The International Tin Cartel

For most of the twentieth century, tin was the site of new forms of international regulation which became a model for other commodities. The onset of the depression of the 1930s saw a collapse in commodity prices, and governments of tin producing countries decided to form a cartel to return the industry to comparative prosperity. This is a detailed study of how the tin industry found itself in difficulty and how the cartel developed its policies of control over production and stocks, together with its enduring legacy after World War II.

These policies proved very controversial and the ensuing conflicts provided many economists with arguments to support a critique of the very principle of cartelization. Examination of the sources of these conflicts shows that they had little to do with economic issues, but reflected deeply held ideological conceptions. They also show that the economic criticisms are without any empirical foundation. This study of a cartel brings together two levels of analysis that are normally kept separate; international co-operation, and national organization, and demonstrates how each affected the other. It is based on a comprehensive review of a wide range of archival sources which are sufficiently rich and frank that they provide an insider's sense of how a cartel actually worked.

The book covers topics including the tin industry, depression of the 1930s, commodities, British imperial economic policy and international trade relations. Its discussion spans a range of countries including Bolivia, Malaya, Siam, the Dutch East Indies and Nigeria. It will be of particular interest to higher level students and researchers in these subject areas, as well as professionals who monitor commodity markets.

John Hillman taught Sociology and International Development Studies at Trent University, Ontario, from 1968 to 2004 where he is now Professor Emeritus.

Routledge Explorations in Economic History
Edited by Lars Magnusson, Uppsala University, Sweden

1 Economic Ideas and Government Policy
Contributions to contemporary economic history
Sir Alec Cairncross

2 The Organization of Labour Markets
Modernity, culture and governance in Germany, Sweden, Britain and Japan
Bo Stråth

3 Currency Convertibility
The gold standard and beyond
Edited by Jorge Braga de Macedo, Barry Eichengreen and Jaime Reis

4 Britain's Place in the World
A historical enquiry into import controls 1945-1960
Alan S. Milward and George Brennan

5 France and the International Economy
From Vichy to the Treaty of Rome
Frances M. B. Lynch

6 Monetary Standards and Exchange Rates
M.C. Marcuzzo, L. Officer, A. Rosselli

7 Production Efficiency in Domesday England, 1086
John McDonald

8 Free Trade and its Reception 1815-1960
Freedom and trade: Volume I
Edited by Andrew Marrison

9 Conceiving Companies
Joint-stock politics in Victorian England
Timothy L. Alborn

10 The British Industrial Decline Reconsidered
Edited by Jean-Pierre Dormois and Michael Dintenfass

11 The Conservatives and Industrial Efficiency, 1951-1964
Thirteen Wasted Years?
Nick Tiratsoo and Jim Tomlinson

12 Pacific Centuries
Pacific and Pacific Rim Economic History since the 16th Century
Edited by Dennis O. Flynn, Lionel Frost and A.J.H. Latham

13 The Premodern Chinese Economy
Structural Equilibrium and Capitalist Sterility
Gang Deng

14 The Role of Banks in Monitoring Firms
The Case of the Crédit Mobilier
Elisabeth Paulet

15 Management of the National Debt in the United Kingdom, 1900-1932
Jeremy Wormell

16 An Economic History of Sweden
Lars Magnusson

17 Freedom and Growth
The Rise of States and Markets in Europe, 1300-1750
S R Epstein

18 The Mediterranean Response to Globalization Before 1950
Sevket Pamuk and Jeffrey G Williamson

19 Production and Consumption in English Households 1600-1750
Mark Overton, Jane Whittle, Darron Dean and Andrew Hann

20 Governance, The State, Regulation and Industrial Relations
Ian Clark

21 Early Modern Capitalism
Economic and Social Change in Europe 1400-1800
Edited by Maarten Prak

22 An Economic History of London, 1800-1914
Michael Ball & David Sunderland

23 The Origins of National Financial Systems
Alexander Gerschenkron Reconsidered
Edited by Douglas J. Forsyth and Daniel Verdier

24 The Russian Revolutionary Economy, 1890-1940
Ideas, Debates and Alternatives
Vincent Barnett

25 Land Rights, Ethno Nationality and Sovereignty in History
Edited by Stanley L. Engerman and Jacob Metzer

26 An Economic History of Film
Edited by John Sedgwick and Mike Pokorny

27 The Foreign Exchange Market of London
Development since 1900
John Atkin

28 Rethinking Economic Change in India
Labour and Livelihood
Tirthankar Roy

29 The Mechanics of Modernity in Europe and East Asia
The Institutional Origins of Social Change and Stagnation
Erik Ringmar

30 International Economic Integration in Historical Perspective
Dennis M. P. McCarthy

31 Theories of International Trade
Adam Klug
Edited by Warren Young and Michael Bordo

32 Classical Trade Protectionism 1815-1914
Edited by Jean Pierre Dormois and Pedro Lains

33 Economy and Economics of Ancient Greece
Takeshi Amemiya

34 Social Capital, Trust and the Industrial Revolution: 1780-1880
David Sunderland

35 Pricing Theory, Financing of International Organisations and Monetary History
Lawrence H. Officer

36. Political Competition and Economic Regulation
Edited by Peter Bernholz and Roland Vaubel

37. Industrial Development in Postwar Japan
Hirohisa Kohama

38. Reflections on the Cliometrics Revolution
Conversations with Economic Historians
Edited by John S. Lyons, Louis P. Cain, and Samuel H. Williamson

39. Agriculture and Economic Development in Europe Since 1870
Edited by Pedro Lains and Vicente Pinilla

40. Quantitative Economic History
The Good of Counting
Edited by Joshua Rosenbloom

41. A History of Macroeconomic Policy in the United States
John H. Wood

42. An Economic History of the American Steel Industry
Robert P. Rogers

43. Ireland and the Industrial Revolution
The Impact of the Industrial Revolution on Irish Industry and Society, 1801-1922
Andy Bielenberg

44. Intra-Asian Trade and Industrialization
Essays in Memory of Yasukichi Yasuba
Edited by A. J. H. Latham and Heita Kawakatsu

45. Nation, State and the Industrial Revolution
The Visible Hand
Lars Magnusson

46. A Cultural History of Finance
Irene Finel-Honigman

47. Managing Crises and De-Globalisation
Nordic Foreign Trade and Exchange 1919-1939
Edited by Sven-Olof Olsson

48. The International Tin Cartel
John Hillman

The International Tin Cartel

John Hillman

Routledge
Taylor & Francis Group
LONDON AND NEW YORK

First published 2010
by Routledge
2 Park Square, Milton Park, Abingdon, Oxon, OX14 4RN

Simultaneously published in the USA and Canada
by Routledge
270 Madison Avenue, New York, NY 10016

Routledge is an imprint of the Taylor & Francis Group, an informa business

Typeset in by Times New Roman by
HWA Text and Data Management, London
Printed and bound in Great Britain by
TJI Digital, Padstow, Cornwall

British Library Cataloguing in Publication Data
A catalogue record for this book is available from the British Library

Library of Congress Cataloging in Publication Data
Hillman, John, 1939–
The international tin cartel / John Hillman.
p. cm.
Includes bibliographical references and index.
1. Tin industry--History. 2. Cartels--History. I. Title.
HD9539.T5H55 2010
338.8'87696--dc22 2009035609

ISBN10 0-415-55412-8 (hbk)
ISBN10 0-203-85661-9 (ebk)

ISBN13 978-0-415-55412-1 (hbk)
ISBN13 978-0-203-85661-1 (ebk)

Contents

Acknowledgements

A study of this scope could only be undertaken in a particularly supportive intellectual environment. A special debt of gratitude goes to Thomas Symons, Founding President of Trent University, who created an institution which encouraged the tackling of large projects. Discussions with colleagues at Trent have sustained this one and they include: Pradeep Bandyopadhyay, Eric Helleiner, David Morrison and Andrew Wernick. Scholars who are familiar with tin have helped refine my position and they include: Gill Burke, Manuel Contreras, John Thoburn and Helmut Waszkis. Some of my students have also been indispensable. Piet Straatman assisted in the clarification of colloquial Dutch. Nadien Godkewitsch and Aqeel Zaman helped in locating and checking some archival sources. Material support for this project has come from grants from the Committee on Research of Trent University, and the Social Sciences and Humanities Research Council of Canada.

Much of the primary research was undertaken with the assistance of my wife, Ruth, who read newspapers, checked references, organized xeroxing and listened to several unrefined nuggets as they were extracted from the archives. Without her help and support, the results of this inquiry would have been far more limited.

Some of the material has been already been published. I am grateful to *Mining History Journal* for permission to use 'The Politics of Equity: Administering Tin Restriction in Nigeria, 1931–1933', vol. 4, 1997, to the Board of Regents of the University of Wisconsin system for the use of 'Chartered Companies and the Development of the Tin Industry in the Belgian Congo, 1900–1939', *African Economic History*, vol. 25, 1997 and to Cambridge University Press for the use of sections from 'Malaya and the International Tin Cartel', *Modern Asian Studies*, vol. 22, pt. 2, May 1988, 'Bolivia and the International Tin Cartel, 1931–1941', *Journal of Latin American Studies*, vol. 20, no. 1, May 1988, and 'The Freerider and the Cartel: Siam and the International Tin Restriction Agreements, 1931–1941', *Modern Asian Studies*, vol. 24, no. 2, May 1990.

Acronyms

AISI	American Iron and Steel Institute
AMM	*American Metal Market*
AOGIT	Anglo-Oriental and General Investment Trust
AOMC	Anglo-Oriental Mining Corporation
ASARCO	American Smelting and Refining Company
ATMN	Associated Tin Mines of Nigeria
BC	Belgian Congo
BM	Billiton Maatschappij
BMC	British Metal Corporation
BMTS	British and Malayan Tin Syndicate
CFL	Compagnie des Chemins de Fer du Congo Supérieure aux Grand Lacs Africains
CGF	Consolidated Goldfields of South Africa
CMO	Compañía Minera de Oruro
CoP	Certificate of Production
CTMB	Consolidated Tin Mines of Burma
CTS	Consolidated Tin Smelters
EIC	East India Company
EMJ	*Engineering and Mining Journal*
EPL	Exclusive Prospecting Licence
ESB	*Economisch-Statistische Berichten*
ESC	Eastern Smelting Company
FIC	French Indo-China
FMS	Federated Malay States
GMB	Gemeenschappeleijke Mijnbouwmaatschappij Billiton
IAMS	*Industrial Australian and Mining Standard*
IRRC	International Rubber Regulation Committee
ITC	International Tin Committee
ITP	International Tin Pool
ITRDC	International Tin Research and Development Council
ITRI	International Tin Research Institute
LME	London Metal Exchange
LTC	London Tin Corporation
LNTM	London Nigerian Tin Mines
MRC	Metals Reserve Corporation

MTRJ	*Malayan Tin and Rubber Journal*
NEI	Netherlands East Indies
NIEO	New International Economic Order
NYT	*New York Times*
PME	Patiño Mines and Enterprises
RFC	Reconstruction Finance Corporation
SAMEJ	*South African Mining and Engineering Journal*
SEK	Société des Etains de Kinta
STC	Straits Trading Company
STS	Siamese Tin Syndicate
TIMM	*Transactions*, Institution of Mining and Metallurgy
TPA	Tin Producers' Association
UFMS	Unfederated Malay States
UMHK	Union Minière du Haut Katanga
VOC	Vereenigde Oostindische Compagnie
VYB	Vivian, Younger & Bond
WCM	*Wild Cat Monthly*

1 Introduction

On United Nations Day, 24 October 1985, Pieter de Koning, Manager of the Buffer Stock of the International Tin Council, informed Ted Jordan, Chairman of the Committee of the London Metal Exchange, that he was no longer in a position to meet his financial obligations.[1] Overnight, £900 million in paper assets vanished and the price of tin dropped from £8,900 to £5,500 per tonne. Bankruptcies of dealers and miners followed in its wake and the economy hardest hit of all, Bolivia, sought relief in the production of cocaine.

International commodity agreements are always fragile but none had ever collapsed in such a dramatic fashion. Over most of the three decades of its effective history (1956–1985) the International Tin Council had been regarded as offering a solution to the difficulties created by markets in primary commodities. With its demise any enthusiasm for such agreements as an integral part of the creation of more equitable and prosperous international economic order evaporated.[2] The 'obvious' lesson was drawn by the British government whose representative, Alan Clark, stated that:

> Experience with the tin agreement and other commodity agreements has shown that market intervention provisions such as buffer stocks and quota arrangements have severe defects: at best they are of doubtful benefit to producers and consumers and at worst they distort markets fundamentally and carry the risk of disastrous failure … The Government's general policy towards commodities is based upon the premise that the best way to assist producers and consumers is to enable markets to work efficiently. We therefore support liberal arrangements and deregulation. We seek to work with our European Community partners to put these principles into practice both in relation to commodity agreements and in the current round of multilateral trade negotiations in GATT which will have an important bearing on wider commodity trade questions of interest to both developed and developing countries.[3]

The International Tin Council was the last in a series of intergovernmental agreements which regulated the tin market over most of the twentieth century. Yet the operation of a deregulated market, which has produced both extremely low and high prices, does not appear to have been a superior method of coordinating the activities of producers and users of this indispensable metal. The failure of the International Tin Council should not be taken as a licence to dismiss the experience of commodity control.

International control of tin began in response to the first economic crisis of the interwar period, 1921–1922 and was resumed in response to the second, 1929–1931. On both those occasions, the problem was sufficiently severe that governments set aside market orthodoxy and co-operated in rescuing their producers from its travails.

The first of these interventions, the Bandoeng Pool, was short-lived, being made redundant by the boom of the mid-1920s and is relevant only to the history of tin. The second, the International Tin Restriction Agreements, 1931–1946, were more durable since they had to address the challenges posed by the depression of the 1930s and World War II. They were also far more significant, not only in the history of tin but in the whole history of primary commodities, especially during the crisis of the 1930s.

Tin was the first intergovernmental agreement which gave its regulatory authority, the International Tin Committee (ITC), the power to control production. This was seen not only as a solution to the problem of tin but also as a model that could be adapted to deal with other troubled commodities. Tin was followed by intergovernmental agreements for sugar, tea, wheat and rubber.[4] Expectations were even higher since rescuing these sections was seen as making a major contribution to the recovery of the overall international economy. It is therefore not surprising that tin has been at the centre of many of the more general debates about primary commodities.

When a small group of individuals determines the overall level of production their decisions are inevitably controversial, especially when they do so in the name of governments. These controversies engaged several different groups. Specific decisions of the ITC directly affected members of the industry, miners, smelters, brokers and industrial users. Their responses, together with the decisions themselves, were subject to extensive scrutiny and commentary by journalists[5] and politicians.

Given the overall significance of the tin experiment, it became the subject of much more comprehensive analyses undertaken by professional economists who became drawn into its controversies. They divide into two camps which reflect the predisposition of their respective intellectual cultures. Anglo-Americans were generally very critical of the ITC,[6] while their continental counterparts were generally supportive.[7] The arguments from the latter group have been ignored or distorted, whereas those from the former have succeeded in shaping the enduring reputation of the ITC.

The dominant critical perspective not only drew on the theoretical power provided by neo-classical economics but it also found supporting evidence in the vociferous complaints from those who considered themselves adversely affected by the ITC, especially where these were echoed in the mining and financial press. Unfortunately, those who were anxious to build a strong case against the ITC relied on a theory which was ill-suited to deal with a commodity such as tin and overlooked the evidence that would provide a better test of its conclusions.

On occasion, those who were responsible for the administration of the ITC provided a public defense of their actions but they were well aware of the limitations of such a forum as a means of convincing or silencing their critics. Fortunately, they have left behind an extensive record of the way in which they actually came to their decisions which now permits the whole experiment in tin control to be reconsidered.

Analysis of a regulated commodity market is inevitably complex since it must encompass three separate but interrelated dimensions: the structure of the industry, the form of regulation and the normative. The first, industry structure, requires answers to three questions. How is the product used? What are the conditions under which it is produced and how is it marketed? Those answers must take into account the ways in which the patterns of consumption, production and exchange have evolved over time. In the case of metals, such as tin, which have a large number of different end uses and a large number of different kinds of producers and technologies, those patterns of historical evolution are inevitably uneven. At any particular moment in time, each pattern is made up of several features, many of which first emerged millennia ago.

The second dimension, regulation, also requires answers to several questions: who are the regulators, how are they connected to the industry, what are the instruments at their disposal and how are decisions made about their use? Discussion of these issues raises two further major questions. What is the nature of the problem with the industry that requires regulation? How does regulation shape the direction of the overall evolution of the structure of the industry? The last of these turns out to be an iterative one. How does the industry respond to regulation in such a way as to shape regulation itself?

The normative dimension emerges by virtue of the assumption that regulation is an aberrant condition and it addresses the following questions. How well did regulation solve the problem that prompted its formation? Who benefited and who lost? This assumption is not simply one made by those attempting to understand the operation of commodity agreements; it is built into the regulatory process itself, especially into the iterative interrelationship between an evolving industrial structure and the regulatory authority.

Constructing a comprehensive account which addresses all these questions will be done in the following way. Chapters 2 and 3 review the evolution of the structure of the tin industry from its inception to the end of World War I. Chapter 4 examines the way in which the industry developed in the 1920s with a view to identifying the source of the problem it faced in 1929. Chapter 5 continues this discussion and examines the impact of the onset of the depression. These four chapters have a primary focus on the first dimension.

The second dimension is introduced in Chapter 6 which considers the preliminary moves towards production control and the formation of the ITC and is sustained through the next eight chapters. They consider how the ITC and its members developed the instruments of control (Chapter 7), how these shaped policy under varying economic conditions of both peace (Chapters 8 to 12) and war (Chapter 15) and how the industry developed within its constraints (Chapters 13 and 14). The empirical and analytical issues raised by the normative dimension are clarified in Chapter 16. Throughout the entire study there is a continuous underlying tension between ideology, politics and economics, one that is resolved in Chapter 17 which demonstrates how governments decided to terminate the ITC. Chapter 18 offers an overview of the history of the tin industry from 1945 to the debacle of 1985 with a view to demonstrating the legacy of the controversies about the ITC and the ways in which its successor, the International Tin Council, attempted to address the problems it had to face.

The experience of the ITC provides an opportunity to develop a fuller appreciation of the strengths and weaknesses of this form of control over commodity markets. Without it, the whole history of tin control may be prematurely written off as an unfortunate and painful lesson on the road to the current 'liberal' wisdom which attempts to ensure that global markets operate on their own logic, unshackled by politics and indifferent to the fate of those whose lives are dependent on them.

2 Tin: the foundations of an industry

Quite apart from its site as the commodity which served as a test case for the solution of the one of the major economic problems of the twentieth century, tin has a remarkable history. It is one that ultimately rests on two material features, the very limited geological distribution of its mineral form and the extraordinary versatility of the metal which permits the fabrication of a wide range of products. Since many of these became an integral part of the daily life of several civilizations, they made the mineral comparatively valuable. The history of the metal is therefore one of the changing ways in which that value is realized and distributed.

Tin stands at a critical point in the history of metallurgy. The first distinctive metallurgical skill emerged in the ninth millennium BC as the annealing (alternate heating and hammering) of native copper resulted in a wider range of more reliable artefacts.[1] By the middle of the fourth millennium, a second set of skills had emerged as experience with the application of high levels of heat required to turn clay into pots was adapted to smelt copper from its mineral form and work the molten metal.[2]

In its pure form, copper is both soft and difficult to cast and the search for ways of overcoming these limitations led in two directions, both of which produced a similar result, bronze. One was to smelt copper-arsenic ores, either found as such or mixed from separate sources.[3] The second was more metallurgically significant and that was to alloy two metals, copper and tin.[4] Only small quantities of tin were required (6–10 per cent) but they provided three major advantages. The resulting product was harder and stronger and much easier to cast. At some point, this paradoxical feature by which the tin makes the molten alloy more fluid but the resulting product more durable was attributed to an unnatural force and tin became the 'Devil's metal'.[5] That reputation of a metal about which there is something quite odd has persisted, expressed less about its metallurgy but rather about its market behaviour. 'It's a bloody perverse metal', summed up the frustrations of at least one metal merchant.[6]

Tin also stands at a critical point in the history of many human societies. Bronze permitted a considerable expansion in the range of ornamental artefacts, together with more serviceable tools and weapons.[7] Just what weight can be attributed to such tools in raising overall levels of productivity is uncertain[8] but the effects of the growing cultural and political significance of bronze are unmistakable. Given the limited distribution of the mineral, long-distance trading networks had to emerge to support many of these civilizations of bronze. The history of tin is one of the interplay between three spheres, the social construction of consumption, the political and economic organization of trading relationships and the capacity of producers to locate and extract tin-bearing ores and then turn them into metal.

Since the focus of this study is on a particular form of organization of trading and production, it will be useful to situate that both historically and comparatively. The ITC represented an attempt on the part of producers to regulate a uniform international market, one that arose in the mid-nineteenth century. Of the regions relevant to the operation of the ITC, only two originated in response to the incentives offered by that market. Three others emerged in response to incentives offered by more segmented markets operating around the Indian Ocean from the first millennium AD. Most could trace their formation back to much earlier forms of exchange relationships which were incorporated into the political organization of the civilizations of bronze and which date from the fourth millennium BC.

Over this long sweep of history the Southwest of England was consistently an important, and normally the most important, producer. Such regional continuity is unparalleled in the history of mining which is generally characterized by dramatic shifts as one ore body is depleted and another developed. The dominance of Cornwall was such that, even as production in the county itself declined at the end of the nineteenth century, Cornish workers, smelters and, above all, capitalists continued to shape the development of the industry internationally until well into the twentieth century. The history of tin is therefore in large measure bound up with Cornwall.

Not only are tin deposits scarce but they generally lend themselves to small scale exploitation.[9] That allows for an extremely diversified pattern of production and has resulted in a technologically conservative industry. When the Hoovers visited the Saxon tin mines at the beginning of the twentieth century, they found the same techniques in place that were described by Agricola three and a half centuries earlier.[10] Had they visited other parts of the world, they would have seen techniques that stretched back over four millennia. This combination of technical conservatism and regional continuity made adaptation to the new political and economic conditions of the interwar period extraordinarily difficult. The tensions that will be examined around the operation of the ITC took their particular form by virtue of the heavy legacy of history.

Throughout that entire history, virtually only one mineral has lent itself to economic exploitation and that is cassiterite (SnO_2) with a 78.6 per cent metal content, often called 'tinstone' or 'black tin' on account of its appearance.[11] Cassiterite is found in both primary and secondary forms. Primary or lode ores are veins embedded in hard rock, often mixed with other minerals and impurities. The secondary forms, generally called 'placer' or 'alluvial', have been eroded and redeposited away from their original formation.[12] Thanks to its high specific gravity at 6.8, the process of erosion has separated the cassiterite from most impurities and deposited it as particles and pebbles in river beds.[13] The same technique of panning that removed and concentrated particles of gold could be applied to cassiterite and it is one that has remained unchanged throughout the entire history of the industry. As soon as pyrotechnology permitted a furnace to reach a temperature of around 1,100–1,200°C, the metal could be freed from its oxide.[14] While smelting involved considerably more skill than extraction, it remained well within the capacity of most miners.[15]

The emergence of a highly differentiated industry from these simple beginnings was a function of three separate processes: (1) extraction of alluvial deposits from less accessible sites, which rested on the development of techniques of prospecting and of more complex forms of mining; (2) separation of extraction from smelting, both in location and organization; (3) extraction of underground lode deposits together with methods of concentrating them to make them amenable to smelting. In virtually all of these spheres, the techniques implemented were adaptations of those already developed for other minerals. Not only does the industry change slowly but its pace is a function of innovations occurring elsewhere. Moreover, as newer techniques are implemented they tend to overlay rather than obliterate older ones. The conservatism of the industry is therefore a function both of the extent to which older techniques are preserved and of the lagged implementation of newer ones.

These technical developments occurred within the framework of the development of capitalist institutions. With the formation of a competitive international market the dominant form of capital associated with the industry became that of investment in the process of extracting cassiterite. The emergence of this form of capitalism rested on a long period, stretching over nearly four millennia, in which the metal only became a commodity by virtue of the dominant role of mercantile capital. The precise form of that dominance was determined within an elaborate system of property rights which established several other claims on the value of the mineral as between prospector, working miner, mine owner, smelter, landlord and state. These proto-capitalist developments evolved from a world where the products fashioned out of tin never took on a commodity form, at least not at the point of final consumption where they were incorporated into a prestige goods economy. These three stages will serve to organize the remainder of this overview.[16]

Prestige goods economies

Bronze metallurgy could only arise within societies which produced a sufficient agricultural surplus to support the allocation of the substantial amounts of social labour required both for the fabrication of artefacts and the prior processes of smelting and extraction of minerals. Such conditions are normally associated with a distinctive class structure and this shaped the overall course of metallurgical evolution. Ideological and political domination by the ruling elites could only be sustained with goods that were specifically fabricated for their exclusive use and which therefore represented the symbolism of their prestige in material form.[17] Such goods were primarily items of personal adornment and display and therefore markers of social position. But they could also be items provided by these elites to serve the whole community such as bells to drive away evil spirits or release the productive powers of the earth.[18] Once bronze had become the successful basis for the communication of the symbols of social power, it could be used for the fabrication of tools. At first these simply

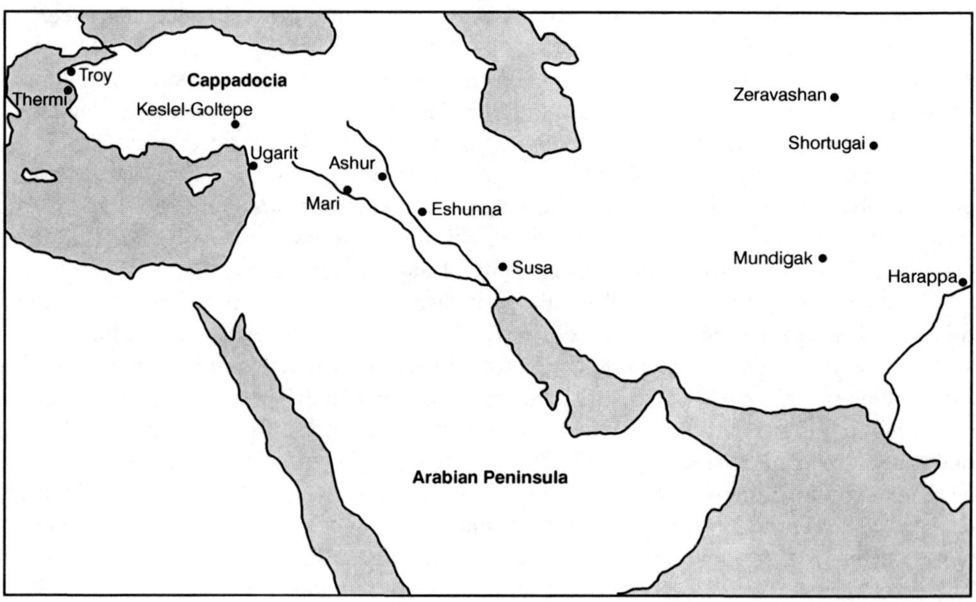

Map 1 Southwest Asia

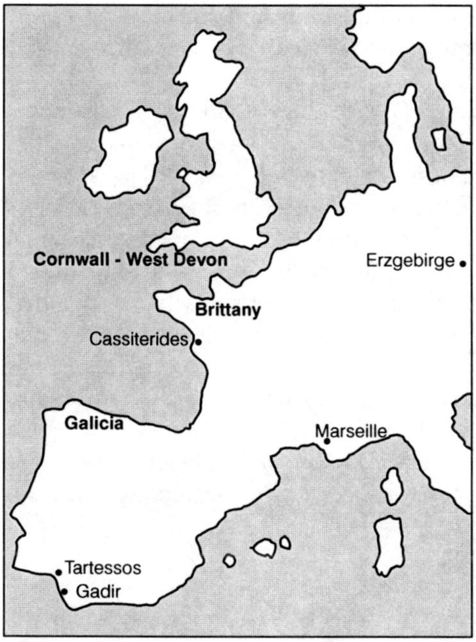

Map 2 Europe

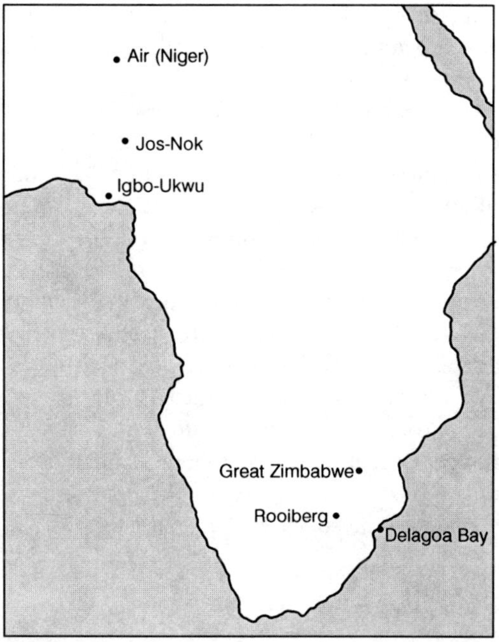

Map 3 Africa

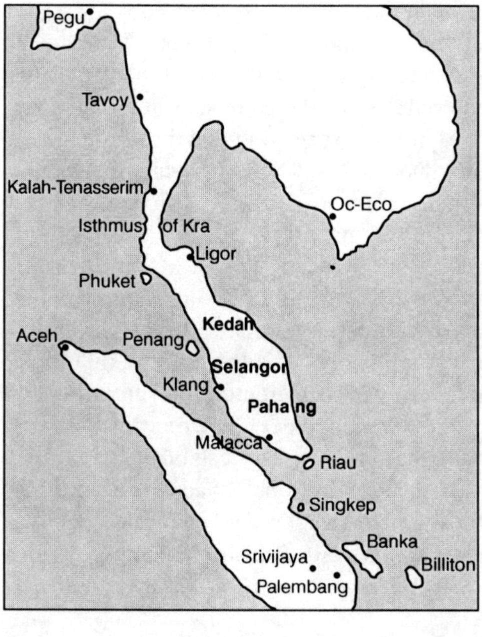

Map 4 Southeast Asia

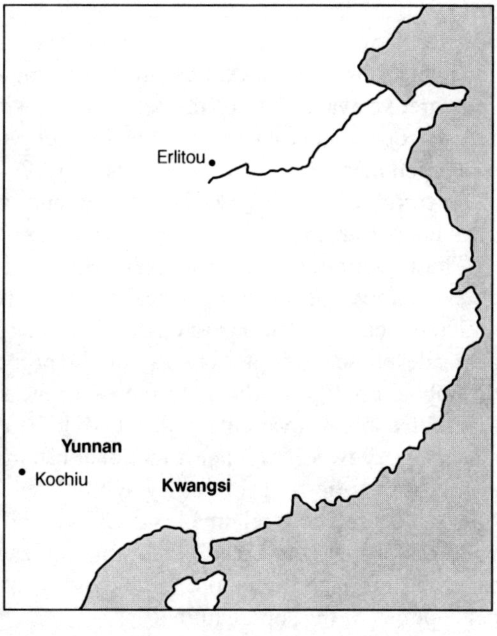

Map 5 Northeast Asia

improved the level of handicrafts to meet other needs of the elite but eventually tools were developed that raised the overall level of food production which always remained the basis of these societies.[19]

Conceptualizing these processes as constituting a Bronze Age as part of a linear model of historical evolution[20] has resulted in several distortions. The most obvious emerges from the empirical observation that the model does not apply to the early class societies of sub-Saharan Africa and the Western Hemisphere, which relied on other material bases to symbolize social domination and for the fabrication of useful tools. While tin played an interesting role in both of these regions, it was quite different from that which it came to occupy in Europe and Asia. More important from the perspective that will serve this inquiry, the model tends to focus on the technical properties of materials at the expense of the social context within which they are produced and used.

Clarification of the significance of this distinctively social context has come from studies of the role of tin in both South and Meso-America. This is a result of a combination of two fortuitous features: an autonomous metallurgical tradition which had embraced bronze but not moved to an iron stage and conquest by a force interested in constructing a literary account of the societies it encountered.

South and Meso-America

The concept of bronze is ambiguous since the hardening of copper can be accomplished with the addition of either tin or arsenic. Much of the discussion about the importance of tin in bronze has rested on the assumption of its technical superiority over copper-arsenic bronzes. However, systematic testing of the two types has not confirmed this.[21] If technical considerations alone prevailed, then the type of bronze in use would be a function of the comparative availability of the respective raw materials. Since there are substantial deposits of both copper and cassiterite in the Southern Andes, it is not surprising that tin bronzes were developed there, probably as early as 200 AD. But, once developed, they eventually became a ‘powerful symbol of Inka hegemony’ and, as that empire expanded, tin bronze replaced local copper-arsenic bronzes ‘because it was a marker of political power’.[22]

The importance of such markers was particularly pronounced in West Mexico as local elites attempted to fill the political vacuum created by the demise of Teotihuacan. Contact with the Incas had demonstrated the technology of tin bronze and from the thirteenth century it was developed to express two symbolic properties, colour and sound. As colour, tin bronze resembled gold in its ability to represent the sacred property inherent in the sun but it had several technical advantages since it could be used for a much wider range of artefacts, such as depilatory tweezers. Their functional use in plucking out facial hair established the ritual purity of the priestly offices they symbolized.[23] The sounds from bells had many symbolic functions in relation to human and agricultural fertility and social order but when combined with colour they expressed ‘a shimmering sacred paradise full of lustrous beings’.[24] Only tin could produce the colour and sounds that served to connect new elites to established conceptions of the supernatural.[25]

Sub-Saharan Africa

Although sub-Saharan Africa has several important cassiterite deposits, bronze played a very limited role in its metallurgical and social history.[26] Deposits at Air in Niger must have been the basis of the limited local manufacture that arose in the middle of the seventh century BC.[27]

A century later, the Nok civilization emerged further south. As one of the earliest centres of iron metallurgy and located near the Jos plateau with easy access to extensive cassiterite deposits, it probably learned how to turn them into metal.[28] However, there is a millennial long gap before they became the basis of the flourishing, albeit limited, bronze metallurgy, reflected in the major collection of ornamental artefacts excavated at Igbo-Ukwu in Southern Nigeria.[29] There is another major gap in both time and space before ornamental bronzes emerged again when Great Zimbabwe produced them from locally smelted copper and tin in the thirteenth century.[30] While all these artefacts are a tribute to the creativity of local craftsmen, the fact that they are so isolated in both time and space raises some important, though as yet unanswered, questions as to why the foundations of this aspect of an otherwise thriving metallurgy were so fragile.

Southeast Asia

The position of tin in the metallurgical history of Southeast Asia is particularly puzzling.[31] Some of the earliest artefacts, many of which are of pure tin, are to be found in northeastern Thailand. By the early second millennium tin bronzes were appearing in Vietnam, by the early first in both Yunnan and Upper Burma.[32] By the late first millennium, bronze artefacts, especially Dongson drums and large bells from North Vietnam, together with bowls from Western Thailand, had become major items traded throughout the region.[33]

The structure of regional trade became intimately connected with a distinctive form of social hierarchy, the port-polity.[34] A strategic coastal location, especially one linked to a hinterland by river, provided an opportunity for accumulation of power through external trade. Alluvial cassiterite deposits were found in river valleys running into both coasts of Peninsular Malaya and tin became its major source of export income.[35] Sustaining power in a port-polity rested on redistribution of a substantial portion of the benefits to local elites who ensured the supply of trade goods and who could be expected to provide the resources required to subdue rival polities.

As an interrelated system, it proved extraordinarily stable over two millennia but the position of any port-polity within it was inherently fragile. There was a tendency on the part of one polity to expand its influence at the expense of others[36] but any success could be undermined as local elites shifted alliances in response to new opportunities, especially those created by changing patterns of trade. This will be clearly demonstrated when that trade becomes penetrated by extra-regional mercantile capital from India, China and then Europe.

The importance of tin on the consumption side is most evident in the Khmer empire of Funan. This was the first of the many Indianized polities in the region and was formed in the first century AD. Its power soon extended across the Gulf and drew tin from the east coast of Siam. By the third century, its main port of Oc-Eo in South Vietnam had become extremely wealthy, much of which was expressed in tin rather than bronze artefacts.[37] Although the tin was not yet part of a mercantile trade, its significance would have been evident to the many traders attracted from both India and China.

Northeast Asia

Northeast Asia presents a more coherent, albeit quite distinctive, picture. Chinese thought distinguished between natural materials such as stone, bone and jade and derived materials such as metals, regarding only the former as suitable for personal ornamentation.[38] Bronze artefacts would, therefore, have a much more circumscribed role than in most other civilizations but a far more important one.

As the Erlitou culture on the Yellow river developed in the mid-second millennium, bronze vessels (ding) became 'the most important symbol of political, religious and economic power'.[39] The position of those entitled to the highest number of dings was then reinforced by the acquisition of sets of musical chime bells.[40] Both dings and bells were used in ritual communication with ancestral spirits and their production required a complex casting technology, reflecting a sophisticated level of industrial organization.[41]

As in Meso-America, control over bronze represented control over access to heaven and was 'so essential that the Great Bronze Enterprise became the focus and core of ancient Chinese civilization'.[42] Here, too, the sources of both copper and cassiterite were found locally and their indispensability forced successive monarchs to move their capital cities to remain close to new mines in the face of the depletion of old ones.[43] As Chinese civilization expanded from its northern centre, the early dings were then collected by subsequent dynastic orders and used to symbolize their continuity and legitimacy.[44] The Great Bronze Enterprise therefore only came to an end in 1911.

Southwest Asia

When the metal artefacts unearthed by Schliemann at Troy in the 1860s were identified as bronze, the question was immediately asked, where did the tin come from? Each stage in the progression of Old World archaeology which uncovered the lost civilizations of southwest Asia (Anatolia, Mesopotamia, Indus valley), the eastern Mediterranean and the Nile valley, encountered similar artefacts. As at Troy, there were no obvious local sources and the question therefore continued to be re-posed. On its answer rest important clues to the interrelationship between these civilizations and the way in which they were constructed in relation to long distance trade, especially with Central Asia.

The earliest known bronze artefact is an axe found at Mundigak in Afghanistan, dated to the late fourth millennium. There is a clear continuity with the copper metallurgy that flourished in Iran which was then taken to a higher stage at Mundigak thanks to the availability of local alluvial cassiterite deposits. To this extent, it fits the pattern already described for the non-Western world. However, it developed in a fundamentally different direction, since Afghanistan was an integral part of trading networks that connected it to both Mesopotamia and the Indus valley. These networks could then transmit both the technology of bronze and the tin required to manufacture it.

Whether these networks remained intact, or were replaced by others, is uncertain. The excavation of a large group of lode mines in the Zeravshan valley between Uzbekistan and Tadjikistan has excited considerable attention.[45] While it may have simply served the needs of the Andronovo further north, some have been tempted to make it a central part of the solution of the regional problem of ancient tin.[46] Among the possible trade routes, the most interesting is to the Indus valley since there was a well established pattern of trade in other raw materials controlled by Harappa from its outpost on the Oxus at Shortughai.[47] It could then have been shipped across the Gulf to Mesopotamia. However, from the perspective of the valley itself, there is no clear-cut case, since it could just as easily have continued to draw on Afghanistan, or even from sources on the subcontinent both to the south and east.[48] Fortunately, none of these unresolved issues affect the major conclusion that the primary source of the tin for both Mesopotamia and the Indus Valley throughout the third and second millennia was in the East. What is much less certain is the pattern of consumption. Tin bronzes are not particularly common and while they increase by the middle Bronze Age, they remain less frequent than artefacts from arsenical copper.[49] Whether the scarcity[50] was a function of economic constraints on supply, or of the social organization of consumption, remains quite indeterminate.

As the focus of discussion moves further west to Anatolia and the Aegean, both these issues of sources and consumption remain unsettled. The region, however, reveals one important artefact and that is a pure tin bangle from Thermi on Lesbos dated to the mid-third millennium which indicates early experimentation with the metal. Surveys of other artefacts found provide widely different impressions both of the ratio of bronze to copper-arsenic and of the change between the Early and Middle Bronze Ages.[51]

A more consistent picture emerges with respect to Anatolia. Bronze was found much more frequently there than in Mesopotamia and that is probably the result of the exploitation of local sources, one of which has been excavated at Kestel-Göltepe.[52] This was also a lode mine and is dated from around 3000 BC and probably continued for another thousand years. It is estimated that both underground and surface operations produced around 200 tons of metal during this period.[53] With tin included among the valuable minerals exploited in the Taurus mountains, a comprehensive trade network emerged which encompassed the Aegean and provided the basis of a set of political alliances and corresponding class structures.[54] However, by the beginning of the second millennium, local sources were exhausted and Anatolia then had to turn to imports from the East and that marked the transition to a trade organized by mercantile capital.[55]

Central and Western Europe

Exploitation of the extensive cassiterite deposits in Europe is likely to have commenced in the Erzgebirge region of Saxony/Bohemia at the end of the third millennium,[56] possibly in response to demand from the Aegean. Recognition of the importance of such deposits soon reached Brittany and the Southwest of England.[57] A fourth major source was also located in Galicia. While it may have been stimulated by the discoveries further north, the possibility that it was a response to the development of an autonomous bronze technology based on minor cassiterite deposits in the south and centre of the Iberian peninsula cannot be discounted.

The development of bronze followed a very different course than that in Southwest Asia. Arsenical copper was rapidly displaced and during the course of the second millennium bronze artefacts became widely distributed throughout Europe. It also saw more experimentation with tin as the malleability of the metal was discovered and tinfoil was then used to decorate pottery to give it the appearance of silver.

Of the various European centres, the one that has attracted most archaeological attention is the Southwest of England which is considered the primary source for a long distance trade that extended to the eastern Mediterranean.[58] British and Breton tin was brought to islands off the Loire estuary, known to the Greeks as the Cassiterides,[59] and then shipped across France, mainly by river, to Marseilles. By the first millennium that trade was firmly under the control of mercantile capital.

Virtually all prestige goods economies found tin artefacts to be an indispensable feature of the ways in which power and hierarchy were expressed and sustained. While it was only in China that the artefacts themselves continued to be cherished symbols of a glorious past that had long disappeared, the skills that lay behind them did not vanish. It is these skills that are the enduring legacy of the prestige goods era since they provided the basis on which many other features of the metallurgy of tin would be explored. However, more extensive technical development would depend on a shift in social structure which permitted mercantile capital to play the crucial role in organizing supplies.

Mercantile capital

Prestige goods economies sought to transform the social surplus into symbolic artefacts and ensured that consumption of metals such as tin directly expressed an established social hierarchy. Mercantile capital emerged within this system by expanding the way in which the metal was acquired, thereby transforming part of that surplus into liquid wealth. At first, markets were simply supplementary sources of supply but changing political configurations allowed for their expansion and the formation of competitive prices. As long as consumption remained subject to the restrictive principles of a prestige goods economy, mercantile capital could only operate within a limited sphere. That would expand considerably once consumption itself was mediated through the market. Relaxation of the constraints on the utilization of the metal allowed for an increase in demand and the tin trade offered the prospect of very attractive profits.

The actual level of profits was a function of the conditions under which the tin was acquired and sold and various fractions of mercantile capital sought to maximize them with monopolist and monopsonist policies enforced through the state. Given the number and distribution of deposits that had been actively exploited during the preceding period, it was far more difficult to accomplish this objective in tin than in many other commodities. The history of mercantile capital is therefore primarily one of contending fractions,[60] often linked to contending states, striving to eliminate the competition that would devalue the price of tin. This section will review the available evidence in the following way: (1) Southwest Asia, Eastern Mediterranean and eastern tin; (2) Southwest Asia, Eastern Mediterranean and western tin, (3) Rome, (4) Medieval Europe, (5) Asia, (6) Africa, (7) Early Modern Europe.

Southwest Asia, Eastern Mediterranean and eastern tin

On the western side of the Zagros mountains, there were two main points at which tin from the east entered Mesopotamia: Eshnunna and Susa. Two main centres, Ashur on the upper Tigris and Mari on the upper Euphrates, not only consumed it but distributed it further west, to Anatolia and the eastern Mediterranean.

Eighteenth century Mari provides evidence of the way in which the mercantile capital operated within a prestige goods economy. The royal court exercised a monopsony in tin and what was not needed for its own use was mainly used as part of a gift-exchange relationship with other monarchs and cities.[61] Much was acquired through other gift-exchange relationships but some was bought and at varying prices that reflected a degree of competition. In many of the commercial transactions, merchants acted as agents of the court but in some they acted on their own account.[62] These provide an indication of just how valuable tin had become since the normal tin/silver ratio at Mari was 10:1.[63]

Whether mercantile capital played an important role in the acquisition of metal at Ashur is uncertain but it became extremely significant in its onward distribution to Anatolia. A colony of Assyrian merchants in Cappodocia conducted a flourishing trade in the early second millennium in which tin was a major item. Over 50 years 80 tons of tin were carried from Ashur by donkey. At a tin/silver ratio of 14:1, the metal was cheaper at Ashur than at Mari but it was sold at a much higher price, at 6–8:1, with profits ranging between 75–100 per cent.[64]

This trade which linked Central Asia with the eastern Mediterranean became much more extensive, as is evident from the shipwreck at Uluburum in Southwest Turkey in 1300 BC which contained one ton of very pure tin, cut into pieces suitable for trading throughout the region.[65] Supplies, however, increased at a much faster rate and by the end of the second millennium prices had dropped dramatically, reaching a tin/silver ratio of 200–277:1.[66]

These trading networks allowed for the formation of port-polities and those controlled by Phoenicians gained the experience that enabled them to adapt to a major change in the configuration of political forces at the beginning of the first millennium which cut off eastern supplies and made much of the Levant dependent on Assyria. As the Phoenicians turned west in the service of their new masters, they would reorient the whole structure of the tin trade.[67]

Southwest Asia, the Eastern Mediterranean and western tin

Phoenician cities on the Levant owed their continued existence to the role they played in meeting the needs of Assyria. It was one that met the classic features of a prestige goods economy: political protection in exchange for tribute. Their westward expansion rested on a major innovation: the establishment of a series of colonies and trading posts across much of the Mediterranean and into the Atlantic. Within this sphere, the Phoenicians were then able to establish a trading monopoly, one backed by an appropriate level of military force.

Control over access to the Atlantic ensured control over Galician tin in northwest Iberia. Trading posts at Tartessos and Gadir collected production from the southern part of the peninsula, while one in northern Portugal tapped the erstwhile flourishing trade that had emerged in the northwest.[68] By the middle of the first millennium, the Phoenician model had been adapted by the Greeks who established a colony at Marseilles and that ensured access to supplies from Cornwall and Brittany.[69] Athens alone consumed large amounts of tin, with single sales reaching almost 4 tons. By this point the comparative value of the metal had begun to recover, with a tin/silver ratio of 27:1.[70]

Technical innovation expanded demand in both Greece and the Levant. Water supplies to many dwellings now came through pipes made of bronze.[71] Hydraulic engineering also relied on lead and as the properties of that metal were explored, a new tin alloy emerged, pewter. While iron may have displaced bronze from some of its older functional roles in tools and weapons, new ones were taking their place.

Rome

Roman expansion ended the division of the western Mediterranean between Phoenicians and Greeks and, by the end of the first century BC, the empire had direct control over the Iberian tin deposits. Roman administration ensured that northwest Iberia became the sole source for the Mediterranean and, by the mid-first century AD, Iberian tin was being exported as far as India.[72] Of the remaining European producers, only Cornwall survived but on a very reduced scale to serve a local market.

Roman consumption saw few innovations, though one has persisted and that is to coat copper utensils with a thin layer of tin to improve the quality of both food and drink.[73] Bronze continued to be used for statues and other ornaments and Roman hydraulic engineering required bronze fittings, especially for taps and valves. While lead replaced bronze for pipes, they were all sealed with a tin-based solder. The versatility of pewter was also explored, since it was used not only for tableware but also as a container for medicinal compounds.[74] With a tin/silver ratio of 13:1,[75] the price of tin had recovered to levels not seen for nearly two millennia.

The Roman conquest of Britain resulted in a revival of Cornish mining. By the mid-third century, the Iberian deposits were being exhausted and Cornwall then assumed the role of primary supplier to the Mediterranean. Roman Britain also provided a local market for tin, especially for pewter. As supplies of glazed pottery from the continent were disrupted, the Romano-British elites turned to pewter for the distinctive tableware required to mark their social position.[76] That function would remain for another 1400 years.

Trading patterns dissolved with the collapse of the Western Empire and Cornish production again became very marginal. When they revived, there would be two separate spheres within which mercantile capitalists operated: Europe and the Mediterranean, with its main production centre in the Southwest of England, and the Indian Ocean and the China Sea, with its main production centre in Southeast Asia.

Medieval Europe, seventh to seventeenth centuries

The distinctive legal forms governing social relations in medieval Europe produced a comprehensive system of property rights which were particularly elaborate in the case of mining. Codified in the form of legislation and enforced through specialized courts, this system ensured that the political constraints and incentives that had characterized the earlier phases of the tin industry would now operate through the sphere of law. Not only did that represent an extraordinarily important institutional innovation but its operation has left an extensive body of documentation which has permitted a comprehensive reconstruction of the relationship between consumption, exchange and production.

Consumption rested on several sources. With the Christianization of Europe came a new demand, at least in the west, for large bells and pipes. Bells were first introduced by the monasteries but were soon added to churches.[77] Cathedrals were only considered complete with the installation of a collection of enormous bells.[78] Organs became fixtures in religious institutions and the sonorous quality of their pipes was provided by tin. The poverty of many churches exempted them from the obligation to serve communion on silver and they were then free to use pewter. Many other ceremonial functions also required artefacts best made from pewter or bronze such as candlesticks, ewers, incense burners and the eagles that supported lecterns.

While always inferior to silver, pewter tableware, plates and drinking vessels served as a universal marker of social superiority to those whose poverty confined them to wooden platters and utensils.[79] Pewter also served more mundane domestic functions such as candlesticks, basins, pitchers, etc. As the Renaissance recovered the artistic heritage of antiquity, bronze statues again adorned wealthy households whose servants were summoned by bronze bells. The new technology of printing rested on a tin alloy for its type metal.

The technology of casting large bells was adapted to cast the bronze cannons that became an essential feature of land and naval warfare from the fifteenth century.[80] Although it was eventually modified to cast the much cheaper iron, bronze cannons were lighter, more reliable and remained the weapon of choice,[81] especially for field artillery, until the development of steel in the mid-nineteenth century.[82]

All of these uses occupied a very specific role in the overall pattern of consumption. They were durable and when they finally wore out, their metal contents were easily reclaimed and recycled.[83] The market was therefore subject to clear limits and the following figure, showing the overall course of the European tin market, suggests that these were reached by the early sixteenth century.

Until that point, production and prices generally rose, though with two important breaks in the mid thirteenth and the early fourteenth centuries. From that point both declined, presumably hastened by the turmoil that beset Europe from the mid-sixteenth to the mid-seventeenth centuries. With the return of political stability in the second half of the seventeenth century, production, though not price, recovered. In England that recovery was driven by the replacement of the church organs destroyed by the Puritans and by a decision of the Board of Trade to help the industry by minting tin coins.[84]

Throughout this period there were three sources, northwest Iberia,[85] southern Germany and the southwest of England. Apart from a brief period in the sixteenth century, when she was the largest producer, Germany played only a regional role, albeit an important one, since it sustained a monopoly on tinplate.[86] The overall market was dominated by England with production from both Devon and Cornwall.

Both counties contained large areas of granite moorland with extensive alluvial tin deposits. With the collapse of Roman authority, individuals were free to prospect and exploit them. That right of free mining with the concomitant right to market the metal remained the foundation of mining law as it applied to tin. Specification of these twin rights began in the twelfth century and while it underwent several important modifications, the fundamental principles remained.[87]

Prospecting freedom extended to all unenclosed land together with most of the land enclosed for the manors held by the crown for the Duchy of Cornwall. Operations on privately held land, as long as it had not previously been mined before enclosure, required the de facto permission of the owner. Successful prospecting led to formal delimitation of the boundaries of the area to be worked and once these bounds were registered, the owner had a legally defensible right to extract the tin. That right was contingent on actual exploitation and since it was divisible and transferable, it would come to support capitalist forms of organization of production.

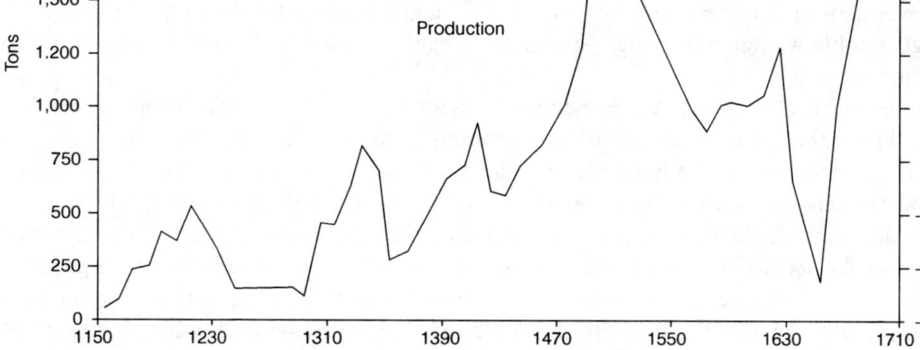

Figure 2.1 European tin production and prices, 1200—1700

Extraction rights were reinforced by granting rights over the water required to dress the ore and over the timber required to smelt it. Working miners were granted a special legal status as tinners which exempted them from all local feudal obligations and over time this was extended to all those associated with the industry.[88] With the exception of major crimes, all disputes involving tinners, regardless of whether they concerned tin, were subject to adjudication in Stannary courts.[89] Stannary Parliaments were convened when required to clarify the general principles governing the industry.[90]

While this was a system designed to maximize the economic space within which the tin industry could flourish, it was subject to several constraints. The primary one, which served as the basis of these extraordinary concessions, was the fiscal interest of the crown. No metal could be legally marketed until it had been weighed and assayed by royal officials who visited designated towns at specific intervals. Assaying was undertaken by striking small pieces from each corner of a large block of metal and the tax was therefore known as 'coinage'.[91] Coinage provided the crown with a substantial annual income of between £2,000 and £3,000 in the early fourteenth century which could be supplemented by exercising the right to speculate on the metal by pre-empting all private purchases.[92]

By the medieval period, most forms of alluvial mining, especially in Cornwall, required extensive work to remove a sterile overburden, bring the deposit to the surface and then dress it for smelting. While the principle of free mining granted individuals the right to extract the mineral, few could do so unaided and a variety of forms of co-operation emerged, ranging from simple working partnerships to proto-capitalist relations between non-labouring owners, or 'adventurers', and hired labour.[93]

Capital expenses in opening up a deposit and then building the necessary waterworks to operate the equipment needed to raise and process it were substantial and many mining operations were dependent for working capital on advances secured against delivery of the metal or the bounds themselves. The additional costs and delays imposed by the method of collecting coinage simply intensified this chronic problem.

Bridging the financial gap between the time at which production was initiated and that at which its final results could be marketed was the responsibility of mercantile capital. Miners received advances in the form of cash or goods, for which privilege usurious rates of interest were demanded. In addition, most miners found themselves compelled to sell their tin before coinage at a substantial discount of between 25 per cent and 50 per cent.[94] Several different local interests, officials, landowners, tenants, clergy, as well as merchants, were attracted by the high rates of return on mercantile capital and that led to a considerable widening of those participating in financing tin.[95] As some tinners defaulted on their debts, merchants took over their bounds and a further incentive to acquire a direct stake in production came with greater knowledge of mining, especially of the differential rents to be won from higher grade properties.

By the fourteenth century, miners no longer smelted their own ores and became dependent on the skill of those who operated separate smelters or blowing houses. These had their own capital requirements especially for the leats and water-wheels required to drive their bellows and the efficiency of their scale of operations granted them a de facto monopoly within a particular region. At first, the smelters treated each parcel of ore on behalf of the miner on toll but as the technology of smelting developed, it became more efficient to organize a more comprehensive treatment of the various parcels submitted. Each parcel continued to be assayed but now the miner received not the metal as such but a bill which entitled him to the appropriate share of the blocks when presented for coinage. The simple toll was then replaced by a more complex formula which guaranteed the smelter not only the costs of processing but a share in the value of the metal itself. These tin bills served to lubricate the overall pattern of financing. Given the more secure position of the smelters as opposed to the miners, it is not surprising to see tin merchants extending their interests to include these blowing houses.

While the law created a distinctive space for mining, it was intimately bound up with the overall pattern of development of local agriculture. On occasion, acute labour shortages reduced wages to the point where mines were a more attractive option than farms but otherwise working miners led a particularly wretched existence, one to which they were only driven by dire necessity.[96] The labour contract itself took several forms: wages, piecework and tribute. Tributing divided the risk of mining but any benefits were as short-lived as the contract itself and shortfalls meant that the tributers remained indebted for the advance or 'subsist' provided by the owners.

Mercantile capital operated in another sphere connecting the marketable tin with final consumers. The pewter industry of London was the largest single source of demand and the city also served as a centre of redistribution both throughout the rest of Britain and to parts of Europe. Both Cornish and London merchants shared the trade and that enabled the former to draw on credit from the latter. The same tin that secured the circuit of mercantile capital within the industry served to reinforce it within a broader network. However, since the Londoners ran fewer risks and dealt with powerful merchants, they were unable to share in the usurious rates of interest paid by the miners for their financial advances.

The overall external trade was extremely important, amounting to perhaps half the total production.[97] Cornish merchants dealt directly with France[98] but the whole of trade beyond Western Europe was in the hands of Italians. This encompassed not only the Mediterranean but also the Black Sea, Syria, Persia and probably the eastern part of the Indian Ocean. Although the Italians often bought at coinage, they played no role in the financing of production. Theirs was the classic form of mercantile capital which took advantage of price differentials in segregated markets. Once the Genoese opened up a direct sea route to England which reduced transport costs to very low levels,[99] the profits on this long distance trade were substantial.

Italian control over the Mediterranean and eastern tin trade came to an end as the sixteenth century closed. Following its defeat at Lepanto in 1571,[100] the Ottoman Empire was anxious to rebuild its military position and once England had irrevocably broken with Rome, she was no longer subject to its interdiction of strategic goods destined to help a religious enemy. Soon after the Levant Company received its charter in 1581, it started shipping tin east.[101] This was followed at the beginning of the seventeenth century when another English chartered company, the East India Company (EIC), started shipping large quantities of tin to Safavid Persia.[102] Its eastern trade in the metal was then expanded to include Mughal India[103] and continued until the 1720s.[104]

As eastern markets were developed, English production, now almost entirely confined to Cornwall, recovered. In this it was aided by several technical improvements. Blowing houses caught up with the German practice of capturing the volatilized tin particles in special chambers.[105] They also learned how to treat varying grades of ore with different fuels. The most important developments occurred in preparing ores for smelting. Impurities were eliminated through roasting or calcining and these, together with other techniques, allowed for the treatment of previously discarded ores.[106] Of longer term significance was the development of the techniques that would allow for the exploitation of the much more complex lode deposits. They would require new technologies and organizational forms but the foundations which enabled both to flourish had been firmly laid over this half millennium.

Mercantile capital in Asia, eighth to eighteenth centuries

Consumption in the main Asian centres had some parallels with the pattern established in Europe. While bells violated Islamic prohibitions against disturbing the spirits, they were widespread in both China and Japan. Domestic utensils were made of bronze or tinned copper. Bronze statues became a prominent feature of both Hindu and Buddhist religious observances and pewter candlesticks and vases were found on all Chinese ancestral altars.[107] China developed its own form of the bronze cannon in the early fourteenth century[108] and Ottoman influence ensured its adoption in India and Southeast Asia.[109]

Two technical developments are worth noting. In the eighth century the chemical properties of tin were explored in Iraq and tin oxide was used to produce an opaque white glaze on earthenware.[110] This provided a base for additional colours which made pots and tiles particularly attractive. Muslim expansion brought this technique to Europe where it remained very popular and was known as majolica.[111] By the eighth century the Chinese had discovered how to beat the pure metal into tin foil, which was then attached to sacrificial paper on which prayers were inscribed and then burnt. The practice of burning this joss paper became extremely widespread and lasted until World War II.[112] This was the first use of tin on a mass scale and since it melted into tiny, irrecoverable globules, it also meant a constant demand.

Until the eighteenth century, Asian consumption levels were considerably below those of Europe but then Chinese demand increased markedly. In part this was a function of a general increase in prosperity but it was also driven by the use of tin in the containers used to ship tea to Europe. This was the first point at which tin served to preserve food.

By the first millennium mining on the southwest-northwest Asian tin belt had virtually ceased. China drew on domestic sources in the south,[113] supplemented by trade with Southeast Asia. The other main centres of consumption, India and Persia, were entirely dependent on transoceanic trade in which mercantile capital inevitably played a dominant role. That trade drew primarily on Southeast Asia, with England and Southern Africa at its margins.

While production in China may have followed a route similar to that discussed in the case of England,[114] Southeast Asia developed in a very different direction. Expansion of demand meant a quantitative increase in the number of production sites and hence of the port-polities that controlled them but, with one exception, there was no qualitative change in the character of production or of the competitive system that had previously emerged.

The alluvial deposits still permitted small groups of men and women to work shallow pits and smelt their ores which they were then obligated to deliver it to local authorities. The principle of seasonal corvée labour, directly in the case of Malayan polities,[115] indirectly in the case of Siam,[116] served as the basis of this production system and the resulting technical stagnation would only be broken with the importation of Chinese labour in the eighteenth century.

Transoceanic trade was first initiated by the Arabs and by the tenth century traders from Oman were acquiring tin at Tenasserim.[117] They were soon replaced by Gujeratis who had the competitive advantage of supplying the high quality textiles demanded by the Malay rulers both for their own display and to pass on to those who directly controlled tin production. Tin was therefore secured against credit advances in the form of textiles.[118] By then the centre of trade had moved further south, to Kedah and Perak.

The trading networks that flourished during thc prc-European phase were based on several interrelated transoceanic exchanges, which ultimately connected South and East Asia and both of these, together with Europe, with the spice islands of the Moluccas. Since the monsoons prevented direct voyages between these points, the commodities were exchanged at intermediary points, of which the most strategic was in the narrowest part of the Malayan inland sea, known after its most important port, Malacca.

The emergence of Malacca and its subsequent decline illustrates the complex interrelationship between trade and political rivalry. As the Buddhist port-polity of Srivijaya, based in southern Sumatra, flourished between the late seventh and early eleventh centuries, its power extended throughout most of the Malayan peninsula.[119] By the early fourteenth century, it was in decline and by 1365 its place was taken by the Hindu Majapahit polity, based in Java. One Srivijayan response resulted in the occupation of a small fishing village in 1396 which was then developed as a pure entrepôt. To attract trade, Malacca was organized in such a way as to serve the needs of mercantile capital, with low port charges, adequate warehousing space and, above all, a legal code which constrained arbitrary political interference and made for predictable, enforceable contracts. Within a few years it had secured a considerable amount of the business transacted by Arab and Indian Muslims and in 1414 the Srivijayan prince converted from Buddhism to Islam.

Malacca's economic prosperity reinforced the prestige of its sultan. As the legitimate heir to one of the most successful empires in the region, he was in a strong position to reassert its influence throughout both Sumatra and the peninsula. Pahang was conquered and tributary relationships were established with other sultanates, especially Klang, Selangor and Perak. Malacca would, therefore, not only serve as an entrepôt but also as a centre of the trade in tin. While most of the metal came from its new hinterland, even tin from as far away as Pegu in Burma was brought to Malacca for sale.[120] However, its success undermined the role of Siam as a source of tin supplies to China.[121] Siam responded by attacking Malacca, which turned to an alliance with China and the consequent creation of a Sino-Malay population led to a distinctive, cosmopolitan city. Siam had to content itself with the seizure of the region around Mergui.[122]

The Portuguese invasion of the Indian Ocean changed this entire system, especially once it was consolidated with the capture of Malacca in 1511.[123] Although Malacca remained an important centre and one that would make its captains very wealthy,[124] it had to contend with other rivals. Since the Estado da India attempted to regulate the entire transoceanic trade by requiring that all ships carried its passes, the Indian merchants diverted much of their tin business to areas beyond its control, especially to Tenasserim.[125] The Malacca sultanate itself was fractured, since the sultan moved to a new base in Johore from which Malacca would be constantly harassed. In the north, Aceh emerged as a rival to both and conquered Perak in 1575 to secure its own tin supply.

By the early seventeenth-century, the Estado da India was being challenged by two much more efficiently organized trading companies, the EIC and its Dutch counterpart, the Vereenigde Oostindische Compagnie (VOC). The chronic imbalance that each experienced in its Europe-Asia trade made tin an important part of its overall commercial strategy. The more tin that could be supplied in Asian markets, the less the drain on silver bullion from Europe. Of these two, the VOC was the more significant until the mid-eighteenth century when the EIC began to oust its rival.

Over the seventeenth and eighteenth centuries the VOC had five primary bases for its trade in tin: Burma, Siam, Perak, Malacca and Banka. The overall commercial policy of the VOC was to secure exclusive rights to purchase tin at a fixed and low price,[126] though in the case of Burma[127] and on occasion in other places, it traded in a competitive market. With the assistance of Johore, the VOC captured Malacca in 1641. Elsewhere, its policy rested on negotiations with independent local rulers. The pattern of negotiations naturally varied according to each specific political configuration and the role that the VOC could play within it but one result was common to them all. Complete enforcement was rarely possible and a black market emerged, supplied both by local officials and those of the VOC itself.[128]

Political weaknesses, both external and internal, led rulers to grant these extraordinary commercial concessions to the VOC. As a European organization, the VOC provided Siam with a useful ally against other European forces. Tin played a particularly important role in reinforcing this relationship since the deposits on both west and east coasts were located in areas where royal authority was weak. Ligor, the east coast centre, was prone to outright rebellion. As the VOC enforced its commercial rights, including the right to confiscate black market tin, it also strengthened that authority.[129] This agreement ran from the 1640s to the 1740s, when the VOC found itself unable to compete with the Chinese who not only exercised greater commercial power thanks to the growth in the trade with China but also greater power in the actual administration of the Siamese state.[130]

Royal power was even weaker on the west coast of Siam where the centres of tin production were Phuket and Pagngna.[131] Here the VOC negotiated agreements through local magnates in 1643 and 1644 but these lasted only for a short time. As a frontier area, the region allowed free mining and the miners simply paid a royalty to the crown in the form of a share of their tin.[132] However, by the early eighteenth century, the extension of Chinese influence at Ayuthaya had led to the appointment of Chinese governors whose oppressive administration reduced the incentive to produce tin.[133] That was eventually relaxed and tin became part of an extensive trading network largely under the control of Malays and Bugis.[134] They brought textiles from the Celebes, together with Chinese goods acquired at southern entrepôts and exchanged them for tin, which in turn was exchanged with English country ships who brought opium from Bengal.[135] As this activity grew, so too did the concern of the central government about the need to assert its power in the region and the implications of the way in which that problem was solved will be considered later.

The initial agreements signed by VOC, first with Aceh, as the overlord of Perak, in 1639 and then with Kedah in 1642, failed to secure much of the regional tin trade which remained in the hands of the Indians.[136] Although a blockade managed to expel them for a short period, Perak responded with violence and the VOC soon found itself rebuffed in Aceh and Kedah as well. Unwilling to modify its commercial policy and unable to enforce it because of insufficient firepower and discipline on the part of its servants, the VOC had failed in its ambitious attempt to control the bulk of the seventeenth century Asian tin trade.

Failure did not lead to a revision of the overall commercial strategy and the eighteenth century saw it successfully applied to both Banka and Perak. Banka was part of the Palembang sultanate based in southern Sumatra where the VOC had long been buying pepper. Soon after tin was discovered in 1710, a violent succession dispute erupted and was only settled thanks to armed support provided by the VOC.[137] In 1722 it was rewarded with an exclusive contract for the purchase of tin supplied to the sultan.[138] VOC military support continued to be essential and it repulsed the attempt of the Bugis to capture the island in 1731.[139]

At first, the tin was worked by the indigenous Bankanese and received as tribute but it soon became evident that the whole island contained extensive deposits and a coherent policy was introduced under Chinese from Palembang. Chinese were then recruited on a large scale from Yunnan and the technology they developed in both mining and smelting permitted a substantial increase in production.[140] What was equally important was the development of a systematic organization of the industry. Key to this was the appointment of ethnic Chinese, ‘Tikos’, as intermediaries between the sultan and the miners/smelters who were then organized as autonomous work teams, or kongsis. The VOC advanced credit to the Tikos who then met the capital requirements of the kongsis.[141] The kongsi received a fixed price for each picul of metal delivered, around half the amount paid by the VOC, with the district Tiko and the sultan sharing the difference. While this provided an incentive to improve efficiency, it also ensured that deposits would be abandoned as soon as more lucrative ones were located.[142] By the 1760s, Banka was responsible for an annual production of over 3,000 tons,[143] outstripping Cornwall, of which around 10 per cent went to Amsterdam and the rest to China. With occasional interruptions, Banka would always remain the largest single producer.

The levels reached in the 1760s could not be sustained throughout the rest of the century. In part that was the result of the exhaustion of the very high grade deposits but it also reflected two administrative weaknesses. The sultan replaced Chinese Tikos with his court favourites who found it difficult to communicate with the miners. The VOC itself was moving towards bankruptcy and was no longer in a position to provide security for the island. Banka's future would therefore rest on the development of a new administrative system. However, it had shown how new technologies, developed within an ethnic enclave and supported by interrelated chains of mercantile capital, could be the basis of a very lucrative industry.

The market conditions that had allowed Banka to flourish also attracted the interest of the Bugis who extended their political control over large sections of the Malayan peninsula to secure tin supplies. Control over Johore allowed them to establish an entrepôt on the island of Riau where tin could be exchanged for Chinese goods and Bengal opium which were then redistributed throughout the peninsula and archipelago. Given the proximity to Banka, Riau was also the destination for much of its smuggled tin.[144]

Bugis influence extended much further north to another major tin producing area, Selangor, where Indian and English traders predominated and this exacerbated the weakness of the Perak sultanate. Perak was a small state, surrounded by much more powerful neighbours and suffered from the chronic factionalism that beset Malay polities. Since they could find allies within the royal court, the Bugis were both an internal and external threat and actually mounted an attack in 1743. The sultan therefore turned to the VOC to shore up his position. A treaty was signed in 1746 which granted the company monopoly rights over all tin brought for sale by the sultan and other members of the aristocracy. Although Perak compensated for the loss of Ligor, it only served to supplement supplies from Banka and the terms of the treaties were the source of constant friction. The VOC paid more than it did at Palembang and less than the price available in neighbouring Kedah, to which large quantities were smuggled.

A severe drought and famine hit Perak in 1776 which made it impossible to enforce the corvée and the VOC advised the sultan to recruit Chinese miners.[145] The advantages of a permanent commitment were soon obvious. Corvée labour was not only less efficient but it was also sporadic. Banka had shown how the Chinese, as a permanent and specialized labour force, could expand production and they would eventually transform not only the economy but the sultanate itself.

As long as smuggling continued on a large scale, increasing production in Perak did not make the overall strategy of the VOC more secure. As it attempted to do so, the VOC inadvertently dealt itself a fatal blow. In 1783 both Riau and Kuala Selangor were attacked. With the support of a Netherlands naval force, the VOC succeeded in its immediate aim of destroying the base of Bugis power but it had two unanticipated consequences. Perak no longer needed the VOC and refused to renew its treaty; the British felt that they needed a secure entrepôt to replace Riau.[146] In 1786 the EIC was able to take advantage of Kedah's vulnerability in the protracted Burma-Siam wars to secure the concession of the island of

Penang.[147] Under the aegis of the EIC, Penang then became not only the main port at which tin from Kedah and Perak was sold, but one in which all traders, Indians, Chinese, Bugis, could operate freely. There was now an entrepôt freed from broader political and commercial rivalries and it would provide the base from which new forms of mercantile capital would propel the Malay states to a leading role in the world tin industry.

By the end of the eighteenth century the European and Asian sectors were beginning to converge. In 1787 the VOC shipped an extraordinary amount of Banka tin to Amsterdam[148] and prices for Cornish tin declined as a result. Cornwall then turned to the EIC for assistance. An agreement was then struck between the smelters, organized as Associated Tinners, and the EIC whereby metal would be bought at a price fixed at a discount of around 20 per cent on the domestic market price and then shipped to Canton. From 1789 to 1797 one third of Cornish production was disposed of in China and domestic prices rebounded accordingly.[149] The success of Associated Tinners in manipulating the market was remembered for a long time. It was also clear that these two sectors could not long be kept apart.

Africa

Southern Africa played a modest role within the Indian ocean trade. In precolonial times, the mines at Rooiberg in the Transvaal produced around 1,000 tons of metal, much of which appears to have been destined for India.[150] This was probably the same source that supplied Delagoa Bay where the VOC was buying limited quantities in the 1750s.[151] Further north, Rhodesian tin was also being shipped to India, at least for a period during the early sixteenth century.[152] However, by the late eighteenth century, trade with Southern Africa had come to an end.

Nigerian cassiterite deposits were rediscovered in the sixteenth century by Sudanese or Arab merchants and incorporated into a trans-Saharan trade.[153] By the eighteenth century, the cassiterite was not only being smelted locally but the metal also supported a flourishing industry which manufactured household utensils and religious artefacts. This is one of the few cases, after the end of the bronze age, where mining, smelting and fabrication were integrated within the same region.

Early Modern Europe

During the course of the eighteenth century, Cornwall experienced a major shift in both mining and smelting. While alluvial deposits remained an important source, they became subordinate to the extraction of veins located underground. Exploitation of lode deposits had begun in late medieval times and drew on techniques of crushing and preparing ores developed by other non-ferrous metal mines.[154] Gradual growth led to an entirely new technology of smelting, based on the reverberatory furnace. That operated on a much larger scale than the older blowing house which treated alluvial tin and relied on imported coal rather than local charcoal.

As underground mining operations became larger they also worked deeper deposits and required much larger, more specialized and more professionally managed work teams. They also required substantial infusions of capital in development work, in the pumps required to keep the mine free of water, in the batteries of stamps that crushed the ore and in the equipment needed to clean it.[155]

Even when cleaned, concentrates from lode veins contained at best 65 per cent metal, whereas those from alluvial sources were around 70 per cent. The reverberatory furnace was unable to refine them to produce metal with the same degree of purity as that which came from the blowing houses. The existence of two production systems therefore meant two sets of prices. Throughout the eighteenth century there was considerable improvement in the technology of smelting and the price differential would be reduced to the point where further innovation effectively ceased, at least throughout the whole of the nineteenth century.[156]

At the end of the seventeenth century, the recovery of Cornish production was associated with a major shift in the financing of the industry. London merchants were being displaced by local smelters who now dealt directly with consumers. That trend was completed with the emergence of the reverberatory smelter. As there were comparatively few such smelters, each of which would have several furnaces, they could impose a common set of returning charges. It was the existence of this routine collusion that permitted the EIC contract.

Although such a system guaranteed substantial profits to the smelters, an incentive was still provided for investment in mining and processing since the mineowner would capture the differential rents as a result of working higher quality ores and benefit from improvements to the organization of production.[157] The owners of the smelters, in turn, were anxious to secure their source of feed[158] and also benefit from the profits available at the point of production and therefore took shares in many of the mines. In addition, many of them were also suppliers of the materials required. The result, therefore, was a hybrid system of capitalist smelters and quasi-capitalist mines. The smelters worked with wage labour[159] and bought most of their inputs in a competitive market. The mines were owned by shareholders who had multiple points at which they took their profits.

The partnership structure that had served until the seventeenth century was now modified into the cost-book company, which had juridical personality and whose shares, while freely transferable, contained the obligation to meet the costs of the mine, together with the right to its profits.[160] Such a system required a professional accountant, the purser, and regular meetings of the main shareholders at which they would distribute any profits or decide whether to put in additional funds to continue the mine. It did not therefore allow for the accumulation of capital in the hands of the company and limited the capacity to engage in long term development. One source of the technical conservatism of the industry lay in its distinctive corporate form. Another lay in the organization of labour. Development or tutwork was undertaken by wage labour but the actual extraction was done by tributers who were invited to bid for the right to work specific sections of the mine. Their income was entirely dependent on their own skill and energy and few were able to avoid relying on the purser for advances or 'subsist'. Miners remained in a condition of chronic poverty and in many case their wives and children became a source of cheap labour employed in processing the ores on the surface.[161] Just as the adventurers looked toward immediate profits, so did the tributers which inevitably led to unsystematic development of the mine, known as 'high-grading'.[162]

The development of lode mining at first meant a slow increase in overall production but by the middle of the eighteenth century, it was around twice the level reached at the end of the seventeenth. Until the 1780s there was a continuous increase in the exports of both tin and pewter, which showed that Cornwall maintained its overall preeminence in spite of the increases occurring in the East. In addition, an entirely new source of demand emerged, tinplate. However, only the better quality grain tin met its needs and the division between the two kinds of metals would be reinforced.

Coating sheets of iron with tin from which a wide range of artefacts could be fabricated was a skill that dates back many centuries but it was the emergence of the Bohemian/Saxon mines that was to make it a flourishing industry.[163] By 1670, fabrication with imported

German tinplate was being undertaken on such a scale in London that the Worshipful Company of Tin Plate Workers was incorporated as one of the city's livery companies.[164] A tariff was imposed in 1706 to encourage domestic manufacture and by 1750 there were four tinplate works, another seven by 1800.[165] The foundations for the industry that would support a major growth in tin production for the following 150 years were laid.

3 Tin and industrial capitalism, 1815–1918

Tin shared in the upheaval that occurred in all non-ferrous metals industries as new needs and producers emerged in response to the imperatives of industrial capitalism. However, several features distinguish the way in which new forms of competition affected the position of tin in the period from the end of the Napoleonic wars to the end of World War I. Although many important new uses were introduced, consumption grew much more slowly than in other base metals.[1] Nonetheless, these new uses supported a major increase in price, illustrated in Figure 3.1, which restored tin to the role occupied during the Bronze Age, as the most valuable of the base metals. While changes in the pattern of both consumption and production made tin the most dependent on the operation of the international market, extraction under the direct control of capitalist corporations proceeded far more slowly. Even by the end of this period, tin remained a comparatively backward industry still shaped by its distinctive history.

The major shifts in the development of the world tin market are presented in Table 3.1

As the initial centre of new forms of capitalist production, it is not surprising to see that British share of world consumption grew until the 1860s. As others emerged, that share inevitably shrank and by the 1890s the United States had outstripped Britain as the

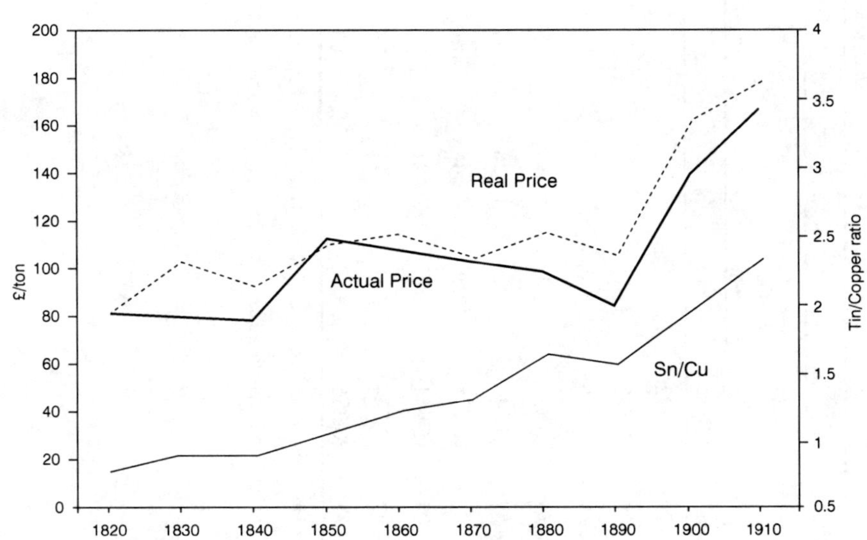

Figure 3.1 Tin prices, 1820–1913

Table 3.1 World tin market, 1820–1913

	United Kingdom								Rest of World			
	Consumption					Production		Trade	Consumption			UK smelter exports RoW
		Increase		Imports		Ore Imports	Exports	World		Increase		
Annual average	tons	tons	%	% cons	% world	% prodn	% prodn	%	tons	tons	%	%
1820–29	2,492				21			37	9,618			19
1830–39	3,520	1,028	41		25		42	41	10,720	1,102	11	10
1840–49	4,580	1,060	30	4	28		23	42	11,670	950	9	14
1850–59	6,821	2,242	49	28	36	3	27	46	11,989	318	3	14
1860–69	8,746	1,924	28	39	36	4	44	58	15,794	3,806	32	26
1870–79	12,776	4,030	46	61	34	7	56	62	24,504	8,710	55	24
1880–89	15,464	2,687	21	74	27	8	56	61	41,006	16,503	67	13
1890–99	17,224	1,760	11	79	22	33	63	53	62,019	21,013	51	9
1901–13	20,236	3,012	17	65	19	81	53	64	85,106	23,087	37	1

Sources: House of Commons, Papers, *Accounts relating to Tin, 1820–1854, Annual Statement of the Trade of the United Kingdom, 1855–1913.*

Note: Consumption is calculated as the residue of metal production, imports and exports.

largest single consumer. Reduction and then elimination of tariffs on imported ore and metal uncoupled the link between British production and consumption. Both consumers and smelters drew increasingly on supplies from abroad and the latter became more dependent on exports for a market for their production. Beneath the surface of an overall world market there were several specialized ones which cut across national boundaries. As will be seen, the main trading links of three of the largest producers were with Britain and that meant a large proportion of world trade, regardless of final destination, touched her shores.

Consumption

The pattern of consumption that had been established in the eighteenth century remained more or less stable throughout the nineteenth and that impression is supported by the very slow rate of growth outside the United Kingdom until the 1860s. What drove growth first there and then in the rest of the industrializing world were new uses directly tied to its major features: mechanization of production and transport, imperialism and urbanization.

Hitherto, tin had entered into household or institutional consumption primarily by virtue of desirable qualities that only it, at least within the range of the available technologies, could provide, such as the sound of bells and pipes and the appearance of pewter and bronze. Fashion dictated that desirability but not the way in which it was reached. While it had also played an important role in various manufacturing processes by virtue of its purely technical qualities, such as solder for joining metals and as a mordant for fixing dyes it was a subordinate one. As that role became predominant, the link between artefact and metal was broken and the consequences were far reaching.

Expansion of this role of tin as an element in the production process rested on technologies whose fundamental features had already been developed in the eighteenth century but which were capable of considerable innovation. As industrialization proceeded, new problems emerged for which tin often offered the best solution. Two examples illustrate the new role for the metal.

Metal axles revolving at high speeds can overheat the unit in which they are contained. Tin proved to be the basis of antifriction metals, the best known of which is associated with an American, Isaac Babbitt. Developed in the 1840s, it solved the problem encountered not only by railways but also by many forms of machinery.[2] The second example is the familiar one of tinplate which solved the problem of how to package food that could be transported over long distances and by the end of the century it was an indispensable feature of several aspects of the food industry. Tin had therefore ceased to be merely a desirable metal and had become one which was essential to modern civilization.

Although essential, tin was, at best, only a minor part of the final product and that had two major consequences. It made demand price inelastic and the market was therefore particularly prone to price fluctuations. It virtually eliminated any basis for the kind of organizational linkage between producer and fabricator that was emerging in other non-ferrous metals industries. That relationship would be entirely mediated through a competitive market.

British domination of tin consumption was a function of the competitive advantage it enjoyed in the production of iron and steel which enabled it to establish an effective monopoly of tinplate.[3] By the 1880s well over half of British tin consumption went into its manufacture. Although a few tinplate works operated in Europe, none produced sufficient to meet even their domestic demand and until the 1890s Britain supplied virtually the entire world. In 1805 production was 4,000 tons of which 63 per cent was exported; by 1891 production reached 586,000 tons of which 76 per cent was exported and of these exports 73 per cent went to the United States.[4]

Behind this remarkable expansion lay a series of technical developments. It is the iron and steel that give tinplate the qualities that make it extraordinarily useful but they need a coating of tin to prevent oxidation. Technical innovation occurred in three different spheres: fabrication of the plates, coating them with tin and manufacture of containers and utensils. Innovation in all three spheres was continuous and gradual and would continue well into the following century.

Of the many innovations, the one that had the greatest effect was the replacement of iron with steel in the 1870s. The smoother surface made bonding with tin easier and the stamping machinery could produce containers in a single process. Both of these developments reduced the amount of tin required but since they made tinplate cheaper and more versatile they increased its demand.[5]

As long as tinplate was comparatively expensive, it was mainly confined to durable products. The technology that had manufactured domestic utensils was adapted to make a wide range of products, as diverse as gas meters, milk churns and equipment for refining sugar. From the 1850s whole sheets were being manufactured for use as roofing material, known as terneplate, where the tin served to bond a lead coating to the iron base. Its light weight made this an ideal solution to the problem of the construction of dwellings in tree-less frontier areas.

As the cost of tinplate dropped, its role in the demand for disposable containers increased. Naval interest in tinplate as the basis for a container that would preserve food for a long period goes back to the mid-eighteenth century[6] and the Napoleonic wars provided further stimulus to food preservation. But the 'heroic age' of the tin can began with the British expeditions to the Arctic where members of the mess marvelled at their ability to enjoy the taste of exotic fresh food in the most inhospitable of environments.[7] Canned food therefore not only overcame a material obstacle to the extension of European control over the world but, since Europeans could maintain their lifestyle anywhere on the globe, it also served as a symbol of that control.

Canned food also served as a marker of social position at home, since it enabled foods to be consumed out of season and it was celebrated in the Great Exhibition of 1851 as a symbol of all the Victorian values of 'healthiness, hygiene, modernity and progress'.[8] As cans became cheaper and easier to open, the packaging of food on a large scale could begin. In the 1870s Australian meat set the pattern, since an otherwise worthless product could now be commercialized and sold to those unable to afford domestic beef.[9] By the end of the century entirely new foods, such as salmon from Canada, fruit from California, sardines from Portugal and pineapples from Malaya, had become a staple of many urban diets.[10]

Tin containers were to be found in many other areas of domestic consumption. From the 1850s petroleum products were all marketed in tins and a major expansion occurred in the 1870s when the technical problems of imprinting designs were overcome. This served to promote a wide range of branded products such as mustard, tobacco and cigarettes.[11] Of this kind of container, the most important was the biscuit tin. It served as a suitable gift item, especially at Christmas and one that continued to have value as a practical, attractive container or toy long after the contents had been eaten.[12]

Most of these developments were disproportionately concentrated in the United States and the rapid growth of its economy following the Civil War placed it among the leading consumers of tin products. By the 1880s the American iron and steel industry was strong enough to press for the domestic manufacture of tinplate and, thanks to the initial protection offered by a tariff for an infant industry, it quickly flourished. By 1912 it was not only producing more than Britain but was competing effectively in many of her other export markets. Where the British industry was made up of a large number of small, conservative firms, the American one soon became far more concentrated both in the manufacture of tinplate and of the cans and that laid the basis for continued technical innovation.

As the century came to an end a new product appeared, the automobile, whose production was even more concentrated in the United States. That also used considerable quantities of tin, especially in solder for radiators. By the 1890s the United States had become the largest single consumer of tin metal and one of the central features that would shape the development of the global tin industry throughout the twentieth century was now in place.

Exchange

The international market became integrated in two separate phases. The first, which began in the 1830s, saw the emergence of a world trade almost entirely untrammelled by tariff barriers. The second, which began in the 1870s, saw the emergence of a single set of interconnected prices which governed that trade.

It was in the 1830s that the price differential between China and Europe came to an end and increasing quantities of tin from Southeast Asia flowed westwards. Competition in her most important foreign markets reduced Cornish exports to less than half the level reached in the 1780s and relief was sought in the abolition of coinage which was granted in 1838. At the same time duties on imports of foreign tin were reduced in response to pressure from the tinplate trade and were finally abolished in 1852.[13] A few countries imposed modest revenue tariffs but none served to generate any significant distortions.

The second phase rested on the telegraph as the basis of instant distribution of information. In 1871 Southeast Asia, China and Australia were connected to London; Bolivia followed in 1875.[14] This not only ensured that virtually all market participants had the same information but it also allowed for the rise of a new marketing system organized by the London Metal Exchange (LME).

The problem addressed by the LME was one of the ways in which market risk is allocated and it was one that became particularly manifest with the rise of Chilean copper and Australian tin. The telegraph solved one part of the problem by permitting a sale prior to the delivery of the metal. The LME was formed in 1877 to solve the other, a guaranteed market for the future delivery. This was accomplished by changing the basis of trade from that in physical metal to a contract for its legal title, or warrants, against which metal held in an authorized warehouse could be delivered. The LME then specified the brands which were considered as meeting the terms of the contract and delivery could be either immediate or at some point in the future, which was standardized as 90 days.[15]

The market in future deliveries not only fixed a price but, since it operated in connection with one in prompt delivery, it allowed for hedging and therefore eliminated the risk faced by traders and others who bought metal.[16] These advantages were secured by a fundamental principle of the LME. Although those who participated in open competitive bidding at the ring sessions, held twice daily, may have been acting on behalf of clients, as far as the LME was concerned they did so as principals. Membership was therefore limited to individuals and firms whose financial standing was strong enough to support their commitments.

However, the advantages of predictability and risk reduction came at some cost which went far beyond the modest brokerage commissions. A futures market in which the contracts are guaranteed[17] invites speculation, both from brokers who manipulate information and from outsiders who control supplies which accentuates price instability.

Tin was part of the most notorious of all corners attempted on the LME. Secretan had formed a successful tin pool in early 1887[18] and in October he started buying Straits tin, driving the price from £105/ton to an unprecedented £167/ton six months later. As is so often the case with such speculations, the overall stock and supply position was misjudged and in April support was withdrawn from the market. Overnight the price dropped to £80/ton.[19] The Secretan corner could be attempted since the LME contract only specified Straits and Australian as brands which constituted good delivery. To widen the market, Banka and Billiton brands were included in 1897 and in 1911 the market was widened much further with the Standard Tin contract. This established two grades: Grade A included all brands that met a minimum of 99.75 per cent tin and Grade B included those with a minimum of 99 per cent tin. The price was established for Grade A but the seller had the option of delivering Grade B at a specified discount. With the exception of a few lower grade metals that were produced for local markets, all of the world's smelters had their brands recognized on the LME, including the English smelters whose refined tin met the requirements of grade A and their common tin, grade B. While widening reduced the extent to which private interests could affect the market, tin retained its reputation as a notorious plaything of speculators.

LME prices were benchmark prices which served as a point of reference in all other markets. Markets also existed in Singapore, Amsterdam, Hamburg and New York but they only dealt in physical metal and all operated within the shadow of London. They were also prices against which specific premia and discounts for particular brands were independently established as the particular needs of final consumers were met.[20] As this system developed, especially in response to the rise in American tinplate production, the pattern of premia changed. Until at least the 1890s, both English refined and common tin sold at a premium over eastern metals but then the pattern was reversed. Although eastern metals remained good delivery under the LME standard tin contract, it was mainly English refined that was delivered under its terms and which was therefore known in the trade as 'standard tin'.

Production – historic centres

The price history presented in Figure 3.1 indicates the existence of two major booms, punctuated by a severe slump. Production history broadly corresponds to these three phases. The first which culminated in the boom of the 1850s was largely confined to the historic centres that had emerged by the end of the eighteenth century: Cornwall and Southeast Asia. The second, the slump of the 1870s and 1880s, was entirely the result of the exploitation of new deposits in Australia whose effects continued to be felt in these centres throughout the third phase, the recovery that lasted until World War I. This section will review how these centres developed throughout the entire period; a subsequent one will consider how new centres emerged towards its end.

Cornwall

The boom of the 1850s stimulated production from two sources. Old mines were reopened, most of which failed, often with enormous losses.[21] Declining copper mines provided a second opportunity. Their ores were being depleted in a market that had turned against them. The largest of these, Dolcoath, took an enormous gamble on the prospect of tin lying beneath the copper and it was one that turned the company into the most important of all the Cornish tin producers. It was distinguished not only by size but also by the scientific basis on which all aspects of mining and processing were undertaken.[22]

Dolcoath's recovery was quick and it had many imitators. Some were reasonably successful but many were flimsy operations designed to generate immediate rewards for the parasitic promoters and share-pushers.[23] From 1862 companies could be formed with limited liability which made external investment more attractive and the Redruth Mining Exchange was formed immediately thereafter as one place where gullible English investors could lose their money.[24] Promoters found mineral lords anxious to co-operate by dividing their properties into a multiplicity of separate operating companies and that would make it difficult to work even viable deposits on a sound basis. Production naturally increased in response to these incentives but it lay on an increasingly insecure foundation. By 1874 there were 163 different lode mining operations, with a median production of less than 60 tons.[25] The property rights system that had enabled Cornwall to flourish was now an obstacle to its long term development.

The position of the smelters was quite different. Expansion of production permitted the entry of two new smelters but they were incorporated within a system of strict allocation of the available concentrates. One of the more recent entrants and who was therefore confined to a marginal position decided to seek fresh sources by turning to concentrates from broadly similar lode deposits in Bolivia.[26]

The long dominance exercised by Cornwall in western markets came to an abrupt end with the discovery of rich alluvial and lode deposits in Australia in the 1870s.[27] By the 1880s she had become the world's largest producer and the sheer magnitude of this new output produced the long slump in price.

The immediate effect of Australian competition revealed an apparently paradoxical feature which would recur in subsequent depressions as mines increased output in order to minimize costs and to extract as much as possible before they closed.[28] Those that survived turned to two new techniques which extended their lives. Rock was drilled by machine rather than by hand and dislodged by dynamite rather than by gunpowder. As a result, the productivity of labour doubled within just five years.[29] New mines would continue to open with some success but from 1878 the industry went into general decline, the rate of which was only slowed by the continued strength of Dolcoath. By 1900 production from the remaining mines was reduced to the level reached in the 1750s.

The 1880s also saw the end of the closed system of smelters. Two new ones were established, Penpoll, partly owned by a metal broker, Strauss, and Cornish Tin Smelting. The newcomers forced a genuine competition for both Cornish and Bolivian concentrates.[30] As Cornish supplies declined, some smelters became dependent on the increasing supplies from Bolivia and two left Cornwall altogether. In 1901 Penpoll moved to the Liverpool region, followed in 1908 by Williams, Harvey. They were then in a position to play a major role in the world market as it evolved in the twentieth century.

Malaya

In the era of mercantile capital, competition had contradictory effects: accentuating the level of political instability among the Malay sultanates but increasing the domestic power of the sultans. The new forms of international competition that emerged in the nineteenth century had a singular effect in destroying their internal political structures and creating the conditions under which the British government felt compelled to assume political responsibility. Under its aegis, Malaya became the world's largest tin producer, a position she would retain for another century. In this process the Malays found themselves largely displaced by Chinese miners and capitalists and, as it was completed, the Chinese found themselves being displaced by European and Australian capital.

As will be seen, the nineteenth century began with the creation of secure entrepôts in Penang and Singapore and a new financial system emerged. As Chinese merchants found themselves increasingly squeezed by Europeans in the overseas trade, they turned to the riskier but more profitable business of financing tin production on the peninsula.

Chinese mercantile capital tended to move directly to the local Malay chiefs, bypassing the sultans, and the focus of power within the sultanates therefore shifted. While the emergence of new centres of power meant that the profits of the tin trade were captured locally, they also created conditions for a struggle between chiefs, especially with those who controlled the transport routes. A vicious cycle was then set in motion. The chiefs preferred to use their Malay subjects for their militias rather than for mining. Financial pressure on the mining industry increased to pay for these militias and that pressure was easy to apply since the chiefs not only controlled the distribution of the tin but also the supplies, especially of opium, needed by the miners. Since the Chinese lived within a social universe shaped by clan associations and secret societies, they had the organizational capacity with which to respond. At first these tensions broke out into massacres of Malays by Chinese and vice versa but in the 1860s they led to open civil war as the rivalries between different clan associations were linked to the rivalries between the Malay chiefs. The structures that had supported the growth of tin production then simply collapsed.[31]

In 1867 administrative responsibility for the Straits Settlements was assumed by the British Colonial Office and it came under pressure to intervene and guarantee the property claims of the Chinese finance capitalists. By the early 1870s the tin trade was dropping off rapidly and the Colonial Office set aside its earlier hesitation and authorized the Straits Governor to intervene in the internal affairs of the tin states.

Intervention was designed to restore the authority of the centre and establish a system of property rights that would allow tin production to flourish. It could then be taxed and the proceeds reinvested not only in the infrastructure of roads and railways that would continue to support mining but also in a wide range of public services. The formula adopted co-opted both the chiefs and the sultans who were provided with the resources required to maintain their symbolic functions. However, they were divested of real power since they were required to accept the advice of a British Resident. In 1874 Perak, Selangor and Sungei Ujong accepted the de facto loss of their sovereignty.[32]

The property rights system evolved over a period of time but it started from the assumption that the mineral in situ was the property of the state which had the right to determine the conditions under which it could be exploited. Those conditions were codified administratively around the concept of the mining lease. The bundle of rights attached to the lease included the clear delimitation of the physical boundaries within which the lessee could mine, the right to sublease and the right to sell. The prospect of the last of these being abused in speculation was pre-empted by requiring that the land be worked at a certain level on pain of forfeit and that incentive was reinforced by imposing two limitations. Leases were granted over comparatively small areas for a maximum of 21 years. Prospecting was subject to complementary regulations. These established an exclusive right to prospect over a defined area within a short period of time and those who found payable deposits had the right to apply for a lease.[33] By deciding how much land was available for mining, who should receive an exclusive prospecting licence (EPL) and the actual conditions of both licence and lease, the state was in a position to establish a coherent mining policy.

This system was ideally suited to the needs of the Chinese miners. It gave them absolute security against both the state and their rivals, together with a property to serve as collateral for loans. It also served the needs of the state. Its main source of income was a royalty on the metal contents of the concentrates at the point of export, based on a sliding scale of prices determined in the Singapore market. This was supplemented by substantial fees representing a portion of the value of the mineral in situ on the issuance of the original lease. The administrative expenses of an elaborate bureaucracy required to issue leases and supervise their workings were met by a rent on the acreage of the land alienated for mining.

Co-optation of the Malay chiefs eliminated the role they had played as intermediaries between the miners and the Straits Chinese. While they could still derive a rentier income from mining operations on land they owned, they became increasingly marginal as new lands were opened up. Their primary role was now taken by local Chinese capitalists, known as 'advancers', and granted the courtesy title of Towkay. While miners were now freed from the political constraints of the Malay chiefs, most were caught in a web of financial commitments to the Towkays. The basis of this relationship was the need for working capital which the Towkays supplied in return for a share of production and the right to buy much of the remainder at below market prices. They then arranged for its smelting and shipping to Singapore or Penang where it was refined. Superimposed on this relationship were many other opportunities for profit: provisions to the mine, gambling, brothels and opium for the workers and revenue farming. The whole web was sustained through secret societies and political connections and many Towkays became very wealthy.

Such a system of multiple opportunities for profit provided a major incentive to production regardless of the returns at that stage but it ultimately rested on the competitiveness of the metal produced by Chinese smelting. That had several technical weaknesses: certain types of ores could not be treated. Recovery rates were low and the metal had only a 95.66 per cent tin content and it rested on supplies of charcoal which were being depleted. These would only be overcome by European technology.

Following the establishment of protectorates over the west coast states, the new administrations recognized the technical inefficiency of Chinese mining methods and were anxious to see the introduction of western capital and technology. Incentives were offered on the naive assumption that the simple application of western forms of organization and equipment would serve as a demonstration effect for the Chinese miners.[34] Most of the early companies quickly foundered on a fatal combination of misleading prospecting, highly paid administrative staff, reluctance on the part of Chinese miners to work for Europeans and unsuitable equipment.[35] In the 1880s there were only two successes, both of which were the result of French mining engineers originally invited to conduct scientific expeditions. In 1884 de la Croix and de Morgan merged the companies they had first formed on their return to Paris into Société des Etains de Kinta (SEK) and this formed the nucleus of a group of companies which continued for another century.

The grouping principle was followed by much of the subsequent capitalist investment that flowed into the industry. The operating companies were 'free-standing'[36] and only mined tin, normally in just one location. Rather than expand the scope of an existing company, its directors found it more advantageous to raise the capital required by floating a separate company but keeping control through cross-appointments and an equity interest. The prospect of returns directly linked to fresh deposits may have made it easier to raise the capital required and the system allowed existing shareholders to decide whether they wished to share in the risks.[37]

SEK was also a model of technical adaptation. Steam engines were introduced to extract and treat the alluvium, followed by a Vlaanderen smelter capable of producing a metal which required no further refining. Not content with the substantial profits made out of a large production, SEK pioneered the use of electricity and in 1906 built its own hydro-electric plant.[38]

On the east coast, there was an entirely different development. Mining in Pahang had languished since the 1850s but the reputation of the state as one of unbounded riches remained. Here the sultan continued to exercise effective authority and decided to take advantage of the gold fever of the early 1880s to grant extensive concessions. One of these covered tin deposits which had already been proved by Chinese and Malay miners and in 1886 it was acquired by London interests who floated it as Pahang Mining Company. A Cornish mining expert provided a very favourable and honest report which would see the property developed as a large lode mine. This would require a more secure title, one that only a British administration could provide and in 1888 the sultan agreed to accept a British Resident. That prompted a rebellion on the part of some Malay chiefs which required British support to subdue.[39]

While the new administration established tighter regulations for mining operations, it was forced to recognize the initial 77 year concession over 200 square miles. Development work and the low prices of the 1890s prevented much production but by 1906 the whole property was being exploited by Pahang Consolidated which would become one of the world's largest lode mines.[40]

The 1890s saw a major technical innovation. From the 1850s the gold mining areas of the American West were being worked by a method known as hydraulicing.[41] It relied on the pressure of water to break down the alluvium and the resulting slurry was then processed through a series of sluice-boxes to capture the metallic particles. Conceptually simple, it was a method that required 'large initial investment, and skilled management and engineering'.[42] An Irishman, Osborne, had witnessed the success of hydraulicing in California and started experimenting with its application to tin in Perak.[43] In this he was assisted by a Cornishman who advised that they turn to the Redruth capital market for the resources required.[44] Gopeng Tin Mines was floated in 1892 and was eventually reconstructed as Gopeng Consolidated; it became one of the largest, longest-lived and most profitable of all tin mines in Malaya.[45]

Gopeng's demonstration of the viability of hydraulicing had two major consequences for the subsequent development of the Malayan tin industry. It showed that the future role of Western capital would lie in opening up fresh areas that were too poor to be worked by the Chinese.[46] It was a role that Cornish capital was only too anxious to play. Eight similar, interrelated companies were formed in Redruth before 1914.[47] At the same time, Osborne formed a partnership with another Cornish engineer, Chappel and this would serve as consultants not only to the original Redruth group of companies but to several others that were subsequently floated with wider capital support.[48]

One of these other companies had its origins in the discovery of a rich layer of tin ore but covered by an overburden too deep to be excavated by existing Chinese methods. In 1898 Foo Choo Choon brought in an Australian engineer to organize production; three years later he turned to Redruth to raise fresh capital and organized Tronoh Mines. This rapidly became one of the largest producers and it would serve as the basis for a second group of companies under Cornish control.[49]

Both the Redruth and the Tronoh groups were headed by strong and articulate Cornishmen, the former by Frank Mair and the latter by Charles Thomas, and they saw themselves as pioneers not simply of very successful companies but indeed as founders of a great British industry. In the case of at least the former it had something of a religious coloration:

> It was very difficult for men in public life to be a strong witness to the Christian faith but Frank Mair was one who did. He was one of those who belonged to that fast dying generation to whom the building up of our wonderful Empire was a personal and devout thing.[50]

Many other large Chinese miners turned to western management without losing control and most of the smaller ones adopted a modification of the suction dredge in the form of the steam-driven gravel pump.[51] This was a method[52] which could work many deposits that did not lend themselves to large scale exploitation and it also fitted well with the established system of advancers. Modern technology allied to traditional forms of finance and entrepreneurship ensured that the Chinese mining sector would be sustained well into the twentieth century.

Where European capital was only beginning to make inroads into mining, the reverberatory furnace enabled it to rapidly take control over smelting. In 1886 two individuals associated with agency houses based in Singapore[53] persuaded the governments of Selangor and Sungei Ujong to drop their ban on the export of concentrates and give it a monopoly; with a secure source of concentrates, they formed Straits Trading Company (STC).[54] In order to draw on imported coal and a wider source of concentrates, they moved to an island off Singapore and by 1890 two reverberatory furnaces were in operation.

STC defined itself as a specialized smelter capable of applying the most advanced forms of western metallurgical practice. That enabled it to produce a very high quality metal, 99.97 per cent tin which would eventually command a premium as Straits tin. Of more immediate significance were the implications for the system of marketing. Cash was paid for the concentrates whose contents were determined on the basis of scientific assays and that solved the problem of financing experienced by many miners.

More competitive purchasing practices provided one support for continued expansion of production in the 1890s but at first much of this was captured by existing Chinese smelters who found ways of cutting their own costs. However, one of these smelters decided to meet STC on its own terms and in 1898 Lee Chin Ho built four reverberatory furnaces in Penang, drawing ores not only from his own mines in Perak but also, thanks to his close connection with the Khaws, from Siam.[55] STC then responded to this competition by building its own smelter with eight large furnaces across from Penang at Butterworth in 1901.

This competitive pressure led the Resident of Perak, Sir Ernest Birch, to persuade several large Chinese mine owners to take over the Lee Chin Ho smelter and enlarge it and in 1907 they formed Eastern Smelting (ESC). As a Chinese company, it enjoyed a competitive edge and by 1910 it was producing 29 per cent of the total shipments from the Straits.[56] However, it was not strong enough to continue to meet the response from STC. In 1911 the company was reconstructed in London in order to raise further capital. ESC then became a British company, with a minority Chinese interest. Birch, now in retirement, together with Cecil Budd, Chairman of a metal broking firm, Vivian, Younger, Bond (VYB) and of the LME, took control.[57] With ESC and STC now both powerful modern smelters competing for concentrates, the fate of the local smelters, not only in Malaya but also in Burma and Siam, was sealed.[58]

It is against the background of growing European mining interests that the final moves towards consolidating British rule over the whole peninsula were made. A confederacy of several autonomous states, including Sungei Ujong, was constructed as Negri Sembilan and made part of the Residency system. Repression of the Pahang rebellion had left that state bankrupt and in 1895 its sultan, together those of Negri Sembilan, Selangor and Perak, signed a Treaty of Federation and the Federated Malay States (FMS) was brought into being.

The administration of mining policy remained a state responsibility but its formulation became a federal one. Federation made the establishment of a professional civil service possible[59] and in 1906 the final feature of a conventional British colony was put in place. A Federal Council was created with legislative authority and, while officials retained a veto, representation was granted to various business interests, including both Chinese and European miners.

As European mining became more successful, there were fears that the northern Malay states, which acknowledged Siamese suzerainty, would succumb to the temptation to issue concessions to non-British interests and draw in other European powers. To pre-empt that possibility, Britain managed to detach Perlis, Kedah, Kelantan and Trengannu from Siam in 1909 and they each accepted another modified form of Residency. These, together with Johore,[60] were then aggregated for statistical purposes as the Unfederated Malay States (UFMS). A formal coherence to this political geography was provided by making the Governor of the Straits Settlements High Commissioner to the FMS and each of the unfederated states.[61] While the term 'British Malaya' attempted to express this coherence, it always had to be qualified to recognize the real political divisions between those parts which were a crown colony, those which were not but which had many of its features and those which only had a British advisor.

The only external threat to the Malayan tin industry came from the United States. A large tinplate industry had emerged after the McKinley tariff of 1892 and in 1903 Standard Oil, the largest consumer of tinplate, decided to build a smelter in New Jersey, planning to feed it with concentrates from Malaya.[62] In Malaya, the fear was that, although such a smelter was unlikely to prove economic,[63] the Americans would protect it with a tariff on imported metal and thereby undermine the competitive position of the Straits smelters. If their supply of concentrates were reduced substantially, they would be forced to raise their returning charges. To pre-empt that possibility the FMS government imposed a prohibitive duty on the export of concentrates outside the Straits.[64]

Some important political consequences followed from this decision. One was the feeling in the United States that Britain had acted to prevent American competition and to maintain control over the international market in the metal against her interests. That resentment would resurface on several occasions.[65] As the major beneficiary, STC considered it had a particular moral obligation to remain British and support the Malayan industry.[66]As will be seen, that industry became increasingly British in composition, especially after Siam had demonstrated the viability of the new technology of dredging.

Netherlands East Indies

The beginning of the nineteenth century saw a continuation of the old English-Dutch rivalry for control over the tin trade of Southeast Asia. War in Europe provided the EIC with the opportunity to occupy Banka in 1812 and the company was interested in rebuilding production to supply the China market. Its agent, Raffles, kept the basic system established by the VOC in place but modified it in two ways. He dealt directly with the kongsis, thereby eliminating both the Tikos and the sultan, and gave them an adequate set of price incentives.

Unfortunately for the EIC, the same war changed the overall political context within which it could operate. The British government was anxious to let the Netherlands exploit its eastern possessions in order to strengthen its position in Europe as a bulwark against any resurgence of French expansionism and the EIC was therefore compelled to relinquish its claim.[67] However, the Anglo-Dutch treaty of 1824 granted the company a major concession when objections to its continued occupation of Singapore were waived.[68] As the long Anglo-Dutch contest for control over the Malayan tin trade came to an end, the inland sea of the Malay world became an international boundary and the Straits of Malacca served to divide their respective spheres of influence.[69]

Banka

When the Dutch returned to Banka they continued to operate the system established by Raffles and the profits became an important part of government revenues. However, by the 1850s it was experiencing three kinds of problems: labour, management and technology.

The labour problem lay in the fact that working conditions on Banka made it difficult to recruit sufficient fresh workers from the coastal ports of China, especially since the demand for coolies was rising not only in Southeast Asia but also in North and South America. That was overcome when the contracts were handed to wealthy Chinese who had settled on the island and who were able to recruit from the interior.[70]

The management problem lay in the autonomy granted to the kongsis who were free to decide where they would work and which had led to an unsystematic development of the deposit. In 1858 a Dutch mining engineer developed a new kind of drill which provided a more accurate assessment of the nature and extent of the deposit. Considerable investment was made in prospecting which identified several new and highly productive areas and the engineers could plan just where the kongsis would work. Control of production was now in the hands of the professionals and that permitted further technical improvements. Railways replaced wheelbarrows and steam pumps compensated for the deficiencies of water-driven chain pumps. Steam pumps allowed for deeper workings and by 1900 the large amount of labour required to lift the ore to the surface was reduced with the introduction of steam-powered trolleys which rode on rails up an incline.[71] With a solution to the labour problem and control in the hands of mining engineers, production virtually tripled from 1880 to 1900. Banka was now the world's largest production unit.

By mid-century Chinese smelting technology had run up against the limits of the local supply of charcoal. A Dutch metallurgist, Dr Vlaanderen, then developed a new kind of furnace which both used less charcoal and produced more metal and that meant further European control over production.[72] Forests were preserved by encouraging farmers to abandon slash and burn agriculture and cultivate pepper. Banka could then count on an adequate supply of charcoal to continue to produce a premium brand of tin.

With mechanization and the presence of a considerable body of Dutch technical and supervisory personnel, Banka was formally reorganized. In 1913 Banka Tinwinning was separated from the civilian authority that governed the island and made responsible to a Department of Government Enterprises.[73] That would allow for greater technical supervision over its operations, together with greater political control.

Billiton

In 1850 the Dutch government invited private interests to develop the mineral resources of the archipelago and a group, including Prince Hendrik, secured a forty year concession for Billiton.[74] The group committed considerable resources to prospecting[75] and soon found itself without the capital required for development. It then formed Billiton Maatschappij (BM) in 1860 which raised the additional capital. The island was worked along the same lines as at Banka and soon overtook its rival. However, pressure to extend the concession on the existing terms which granted 10 per cent of the substantial profits[76] to the NEI government led to a major political debate about the role of private capital and it was only on the eve of the expiry of the concession in 1892 that the Dutch Parliament consented to renewal on condition that the share was raised to 62.5 per cent.[77]

This proved to be a heavy burden[78] and production began to stagnate. The administration that had been responsible for the initial success of the company was now very pessimistic about its future, especially since it was unable to see any merit in mechanization.[79] By 1904 a new administration was installed which resumed extensive prospecting and started a comprehensive programme of technical modernization.[80] It was particularly interested in the Australian experience and in 1909 a suction cutter was ordered from Melbourne.[81] By 1914 another three were in operation which together were responsible for around 17 per cent of the total output. This administration also turned its attention to the lode deposits. Diamond drilling suggested that they would be even more profitable than the alluvial ones but the systematic development of deep mines would require a capital commitment that could not be justified unless the concession were extended beyond 1927 and its terms revised.[82]

As extraction techniques changed, Billiton revised its smelting plans. Rather than treat the increasing proportion that came from less pure concentrates, it decided to send them to STC which smelted them on toll. By 1905, half of all its production was being shipped to Singapore and marketed as Straits tin.

Singkep

Singkep is a much smaller island and lay within Singapore's sphere of economic influence.[83] It remained under the control of the Sultan of Lingga and by the 1860s its tin deposits were being worked by a few Chinese. In 1886 the sultan accepted the recommendation of various mining engineers that the deposits be worked by a private company and in 1889 Singkep Tin Maatschappij was formed. By 1900 it had become an important producer though never of the stature of Billiton or Banka.

In 1907 the NEI government granted its own concession to Singkep which was now in a position to work the offshore deposits. By 1913 two bucket dredges were acquired to treat them and, within two years, they were providing nearly 60 per cent of the company's production. All of this was also smelted by STC.

By the early twentieth century the NEI was in a very strong position. Banka had developed a system which combined European and Chinese expertise and which ensured that alluvial deposits were extracted far more efficiently than on the Malayan peninsula. Billiton and Singkep also had the advantage of working whole islands were introducing even more mechanized techniques. This strength would continue to hold through the rest of the century.

Siam

The destruction of the west coast of Siam in the wars with Burma at the end of the eighteenth century created a demographic, political and economic vacuum, one that was recognized by a Chinese peddler, Khaw Soo Cheang, who arrived in 'the sleepy little village' of Ranong in the 1820s.[84] It was filled by rebuilding the tin industry on the basis of Chinese immigration and by the end of the century the Khaw family controlled both the political and economic life of the entire region. The patriarch had arrived by way of Penang and the connection was maintained, so that the importance of the family extended well beyond Siam itself. Its base, however, lay in the relationship cultivated with the Siamese court which depended on the Chinese to provide effective local administration, especially through tax farming. In the case of the Khaws, that relationship became of strategic significance, as Bangkok found them indispensable in protecting its interests on the west coast from British expansion and in negotiating the terms on which foreign capitalists were eventually allowed to exploit the tin deposits.

The system of tax farming provided both an incentive and a constraint on the development of the tin industry. On the one hand greater immigration and production meant higher revenues but, while the system of competitive bidding maximized the amounts going to Bangkok, it forced the farmers to increase the rate of taxation. The effect is most clearly seen on Phuket which grew after 1850 to the point where it was producing around 5,000 tons by 1889, but the high tax rates of around 40 per cent meant that by 1900 it was down to only 2,500 tons.[85]

Modernization of the overall administrative structure of the Siamese government led to the abolition of this system and the development of a mining code which generally followed the principles established in Malaya. This decision to place mining on a new foundation coincided with a singular event that led to the introduction of a new technology that would not only transform Siam but the entire region. It was one that built upon the experience of Australia in dredging.

Tin dredging in Australia emerged in the late 1890s and was adapted from the experience in New Zealand in extracting alluvial gold deposited from river beds. Dredges were designed to float on water and they varied according to the method used to break up and excavate the cassiterite-bearing alluvium together with any overburden. The suction cutter loosened the material and sucked it to the surface where it was then treated, whereas the bucket dredge used a continuous chain of buckets to dig into the material which dumped the alluvium into a revolving trommel and which then saved the particles of cassiterite. Both types were in use in Tasmania from 1897 and were soon adopted in New South Wales and Queensland.[86]

The early versions of the suction cutter and bucket dredge were not particularly reliable but those that succeeded demonstrated the basic viability of these mechanical means of extracting the alluvium.[87] Two fundamental advantages were opened up. Deposits whose grade was too low to be worked by any other method could now be treated and that gave a new lease on life to those that had been abandoned. Dredges could also reach deposits that were otherwise inaccessible. It was this feature that provided the point of entry into Siam.

One of the business interests of the Khaw family was shipping and this brought them into contact with an Australian, Captain Edward Miles. Since he was familiar with dredges operating in Tasmania,[88] he was invited to Phuket to review the prospects of applying the technology there. By 1906 their negotiations resulted in the flotation of Tongkah Harbour Tin Dredging on the Melbourne Stock Exchange.[89] Five years later Tongkah Harbour had five bucket dredges working the concession, all of which were financed from the early profits and by 1914 the company had also paid dividends of 150 per cent on its issued capital.

This extraordinarily successful initial venture stimulated further Australian investment which set a distinctive pattern. It was one that involved co-operation between Khaw Joo Tok, members of the Miles family who administered new dredging companies and Achelan Palfreyman who raised the necessary capital in Melbourne. By 1913 this network had floated three other companies with four dredges operating on the mainland. In addition, Palfreyman became part of another network which included Ambrose Pratt, editor of the *Australian Mining Standard*, and who provided considerable publicity for the whole dredging industry. In 1910 this group floated Tongkah Compound to work in Phuket itself.

These initiatives were part of a much broader picture. In 1906 two British companies were formed both of which enjoyed official support and were granted extensive leases inland. By 1913 Siamese Tin Syndicate (STS) had three dredges operating and was closely linked to Renong Tin Dredging also with three dredges.

Dredging

When Miles visited Phuket he saw Chinese miners extracting alluvium from the harbour at low tides. Sensing that the deposit reached much further out into the seabed, he developed a method of boring through the overburden and was extremely fortunate in the initial results. Subsequent initiatives would require much more systematic and expensive methods to determine the extent and value of the deposit in order to justify investment in a dredge. The level of that investment was not simply in the cost of building the dredge but in towing it from Europe or Australia and installing it. Those costs were particularly high for inland dredges as were those of dismantling and moving it should the deposit not prove to be viable.

The more that was known about the character of the deposit, the potential obstacles it contained in the form of clay, boulders and buried timber, the underlying character of the bedrock to be scraped and the actual quality of the alluvium, the easier it was to design and finance the dredge. As a result, considerable sums had to be invested in boring the deposit and they established a reasonable estimate of the total value of the mineral in situ.

Both of these aspects were reflected in the capital structure of a dredging company. Since the property had already been proven prior to its exploitation, the owner was entitled to claim the value of the mineral it contained, together with the prospecting expenses as the basis of transferring it to an operating company. Expenses were normally covered in a cash payment from the capital raised, while the value of the mineral itself was reflected in the vendors' shares.

A rather paradoxical feature therefore emerged: the higher the investment in the physical equipment, the higher the proportion of the initial capital represented by the value of the property prior to that investment.[90] As a result, it was extraordinarily difficult to assess the comparative competitiveness of dredging with lode mining. A lode mining company developed the property gradually in relation to the process of actual extraction and the costs of doing so were included in its operating expenses. It therefore had comparatively high operating costs but comparatively low capital invested in relation to each ton of mineral prepared; dredging companies had comparatively low operating costs but a high level of capital invested. This would bedevil any attempt to define a price at which even the capitalist sector of the industry could find an equilibrium.

The system of property rights provided an incentive for the investment of capital in the prospecting stage as well as in the extraction stage. The EPL gave the prospector the basis on which extensive boring could be undertaken. Since it also normally granted the right to apply for a mining lease, the capital invested in prospecting received its main return from the dividends paid on the vendors' shares once an operating company was formed and became profitable.[91]

Dredging not only affected the capital structure of operating companies but reinforced their common financial linkages. Dredges required extensive maintenance and that was facilitated by a common pool of supplies. Should production be interrupted, as the dredge was moved around the deposit or needed extensive repairs, then parent or sibling companies could offer temporary financial assistance.

The most sophisticated form of grouping adopted a South African institution, the mining finance house which was designed to balance diversification of risk with consolidation of control. The house had its own corporate identity and undertook three functions: prospecting and development of new properties, flotation of companies to work them and over which it kept a controlling equity interest and their management. Each of these provided an income stream which could finance further expansion.

Another Australian, Herbert Pratten from Sydney, was responsible for creating the first such house in the region. Following his visit to Malaya 1908, he formed Austral-Malay which developed a large property at Kamunting. Kampong-Kamunting was floated to work one section, while the remainder was transferred to an English company, Kamunting. Austral-Malay was part of a more extensive Australian commitment to the region. A Melbourne company brought a suction cutter to Burma, while Charles Kerry, from Sydney, floated Ratrut Basin Tin Dredging to work in Siam. In 1915 Pratten left Austral-Malay and formed a similar finance house, Alluvial Tin, which would become far more important.

English capitalists also responded to the opportunities demonstrated by dredging in Siam, though these were confined to Malaya. Cornelius Stephens floated Malayan Tin Dredging in 1911 and was followed by the established Cornish interests. Tronoh started its first dredge in 1912 and the Redruth group floated Tekka-Taiping in 1913. The same year saw other newcomers: Addinsell with Kamunting Tin Dredging and Chenderiang Tin Dredging and Pawle with Ipoh Tin Dredging.[92] All of these would be prove to be strong, profitable companies.

Although it was only in Siam that dredging amounted to a substantial proportion of production before the war, it was clear that development of this technology held the key to the future role of Southeast Asia in the international tin industry.

Production – new centres

China

Tin production in China had traditionally served domestic needs and it was only towards the end of the nineteenth century that its most important region, Yunnan, began to play a role in the international market. However, it was one that was constrained by three factors: property rights, historical legacy of violence and the character of the lode deposits.

The system of property rights in China rested on the principle of free mining which produced a large number of separate, small scale workings, all of which were dependent on smelters in Kochiu. These, in turn, were controlled by the local merchant-gentry[93] and the Chamber of Commerce had the power to seize any ore moving outside the district and impose the death penalty for a repeated infraction.[94] This was part of a broader monopoly, since smelting was only one interest of the merchants who brought in the food and other supplies required in this arid region and the tin shipments served to finance this wider trade.[95] Local prices for tin metal could therefore be kept high and support a very heavy burden of taxation.[96]

Yunnan still retained many features of its early history as a frontier area with a tradition of independence and rebellion. Conflict between Muslim and Chinese miners was the root cause of the uprising of 1855–1873[97] and banditry remained an endemic problem. More routine was the resort to lethal violence in settling disputes and foreigners could only visit the mining district under the protection of a large armed escort.[98]

Although the deposits were often rich, they contained high levels of iron and lead which made smelting particularly difficult. One response was to produce two grades, one of 84.4 per cent which was sent to Hanoi for refining and a higher quality one of 99.49 per cent tin which was sent inland.[99] By the 1890s miners were producing concentrates with higher impurities which made it impossible to maintain the better grade and all metal was then reduced to a grade of 90 to 92 per cent. This was then shipped via Haiphong to refiners in Hong Kong who raised the grade but only to 95 to 99 per cent.[100] These inferior brands were quite suitable for Chinese domestic consumption and they also found important new markets in Europe and the USA where they sold at a discount on standard tin. That incentive was particularly pronounced as the price of standard tin rose in the 1900s and at the same time the industry was provided with other sources of support.

Fresh support for the industry came from Peking which introduced a new mining policy which reduced taxes to around 13 per cent of the export value of the metal[101] and from Hanoi which financed the construction of a railway which would ultimately connect the mining district to Haiphong.[102] That also opened up the prospect of developing a modern smelter that could use imported coal and overcome the limits imposed by reliance on diminishing local sources of charcoal.

Of longer term significance was the decision of the Yunnan provincial government to form Yunnan Tin Trading in 1909 which built a smelter with German technical and financial support.[103] Unfortunately, that support was exhausted in the installation of the smelter; unable to secure adequate working capital, it soon closed.[104] However, the government was left with the plant and it was only a matter of time before it would be restarted.

No discernible change in the extraction of ores or in their processing occurred.[105] Chinese mining technology had developed over a long period of time and there was no particular pressure to change it. The unsystematic way in which the mines had developed meant narrow tunnels where the ventilation was very bad and the 'atmosphere indescribably foul'.[106] While productivity of labour remained low, it was cheap with the result that China did not experience the processes of technical and institutional modernization that were occurring elsewhere.

By the 1910s China had emerged as a major producer, the world's fourth largest, but one that occupied a distinctive role with inferior metal and techniques of mining and smelting that were universally regarded as extraordinarily primitive. However, at the point at which both problems could be addressed she would escape her marginal position and become a far more important force.

Bolivia

The increase in price had a more dramatic impact in Bolivia than anywhere else. Her mines faced three sets of heavy transport costs, from the mine to the concentrating mill, from the mill to the railhead and from the railhead to the coast. With large fixed external costs,[107] a doubling of prices in London could easily mean a tripling of income at the minehead.

Fortuitously, this bonanza in tin occurred within a critical economic and political conjuncture. Silver had been the primary base of the commercial sector of the Andean region of Bolivia and from 1895 to 1905 production declined by a precipitous 85 per cent. Tin emerged at exactly the right moment and from 1901 it became the country's leading export. It was helped by a new political regime which emerged from the civil war of 1898–1899 and which was committed to establishing an infrastructure, especially in the form of expansion of railways, that would serve the mining districts.

Capitalist responses to this new incentive took many different forms. One relied on substantial fresh investment to take advantage of the most advanced forms of mining technology and administration to build a very competitive industry. At the other extreme were those captivated by the opportunities for immediate returns, thanks to the existence of pockets of high grade ores, together with a gullible set of investors, especially in Santiago, prepared to gamble on new company flotations. In between were many companies which failed to develop a technically efficient industry and whose financial viability was entirely dependent on working high grade deposits.

Of the first set of cases, two proved to be of enduring significance. One was the Aramayo family firm whose mining interests in Bolivia dated back to the 1830s.[108] By the end of the nineteenth century it produced a wide range of minerals and had established good financial connections in Europe. While that diversity of interests remained, it expanded its tin production for which fresh capital was needed. In 1906 the firm was incorporated in London as Aramayo, Francke & Co[109] and it soon acquired the whole of Chorolque mountain which became its most important single source of tin.[110] The firm's tradition of operating with the most advanced technologies was sustained with the appointment of a British mining engineer as its administrator.

The other case would lead to a fundamental transformation of the structure of the tin industry at first in Bolivia and then eventually of the whole world. From 1895 Simon Patiño had been working a small, marginal deposit, La Salvadora, high up on Llallagua mountain, with the most primitive equipment and with indifferent results until a rich vein of pure cassiterite was struck in 1900. Patiño then set about developing the mine with modern, mainly German, equipment. As he began to market his concentrates, first through Aramayo and then the Anglo-South American Bank, Patiño secured a credit standing which enabled him to borrow the funds required for further investment both in La Salvadora and in the acquisition of other mines.

By 1906 Patiño had £1 million of his own money with which to found a new bank, the Banco Mercantil, with a capitalization twice that of all the other domestic banks combined.[111] Just as he moved to provide a domestic financial infrastructure, so he also moved to take more control of his own marketing by opening an office in Hamburg in 1911. That in turn led to taking an interest, shared with a major American tin consumer, National Lead, in a German smelter, Zinnwerke Willhelmsburg, to which La Salvadora had been shipping some of its concentrates.

By 1908 La Salvadora had become the largest single tin producer in Bolivia and in 1910 Patiño used its profits to buy out two neighbouring properties. The following year two other important mines in another region, Huanuni, were acquired.[112] This expansion made Patiño the world's leading tin capitalist and with it came a fresh set of commercial relationships. As he became one of the most important clients of the leading British merchant houses of Antony Gibbs and Duncan, Fox, he could draw on fresh sources of credit with which to engage in a further round of major acquisitions in the postwar period.

The process of installation and operation of the most advanced forms of mining technology in one of the most inhospitable regions of the world was not at all straightforward.[113] Moving equipment from the railhead to the mine by llamas was a slow process and smaller items were always subject to breakage and pilferage. Maintaining that equipment was particularly complicated given the long delays in securing spare parts and the lack of technical training on the part of many of those responsible for ensuring that it functioned smoothly. Administration was an additional challenge since it required attention to a mass of detailed issues, few of which could be settled in any straightforward manner.[114]

Although most companies failed to master the technical and organizational challenges, they continued to produce. Their concentrates served as security against loans that commercial houses were only too willing to advance. Three of them became quite large. The most important was the Compañía Estañifera de Llallagua, Patiño's neighbour and rival. The mine had been developed by a Bolivian politician who was anxious to retire in Europe and he eventually found a buyer in a Chilean consortium in 1906. Modernization of the plant required an injection of £100,000 bringing the capital to £450,000 and the shareholders insisted on distributing most of the profits. The high grade of ores treated ensured that these were substantial, though far less than could have been secured with responsible administration. These, in turn, were much lower than consolidation of all of the mines on Llallagua mountain could provide.

Both of the other large companies were also Chilean. One was an old established silver mine, the Compañía Minera de Oruro (CMO), whose veins also contained a small proportion of tin. This was a mine that appeared to be coming to the end of its life, though there would be several attempts at resuscitation. The second was the Compañía Minera y Agricola de Oploca. This too had been an old silver mine but under Bolivian control and was reorganized as a Chilean company in 1906.[115] It experienced extraordinary difficulties in installing a new concentrating mill, thanks primarily to the congestion in the port of Antofagasta.[116] Bank loans enabled the company to survive and it only emerged as a large producer on the eve of World War I.

The one sector that experienced no technical innovation was that of small miners. While they continued to rely on advances from commercial houses, local houses found themselves facing stiff competition from the established British ones, especially as they built up their business with the both the medium and large companies.

The rapid growth of the industry established a pattern that would remain intact for many decades. It had become a highly differentiated industry, along three separate axes: size, nationality and viability. Seventy-five per cent of production came from six large companies producing concentrates with over 1,000 tons of metal; 6 per cent came from mines producing less than 100 tons. Of the large groups, the three under Bolivian control produced 72 per cent of that sector, whereas foreign companies were responsible for 80 per cent of the medium sector. A few large and medium-sized companies were strong enough to meet any competition but most were dependent on a combination of high grade ores and a buoyant market to maintain their viability. However, the marginal operations left behind an important legacy. One measure of their inefficiency was the high proportion of cassiterite discarded in the concentration process. That reinforced the incentive for a subsequent generation of entrepreneurs to raise the funds required to take up the technical challenge required to turn these dumps and mines into profitable operations.

Completion of the railway to Potosí in 1912 reduced transport charges to the point where local production of an inferior grade of metal was no longer viable and thenceforth concentrates were all shipped to Europe where at least eight separate smelters were potential competitors for them. In 1913 Britain took 56 per cent of the total, while Germany took 40 per cent and France the remainder. Bolivian miners therefore found themselves in the unique position of selling into the most competitive market of all but the country was left without any smelter of its own. That would establish a persistent tension between the miners who relied on the competition and the government which was anxious to build forward linkages. It also established the basis on which ore-dealers with international connections played a crucial role in finding specific markets for these concentrates, since only they could determine which smelter was prepared to offer the best terms for particular parcels, with their distinctive pattern of impurities, at any point in time.

Nigeria

When a British Chartered Company, the Royal Niger Company (RNC), signed treaties with Northern rulers in the 1880s they secured an extraordinary set of privileges. These treaties did not simply establish the administrative authority to regulate the trade in tin, they also transferred the mineral rights to the RNC.[117] No actual prospecting or mining was undertaken under RNC auspices but it was able to claim compensation for these rights when the British government took over its administrative responsibilities with the formation of the Protectorate of Northern Nigeria in 1900.[118] That compensation took the form of a half share in whatever royalties and other taxes the government might impose for the remainder of the century, an arrangement denounced by the second governor as a 'fraud on the public that ought to be revoked by Parliament'.[119]

With the revocation of its charter, the RNC became the Niger Company and took on the functions of a mining finance house. It managed to secure extensive prospecting rights and recruited two Australians to explore them. By 1909 the company was shipping sufficient quantities of high grade cassiterite to warrant systematic development of the tinfield. The primary vehicle was a company that had failed on the goldfields of West Africa, Champion Gold Reefs, reinvigorated by Oliver Wethered, Chairman of Dolcoath and Geevor.[120] Champion, in turn, floated six other operating companies. The Niger Company was then poised to control the whole tinfield as the main source of supplies, transport, technical and financial assistance.

Overall development of the new field was shaped by two crucial decisions by the new colonial regime. The first was to establish control over all land which extinguished native mining rights and the second was to issue a mining code which broadly followed the Malayan pattern, with its system of EPLs and 21 year renewable mining leases.[121] However, its specific provisions which required a fee of £5 per acre for an EPL and £2 per acre of annual development work under the lease were designed to severely circumscribe the opportunities for individual miners.

The discovery of a major new alluvial field, especially at the point at which the market was becoming acutely aware of the depletion taking place in Southeast Asia, led to an extraordinary boom. The form it took rested on two institutional failures: companies were not required to provide any evidence of their assets before selling their shares and the officials of the Mines Department were not prepared to eschew the prospect of personal gain.

Anglo-Continental Mines was the vehicle used by a group with substantial financial and mining connections with which to take advantage of a gullible public. The group included Herbert Hoover, Edmund Davis, British, German and Belgian stockbrokers, together with the leading South African mining companies and finance houses, of which the most important was Consolidated Gold Fields of South Africa (CGF).[122] This was far more credible than the Niger Company and Anglo-Continental soon came to an arrangement with the Champion company and started promoting its own subsidiaries. These two groups also began to work together politically, first forming the Northern Nigerian Mines Association in 1910 and then bringing it under the direct control of the Nigerian Chamber of Mines formed in London in 1911.

Development of the tinfield rested on the construction of an extension of the railway to the Jos Plateau. That would not only reduce transport costs considerably but also opened up the prospect of importing the machinery required to make the tinfield extraordinarily profitable. In early 1911 lobbying efforts were successful in persuading the British government to finance the line.[123] By the end of the year around 250 companies had been formed with an issued capital of £4.4 million and puffed up to a market value of £10.7 million.

Anglo-Continental then made an even more dramatic assault on the stock market by announcing it held an EPL over a large property assaying 25 per cent tin. Davis was able to manipulate the market for the company's shares, selling them at a premium, buying them back when they fell and then selling them again when CGF lent public credibility to the project. Within a few weeks, an independent investigation made it clear that the earlier reports were largely fictitious and the market for Nigerian shares quickly collapsed.[124] The memory of these blatant market frauds lived on for a long time, especially in government circles. Officials had little sympathy with an industry which they saw as much more concerned with stock exchange manipulation than effective production.

While estimates of potential yields may have been grossly exaggerated, the field was both rich and extensive. This proved to be sufficient to enable many of the early companies to become steady producers. By 1913 Nigeria was contributing 4,000 tons to the world's tin supply of which approximately 60 per cent came from companies floated under the auspices of the Niger Company, Anglo-Continental and CGF. Within this group one individual, Herbert Latilla, with close links to CGF,[125] exercised particular influence and would play a major role in shaping the evolution of the industry in the postwar period.

Two other groups which would also play an important role had their origins in this early phase. Kaduna Syndicate and its affiliate Kaduna Prospectors were something of an anomaly since the capital was mainly supplied by the Livingstone-Learmonth family without much City participation.[126] The second emerged from the prospecting work undertaken by an individual, Davidson, who then turned to Broadbridge to secure the necessary financial support from the City for several companies.[127]

Nigeria also occupied a particular position in the structure of the smelting industry. Local smelters were dismantled in 1911,[128] since they made it impossible to prevent the theft of concentrates on the part of the loosely supervised tributers. Their high quality allowed the British smelters to overcome many of the problems posed by the dirty concentrates from Bolivia and they could then produce a higher proportion of metal at the grade required for standard tin. Control over Nigerian tin was seen as the key to overall competitiveness of British smelting.

South Africa

Tin deposits were exploited in several parts of the Union of South Africa, Swaziland and German Southwest Africa but only those in the area of northern Transvaal that had supported earlier tin workings became significant.[129] British interests had taken advantage of the low cost of land to make speculative purchases on the assumption that it contained minerals and they were quickly rewarded with the discovery of tin at Rooiberg. In 1910 the mine was taken over by a major South African mining finance house, Anglo-French Exploration, with close links to Dolcoath.

The only other mine of importance was some distance away at Zaaiplaats. This had its origin in the decision of the government to operate a mine as a way of maintaining skilled white labour but in 1908 it was sold to a company affiliated with another mining finance house, Wernher-Beit. Although the early production levels were not sustained, both Zaaiplaats and Rooiberg continued to be very profitable and together they accounted for around 70 per cent of the total production of South Africa. Given the erratic character of these lode deposits, strong financial support was essential to success. However, it was not always sufficient even when 'the last word in tin mining equipment' was installed as in the case of Leeuwport,[130] supported by a third mining finance house, South African Townships. It was more successful with McCreedy Tins which worked an alluvial property in Swaziland.[131] The remainder of the industry in the Transvaal was made up of 11 small independent companies, of which only two survived into the postwar period.[132]

Belgian Congo[133]

Leopold II's strategy for the development of the mineral wealth of his colony rested heavily on Chartered Companies, of which the most important were the Comité du Katanga (CK) formed in 1891 followed by the Compagnie des Chemins de Fer du Congo Supérieure aux Grand Lacs Africains (CFL) in 1902. Each enjoyed concessions over whole provinces, Katanga and Manièma-Kivu, which would run for 99 years and they naturally included all the mineral rights. Each was formed with strong financial support from major sectors of Belgian finance capital, the CK from the Société Générale de Belgique (SGB) and the CFL from the Empain group. Although the Belgian state assumed overall administrative responsibility for the colony in 1908, the companies retained their charters intact. They would provide the framework for the systematic exploitation of the tin deposits.

To pre-empt British imperial designs on Katanga, the CK co-opted a British company, Tanganyika Concessions (TCL) and granted it prospecting rights over the southern portion of the province. The TCL team located deposits of copper and tin which were then transferred to an operating company, Union Minière du Haut Katanga (UMHK), created in 1906 as a joint venture between TCL and SGB. Initial exploitation of the copper was not easy and UMHK turned to the more easily worked tin deposits which could generate immediate profits.

In 1909 the region was opened up to other prospecting companies and by 1914 they had formed Simkat which started exploitation on a modest scale of its substantial tin deposits. It, too, enjoyed SGB support. Concern about the dominance of the SGB group in its new colony led the Belgian government to encourage new groups to apply for concessions, one of which became the basis of the Géomines, formed in 1910.

Several factors combined to place these operating companies in a very strong position. One was their financial support which was reinforced by the fact that the Comité Special du Katanga (CSK), as the successor to the CK, was entitled to a third of the capital issued by the operating companies. As the CSK took its return on the mineral rights through the profits, rather than an ad valorem royalty, the cutoff grade below which exploitation was not considered viable was correspondingly reduced. That permitted a more systematic and comprehensive extraction of the deposit. Most important was the fact that the concessions granted to the companies would only expire with the master concession enjoyed by the CSK. Since they were also granted exclusive rights over large areas, they were in a position to devote the bulk of their resources to prospecting and development rather than immediate extraction. While these features would allow the companies operating in the Congo to become extremely important, they simultaneously limited the basis on which the Belgian government could formulate an independent policy.

Minor producers

This period also saw the foundations laid for two other minor producers whose full importance would only become evident in the postwar period. In Portugal at least one English company was floated and California interests sponsored the formation of Portuguese-American Tin Dredging which started in 1913.[134] French Indo-China saw the first modest European investment as early as 1896 in Laos.[135] Tonkin was opened up in 1904 which saw more systematic development especially after the formation of Société des Étains et Wolfram du Tonkin in 1911. Until adequate transport facilities were constructed, production would remain very modest.

Secondary tin

The process of manufacturing containers from tinplate resulted in a large quantity of clean scrap. For the manufacturers, this posed a problem of disposal and was therefore available for collection. If the tin coating could be removed from the steel base, both could be recycled. Karl Goldschmidt, a German metallurgist linked to Metallgesellschaft, started experimenting in the early 1880s and by the end of the century had overcome the substantial technical difficulties.[136] Germany then started importing scrap from across Europe and a large detinning industry also emerged in the United States.[137] Since the technology could be modified to treat a large number of otherwise useless tin products, secondary tin came to supply a significant proportion of final consumption.[138]

It was always tempting to aggregate consumption figures and treat the secondary industry as in competition with primary producers. While there were some spheres where this was the case, they did not include tinplate which always used virgin tin. As far as the others are concerned, secondary tin occupied a very specific role. It was one dictated by the forms in which the tin was recovered, a high proportion of which were chemicals or alloys, rather than pure metal[139] and which were only suitable for a limited range of purposes. While many were not themselves recyclable, the tin in others was used over and over again, so that secondary tin became a constant stock rather than a flow.[140]

The precise role of price is also difficult to determine, though it is unlikely to have been particularly important. Tinplate scrap was sought for its steel rather than its tin[141] and no price for virgin tin would have been low enough to reduce the secondary supply. However, in the case of other sources with much higher tin contents, the price of tin may have been the factor determining the extent to which detinning was profitable. Even here, price competition would only have affected a few manufacturers capable of switching between different kinds of metals.[142]

Although secondary tin played a largely passive role in the functioning of the tin market, it was virtually the only domestic source for two of the world's largest consumers. When Germany and the United States became concerned about their overall position in the tin market, they would give considerable attention to the way in which this role could be expanded.

Summary

As the twentieth century opened, the tin industry found itself both more united and diversified than ever before in its history. Unity prevailed in the spheres of consumption, marketing, finance and smelting; diversification in its fundamental base, mining.

The overall course of western economic development, now paced by the United States, was characterized by the interrelationship between industrialization, urbanization and rising living standards. All of these generated their own demand for tin and, while the metal had lost its glamour, it had become indispensable to the maintenance of that form of civilization.

Only a very small proportion of this demand, around 5 per cent, was met by domestic sources. As the United States, otherwise well endowed with minerals, became the largest consumer, tin became the metal that was most dependent on the operation of an international market.[143] It was one governed by a uniform, albeit volatile, set of prices established on the LME.

London also served as the centre of finance and the capital it raised could flow freely throughout most of the world of tin. While other capital markets played an important role, none of them created any obstacles for external investors. It was only China and Burma that explicitly blocked foreign investment. Financial unity also emerged as declining silver prices forced most Asian currencies to switch to gold.[144] Again, China was the exception where a complicated system of regional silver currencies was retained.

Smelters also operated in a competitive world market with a similar technology, the reverbatory furnace. Insulation from that market in the case of China and the ability to produce a superior brand in the case of Banka were the basis of the main exceptions. Competition produced a a high level of concentration, among the highest of any of the base metals. STC was capable of treating 26 per cent of world production; the next seven could treat another 50 per cent.[145] Such a degree of concentration carried with it an apparently high level of market power. However, it was not one that could be easily exercised against producers, at least not over the long term.[146] Capital costs of a smelter were comparatively low and they posed no barrier to new entrants.[147] If returning charges were too high, miners could always threaten to sponsor their own smelter.[148] Nor could such power be exercised against consumers for lack of sufficient finance and tolerance for market risk.

Diversification on the extraction side occurred along three separate dimensions, geographical, technical and organizational. The extent of geographical diversification is presented in Table 3.2.

Table 3.2 World tin production, 1901–13

			Increase		*Market share*		
	1901	*1913*	*1901–13*		*1901*	*1913*	*% share*
	tons	*tons*	*tons*	*%*	*%*	*%*	*increase*
Historic centres							
Cornwall	4,600	5,290	690	15	5	4	2
NEI	15,780	20,920	5,140	33	17	16	12
Malaya	46,740	51,380	3,940	8	51	38	10
Siam	3,200	6,710	3,510	110	3	5	8
Total	71,020	84,300	13,280	19	77	63	32
New centres							
Bolivia	12,920	25,940	13,020	101	14	19	31
China	3,000	8,300	5,300	177	3	6	13
Australia	5,000	7,780	2,780	56	5	6	7
Nigeria	0	4,010	4,010			3	10
Other	200	3,800	3,600			3	9
Total	21,120	49,830	28,710	136	23	37	68
World	92,140	134,130	41,990	46	100	100	100

Source: ITRDC, *Statistical Yearbook*, 1938, The Hague, 1938

Table 3.3 Major tin producers, 1913

	tons	*% world*
Banka	15,484	11.5
Patiño	10,714	8.0
Billiton	4,378	3.3
Redruth/Tronoh	4,230	3.2
Llallagua	3,538	2.6
Anglo–Continental	2,400	1.8
Aramayo	2,341	1.7

All continents, with the exception of North America, were important centres of production. While that was not sustained in the case of Europe and Australia, South America and Africa became even more significant. Historic centres, however, retained their predominant position. While two thirds of the increase between 1901 and 1913 occurred in the new centres, their overall market share only moved from around a third to just over half that of the historic ones. However, since the largest historic centre of alluvial mining, Malaya, expanded at the slowest rate and the bulk of the production from the new centres came from lode mines, those attempting to extrapolate long term trends became increasingly sceptical about the ability of the historic centres to keep their position.

Technical diversification was not simply between lode and alluvial mining but increasingly within the latter. It was one of different forms of mechanization of extraction with dredges, gravel pumps, excavators and hydraulic monitors. With that went different organizational forms, kongsis and capitalist corporations, each tied to complex webs of ownership and control.

The most important manifestation of those webs was the emergence of corporate groups with recognizable international stature. Table 3.3 identifies those that were responsible for at least 1 per cent of the world market.

The combined market share of these groups was 32 per cent, a very low figure in comparison with other base metals.[149] What is also striking is that in two of the new centres, Bolivia and Nigeria, such groups controlled well over half of their respective national output. By contrast, Malaya supported only one such group and that accounted for just 8 per cent of its production. Organizational diversification meant qualitatively different corporate sizes and of the degree to which the larger ones dominated their respective countries. This would have important political consequences since the asymmetry between size and concentration made it difficult for any one group to speak authoritatively on behalf of the world's largest producer.

The twin processes of capitalist penetration and technical modernization were far from complete, leaving tin in a comparatively backward position. Only in the NEI and the new African centres were small miners insignificant. In Malaya, Siam and China they were the dominant force and had a substantial presence in Bolivia. Between the large and the small stood numerous medium-sized companies and the institutional framework within which the industry had grown was becoming a serious impediment to its future. Its inability to keep pace with rising demand indicated a chronic weakness, one that was reflected in the continuing rise in price which by early 1913 had reached £238/ton. After the hiatus of World War I, the question of how, when and under what conditions could tin overcome its backward state became extremely pressing.

World War I

The industry that had developed to 1914 emerged in a rather different condition in 1919. Some of these changes were simply continuations of trends already evident, though in many cases their precise rhythm can be traced to the impact of the war. The immediate effect of the war itself was to cut off the Central Powers from all external sources of the metal and that meant severing the Bolivian-German smelting relationship. Towards the end of the war, government controls over marketing were established and as they were dismantled the industry found itself in one of the worst crises in its history.

On the consumption side, the most important feature is the continued growth in demand in the United States. Whereas in 1913 the USA had consumed 45,000 tons, 34 per cent of world consumption, by 1917 she was using over 68,000 tons, 52 per cent of the world. This was disproportionately based in personal consumption; from 1914 to 1917, tinplate production increased at an annual rate of 13 per cent and automobiles even faster at 47 per cent.[150] The war itself had little immediate impact,[151] though towards its end government allocation constrained civilian consumption so that tin could replace metals of direct military significance. From that point onwards, there would be a continuous incentive to find substitutes for tin itself.

The war itself hastened the decline of Cornwall as conscription reduced the number of miners and those that remained were unable to undertake development work. The county would be reduced to a very weak condition by its end.[152] More worrisome was the decline in Malaya, by 16 per cent. This was a new trend and one that marked a major transition. While this would not end her position as the world's leading producer, it was now to be one based less on high grade deposits worked on a small scale, than 'exploitation of ground of lower grade which can only be worked economically and profitably with up-to-date machinery'. The differential rents had gone and Malaya would have to cut production costs to 'compete successfully in the world's markets'.[153] The question would soon be raised as to whether the Malayan pattern would eventually be followed by the other alluvial deposits.

There was no immediate sign of any such problem. The NEI simply remained steady and Siam enjoyed an increase of 36 per cent. Nigeria grew at the fastest rate of all at 50 per cent which enabled her to overtake both Cornwall and Australia and emerge as the world's fifth largest producer. In the case of Burma, the war itself played an important, albeit indirect, role. Demand for wolfram led to extensive prospecting in regions which also contained cassiterite deposits. That laid the foundations for the emergence of a significant tin producer as soon as the war was over and the market for wolfram collapsed.

British smelting capacity was expanded to treat the concentrates that had previously gone to Germany. Williams, Harvey reopened its Cornish plant and two new smelting companies were founded.[154] However, since the United States continued to import the bulk of its metal across the Atlantic, there were, at least potentially, considerable savings to be realized by treating Bolivian concentrates there. American Smelting and Refining (ASARCO) and Williams, Harvey both responded to this new opportunity.[155]

What prompted ASARCO's interest is unclear, especially since the Guggenheims had no previous connection with the tin industry.[156] Whatever its source, it was one that led them to establish a plant in Perth Amboy, New Jersey. This introduced a novel electrolytic method of refining designed to produce a metal of 99.98 per cent purity, equal in every respect to Straits.

Although Williams, Harvey had been able to negotiate a contract with Patiño for the supply of his concentrates by the end of 1914, the company was anxious to start a new plant in the United States. In early 1915, serious planning began and National Lead was approached for its co-operation.[157] National Lead was the second largest American consumer and was a reliable customer especially for the lower grades of tin. The initiative had two advantages. It eliminated the possibility that National Lead would sponsor a smelter of its own and placed Williams, Harvey in a position to compete with ASARCO. As the project developed, it was seen not simply as a wartime measure but one which would give Williams, Harvey, with control over two smelters on two continents, the flexibility to remain internationally competitive in the postwar period.[158] Both National Lead and Williams, Harvey pre-empted each other's freedom of independent action through cross-investments. National Lead bought half the shares in Williams, Harvey and transferred half the shares in the American subsidiary to the British company.[159]

By mid-1917 plans for the American smelter were well advanced and the manager of Williams, Harvey, Pearce, was anxious to secure the Patiño concentrates since otherwise he would have to negotiate for a number of smaller tonnages. His proposal found Patiño thinking more about the shape of the postwar tin world and received the following response:

> At present, our only tie is a contract similar to that with any other Smelters. If, after the war is over, I should still hold to the proposal for a combination which would permit of the Smelters being, at one and the same time, Tin Producers and consequently directly interested in dominating and fixing market prices, it naturally follows that the Company which has the preference should be the one in which I am already interested.[160]

National Lead and the original Williams, Harvey shareholders then reduced their share of both the American and the British companies so that Patiño could take a third share in each.[161] In 1918 the first slabs of tin were poured at the Jamaica Bay plant in Brooklyn.

ASARCO lost a considerable amount on its experiment, largely as a result of the additional costs incurred to refine the metal to the standard it desired. By contrast, Williams, Harvey, which worked with an established technology, was very successful.[162] However, the combined production of these two smelters only accounted for at most 25 per cent of American consumption and 40 per cent of Bolivian exports. In spite of the changed circumstances of the war, American consumers still preferred Straits and many Bolivian producers preferred to preserve their position within familiar networks rather than risk developing new relationships.[163] At first, ASARCO could only sign contracts with small and medium producers and even when it managed to land a large contract with Compañia Estañifera de Llallagua in 1918 it was not renewed. In 1924 both smelters closed but the legacy of this episode would continue to shape much of American thinking about tin.

The tin market functioned reasonably smoothly until 1917, although price increases failed to keep up with inflation. But then the war began to put specific pressure on consumption, as tin was substituted for metals in scarce supply and attempts were made to improve shells with a coating of tin.[164] Increases in consumption and uncertainties about shipping inevitably brought about a disproportionate increase in manufacturers' demand for stocks, resulting in a runaway price. By August 1918 it reached £380/ton, around twice the level of 1916 which provoked the American, British, French and Italian governments to establish an Interallied Tin Executive and regulate the distribution of available supplies.[165] It also permitted the US government to accumulate stocks of 10,000 tons and the rapid rate at which the war came to an end meant that these, together with stocks in private hands, would depress the postwar market.[166]

Although the Interallied Tin Executive was short-lived, it proved to be an important turning point. For the first time, the international market was constrained by international political intervention. The dislocation it caused would be corrected by a different form of international co-operation. For the rest of the century, whenever tin ran into problems, whether they were from the perspective of producers or consumers, solutions were always sought through international collusion.

4 The problem with tin, 1919–1929

The aftermath of World War I fundamentally changed the world within which tin had evolved. Hitherto, decisions that shaped the industry had been made on a purely pragmatic basis, divorced from any general form of economic analysis. Indeed, such a divorce was an inevitable consequence of the reigning orthodoxy whose three moral principles, unfettered competition, free trade and the gold standard, sustained the institutions within which such pragmatic decisions were encouraged. War had now destroyed those institutions and the old orthodoxy looked increasingly irrelevant.

While it required the collapse of the international capitalist system in the early 1930s to expose the theoretical limitations of that orthodoxy, much of the economic discussion in the 1920s, especially in Britain, as the erstwhile centre of that system, revealed the extent to which its moral hold had evaporated. Imposition of tariffs were defined as 'reform'. Competition was often dubbed 'ruinous'. Schemes of industrial consolidation were promoted as 'rationalization' and the case even had to be made for returning sterling to gold. Capitalism had now become problematic for capitalists and decisions about tin would be affected by the ways in its problems were defined and solutions were sought.

Particularly problematic was the level and structure of international trade. It remained one organized around the exchange of manufactures for primary commodities, making primary incomes a major determinant of the level of demand for manufactures.[1] Unfortunately, many primary commodities were in trouble as increases in productive capacity were translated into low prices and incomes. In the 1920s governments often responded by manipulating markets to raise prices[2] and the intergovernmental schemes of the 1930s simply expanded the scope of intervention to deal with a more severe crisis.

Fresh ideas about primary commodities began to percolate. One set defined the special features of their markets; another, those of minerals. Both were concerned about optimal pricing policy over the long term.

A distinctive problem in primary commodity markets emerges from the time lag between the point at which a rise in consumption signals the need for fresh investment and the point at which new production comes on stream. During the course of that lag, prices rise above the level required to generate the actual new capacity needed resulting in excess investment. There inevitably comes a point where excess production forces prices down. Many factors, such as heavy level of fixed costs, immobility of capital and labour, prevent immediate adjustment, so there is a lag between the onset of a decline in price and the point at which it is so low that many producers are forced out. This is the familiar boom-bust cycle. Too many producers are pulled in on the upswing and too many expelled on the downswing. An unfettered market is condemned to repeat the commodity cycle indefinitely and it is simply exacerbated by the overall business cycle. Since violent oscillations were considered injurious to both producers and consumers, price stability over time becomes the dominant economic desideratum.

The distinctiveness of minerals lies in the fact that they are non-renewable. That raises the question of the optimality of the rate of depletion both from the standpoint of the owner of the resource and from the standpoint of the society which uses it.[3] The experience of the oil industry in the United States in the 1920s showed that competition could not only be 'ruinous' but also 'wasteful'. Since waste meant an acceleration in the rate of exhaustion, the adverse consequences for consumers were obvious. This observation had two important implications. One was yet another recognition of a point of convergence between the interests of producers and consumers. The second was to charge producers with the moral responsibility of conducting their affairs in such a way as to avoid this waste; the scarcer the resource, the greater the moral imperative.

The 1920s represented a new phase in the evolution of the tin industry. Throughout the decade fresh conceptions of economic morality were taking root. A new actor emerged with an ambitious programme to be implemented in many different countries. It ended with a problem which was quite unlike any that the industry had experienced hitherto. The following discussion will consider the way in which these three features were connected.

The tin market

The problem of the 1920s is best grasped against a picture of the price history of the metal, which is presented in Figure 4.1.

Tin shared in the short-lived postwar commodity boom of 1919–1920 but dropped further than other non-ferrous metals in the depression of 1921 that followed in its wake. It made a strong recovery in 1926–1927 while prices for other metals remained steady. At the end of the decade prices went in opposite directions, with tin experiencing severe declines while some of the others continued to rise.

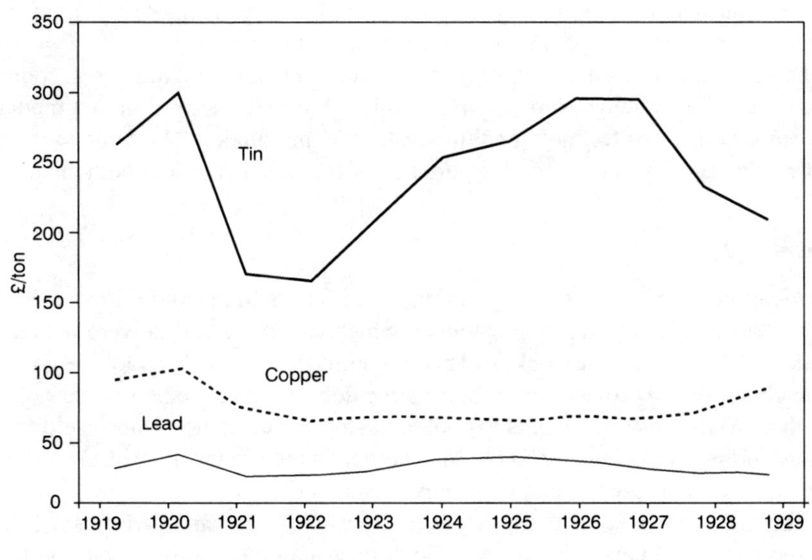

Figure 4.1 Non-ferrous metals prices, 1919–1929

This price history was read in two different ways. The boom of 1926–1927 could be seen either as establishing an appropriate norm for tin, or as a temporary deviation from the long term trend as established by other non-ferrous metals. The prices of 1929 could therefore be regarded either as a manifestation of a temporary problem, or as a return to a more appropriate norm. The fate of the industry would largely turn on the answer to this question. If the long term norm for tin were to be established around the levels reached in 1926–1927, that would set a base for the profitability of new investment and the maintenance of the new productive capacity that it engendered. Any departure from that base would be just a short term aberration for which corrective measures to control the market would be justified. If, on the other hand, these price levels themselves were the aberration, then all such interventions would simply amount to short-term palliatives, leaving the long-term problem at best unresolved if not actually worsened.

The price boom of 1926–1927 has raised several important, interrelated questions, both about its origins and consequences. When critics of the ITC were trying to diagnose the problem of tin in such a way as to point to the artificiality of cartelization as its solution, they focussed on this issue. The argument they laid out has a neat theoretical elegance. Prices serve as signals for producers which enable them to make investment decisions, permitting an equilibrium between production and consumption. Since the prices of 1926–1927 were artificially high, they stimulated excessive and financially unsound investment which created the problem of surplus capacity. They were artificially high because of an earlier intervention in the tin market through the Bandoeng Pool which withheld metal from the market in the slump of 1921–1922. When it was released on a rising market in 1923–1924 it prevented prices from signalling the investment that was needed then to meet the continued rise in demand.[4] As demand ran ahead of supply, a price boom was the inevitable consequence. The lesson to be drawn was clear: artificial interventions provide only short term solutions and compound long term problems.

Those who felt that 1926–1927 represented a long term trend were also conscious of the significance of price. However, they were struck by the extent to which it revealed the degree of price inelasticity since high prices had no obvious adverse effect on consumption. More important was the extent to which they revealed an inelasticity in long-term supply, since the strong incentive given to prospecting revealed no new tinfields.

Price however does not tell the whole story, at least not without taking into account at least two other factors: the system of property rights through which price signals are mediated and the autonomous effects of technological innovation on production. Nor is price the only the source of market signals as the actual experience of the Bandoeng Pool demonstrates.

The Bandoeng Pool

The Bandoeng Pool had its origin in an initiative of STC's Managing Director, W. F. Nutt, which was designed to alleviate the extent to which excessive stocks were depressing the industry in 1921.[5] Most of these stocks had been acquired by the FMS and Johore governments as they sought to try and provide a temporary price floor for their producers. Nutt was able to persuade these Malayan governments to pool their stocks with some of those held by the NEI government adding stocks held by STC and the other Dutch producers, Billiton and Singkep. The pool would then freeze these stocks until the market recovered.

On its formation, the pool held around 19,700 tons or 34 per cent of world stocks. Billiton and Singkep were not financially strong enough to sustain their holdings but the rest of the pool held together and released its metal in an orderly fashion during 1923–1924. Contrary to many assertions, this pool was a financial failure.[6] Although the prices it received were substantially higher than those it initially paid, such profits were more than wiped out by the financing and warehousing costs of the metal.[7] For STC there were no compensating gains[8] and Nutt was dismissed.[9] The experience left an indelible lesson on his successor, John Bagnall, that such speculative activity was far too risky to be undertaken.[10]

Otherwise, the Bandoeng Pool was a success. While prices continued to fall after its formation, the fact that this large quantity of stocks was held in strong hands almost certainly reduced the full impact of the 1921 depression. It also demonstrated that it was possible to release a large quantity of stocks on a rising market without undue disruption. That was also a lesson that would be remembered for some time.[11]

Since the size of the pool was public, as was its release policy, the tin market could easily see the extent to which rising consumption was outstripping available production. Although the releases inevitably reduced the extent to which consumption was putting pressure on prices, they certainly did not deceive the market. To suggest that they did reveals a naive confidence in the significance of prices in shaping decisions on mining investment. Contrary evidence is easily available since annual meetings of companies provided an opportunity for Chairmen to reflect on the overall state of the market and throughout 1923 they regularly commented on its bright prospects once 'the artificial position caused by the "Eastern Pool" is got rid of'.[12]

Whether the Bandoeng Pool in fact created any distortions in the operation of the tin market in such a way as to contribute to the problem of 1929 is less important than a consideration of the nature of the immediate problem it was designed to solve. Excess stocks have to be carried by someone and few producers had the financial resources with which to do so. There are only two other possibilities: professional speculators and industrial consumers. Since the former can never guarantee the financial support required, not least as a result of other commodity speculations, there is always a danger that they will be forced to suddenly release their stocks and that risk will be discounted in price. If stocks are carried by industrial consumers, they will draw them down during the recovery phase and when they return to the market a runaway price can easily follow. Any alternative solution would simply have intensified the swings of the commodity cycle, placing the industry in a far more difficult position as the downswing began in 1928.

The tin cycle of 1923–1929

By 1923 world consumption had returned to the level reached in 1913 but with a very different pattern. Europe was reduced from 55 per cent to 35 per cent, while the United States increased in exactly the same proportions.[13] For the remainder of the decade, consumption steadily increased; on the upswing of 1923–1926, at annual rate of 2.9 per cent, on the downswing of 1927–1929, at an even faster rate, 10.3 per cent.

The strength of this consumption trend was not a function of any major new uses but of the expansion in the demand for established products, especially automobiles, electrical goods, silk and cans[14] which ensured a 'bright sunlight for the future'.[15] Projections would make that light quite dazzling. In 1929, the United States consumed 23 ounces of tin per capita, closely followed by the United Kingdom at 19 ounces. Even if these represented the outer limits of consumption, they reflected a pattern to which much of the world was moving. When the rest of western Europe caught up to Britain, demand would increase by another 20 per cent; when the newly industrializing countries of Japan and the USSR caught up, demand would double. While there were some signs of displacement in virtually all the spheres where tin had been strong,[16] they were not sufficient to undermine such optimism.

Table 4.1 Tin market, 1923–1929

	Production	Consumption	Change in stocks		Stock support Days consumption		
	tons	tons	Visible	Other	Visible	Total	£/ton
1923	126,300	139,000	–4,300	–8,400	66	150	203
1924	141,300	144,500	4,100	–7,300	53	112	250
1925	146,100	154,500	–7,100	–1,300	59	97	262
1926	143,700	146,700	–1,700	–1,300	45	82	284*
1927	158,700	150,900	–600	8,400	39	72	282*
1927	180,300	170,600	8,800	900	34	81	226
1929	196,800	183,600	3,800	9,400	49	94	204

Notes: stock support calculated on basis of average annual consumption divided by stocks at previous year end; £ forward price; * backwardation.

In spite of the general interest in the operation of commodity markets, only one source attempted to look at the interrelationship between production, consumption, stocks and price in a comprehensive manner.[17] It was well aware of the limitations of the data on the first three of these variables and only a general picture of the tin market can be constructed from them. It is presented in Table 4.1.

The rise in consumption that sustained the upswing of 1923–1926 was met both by an increase in production and by running down excess stocks. For the first three years, the market was working well and the rise in price reflected both the increase in consumption and the decline in stocks.

1926 saw a major problem, one that was only kept within manageable proportions by the fortuitous decline in consumption thanks to the slump in Germany and the prolonged coal strike in Britain. Important declines in production were experienced in China and Bolivia; modest ones in Malaya and Siam. While these were partly offset by increases in Nigeria and Burma, overall production therefore actually dropped even in the face of a continued rise in price.

The immediate result was a reduction in stocks to suboptimal levels. Prices therefore continued to rise, coupled with backwardation. Throughout the whole of 1926 and 1927 the average premium paid for spot over forward metal was around £6/16/–, rising on occasion to as much as £17. The higher the level of backwardation, the higher the cost of carrying stocks paid by consumers and a vicious cycle was then set in motion. It was also a problem that producers perceived, since they considered that if only they could organize themselves and secure adequate financing, they could capture this premium by selling into the spot rather than the forward market.

Both production and consumption recovered in 1927 and the stock problem remained. However, the momentum behind the continued increase in production was much stronger than that behind consumption. Excess production therefore resulted in a return to excess stocks and prices declined in sympathy. As soon as that point was reached, around late 1928, the industry was again in trouble. When productive capacity continued to increase in 1930–1931, while consumption again dropped away, trouble turned into crisis.

The fundamental difficulty revealed by the cycle of 1923–1929 was one of coordinating changes in production and consumption around a normal level of stocks that would permit price stability. That was inherent in the operation of the tin market but it would only become evident once the crisis itself was solved.

What was distinctive about the crisis was its location in a production momentum firmly tied to one in consumption. The next section will examine the way in which that momentum developed. A final section will consider some of the difficulties in reading the tin market on the downswing which exacerbated the problem with tin.

Evolution of production

The growth in consumption in the second half of the decade led many to extrapolate this as a long term trend made inevitable by the progress of industrial civilization.[18] Given the apparent limitations on production, the conclusion that the world faced a serious shortage was inevitable.[19] In many circles it was tempting to characterize such a situation as a tin 'famine'.[20]

In one form or another, the famine thesis came to dominate analysis of the industry throughout the decade and it had its origins much earlier than the heady days of 1926–1927. At the end of 1919, an official enquiry was held into the state of non-ferrous mining in Britain. In order to assess the future of Cornish tin mining, it was particularly interested in the likely course of production among the major producers. Cecil Budd considered that the FMS had 'passed its zenith' and would continue to decline.[21] A senior partner in Osborne & Chappel confirmed this assessment, adding that it would not be compensated with the development of dredging and that by 1929 it would start to drop off again.[22] The manager of the Williams, Harvey smelter, which now treated the bulk of the Bolivian ores, was sceptical not only about the prospects of substantial increases there but also in Siam and the NEI.[23]

Five years later, W. R. Jones, a geologist with some 14 years experience in tin mining throughout Southeast Asia, provided the most explicit public statement of this position. Writing in 1924, he confidently made the following predictions:

> A few of these areas [worked by bucket-dredging] are already almost worked out, some others are fast reaching that stage and 10 years hence operations on the majority of the remaining areas will have ceased; and although … improved methods of recovering the mineral … will enable … these properties to be re-treated, the fact remains that the exhaustion of the known areas is a matter of a few years. Not many new areas, suitable for bucket-dredging, are likely to be discovered … so that tinfields like those of Bolivia and the West of England … are likely to become increasingly important as sources of supply in the future.[24]

Such an assessment had an inherent plausibility and the notion that the alluvial deposits of Southeast Asia would soon run out became a commonplace. The condition of the world's largest producer seemed to reflect a more general tendency. In Australia, alluvial tin deposits had been generally depleted between 1870 and 1920 and gold had seen a major shift from alluvial to lode exploitation in the nineteenth century. Jones could also find confirmation in the halving of production in China between 1922 and 1924. While Jones' judgement was challenged by the Chief Geologist in the FMS, he was not prepared to provide an alternative timetable for the depletion for Malaya's tin resources.[25] Without any hard and expensive evidence about the quantity and distribution of reserves, the exhaustion thesis could not be refuted.

From the early 1920s all the major figures in the industry were acutely concerned about the need to stimulate production without being clouded at all by any temporary fluctuations in the level of price. Indeed, even in 1919 they were quite prepared to contemplate price levels in the range of £350–£400 as both necessary and inevitable. With the price boom such projections became common. While few went so far as Askwith, who predicted that in 20 years' time 'tin would be almost as expensive as silver or even gold',[26] even someone with considerable experience such as G. Temple Bridgman, chief mining engineer for Guggenheim Brothers, could make the following assessment to the Mining and Metallurgical Society of America: 'Constant, fairly rapid, and virtually complete approach to exhaustion of world tin resources with no limit to the levels to which prices may rise.'[27] A more precise assessment of the inherent constraints on tin supply was supplied by Ambrose Pratt. He interpreted the low rate of expansion in the NEI as indicating a life of only 10 years, doubted whether either Nigeria or Bolivia could expand and considered that 'the utmost we can expect from the experience of the dredging industry is to compensate for the decline of other forms of production'.[28] With such limitations on supply, Pratt prophesied that never again would the world see tin under £250 and that £400 was its likely future level.[29] Even such an authority as the *Engineering and Mining Journal,* with no particular axe to grind, predicted that 'all indications are that the world's output soon will begin to fall off'.[30]

Once the exhaustion thesis had taken hold it proved extraordinarily difficult to shake off. A mining engineer with considerable experience in Malaya presented it without challenge at a professional conference August 1927 and considered that the future of the industry would henceforth lie in Cornwall and Australia.[31] Even a year later, when fresh capacity from Southeast Asia was already causing concern, the region's long-term future could still be written off, auguring well for Bolivia.[32] Long after the depression had demonstrated chronic excess capacity, the term 'famine' was still being used to characterize the future of tin.[33]

While projections of both price and sources proved to be wildly inaccurate, they pointed to a major problem with tin. The industry was still organized in a way that was far too archaic to meet its responsibilities. The deficiencies were particularly evident by comparison with other non-ferrous metals industries. Much of the metal was wasted and there was no research to support more efficient use. Producing units were generally too small to support the kind of capital needed to apply fresh technologies. Price instability and excessive speculation made the environment within which they operated too uncertain. Reliable statistics required to clarify the sources of that uncertainty were unavailable. In short, tin was an irrational industry in need of a comprehensive programme of 'rationalization'.[34]

Rationalizing the production side of the industry involved four interrelated processes. The most important was mechanization which led to larger units of production, with proportionately larger capital requirements. Associated with this was a process of consolidation into fewer groups, backed by influential mining finance houses. As these groups became stronger, they were in a position to establish closer relationships with smelters and the industry therefore emerged much more tightly integrated both vertically and horizontally.

This process of reshaping had two important consequences for the politics of the industry. As new groups entered the industry, they began to displace established interests that had been responsible for laying its foundations before the war. Of these new groups, the most important was Anglo-Oriental. In some cases it could either absorb or form alliances with older interests but it also created serious rivalries. At the same time, the growth of Anglo-Oriental entailed a fundamental realignment of power at the international level, so that domestic rivalries also became international rivalries.

The overall pattern of growth is presented in Table 4.2.

Over the period from 1925 to 1927 world production increased by some 9 per cent and this was very unevenly distributed. Many countries experienced an actual decline and the

Table 4.2 World tin production, 1925–1929

				Change			
	1925	*1927*	*1929*	*1925–27*		*1927–29*	
	tons	*tons*	*tons*	*tons*	*%*	*tons*	*%*
Burma	1,660	2,520	2,660	850	51	140	6
Siam	8,060	7,530	10,520	–530	–7	2,990	40
NEI	31,240	33,740	35,920	2,500	8	2,190	6
Malaya	48,070	54,320	69,370	6,250	13	15,050	28
Southeast Asia	89,030	98,100	118,470	9,070	10	20,370	21
Nigeria	6,260	8,060	10,730	1,800	29	2,680	33
Bolivia	32,220	35,810	46,340	3,580	11	10,530	29
Other	18,570	16,690	17,170	–1,880	–10	480	3
Total	146,080	158,660	192,710	12,570	9	34,060	21

Sources: ITRDC, *Statistical Yearbook*, The Hague, 1938, Siam, *The Record*.

largest rate of growth occurred in two smaller producers, Burma and Nigeria. From 1927 to 1929 world production increased by a further 24 per cent. While virtually all countries shared in this increase, some, such as Burma and the NEI, slowed down their rate of growth, while Siam, Malaya, Bolivia and Nigeria increased theirs by a considerable margin. It was this asymmetry in the pattern of growth that would make finding a solution to the problem it posed extraordinarily difficult.

Since the problem of the late 1920s proved to be one of chronic excess productive capacity, the critical question became one of who bore responsibility and therefore who should pay the highest price in its solution. There are three different, though overlapping, ways of looking at this issue. One is to focus on the location of new investment and since it determined the terms on which the ITC was formed, it will be considered in more detail later.

A second approach is to focus on the sources of capital investment and comparative cost conditions and that has been favoured by those who look to market solutions. They have constructed a theoretically neat model. Consistent with the claim that the Bandoeng Pool distorted the market by deferring investment decisions and generating an artificial price boom in 1926–1927, it is also argued that this boom induced excessive investment over the period from 1928 to 1930. This excessive investment took the form of properties that could only be profitably worked at high prices and of companies that were overcapitalized. What makes the argument neat is not only its consistent reliance on price as the ultimate determinant of market structures but that it also offers an obvious solution to the problem of excess capacity. The market can be left to regulate itself since a new set of stable prices will quickly eliminate the excess. There is a moral lesson to be drawn here too: those who should not have been drawn in deserve to be expelled.

The most forceful advocate of this position, May, argued that 'the threat of shortage precipitated a speculative fever which made it too easy to get capital and resulted in overbuilding'.[35] There is one clear culprit, Anglo-Oriental, which 'opened enormous new areas'[36] and which manipulated the market by watering the stock of the companies it promoted very heavily.[37] Eastham added a further twist to this argument by suggesting that this kind of new investment by Anglo-Oriental continued throughout 1928–1929 as it maintained prices at an artificially high level by holding stocks of metal off the market.[38]

A third approach is to focus on the form of new investment, especially in new technologies which both make new deposits economically viable and reduce costs and thereby make older technologies obsolete. The primary advocate of this approach, Rowe, also considered that the market offered the only solution, since it alone had the capacity to force the amputation of the now economically redundant capacity.[39] However, such a solution could easily be long drawn out.

In looking at the precise way in which an increase in investment generated an increase in capacity, there are three dimensions to be considered. One is the balance between small scale and large scale production. In the case of Burma, Siam and Malaya, the former sector was almost entirely Chinese and the latter European.[40] A second is the date of the entry of the various corporate groups into the tin industry. In order to address some of the above issues, it is useful to divide these groups into three categories: those that entered before the war, those that entered following the war but before 1927 and those that entered after the boom was underway. A third is the balance between dredging, as the most important and easily identifiable new technology, and other techniques of production.

The detailed pattern of growth will be considered in relation to all tin producing nations in the following order: (1) the British sphere, within which Anglo-Oriental played an important role, Burma, Siam, Malaya, Nigeria, Cornwall, Australia and Japan; (2) other large producers, Bolivia, NEI and China; (3) other small producers, Portugal, Spain, Belgian Congo, South and East Africa and French Indo-China.

First, however, some attention must given to the way in which the Anglo-Oriental group itself emerged under Howeson's leadership.

John Howeson and Anglo-Oriental

The reshaping of the industry brought one individual, John Howeson, to prominence with an all-embracing programme for the rationalization of tin. Some elements were part of a comprehensive strategy for corporate growth and included:

1 flotation of new producing companies
2 acquisition and consolidation of existing companies
3 expansion and consolidation of smelters
4 formation of a mining finance house.
5 stock control

Others were undertaken on behalf of the industry as a whole:

1 promotion of research to sustain demand
2 improvement in the quality of the statistical information
3 political leadership to find an effective solution to the industry's problem.

Howeson was firmly committed to the famine thesis and its basic tenets were systematically promoted throughout the industry, especially with the publication of the Anglo-Oriental house journal, *Tin*.

As the industry found itself in greater and greater difficulty and Howeson became more and more public, two issues emerged: first, the extent to what was he responsible for the problem with tin, especially with the flotation of overcapitalized, weak companies that had brought on the excess productive capacity? ; second the extent to which his political activity designed to rescue his private interests as opposed to those of the industry as a whole. The first of these issues will be considered in the light of the experience of Howeson's companies in each of the producing countries, while the second belongs to subsequent chapter which examines the political role he played in the late 1920s. Before turning to production, the overall evolution of Howeson's tin empire will be reviewed.

John Howeson was the youngest of four children of Baron Oscar von Ernsthausen. Oscar was born in Rhenish Prussia but emigrated, first to London in 1849 and then to Calcutta in 1855 where he established an agency house, Ernsthausen, Ltd. which operated both in London and India.[41] In 1904 John left for Calcutta and joined the family firm which not only served German commercial interests but was also active in the jute trade.[42] The house enjoyed an excellent reputation throughout the East and John founded the Calcutta Philharmonic and served as its conductor.[43]

Howeson brought two resources to the tin industry. He could be personally very impressive. He stood 6′4″ and was a 'clever linguist and subtle negotiator blending in nice proportions the tone and manner of Oxford with a cosmopolitan experience and a financial acumen acquired in various countries'.[44] In the City he was known as 'Uncle John' with 'a bedside manner and an avuncular appearance'.[45] The financial acumen rested on a background in the jute industry,[46] providing a distinctive perspective which was both the source of his strengths and weaknesses. But, with no practical mining experience, many industry circles considered him an outsider, a position to which he could be the more easily confined on account of his purely nominal German origin.[47]

From his business background in India, Howeson brought not only experience but also a network of associates. They included Sir William Henry, W. H. Edwards, Sir Godfrey Fell,[48] and togther with a clerk in the London branch of his father's firm, Louis Hardy. Among Howeson's eastern interests was a holding company, Indo-General Corporation which served as a vehicle for several tin initiatives. In 1920 the group formed a private British company, Anglo-Oriental and General Investment Trust (AOGIT) and for many years it did little independent business.[49] Howeson returned to Britain around 1923 and AOGIT then became active in tin when it served as the conduit for shares to be sold to the public in two other companies floated through Indo-General. At the same time AOGIT became secretary to these companies, a service that it would provide for all subsequent members of the group.

Indo-General's tin interests were divided into two groups and in early 1925 they were sold to two newly formed public companies, Tin Selection Trust (TST) and Eastern Tin Trust (ETT). Both then created a subsidiary, London Tin Syndicate (LTS). ETT was an immediate failure. Its 20 shilling shares soon dropped to 18 shillings and it is one case that lends some credence to the charge that Howeson was guilty of shady company promotion, since all the money raised went back to himself and his associates through Indo-General. However, the shareholders were not permanently disadvantaged since the company was soon absorbed by TST on generous terms. TST went on to raise further funds to become a large mining finance house, with a capital of £1.8 million. LTS played a parallel role with a similar capitalization. In 1928 the final piece in the construction of the financial shell for Howeson's projects was put in place with the creation of a public company, Anglo-Oriental Mining Corporation (AOMC), which held the share capital of AOGIT and then raised a further million pounds.

While there was some overlap between these three public companies and considerable cross-shareholding, there was a broad division of responsibility along the following lines: TST defined new opportunities; LTS developed them in Cornwall and Nigeria, while AOMC focussed on the East. By 1929 it was evident that the era of flotation of new companies was at an end and in early 1930 TST was merged with LTS to become London Tin Corporation (LTC). At this point Howeson was now in control of a complex financial system, responsible for over £4 million of investors' funds; quite a remarkable accomplishment given his rather obscure background. However, the whole edifice has generally been regarded as being built on sand[50] and the extent to which such a charge is justified will be assessed once the experience of the various production companies has been reviewed.

In creating this mining finance house, Howeson was able to establish several important linkages. The first was to the world of mining engineers with the appointment of the experienced and respected E. T. McCarthy[51] as a technical adviser and the firm of Pawle & Brelick as consulting engineers. In 1927, the technical management of the whole group was delegated to Gerard Hutton, of the American firm, Yuba Associated Engineers, with extensive experience in building high quality dredges.[52] A second linkage was to the world of British industry through Lord Askwith, ex-President of the British Federation of Iron, Steel and Tinplate Merchants and ex-secretary to the Board of Trade, who was made Chairman of LTS. More important were the connections to various financial circles. They included Reginald McKenna, Chairman of the Midland Bank, who was particularly interested in schemes of industrial consolidation,[53] and Sir Hugo Cunliffe-Owen, Chairman of British American Tobacco, who became an active partner in many of Howeson's market speculations.[54] Howeson also had access to French finance and TST was listed on the Paris Bourse.[55] But the most significant of all these financial connections was one to Oliver Lyttleton and British Metal Corporation (BMC), an organization at the very centre of non-ferrous metal trading. While initially sceptical about the solidity of the financial structure Howeson had erected, Lyttleton 'formed the opinion that he himself was changing into a serious figure in the metal world and started to finance him and his two companies on a large scale'.[56] Not only would Lyttleton provide financial support but he also opened up a larger set of relationships with other key figures in the industry.

Although Howeson's interests on the production side encompassed virtually all tin producing regions of the world, the most important, at least in the mid-1920s, were in Nigeria. Growth in Nigeria led to a close relationship with CGF which decided to transfer its interests in 1928 in an exchange of shares with LTS. Henceforth CGF would be well represented on its Board of Directors. Not only did that bring access to the financial resources of CGF but several intangible assets. The following year the Chairman of CGF, Lord Harris, commented on the attractions of this new relationship: 'the prudent structure, conservative finances, and able administration of London Tin Syndicate.'[57] Howeson had clearly gone a long way in overcoming his earlier reputation.[58]

The two directorships granted to CGF were held by Lord Brabourne and Oliver Hoare and both had close connections to key members of the Conservative Party. Oliver was the younger brother of Samuel Hoare.[59] Samuel was a good friend of Philip Cunliffe-Lister, President of the Board of Trade, and also sympathetic to a programme of industrial rationalization through protection and reorganization.[60]

As LTS expanded its interests in Cornwall and Nigeria, Howeson decided to buy an old smelter, Penpoll in Bootle, which treated both Cornish and Nigerian ores and which was owned by a metal broker, Strauss.[61] Penpoll was then rebuilt and considerably enlarged but, with a monthly capacity of around 1,000 tons of concentrate, it was a long way from living up to Howeson's boosting as the 'largest tin smelter of the world'.[62] When it was opened to some fanfare by Lord Derby in 1928, over 200 guests, among them many MPs and industrialists, came by special train from London, which provides some indication of the stature Howeson had now acquired.[63]

Burma

Immediately following the war, the drop in the price of wolfram released a number of prospectors to explore the region's tin potential. Mining conditions were generally difficult throughout the whole wolfram-tin belt, from the Mawchi mine up in the Southern Shan States,[64] to the main alluvial deposits on the southeastern coast, an area where 'heavy rainfall, thick vegetation and inadequate means of communication' posed major obstacles.[65]

During the 1920s tin was less than 5 per cent of exports so that it played only a minor role in the overall economy of Burma.[66] Administratively, the country was divided into two sections. Mines in the south were subject to the minerals regulations of the government of India though these did not entail extensive supervision of mining operations. These regulations did not apply to the Shan States which retained the considerable autonomy they had traditionally enjoyed under the Burmese kings. The European sector in the south was organized as the Tavoy Chamber of Mines, while the small miners had no organizational vehicle with which to press their interests.

In spite of its modest size, Burma played an important role since it was there that Howeson established his first and one of his most successful companies. In 1920, Indo-General Corporation floated the Indo-Burma Tin Dredging Company in Calcutta. This acquired the dredge first introduced in 1911 by a Melbourne company together with a number of dredges that had been employed in working gold in Northern Burma and brought them down to Tavoy.[67] One of the first initiatives undertaken was to reorganize Indo-Burma as Tavoy Tin Dredging in 1923. This proved to be an extraordinarily successful company since it worked a high grade deposit at 1.21 pounds per cubic yard (lbs/cuyd) and enjoyed very low operating costs at £28/ton.[68] Growth took the form of acquisition of larger dredges and separating off particular areas to be worked by subsidiary companies. Theindaw Tin Dredging was established in 1925, followed by Northern Tavoy Tin Dredging and Thingandon Tin Dredging in 1926. None of these repeated the success of the parent, in large part because of delays in bringing the dredges to an operational condition and in 1930 they were absorbed back into Tavoy Tin. By 1929 this group accounted for 25 per cent of Burma's production and this is the only country where Howeson established control primarily by the development of new properties rather than by the acquisition of existing companies.

In 1923 Burma again attracted Australian capital and Austral-Malay floated another dredging company, Thabawleik. Although it enjoyed an even higher grade of ground at 1.5 lbs/cuyd than at Tavoy Tin, the heaviness of the clay led to higher costs and less spectacular success. With the price boom came a number of fresh investments of which the most important were sponsored by various South African mining finance houses. Anglo-French Exploration floated Anglo-Burma in 1926; Anglo-American provided much of the capital required to reconstruct Mawchi Mines in 1928; and Central Mining was behind the formation of another large tin-wolfram company in 1928, Consolidated Tin Mines of Burma (CTMB).[69] Two minor British finance houses, British and Malayan Tin Syndicate (BMTS) and Malayan & General, also moved into Burma in the late 1920s but neither emerged as a significant producer.

The extent to which such investments were prompted by the specific level of prices obtaining in 1926–1927 must be a matter of some conjecture. The fact that so much came from gold suggests as much a push effect out of an industry depressed by the revaluation of sterling as a pull effect into tin. The dramatic rise in wolfram prices in 1928 may also lie

Table 4.3 Burma tin production, 1927–1931

	1927	*1929*	*1931*	*Change 1927–29*		*Change 1929–31*	
	tons	*tons*	*tons*	*tons*	*%*	*tons*	*%*
Total	2,446	2,586	3,605	140	6	1,019	39
Small	716	413	446	–303	–42	33	8
Corporate	1,730	2,173	3,159	443	26	986	45
Corporate by Cohort							
A: pre–1920	0	0	436			436	
B: 1920–26	930	1,189	1,595	259	28	406	34
C: 1927–	800	984	1,128	184	23	144	15
B: Anglo-Oriental	546	659	1,007	113	21	348	53
Dredge	930	1,017	1,365	87	9	348	34

Source: Mining Yearbook.

Note: 1931 estimate productive capacity.

behind the investments in Mawchi and CTMB since both of these had as much interest in that metal as in tin. In any case, it is important to note that both major dredging operations were started long before the price boom.

One case serves to illustrate the conventional view of highly speculative investment prompted by the prices ruling in late 1926. In January 1927, BMTS sponsored the formation of Anglo-Scottish Tin Corporation to acquire and develop properties in Burma and Malaya. One of its Burmese acquisitions, Mergui, proved unviable but another served to float a dredging company, Kamounghla Tavoy Tin, in May 1927. The prospectus offered the prospect of a 20 per cent annual rate of return, net of all expenses including amortization, for eight years on a total capital of £100,000, even if the price were to drop from its current level of £300 to £250. Of that capital, £75,000 was raised in cash which in turn was to be distributed between the vendors, £23,000, promotional expenses, £9,500, the cost of the dredge, £32,500, leaving just £10,000 as working capital. However, by the time full title to the properties had been acquired and an appropriate dredge designed, it was early 1930 and with the price approaching £150 the project had to be abandoned. The company's financial resources were exhausted, as were those of its parent.[70]

The overall pattern of production is presented in Table 4.3.

The first trend that is evident is the decline in the role of the small miners. Whereas in 1927, they had been responsible for 29 per cent of production, by 1931 they were reduced to 12 per cent. The second feature is the overall large increase from 1929 to 1931. Of this, the most important single source came from the rehabilitation of Mawchi of the prewar cohort. Otherwise, the growth was primarily the result of continued investment by Anglo-Oriental. Table 4.4 presents the pattern of capital inflows that supported the expansion.

Two measures of capitalization are computed. The equity measure refers to both vendors' shares at their nominal value and the total cash raised from the sale of shares; the cash measure refers to the latter plus sale of debentures.[71] On all measures, new company flotations, equity and cash, the capital inflows were greater before the boom than afterwards. The emergence of CTMB after the boom actually reduced the capitalization per ton of capacity. Of the pre-boom cohort, Anglo-Oriental stands out on all measures but with a significantly lower capitalization than most of the others.

Table 4.4 Burma capital inflow, 1920–1930

	1920 – 1926			*1927 –1930*			*1931*	
	Cos No.	*Equity £000*	*Cash £000*	*Cos No.*	*Equity £000*	*Cash £000*	*Equity £/ton*	*Cash £/ton*
A: pre–1920		20	120		103	103		
B: 1920–26	6	815	430	1	236	211	596	355
C: 1927–				1	405	152	359	135
B: Anglo-Oriental	4	492	239		136	136	624	372
Total	6	835	550	2	744	466	500	264

Source: *Mining Yearbook*.

Note: 1931 measures the inflow into new operating companies against the capacity from table 4.3

Siam

In spite of her role as the pioneer of dredging in Southeast Asia, production in Siam languished for most of the 1920s. Siam was generally a less attractive field for investment than Malaya. With a more extreme climate and more difficult deposits, costs were generally higher.[72] While the schedule of royalties was similar to that in the FMS, Siam added duties on imports, provided far fewer services[73] and local officials annoyed foreign companies with demands for petty bribes.[74]

Prior to the price boom only two of the established groups expanded their commitments. Kerry floated Takuapa Tin Dredging in 1920 and in the same year Siamese Tin floated Bangrin Tin. However, the other Guthrie company, Renong, abandoned Siam altogether and moved its whole operation to Malaya in 1923. Tongkah Harbour also moved one of its dredges in 1925. While the Palfreyman group sustained its Siamese base, it moved Tongkah Compound to Malaya in 1920.

Four new entrants compensated for these departures. BMTS floated Eastern Siam Tin Dredging in 1925; Malayan and General Tin Trust floated Pattani Consolidated Alluvial Tin in 1926, followed by Kamra Tin Dredging in 1927.[75] The other two were based in Australia. In 1920–1921, King[76] secured rights over a large valley and formed Malay-Siamese Prospecting with Chinese capitalists in Penang and eventually floated three companies, Nawng Pet, Haad Yai Tin Dredging and Huey Yot Tin Dredging. Burma-Malay Tin followed by floating Katu Tin Dredging and Renong Consolidated Tin.

From 1927 not only did these earlier groups develop and expand their properties but they were joined by others. That year saw Anglo-Oriental float Talerng Tin Dredging, Austral-Malay float Pungah Tin Dredging and the formation of one of the few independent dredging companies, Muang Tin. In 1928 another independent company was formed, Anglo-Eastern and, even as late as 1930, the Tronoh group floated Phuket Tin Dredging. From 1927 to 1929 Siam experienced the extremely high annual growth rate of 18 per cent, tapering off slightly from 1929 to 1931. That meant a virtual doubling of her productive capacity over the whole period. As is evident from Table 4.5, the whole of this increase came from dredging.

Fresh dredging capacity came from several quarters. Groups formed before the war were largely responsible for the increase from 1927 to 1929 and of these two Australian groups, Palfreyman and Kerry, grew particularly rapidly from 1929 to 1931. They were joined by another Australian group, Austral-Malay, which only entered after the boom but accounted for virtually half the growth in that crucial period. By contrast, Anglo-Oriental's contribution was very modest.

Table 4.5 Siam tin production, 1927–1931

	1927	1929	1931	Change			
				1927–29		*1929–31*	
	tons	*tons*	*tons*	*tons*	*%*	*tons*	*%*
Total	7,530	10,520	14,000	2,990	40	3,480	33
Small	4,880	4,790	4,230	–90	–2	–560	–12
Corporate	2,650	5,730	9,770	3,080	116	4,040	70
Corporate by Cohort							
A: pre–1920	2,507	4,465	5,697	1,958	78	1,232	28
B: 1920–26	144	1,169	1,766	1,025	712	597	51
C: 1927–		100	2,306			2,206	
A: Kerry	428	316	1,305	–112	–26	989	313
A: Siamese Tin	1,314	1,700	1,700	386	29	0	0
A: Palfreyman	332	847	1,727	515	155	880	104
C: Austral–Malay			1,008			1,008	
C: Anglo-Oriental			470			470	
Dredge	2,180	5,200	9,400	3,020	139	4,200	81

Sources: *The Record*; *Mining Yearbook*, J. B Were, *Tin Shares as an Investment*, Melbourne, 1938.

Table 4.6 Siam capital inflow, 1920–1930

	1920–1926			*1927–1930*			*1931*	
	Cos No.	*Equity £000*	*Cash £000*	*Cos No.*	*Equity £000*	*Cash £000*	*Equity £/ton*	*Cash £/ton*
A: pre–1920	8	1,012	764	2	541	610	273	241
B: 1920–26	5	800	565	2	450	408	708	551
C: 1927–				5	1,069	949	464	412
A: Kerry	1	44	44	1	142	138	143	139
A: Siamese Tin	1	176	155				104	91
A: Palfreyman	6	680	453	1	322	396	580	492
C: Austral–Malay				1	430	392	427	389
C: Anglo-Oriental				1	317	234	674	498
Total	13	1,812	1,329	9	2,060	1,967	396	337

Sources: Mining Yearbook, W. K. Garnsey , *Eastern tin dredging companies*, Sydney, 1937.

Note: 1931 measures the inflow into new operating companies against the capacity from table 4.5

The distribution of the capital raised to support this expansion is presented in Table 4.6. The pattern of capital inflow broadly reflects that of the expansion, with the bulk of the fresh capital over the whole period being raised by the prewar groups, while the post-1926 groups were responsible for most raised during the second period. There is no consistent picture of the level of capitalization expressed per ton of productive capacity, though Anglo-Oriental again appears on the high side. One specific feature of the capitalization of many of the new companies needs to be noted. Since they moved into the production phase as the market declined, they were placed in a particularly difficult financial position and were forced to raise working capital through debentures.

The overall picture offers broad support to both the May and the Rowe interpretations of the condition of the industry. Rowe could point to the impressive growth of dredging with its implications for the long term reduction of costs. May could equally well point to the large amount of fresh capital and excessive capacity induced after the price boom, though the data clearly absolve Anglo-Oriental of any particular responsibility and suggest that the financially weaker companies were those that were formed before the boom rather than afterwards.

Neither of these broad positions captures the particular rhythm of the development of the tin industry in Siam in the 1920s which was quite distinctive, stagnant during the first part of the decade and then rapid growth towards the end. If attention is devoted to the supply side of suitable mining land, then dredging should have been expanding much earlier since fresh deposits were being opened up with the completion of the railway link to the FMS in 1922. In explaining the Siam puzzle, primary attention must be given to the system of property rights.

Unless a dredging company initiated its own prospecting, it had to acquire land which was either held under an existing mining lease or under an EPL. Normally the land required to justify the investment in a dredge was considerably larger than that held under a single lease or EPL, so that the company had to negotiate with several owners over the value of the tin in situ. Anticipation of a price rise would reduce the incentive to sell and increase the risk for the buyer but this would change as soon as it became clear that prices were on a downward trend.[77] There were therefore two lags between the point at which price served as a signal for fresh investment and the point at which new production came on stream. One is the interval of at least two years between the point at which a dredging company was formed and the point at which its dredge became operational. The second is the lag as result of the time required to complete negotiations with several different owners, often with unrealistic expectations of the value of their claims.[78]

In Siam these principles operated under additional constraints. Until new regulations were introduced in 1928, holders of EPLs were under little pressure to apply for mining leases and hence they could tie up land for speculative purposes.[79] Although the Department of Mines in Bangkok issued the leases, local politicians also claimed a role in the granting and transfer of leases which added to the delay and cost.[80] In addition, a lease did not carry with it the automatic right to start extraction. Before that could be exercised, those who held even squatter's rights to the land also had to be compensated.[81] Claims could be made on the basis of flimsy evidence and there is at least one case of banana trees being transplanted overnight to force exorbitant compensation.[82] Overcoming all these obstacles required both patience and the stimulus of a drop in price. In light of these considerations, it is not surprising to see fresh capacity being deferred to as late as 1931.[83]

Growth in tin production did not materially change the role of the industry within the overall political economy of Siam. Tin was the second most important export commodity but at 17 per cent of 1929 exports, it was still well behind rice at 70 per cent.[84] As a proportion of overall government revenues, tin was quite modest. In 1928–1929 royalties and other direct fees contributed 2.4 million baht (£220,000) or 2.3 per cent of ordinary revenue.[85]

Growth, however, did stimulate one important organizational change and that was the formation of a Siam Chamber of Mines in 1928. While it attempted to recruit 'reputable' members from the Chinese community, it remained an exclusively European association.[86] At first the headquarters was placed in Tung Song, between the East and West coasts but that soon proved inconvenient and it was then moved to Penang which 'emphasized its alien composition'.[87] Regardless of location, the ability of the Chamber to shape the formulation of mining policy in Bangkok remained very limited.

Malaya

The famine thesis rested on projections of production from Malaya and it was here that it was both confirmed and undermined. While she began to rebuild her productive capacity, output remained well below that reached before the war. No increase at all occured between 1925 and 1926. The most promising feature of that rebuilding, bucket dredging, stagnated for four consecutive years, 1923 to 1926. All of these features provided ample cause for concern and would reinforce the developments, especially in dredging, that would only come to fruition after the peak in price in 1927. As they changed the supply position in the world industry, they transformed the structure of the industry in Malaya itself. Table 4.7 illustrates the extent to which dredging came to dominate Malaya.

While production in the rest of the world grew by some 18 per cent from 1927 to 1929, Malaya grew by 33 per cent. As the largest producer with 37 per cent of world production, such a growth rate produced the greatest quantity that had to be absorbed. Had this production momentum reached its peak in 1929, the attempts to address the problem of tin that will be considered in the next chapter might well have succeeded. The fact that it continued largely

Table 4.7 Malaya tin production, 1927–1931

				Change			
	1927	*1929*	*1931*	*1927–29*		*1929–31*	
	tons	*tons*	*tons*	*tons*	*%*	*tons*	*%*
Total	54,319	69,364	89,900	15,045	28	20,536	30
Chinese	32,284	27,653	27,600	–4,631	–14	–53	–0
European	22,035	41,711	62,300	19,676	89	20,589	49
European by cohort							
A: pre–1920	19,270	31,704	49,863	12,434	65	18,159	57
B: 1920–26	2,764	7,144	7,487	4,380	158	343	5
C: 1927–		2,862	4,950			2,088	73
A: Redruth	2,649	5,177	9,669	2,528	95	4,492	87
A: Alluvial Tin	1,798	4,307	8,009	2,509	140	3,702	86
A: Tronoh	1,547	2,994	6,086	1,447	94	3,092	103
A: Malayan Tin	1,246	3,161	4,535	1,915	154	1,374	43
A: S E de Kinta	2,800	3,000	3,229	200	7	229	8
A: Henggeler	342	842	2,226	500	146	1,384	164
B: Yukon Gold	1,434	2,110	2,110	676	47	0	0
A: Palfreyman	1,115	1,558	2,045	443	40	487	31
A: Guthrie	344	646	1,785	302	88	1,139	176
A: Kamunting	1,053	1,036	1,651	–17	–2	615	59
C:Anglo-Oriental		342	1,570	342		1,228	359
A: Ipoh	437	978	1,564	541	124	586	60
B: David	273	799	1,316	526	193	517	65
A: Miles	1,058	1,031	1,221	–27	–3	190	18
A: Sime	402	693	1,069	291	72	376	54
A: Austral–Malay	986	866	1,026	–120	–12	160	18
Dredge	13,005	26,614	46,220	13,609	105	19,606	74

Sources: *Mining Yearbook, Wild Cat Monthly*

Notes: includes UFMS; Anglo-Oriental only includes properties it developed.

unabated over the next two years made it extremely difficult to diagnose the precise problem until its full magnitude was evident. At that point it would become clear that the problem with tin was largely a problem with Malaya.

The first feature to note is the emergence of several groups which now had some stature in the international market. Whereas in 1913 there was only one which controlled more than 1 per cent of the world market, there were now ten which brought the total share from 3.2 per cent to 22 per cent, from 8.2 per cent to 50 per cent of that supplied from Malaya.

That qualitative shift rested primarily on a quantitative expansion by the groups already established before the war. The pioneering Cornish groups were responsible for the greatest absolute increase in both periods, 1927 to 1929 and 1929 to 1931, but their share of the European sector had declined from 37 per cent in 1913 to 23 per cent by 1929. That dilution was a function of more rapid rates of growth on the part of some of the other prewar groups. They include Malayan Tin Dredging, loosely linked to Tronoh, and the agency house, Guthrie, which shifted its attention from Siam to Malaya. However, the one that grew fastest of all was the Australian Alluvial Tin group and this provided the base from which the dominant position held by the Cornish would be challenged.

Dilution was also the function of the emergence of a number of newcomers. They were responsible for a significant part, 37 per cent, of the increase from 1927 to 1929 but only 12 per cent of that from 1929 to 1931. While there may have been a number of inexperienced operators whose interest in Malaya was prompted by the price boom of 1926–1927, the companies they floated were too small to be a significant source of the problem of Malayan excess capacity.

Two of these newcomers represented established outside interests. The first were the Guggenheims. By the end of the war dredges belonging to Yukon Gold operating in Klondike and California were working low grade properties. In 1917, the company started looking at suitable areas in the FMS to which to move them.[88] Thorough and expensive prospecting proceeded throughout 1920[89] and in 1923 a subsidiary, Ampang Tin, began its successful career.

Anglo-Oriental was the other important newcomer and floated Kampar Malaya Tin Dredging in 1927. This was to be a demonstration of Howeson's plans for technical rationalization, since this was the first Yuba dredge to be built specifically for the Malayan tin industry.[90] With a capacity of 215,000 cubic yards per month (cydm), it was the largest of all Malayan dredges and among the cheapest of all producers. An even larger Yuba dredge, with a capacity of 333,000 cydm, was then ordered for a sister company, Lower Perak Tin Dredging.[91]

Both of these properties had been held by Alluvial Tin but it decided to co-operate with Anglo-Oriental in their development. That association led to the most important acquisition made by Howeson. In late 1927 Anglo-Oriental bought Alluvial Tin's Malayan subsidiary, Anglo-Malayan Tin, which included a minority interest in each of the companies of the group, replacing it with a more comprehensive technical and secretarial service. London Malayan Tin Trust (LMTT) was then floated in October 1928 to raise the £1.25 million required to cover the costs of this acquisition. Unfortunately, the share price of LMTT fell even faster than the price of tin and it was soon absorbed by LTC.

The Alluvial Tin connection changed the character of the Anglo-Oriental empire in three crucial respects. Its focus had shifted from development of new properties and flotation of new companies to acquisition of control and consolidation of established ones. As it moved in this direction, its future would be no more dependent on the continuation of boom prices than other major groups. The terms on which companies were acquired reflected the fact that such prices had long passed. Finally, Malaya had become the most important element in that empire, by a considerable margin.

It also changed the character of Malaya since Anglo-Oriental now controlled the largest number of dredges. Furthermore, with an explicit commitment to take control over the technical development of dredging through association with an American firm, Howeson had broken with the expectations of the established British companies which had always seen themselves as loyal members of the empire with the responsibility of supporting British technology.[92]

Howeson's entry into Malaya would disrupt more than convenient business linkages. His programme of rationalization meant a bid for leadership of the whole industry. The full extent of this was as yet unknown outside very limited circles but as far as Malaya was concerned it would mean a displacement of the position occupied by the Cornish groups. They had been important leaders of the process whereby European technology had ensured the predominance of Malaya in the world tin industry in the face of the decreasing competitiveness of the Chinese but now found it difficult to know how to respond to the challenge posed by Howeson.[93]

Figure 4.2 maps the interrelationships that constituted the basis of Cornish influence. As head of the Redruth group, Mair remained the pivotal figure and maintained close links with the Tronoh group controlled by Thomas.[94] He was also the primary link with both the Inchbald and Addinsell groups. While Osborne & Chappel served part of Tronoh and all of the three other groups, it could not offer the kind of leadership to the entire network that enabled it to match Anglo-Oriental. The core network had close links to groups entirely outside the scope of Osborne & Chappel, to Stephens of the Malayan Tin Dredging group and to Byrne of the Kamunting group. The unfolding pattern of politics would see Howeson detaching both of these ancillary groups as well as absorbing several others which at this stage remained quite independent.

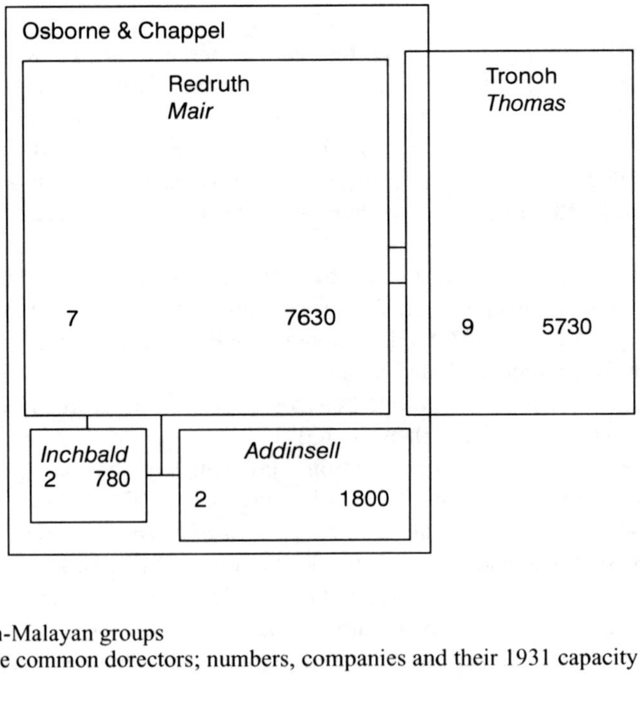

Figure 4.2 Cornish-Malayan groups
Note: Lines indicate common dorectors; numbers, companies and their 1931 capacity

Table 4.8 Malaya capital inflow, 1920–1930

	1920–1926			*1927–1930*			*1931*	
	Cos No.	*Equity £000*	*Cash £000*	*Cos No.*	*Equity £000*	*Cash £000*	*Equity £/ton*	*Cash £/ton*
A: pre–1920	32	4,911	3,577	11	2,922	2,792	157	128
B: 1920–1926	8	2,503	1,955	3	1,210	598	496	341
C: 1927–1930				7	1,060	685	214	138
A: Redruth	3	450	492	3	590	501	108	103
A: Alluvial Tin	10	1,525	1,017	3	654	469	272	186
A: Tronoh	3	366	262		371	394	121	108
A: Malayan Tin	1	326	116		347	422	148	119
A: Henggeler	4	349	233	2	207	184	250	187
A: Palfreyman	2	207	92				101	45
A: Guthrie	2	545	410	1	281	410	463	459
C: Anglo-Oriental				1	270	150	172	96
A: Ipoh	1	200	145	1	180	140	243	182
B: David	2	288	149	1	838	251	856	304
A: Miles	2	130	130	1	112	90	198	180
A: Sime	2	343	308				321	288
A: Austral–Malay	1	175	133		30	30	200	159
Total	40	7,414	5,532	21	5,193	4,075	202	154

Sources: Mining Yearbook, Garnsey, op. cit.

Note: 1931 measures the inflow into new operating companies against the capacity from table 4.7.

Just as expansion of production was concentrated in the hands of groups established before 1920, so was the pattern of inflow of the fresh capital. In that regard, Malaya charted a very different course than either Burma or Siam. On all measures, established groups retained their dominance throughout the period, as is illustrated in Table 4.8.

Just as these data show that established groups raised the bulk of the fresh capital, they also show considerable variation in the extent to which groups invested that capital efficiently in new dredging companies. Fresh entrants could often do much better than established ones. Neither feature provides any confirmation for the Eastham thesis which locates excess capacity in the new overcapitalized companies floated in response to the artificial conditions in the mid-1920s. In the light of the tension that emerged between Anglo-Oriental and the Cornish, it is worth noting that the capital requirements of its flagship company, Kampar Malaya, were comparable to the flotations by Redruth and Tronoh. The results of the comparison between Malaya and Siam should also be noted. Capital requirements of new dredging companies operating in Siam were about 17 per cent greater than their counterparts in Malaya. That result is consistent with the overall impression that the costs of doing business in Siam were higher than in Malaya and companies operating there would therefore find themselves in a particularly difficult position when the market entered its crisis of 1930–1931.

The dynamism demonstrated by the European sector in the FMS has tended to overshadow developments in the other parts of Malaya. The UFMS experienced a very different rhythm. Only one dredging company was formed, Bundi, by an Australian group in 1922 and following the installation of its second dredge in 1927 it became a substantial producer. While production in the FMS in the first part of the decade had dropped considerably below its prewar level, prompting fears about the long term future of supply, production in the UFMS remained stable. Production reached its peak in 1924 and then slightly declined but these were rather minor fluctuations and the UFMS was one of the few areas which was neither booming nor declining.

Although the Chinese sector never recovered its prewar level, it did experience an important pattern of growth towards the end of the decade, tailing off with the price decline of 1929. Since this coincided with the emergence of much fresh dredging capacity, its share of FMS production dropped from 57 per cent in 1927 to 39 per cent in 1929.[95]

The failure of the Chinese to participate in the expansion of dredging suggests a comparative technical and institutional backwardness and several explanations have been advanced: capital requirements of dredging were too high; antipathy towards the more formal structure of the joint-stock company; low levels of labour required by dredges reduced ancillary profits available through control over opium and gambling; lack of familiarity with the technology.[96] The extent of Chinese exclusion should not be exaggerated. Khaw Joo Tok continued to be active in facilitating the growth of the Palfreyman group in both Malaya and Siam. Chinese directors were also to be found among some of the Singapore-based companies and the Chinese certainly included dredging companies among their shareholdings.[97] The growth of dredging placed substantial unearned profits in the hands of the Chinese on the transfer of their mining land.[98] All of the explanations offered are speculative and none survives much scrutiny. There were many wealthy Chinese capitalists who could personally provide the £65,000 required to buy a new medium-sized dredge[99] and profits in dredging companies could far outweigh those derived from any excessive exploitation of labour.[100]

Growth in the Chinese sector occurred in two spheres. The decade saw a considerable extension of gravel pumps[101] and it also saw the formation of the first joint-stock companies under Chinese control, Toh-Allang in 1923 and Hong Fatt (Sungei Besi) in 1926. Hong Fatt demonstrates the considerable potential remaining within the Chinese sector. When it was formed it was a joint Chinese-European venture and managed by a European but by 1929 the Chairman was Chinese as was the manager. New management resulted in an expansion of output to 860 tons of tin-in-concentrate, with costs cut by 42 per cent. In spite of the low prices realized in 1929, it declared a net profit of £64,000 on its capital of £345,000 and was sufficiently strong to withhold a portion of its production in anticipation of a recovery in the market.[102] This was a record matched by few European companies.

In spite of the way in which ethnic categories were used in the administration and analysis of the industry, differences within the two sectors were far more important than the differences between them. That would add further complications to the problem of organizing the industry in such a way as to meet the crisis of 1929–1931.

Tin in Malaya

Although the political economy and demography of the West Coast of Malaya had largely been shaped by the growth of the tin industry prior to World War I, from then on it was overtaken by rubber. During the 1920s the value of rubber exports were more than twice those of tin and the labour force required even larger. However, the greater value added by the tin industry allowed for a much higher level of taxation, so that royalties on tin generated 17 per cent of government revenue, far surpassing the export duties on rubber at 8 per cent.[103]

The FMS government maintained an elaborate administrative structure of support and supervision of its mining industry. The Department of Mines was responsible for issuing EPLs and recommending applications for conversion to mining leases. Inspectors of Mines paid regular visits to the mines both to ensure that the labour conditions of the leases were being met and to monitor compliance with a host of regulations governing safety in mines and their impact on the environment. With that troop of local inspectors, the Department could compile a comprehensive set of statistics, covering not only production but also employment, utilization of machinery and accidents. The government had a strong interest in the welfare of the tin industry as its primary source of revenue and had the administrative capacity which enabled it to shape future development.

At an early stage the European section of the industry organized itself as the FMS Chamber of Mines to both exchange technical information and lobby government. While leading Chinese miners played important roles in the Chamber, the majority of small miners were represented through state organizations, the Selangor Chinese Miners' Association and the Perak Chinese Miners' and Planters' Association. Further complicating this picture were two other European organizations. Southern Malayan Dredging Association served as an important vehicle for exchange of information on various aspects of dredging technology and the Malayan Chamber of Mines, formed in 1918, served as a means with which London-based companies could lobby the Colonial Office. Added to this set of voices claiming to speak on behalf of the interests of Malayan mining were two unofficial members of the FMS Legislative Council designated as mining representatives.

In the UFMS tin played a far less important role. In Johore and Trengganu it was even subordinate to iron ore. Only Johore maintained a separate Department of Mines and in none of the states did the industry have any organization of its own.

The spectre of rubber

The politics of tin in Malaya were conducted within the long shadow cast by the failure of rubber control under the Stevenson plan. Of all commodity agreements this was, perhaps, the worst thought out. It was designed to rescue the industry from the depression of 1920–1921, which was particularly severe in rubber since new production from European estates, originally planted during periods of high prices, was now coming on stream. The British government supposed that, since Malaya and Ceylon provided the bulk of the world's supply, it could easily regulate the market.

The plan was first implemented in 1922 and made three serious mistakes. The first was in its administration which regulated production strictly in accordance with fluctuations in price and the ensuing inflexibility created a great deal of antagonism among consumers. The second was lack of control over other potential sources and a considerable increase in outside production occurred, especially from the native producers of the NEI. Finally, the sponsors underestimated the ability of American consumers to take countervailing measures; they increasingly resorted to reclaimed rubber. By 1928 it was clear that the government had lost control of the market and the scheme was wound up. Whatever benefits Malaya had enjoyed were ephemeral, since the overall market for natural rubber was reduced by the increase in secondary rubber and she had lost her share of the natural market. There was always a fear that tin regulation would be as ineptly managed and the word 'restriction' inevitably conjured up this spectre.[104]

Nigeria

During the initial phase of capitalist control, production relied on comparatively simple, labour intensive techniques, within an equally simple organizational framework. Companies contracted with gangs of African labourers, paying them by results and providing minimal supervision by European mining engineers. While this pattern continued for some time, it soon became evident that the long term future of the minefield lay in much deeper deposits. Since they could be as much as 100 feet below the surface, that would require mechanical excavation of the overburden. The only source of the cheap energy that would be needed was to be found in two sets of falls which could be harnessed for hydroelectric power.

The hydroelectric project at Kwall Falls sponsored by Anglo-Continental's Northern Nigeria (Bauchi) came to fruition in 1925 with a dramatic effect: production doubled and costs dropped by 45 per cent. A similar project at Kurra Falls, sponsored by Latilla's Nigerian Tin and Power, had less success in raising the capital required to bring it to completion.[105] In 1929 Anglo-Oriental acquired Latilla's interests and floated the Nigerian Electricity Supply Corporation (NESCO) to raise £450,000.[106] By July 1930 the plant was finally ready. Given the importance of cheap power in sustaining a programme of mechanization of extraction, Nigeria therefore found herself in a rather similar situation to that already encountered in Southeast Asia. Productive capacity was continuing to expand long after the tin market found itself in serious trouble.

The networks around the Anglo-Continental group that had shaped the prewar organization of the industry became more tenuous in the postwar period. Latilla and Naraguta emerged as autonomous groups. CGF developed one of the largest and most profitable companies, Ropp Tin. Further dilution of the erstwhile dominance of the group came from two other major sources. One was the Davidson and Broadbridge group which formed Barrier and General Finance in 1924 and under its auspices they floated eight new operating companies. The other was Anglo-Oriental. In addition to NESCO, the group established two major mining companies.

In 1926 LTS transferred its Nigerian tin properties to Associated Tin Mines of Nigeria (ATMN) which was formed with a triple mandate. In addition to taking over existing companies, it undertook initial prospecting and development of fresh properties, which could be floated off as separate companies, while retaining others for its own exploitation.

At first, the primary source of income came from the profits on the flotations but as the tin market began to decline at the end of 1927 ATMN was approached by Broadbridge. Attracted by the high dividends initially paid by ATMN and concerned about the increasing cost of maintaining prospecting levels on its own properties, Broadbridge felt that Barrier and General should be folded into ATMN and in early 1928 a share exchange was completed.[107]

London Nigerian Tin Mines (LNTM) was the second of these companies. This was formed in early 1930 from the assets of Ropp and the remaining three Anglo-Continental companies through a similar share exchange arrangement.[108] Since this included the Kwall Falls hydroelectric project, Anglo-Oriental was placed in control, not only of around half of Nigerian tin production but also most of the electrical power required by the remainder.

The precise pattern of development in the 1920s is illustrated in Table 4.9.

It was the established groups that bore the primary responsibility for the increase in production over 1927–1929. Assigning the fresh capacity permitted by NESCO to ATMN shifts the balance to the newcomers for 1929–1931, but it should not disguise the fact that this reflects the strength of a technological momentum that was well underway long before the crisis emerged.

Nigeria, however, differed in three respects. The boom of 1926–1927 attracted no new groups. As the crisis of 1929–1930 intensified, independent miners became particularly active. Most had originally arrived in the service of existing companies but once there, discovered other opportunities to develop fresh properties and strike out alone.[109] As companies began to reduce output, the labour force released became available for these

Table 4.9 Nigeria tin production, 1927–1931

	1927 tons	1929 tons	1931 tons	Change 1927–29 tons	Change 1927–29 %	Change 1929–31 tons	Change 1929–31 %
Total	8,056	10,734	13,036	2,678	33	2,302	21
Small	604	618	943	14	2	305	53
Corporate	7,452	10,117	12,093	2,665	36	1,976	20
Corporate by Cohort							
A: pre–1920	5,620	7,964	7,973	2,344	42	9	0
B:1920–26	1,832	2,153	4,121	321	18	1,968	91
A: Anglo–Continental	1,603	1,640	1,640	37	2	0	0
A: Latilla	1,395	2,463	2,885	1,068	77	422	17
A: Naraguta	502	1,004	1,173	502	100	169	17
B: Barrier & General	1,020	932	930	–88	–9	–2	0
B: ATMN	358	800	2,738	442	124	1,938	242

Sources: 1927, 1929, Nigeria, Department of Mines, Annual Reports; 1931, SNP14034A, 2 July 1931.

miners and their productive capacity continued to rise. Above all was the rapid emergence of Anglo-Oriental which made Nigeria the only British territory where a single group could authoritatively claim to speak on its behalf. Although Howeson's Malayan interests may have been larger than his Nigerian ones, it was the latter that played the more strategic role. They had arisen by virtue of his connections with important London financial interests and they served to cement his position within the world of British smelting.

Some of the aspects of Anglo-Oriental's Nigerian initiative, however, can be seen as confirmation of several features of the charges that have been levelled against Howeson. The three companies floated by ATMN were all weak. Maiangwa Tin Mines never operated. Jarawa Tin Dredging was absorbed back into its parent and while Juga Valley Tin Areas survived, it never paid a dividend. They appear to have been created solely to provide opportunities for stock promotion. However, the flagship company was comparatively strong, especially after its acquisition of the Barrier and Finance group. Consolidation coupled with fresh sources of power placed ATMN in a very competitive position and by 1931 all-in costs had been cut to £130/ton metal. The pattern of capital inflow presented in Table 4.10 confirms the comparative strength of Anglo-Oriental.

Table 4.10 Nigeria capital inflow, 1920–1930

	1920–1926 Cos No.	1920–1926 Equity £000	1920–1926 Cash £000	1927–1930 Cos No.	1927–1930 Equity £000	1927–1930 Cash £000	1931 Equity £/ton	1931 Cash £/ton
A: pre–1920	5	1,250	1,042	1	575	665	302	210
B:1920–1926	15	1,250	1,080	4	1,000	852	418	235
A: Latilla	2	1,038	890		169	381	276	224
A: Naraguta	3	165	105	1	246	184	340	192
B: Barrier & General	6	392	362	2	430	355		
B: Anglo-Oriental	3	460	410	2	523	448	377	184
Total	20	2,500	2,122	5	1,577	1,517	380	227

Source: *Mining Yearbook.*

Note: 1931 measures the inflow into new operating companies against the capacity from table 4.9

These data also show far less diversity in capital intensities than was the case elsewhere and that homogeneity would be reflected in the absence of any division among the companies about the principle of tin restriction. While these capital intensities may have lent some substance to the widely accepted claim that Nigeria was a high cost producer, they were below those for Burma and Siam.

For quite different reasons, the most dynamic forces in Nigeria at the end of the decade were the largest and the very smallest operations. All the established groups that remained outside the Anglo-Oriental orbit were growing at a far slower rate. The differentiated character of the industry would make internal co-operation with restriction very difficult.

In one important respect, the industry was quite homogenous, since all the companies were incorporated in London and policy could be formulated through the Nigerian Chamber of Mines. The Chamber created a local council but granted it no autonomy, using it simply as a conduit to the Nigerian government. The independent miners, however, continued to be excluded from both bodies and had no organization of their own with which to press their interests.

Nigerian tin production was not only about the same level as that of Siam but it played a similarly modest role in the overall national economy. While tin was the most important mineral exported by a considerable margin, it accounted for only 11 per cent of total exports, well behind agricultural products, especially palm-oil, ground-nuts and cocoa. Unable to fully compensate the loss of half the royalties to the United Africa Company, successor to the Niger Company, with increases in land rents, duties and railway tariffs, the government derived an even smaller amount of its income from tin.[110]

As in the case of Siam, tin was extremely important in a particular region, one that was quite remote from the central government. That regional gulf was exacerbated as a result of Nigeria's colonial history which left a great deal of autonomy to the administration of the Northern Province whose civil servants felt a 'pride and prestige [that] will make them unwilling to brook inspection and criticism from that outside world from which they are so largely shut away'.[111] Co-operation at the government level would prove to be almost as difficult as that at the corporate one.

Cornwall

Cornwall entered the 1920s in a profoundly ambivalent position. Suspension of development and maintenance work during the war had made it particularly vulnerable to the depression of 1921–1922. Given the inefficient methods of mining in the nineteenth century, there were a large number of abandoned mines which still contained substantial quantities of minerals and there was still considerable scope for further prospecting and location of new deposits.[112] It was therefore easy to read the recent past as a temporary aberration and conclude that the county had a future as bright as its history.

Given the importance of mining in this remote and impoverished county, Cornwall was never short of local boosters of tin.[113] Their optimism was naturally reinforced by the results of the Enquiry into the State of Non-Ferrous Mining and the decision of the Treasury to grant generous loans to rehabilitate some existing mines.[114] It was sufficient to encourage further investments in both new and existing mines [115] and drew in three outside groups, of which the most important was Anglo-Oriental.

Anglo-Oriental first expressed interest in Cornwall in 1925 and developed three mines, Wheal Kitty, Polhigey and Parkchany, each of which were floated as separate companies. By 1929, the first two had become substantial producers and the whole group was the second largest producer in the county. BMTS lay behind two flotations, Kingsdown and Wheal Reeth. Camp Bird was responsible for Wheal Buller and this is yet another example of a company originally based in gold moving into tin.[116]

None of these new developments, whether by existing companies or fresh ones, were able to reverse Cornwall's historic decline. In spite of Anglo-Oriental's growth from 1927 to1929, Cornwall's production was just over half the level reached immediately before the war. Nor did any of the new companies have any capacity to withstand the downturn in the market and by 1930 they had all folded.[117] Although the new companies were responsible for most of the increase from 1927 to 1929, the bulk of Cornish production continued to come from three established companies, Geevor, South Crofty and East Pool & Agar.[118]

Anglo-Oriental's venture into Cornwall was the most important of all its failures and since the capital was largely provided by the parent house, it was dealt a severe blow. The source of the problem was not uncommon: excessive investment in surface works before the realization of the extraordinary complexity of the ore body.

Australia

In many respects the experience of Australia was parallel to that of Cornwall. It was unable to achieve prewar levels of production and the overall pattern was one of gradual but steady decline. It, too, was the site of a great deal of unproductive British investment.

The Addinsell group floated Derek's Tin Mines in 1920 but it had abandoned operations by 1924. BMTS was responsible for two flotations, Federation Tin Mines in 1926 and St Paul's River Tin in 1928. St Paul's River was also soon abandoned and the promoters at least returned their vendor shares to give the company an opportunity to find a fresh property. Federation continued to struggle until World War II but was never profitable.

Anglo-Oriental also sponsored a company that proved to be very disappointing and its experience is quite revealing. Ten years of experience in the Tingha district led James Symes to develop an elaborate plan for regulating the flow of water from nearby mountains and provide the necessary conditions for effective dredging. Finding no interest in Australian capital markets, he went to London in 1921[119] and eventually Howeson's Indo-General Investment Corporation was attracted to the project. Tingha (NSW) Hydraulic Tin Mines was formed in 1924 and the *Investor's Chronicle* endorsed the flotation, adding that the 'control is experienced and successful in regard to tin enterprises'.[120] In fact, it was only able to operate for a few years and its dramatic failure in 1928 with the loss of nearly £100,000 may also have contributed to Howeson's reputation as a shady stock promoter.[121] Perhaps the most important lesson to be drawn is that even mining alluvial deposits carried substantial risks.

While British investment may have been fruitless, Australians were able to sustain a few successful mines. Aberfoyle, which started operations in Tasmania in 1925, became the most important of these. But with few other corporations and a large number of individual miners[122] scattered across the continent the industry lacked an organization of its own that could speak on its behalf. Smelters, however, did manage to recover their competitive position, so that the whole of Australian production was now treated locally. That placed them in the best position to represent the interests of Australian producers.

Japan

Tin has a long but modest history in Japan and the only mine of any importance was developed by Anglo-Oriental. In 1927 Hans Hunter approached the group to take over an operating mine, Mitate[123] and Toyo Tin was then floated in London to raise the additional capital required. Most of that was directed to expanding the capacity of the mill which only became fully operational until late 1929. However, it soon became not only the largest mine in Japan, responsible for around a third of the country's output but, at least for a short while, a very profitable one.

Bolivia[124]

Before the war, the high quality of the mineral had compensated for the high costs of mining at high altitudes. As these were depleted, it became clearer that large scale production through mechanization was the only basis for the long-term viability of the industry.

Patiño used the strength of the position he had built up before and during the war to become a major force in the world industry. In 1924 he completed his ambition to combine his La Salvadora-Uncía mines with those of the Chilean Compañia Estañifera Llallagua.[125] The new company, Patiño Mines and Enterprises Consolidated (PME), not only became the largest single tin mining enterprise in the world but also a model of technical efficiency in extracting and milling lode deposits.[126] In 1922 Patiño acquired a controlling interest in another Chilean company, Oploca, followed in 1926 by the completion of control over a German-Bolivian mine, Araca. A similar pattern of technical modernization ensured that these companies also became very profitable. By 1929 Patiño controlled over 52 per cent of the total production of Bolivia, making his group the largest in the world.

Throughout the process of expansion and consolidation, Patiño continued to draw on the support of a number of British financial institutions. In this respect, he occupied a position rather similar to that of Anglo-Oriental in Nigeria. Patiño could speak for Bolivia, had good connections with British finance capital and linked his mines with a British smelter. From 1924 he established permanent residence in Paris, serving as Bolivian Ambassador to France, though his diplomatic obligations did not prevent him from retaining detailed personal control over all aspects of his tin empire. This European base would facilitate the process of securing his co-operation in finding solutions to the eventual crisis.

A distant second to Patiño was Caracoles, sponsored by the Guggenheims. This initiative was originally designed to solve the problem of securing a supply of feed for the ASARCO smelter. The deposits at Caracoles were not easily worked and much of the available capital was expended in building a road and it was therefore only at the peak of the boom in 1927 that any profits were realized. At least at first, the Guggenheims were not daunted by the difficulties experienced at Caracoles and extended their interests by taking over the management of the Compagnie Aramayo des Mines en Bolivie.[127]

Caracoles would soon fail[128] and it represents the largest single demonstration of the inherent pitfalls in lode mining; there were many other cases of companies formed with expectations that could not be met.[129] Among them were several floated on the Santiago stock exchange in which not just bad luck but outright fraud played an important role in ensuring their failure. However, none became very important and most were easily driven out long before the crisis of the late 1920s.

Virtually all this fresh investment occurred in the first part of the decade and can be attributed to the generally optimistic view of the role that lode mining in Bolivia would inevitably play in the overall development of the industry, rather than to any specific effect of the price boom. One important development did occur towards the end of the decade and that was the formation an American dredging company, Bolivian International Mining, though it did not become operational until 1930.

A much less visible success was that secured by Mauricio Hochschild. He was German, closely linked to the major German-Jewish metal firm, Metallgesellschaft, and established himself in Bolivia in the early 1920s as an ore-dealer.[130] Metallgesellschaft provided an entry to its German smelter, Berzelius with which Hochschild negotiated an outlet for otherwise worthless low-grade ores. That provided one basis on which he acquired a dominant position in the fiercely competitive business of servicing the small miners of Bolivia. Hochschild began to develop a comprehensive vision of the future of mining in Bolivia which would move it away from dependence on high-grade tin ores towards low-grade ores of all kinds. In 1927 he moved from marketing into mining when he took over the management of a bankrupt silver-lead mine and demonstrated how it could be revived with new techniques based on large scale production. That would become a model for subsequent initiatives in tin.

The overall pattern of growth during the early part of the decade dramatically changed towards its end. The price declines during 1927 were already beginning to expose the marginal character of much of Bolivian mining, especially on the part of the small miners. Otherwise the picture is one of general stagnation, with the singular exception of the Patiño group. While all members of the group showed substantial growth, the one that was truly impressive was the flagship, PME. By 1929 it was producing 20,900 tonnes, 73 per cent over the level reached in 1927, and the whole group now accounted for 59 per cent of Bolivia. That dominant position would become very controversial.

Tin in Bolivia

Not only did tin operate within a very inhospitable natural environment, it also did so within an increasingly inhospitable institutional one. Of all the producing countries, Bolivia was by far and away the most dependent on tin. The mineral was the hub around which the whole of the commercial life of the country revolved. It provided 75 per cent of exports and 30 per cent of government revenues.[131] From the early 1920s, successive governments were anxious to meet the demands of their supporters by imposing three sets of constraints. One was to increase the obligations of companies towards their labour force. A second was to increase the resource transfer from the mining sector through higher taxes, especially on profits.[132] With an expansion of its fiscal basis, the state was then able to contract foreign loans which tended to be dissipated in prestigious but unproductive, projects.[133] The third constraint was on exchange rate policy. The boom allowed for stabilization of the exchange rate in 1927 but at a high level and the consequent reduction in the cost of imports allowed the urban population to enjoy cheap food, at the expense of any stimulus to domestic agriculture. That dependency and its concomitant political and economic distortions would continue to grow.

The mining industry only grudgingly accepted the fact that it was required to play a new role in the political economy of Bolivia, one that would ensure that the mineral wealth was more widely distributed internally in accordance with a set of policies over which it had no control. Each new initiative on the part of government was met with resistance and in 1924 the industry formed the Asociación de Industriales Mineros (AIM) as its lobby group. Since its membership was restricted to companies, it could not claim to speak on behalf of the whole industry.

In this climate of increasing antagonism between state and industry, there was no governmental department to serve as a buffer. Since the concessions were granted in perpetuity, subject only to a modest rent, the state's primary administrative connection with the industry was fiscal through the Ministerio de Hacienda. State institutions were required for the registration of claims, adjudication of disputes and monitoring of safety regulations. But the first of these required no direct contact with the industry and the second was done only sporadically. Without a specialized Department of Mines, Bolivia would find it very difficult to formulate a coherent policy for the industry. The fact that the Patiño group was both large and wealthy served to distract attention from the real needs of the rest of the industry.

Netherlands East Indies

The overall pattern of growth also followed a quite distinctive pattern. While the NEI grew at a somewhat faster rate than any of the other major producers from 1920 to 1927, the rate slowed down considerably from 1927 to 1929. The annual growth rate in this crucial period was just 3 per cent whereas the rate for the rest of the world was over 14 per cent. This slow rate of growth was naturally read as further confirmation of the famine thesis and the NEI was seen as pursuing a policy of restriction in order to make the deposits last as long as possible.[134] In fact, it reflected a distinctive policy of extraction. As effective monopolies over their respective concessions, the companies could plan a strategy which would deplete the entire deposit over the long term. The central feature of that strategy was to work the lower grade sections when prices were high and the higher grade sections when prices were low.[135] Price therefore affected less the quantity of production and more its source. The result was a much more stable level of production, which coincided with the government's desire for a stable revenue stream. With a long term approach to exploitation, both Banka and Billiton could move slowly in generating the capital they needed from internal sources and there was, therefore, no momentum of the kind that existed in Nigeria, Malaya and Siam which continued to generate fresh productive capacity until 1931.

Whereas the dredge had provided the impetus for the surge in production in Burma, Malaya and Siam, it played a much less important role in the NEI. Although the first dredge was introduced at the same time as in Malaya, by Singkep in 1911, it was not until 1920 that Billiton started dredging and Banka waited until 1927. One of the factors behind this lag was the incentive system adopted by both Billiton and Banka.

The method whereby the companies exploited their deposits involved systematic prospecting on the basis of which targets could be set for the Chinese gangs contracted to work them. Such a system generated a particularly high rate of recovery, since the company had the incentive to minimize the difference between the targeted and the actual amount extracted and the contractor had the inverse incentive to maximize the difference. One result was the absence of the kind of high-grading frequently found among Chinese miners in Malaya which left considerable quantities of low grade deposits behind that could only be extracted by dredging. Production methods were distinctive in yet another respect. Most of the labour continued to be recruited from China on the basis of yearly contracts, with criminal sanctions applied to those who broke them. That would create much less flexibility, at least in the short term, in cutting production.

The 1920s saw the formation of three new operating companies but they remained quite insignificant, so the prewar structure was retained intact. An important shift, however, occurred with the reorganization of Billiton following the expiry of its concession in 1924. A new company was formed to work the island, Gemeenschappelijke Mijnbouwmaatschappij Billiton (GMB), with the same share structure as the old and this permitted the reconstruction of Billiton Maatschappij as a holding company.[136] While its main asset remained the three-eighths stake in GMB, it was now free to explore other interests. One direction was diversification into other minerals such as bauxite; the other was to consolidate its position in the tin industry.

Consolidation in tin involved three separate dimensions. One was expansion abroad, especially into Africa; a second was metal trading and the third, and most important, was the construction of a smelter at Arnhem. When this opened in 1928 it treated medium-grade concentrates from Bolivia[137] and eventually it was extended to treat concentrates from GMB itself. While Banka remained the largest producer in the NEI, it was Billiton that emerged as the more dynamic company. It would prove to be a good match for Howeson.

Both Billiton and Banka were sufficiently large and well-financed that they could retain ownership of the metal and choose their own timing for its marketing. In the case of Banka this was essential to preserve the special premium that its metal commanded. Banka operated five separate charcoal burning smelters, though it diverted around a third of its production to STC in Singapore in order to prevent oversaturation of the limited market prepared to pay the premium.[138] Until the expansion at Arnhem, all other producers relied on STC to treat the whole of their output.

Given the size and diversity of the NEI economy, tin played a comparatively unimportant role and accounted for only 4 per cent of exports. However, as a source of government revenue it was much more significant, since it generated around 8 per cent of total income.[139] In most respects the NEI proved a model of how to organize a tin industry. Its essential features included: large concessions to well-financed companies, systematic prospecting, long term extraction policy, close links between industry and government, close links between industry and the market. Had this model been more universally applied there would have been no problem with tin.

China

China is an important exception to the pattern of modernization and mechanization that occurred throughout the rest of the industry. Production had peaked at nearly 12,000 tons in 1917 and generally declined throughout the 1920s, so that by 1929 it was just under 7,000 tons. Banditry and open warfare, especially in 1927, were immediate causes but there were more fundamental problems: several seasons of abnormally dry weather, depletion of local sources of charcoal required for smelting[140] and the system of property rights which served to perpetuate its antiquated mining methods.[141]

With large reserves, China had substantial potential but in the 1920s there were but few signs that she was making the necessary changes in the organization of both mining and smelting to realize it.[142] The one development that offered some promise was the formation of Yunnan Tin Corporation (YTC) which took over both the German smelter and the Malaga mine with which to feed it.[143] By 1923 it was under American management and producing around 60 tons of metal a month.[144] At least there was a modest base on which further modernization of both smelting and mining was feasible.[145]

Portugal and Spain

The success of Portuguese-American Tin attracted some fresh British investment but none of it was at all successful. Mondego Tin Dredging was formed in 1924 but its dredge only operated for a few months since it failed to live up to the boring samples.[146] Camp Bird floated a lode mine, Lagares Tin Mines, in 1927 but it only became operational in 1929 and while it limped along for another ten years it was never profitable. Across the border in Spain, Malayan and General floated another company, San Finx, in 1926 which had some initial success but eventually failed.

Southern Africa

While the tin boom stimulated considerable prospecting activity in Southern Africa, especially in the mandated territory of Southwest Africa, little was found that warranted fresh investment. The industry therefore remained largely unchanged from its prewar position.

Belgian Congo

During the 1920s the Belgian Congo was even less visible than Southern Africa but this would prove to be extraordinarily deceptive. The system of property rights under the Chartered Companies encouraged the commitment of considerable resources to prospecting and this was reinforced by the scarcity of mining labour, especially in Katanga. Extraction could only take place on the basis of extensive mechanization and the capital investment could only be justified on the basis of a comprehensive knowledge of the size and nature of the deposits.

Géomines remained the most important of the companies operating in Katanga. Extraction continued at a modest rate, though sufficient to justify the building of a smelter in Belgium.[147] Prospecting, on the other hand, increased exponentially. Over the whole decade, Géomines produced just under 6,000 tons but its reserves rose from 20,000 to 200,000 tons. It was then in a position to increase its capital and in 1929 the equivalent of just over £1 million was raised, making the company one of the largest in the tin world. Géomines had emerged from the 1920s in an extremely strong position and while the depression would inevitably delay the realization of its plans, they could not be permanently shelved.

The experience of Géomines demonstrates the perversity of the famine thesis and with it any simple reliance on price as the primary factor accounting for production levels. Given the devaluation of the Belgian franc against sterling, prices were even better for this producer than for most others. Géomines was well aware of the famine thesis and this reinforced its decision to take a long term view of its development strategy and concentrate on prospecting. But the absence of any fresh production from a new minefield simply confirmed one of its fundamental assumptions, that there was an inherent limitation to the world's supply of tin.

Development of Manièma under the auspices of the CFL proceeded more slowly, so no extraction occurred but prospecting laid the foundations for another major producer, Symétain, that would come to rival Géomines. Across the great lakes in the mandated territory of Ruanda tin was not located until much later but since the terms of the mandate precluded the same kind of concessions that were granted in the Congo proper, the region was opened up to the pattern of competitive prospecting that had resulted in the formation of several companies in Nigeria. Ruanda therefore quickly moved to the production stage but by then the depression was under way.

East Africa

The tin deposits of Ruanda are an extension of those found in British East Africa. Here prospecting was well established by the middle of the decade and three companies emerged to develop them. Billiton took a large interest in all of them. Kagera became a respectable producer but the other two, Ankole and Bukoba, soon faded away. Anglo-Oriental was involved here too but only for a brief moment while it reviewed an option on Ankole, which was wisely declined,[148] demonstrating that Howeson drew some limits to the kinds of properties he was prepared to turn into stock exchange promotions.

French Indo-China

The prewar initiatives provided a basis for more sustained production in both Tonkin and Laos but the period is important less for the actual output, which only reached 830 tons in 1929, than for its promise. Under the mining legislation of 1912 permanent concessions were granted to individual Frenchmen without any prospecting. Such a system inevitably encouraged speculation[149] and the tin boom provided the concessionaires with the opportunity they needed to take immediate profits through company promotion.

The valley of Nam-Patène in Laos was dubbed the 'Klondyke Indo-chine' and a 'Nouvelle Bolivie' and in this heady environment the Dubost group floated five separate companies with a total market capitalization equivalent to almost £9 million.[150] This was over ten times the market value of Gopeng Consolidated which produced considerably more than the whole of Laos and whose proved reserves could last another 50 years! Such optimism also infected the French Geological Survey which considered that the region contained extensive deposits which could support an annual production of 5,000 tons.[151] At least Indo-China had been recognized as a 'respectable' producer,[152] which showed considerable promise for the future.

In the early part of the decade virtually all concentrates were smelted locally but, as production expanded, an increasing proportion went to Singapore where their abnormally high quality was appreciated as a mixture to be added to somewhat lower grades. By 1930 these local smelters had all folded, though the technical experience gained served as the basis for the development of a refinery in Haiphong which competed with those in Hong Kong to treat the crude tin from Yunnan.

The production problem

The problem posed by the overcapacity evident in 1929 was almost entirely located within the sphere where British capital was invested and the overall distribution of production between various sectors is presented in Table 4.11.

1927 to 1929 saw a particularly rapid rate of growth and while the rate declined from 1929 to 1931, the quantity did not. Underlying this pattern are two consistent trends: an absolute decline in the amounts provided by small miners and an increase in market share taken by the four large groups. Of these, the largest amounts came from the Cornish-Malayan companies. While Anglo-Oriental was the most dynamic group in first period, the momentum that carried on into the second was comparable to the sector as a whole. Such dynamism and its visibility has obscured the quantitative impact of Anglo-Oriental. Of the total increase of 46,000 tons between 1927 and 1931, the group was responsible for 10 per cent. It is therefore clear that even if Howeson had never been attracted by tin, the industry would have still faced the same fundamental problem.

Table 4.11 British sphere production, 1927–1931

	1927	1929	1931	Change 1927–29		Change 1929–31	
	tons	tons	tons	tons	%	tons	%
Total	72,350	96,470	121,520	24,120	33	25,050	26
Small	38,130	33,490	33,220	–4,640	–12	–270	–1
Corporate	34,220	59,710	87,320	25,490	74	27,610	46
Medium	25,640	38,040	49,570	12,410	48	11,530	30
Large	8,590	21,670	37,750	13,080	152	16,090	74
Redruth/Tronoh	4,200	8,170	15,650	3,980	95	7,470	91
Anglo-Oriental	1,340	6,030	9,560	4,950	460	3,540	59
Alluvial Tin	1,800	4,310	8,010	2,510	140	3,700	86
Malayan Tin	1,250	3,160	4,540	1,920	154	1,370	43

Source: Tables 4.3, 4.5, 4.7, 4.9 and Cornwall.

Notes: Large defined as over 4,000 tons in 1931; Anglo-Oriental includes Nigerian acquisitions.

Howeson's ability to play a role in solving that problem rested not on the development that he initiated but on the acquisitions made in 1928–1930. Table 4.12 presents the comparative position of the world's largest groups.

By 1929 Anglo-Oriental's acquisitions of the Alluvial Tin, Anglo-Continental and Barrier & Finance groups had made it the world's fourth largest group. Howeson was therefore uniquely placed to play the key role in finding a co-operative solution to the problem of tin. He was now large enough to be able to talk to the Dutch and Patiño but not sufficiently large that he could so without building further alliances within the British sphere.

The overall pattern of growth lends itself to no simple comprehensive explanation. There is plenty of evidence to support the May-Eastham thesis that there were many fresh properties developed on the unrealistic assumption that boom prices would continue and Anglo-Oriental bore responsibility for a few of these. However, the problem with tin was not primarily located in such a source of excess capacity. The evidence presented above demonstrates that most of the excess capacity, whether it be measured from 1927 to 1929 or 1929 to 1931, was brought into being by the established groups. Flimsy companies, including seven of Anglo-Oriental's, were quickly destroyed long before the ITC took shape. In any case, flimsiness can be equally a function of the inherent risks in applying new techniques to old ore bodies as of the opportunities for stock market promotion. Caracoles

Table 4.12 Major world groups, production 1927–1931

	1927	1929	1931	Change 1927–29		Change 1929–31	
	tons	tons	tons	tons	%	tons	%
Patiño	16,890	26,920	26,920	10,040	59	0	0
Banka	20,670	21,310	21,600	610	3	290	1
Anglo-Oriental	1,340	6,030	17,570	4,690	350	11,550	192
Redruth/Tronoh	4,200	8,170	15,650	3,980	95	7,470	91
Billiton	11,940	13,400	13,430	1,460	12	30	0
Total	55,060	75,830	95,160	20,770	38	19,340	26
Total % World	34%	39%	46%				

Notes: 1931 estimate of capacity; Anglo-Oriental includes all acquisitions, Billiton includes African holdings

and Tingha provide salutary and expensive lessons that mining is a very different industry from others.

Where Anglo-Oriental was uniquely exposed was not in the high cost side of its mining activities but in the low cost side, or rather in the side that had the potentiality of becoming so. Its major investments in new productive capacity were in modern dredges at Talerng, Kampar Malaya, Lower Perak and Northern Tavoy and in NESCO, all of which promised considerable reductions in cost. In addition, it held an undeveloped property at Berjuntai. The capital that flowed into these projects came through the Anglo-Oriental group in its role as a finance house, with a set of connections to the wider world of mining and finance that none of the other groups that operated within the British sphere came close to matching. As the crisis unfolded, Howeson would have faced a simple choice: try and weather the storm with the financial support on which he could draw or try to erect a barrier beyond which some protection could be secured. Anglo-Oriental, at least in some form, could have survived the former and, indeed, may well have emerged in an even stronger position as a result. However, a collective solution was simply the more efficient.

Smelting

While the 1920s saw major changes in the location and organization of extraction, they were largely absorbed within the prewar pattern of smelting. Expansion of production in South East Asia was simply matched by an increase in the number of furnaces on the part of both ESC and STC with no major technical modifications to the way in which the concentrates were treated.[153] The expansion of Nigeria was the basis of the expansion of Penpoll and its emergence as the smelter with the most modern of reverberatory furnaces.[154] But it was with respect to Bolivia that the most important changes occurred.

By 1924 both the American wartime smelters had ceased to operate, unable to match the competition from Britain.[155] At the same time, European smelters began to develop new techniques to process the low-grade and complex ores which were becoming an increasingly high proportion of Bolivian production. Two developments have already been noted, Billiton's smelter at Arnhem and Hochschild's success with Berzelius. This was matched by Zinnwerke Wilhelmsburg in which Patiño continued to hold a large interest.[156] Capper Pass undertook similar experiments, though with electrolytic techniques which produced the world's purest tin at 99.99 per cent[157] and an electrolytic smelter was also completed at Annecy in France. Table 4.13 provides a picture of the distribution of the productive capacity of the world's smelters of tin concentrates in 1929.

International smelters accounted for 87 per cent of the world market and, of these, three were responsible for 75 per cent. While this was a level of concentration considerably higher than that reached before the war, it was still not a source of independent power. However, it did provoke some anxiety in the United States.

The absence of an American smelter was not a matter of concern to industrial consumers since they drew over 75 per cent of their tin from Southeast Asia for which they were prepared to pay a premium. However, it was one that continued to rankle with the metallurgists who had been responsible for developing the wartime smelters, especially since American technology led the world in treating other non-ferrous metals. A convenient explanation was to be found in the protection that the British smelters enjoyed in securing access to the purer concentrates from Nigeria with which to sweeten those from Bolivia. The absence of an American smelter therefore became symptomatic of a system of control which appeared to deprive the world's largest consumer of any market power.[158]

Table 4.13 Primary tin smelters' capacity, 1929

	Local	*International*											
		British											
		United Kingdom					*Straits*		*Continental Europe*				
		WH	*CP*	*CTS*	*PP*	*TM*	*ESC*	*STC*	*ANH*	*ANY*	*BZ*	*ZW*	*HBK*
Australia	2,260												
Belgian Congo													x
Bolivia		x	x	x	x				x	x	x	x	
Burma						x	x	x					
China	7,000												
Cornwall		x	x	x	x								
NEI	15,000							x					
East Africa		x			x								
Indo–China	350							x					
Japan	1,000							x					
Malaya	300						x	x					
Nigeria		x			x								
Portugal	5	x											
South Africa								x					
Siam							x	x					
Total	25,915	45,000	1,000	5,000	9,000	500	40,000	63,000	2,000	700	2,540	5,000	1,000

Notes: Total capacity ca 200,000 tons. Code: WH; Williams Harvey; CP: Capper Pass; CTS: Cornish Tin Smelting; PP: Penpoll; TM: Thames Metal; ESC: Eastern; STC: Straits Trading; ANH: Arnhem; ANY: Annecy; BZ: Berzelius; ZW: Zinnwerke Wilhelmsburg; HBK: Hoboken.

While there was some pressure to level the playing field by imposing a tariff on imported metal, it met with considerable consumer resistance, especially from the tinplate manufacturers who saw it as eroding their competitiveness in export markets. However, when the United States government decided to take a more active interest in the politics of tin, this group of metallurgists would find an opportunity to articulate their vision of a new world order in tin much more forcefully.

Reading the tin market, 1927–1929

Over the course of 27 months, from March 1927 to May 1929, prices dropped from £313 to £198/ton. This trend was interrupted for only one sustained period; from July to November 1928, when they rose from £212 to £230. Throughout the entire period the tin market received considerable attention, as commentators attempted to discern the actual state of the market that underlay these prices. Since their objective was to establish long term projections, commentators had the difficult task of reading the immediate results produced by the operation of a very complex market in such a way as to disentangle the factors causing short term fluctuations.

Financial commentary probably has an inherently bullish bias but it was one that was exacerbated in the case of tin, thanks to the persistence of the famine thesis. A vicious cycle may then have been produced in which the commentary supported the thesis which in turn supported investment decisions which in turn created an audience anxious for reassurance.[159] It is difficult to determine the significance of this process in contributing to the problem of tin but it certainly indicated the real difficulties in the way of understanding it.

One of the best examples of that lack of understanding is provided by *The Economist.* Unaware that March represented the peak of the boom, it declared that 'the position of the metal cannot be considered otherwise than sound'.[160] In September, as the price had dropped by £33, this assessment was cautiously qualified: 'Although the outlook is none too clear, it is possible that tin has seen its worst under present conditions.'[161] When the price continued to fall by another £30, this was defined as a puzzle since the statistical position had not changed and the market fundamentals were sound.[162] This kind of judgement would continue to be made for the next two years. Why then were price signals discounted?

Part of the problem lies in the fact that price changes were not confirmed by other indicators, of which the most important was the level of stocks. It was difficult to reconcile the fact that stocks in the UK had dropped by over a quarter from March to June with the drop in price from £313 to £296.[163] Henry tried to analyse this apparent price perversity in light of the statistics available and concluded that the statistics themselves were quite misleading. They had suggested that production was exceeding consumption by some 12,000 tons, whereas he thought, quite erroneously, that consumption still exceeded production.[164] Two conclusions followed: price fluctuations were the result of speculative manipulation and the tin market could only rid itself of this plague with the kind of comprehensive statistical picture developed by the American Zinc Institute, 'as a speculation of the type which disorganizes the market becomes well-nigh impossible when all the facts of the case are known'.[165]

As the price approached £250 in January 1928 a comprehensive guide to the industry considered it:

> very improbable that the market will have so much tin rapidly thrown upon it as to depress the price below £250 for metal. ... many of the alluvial deposits are rapidly

> becoming exhausted … and there is every reason to believe that the demand for tin … will extend rapidly every year. We must expect, therefore, something in the nature of a world shortage, and … the time may not be far distant when £350 may be a regular figure.[166]

Since it was assumed that the alluvial mines were becoming increasingly marginal, saved only temporarily by the development of dredging, this would set a floor to the price of tin. It could only fall to the point where it drove out the small Chinese producers, at which level it was bound to stabilize. Initial predictions of this level suggested it was around £250. This was a figure initially calculated by Rae and he considered that at such a price, 24,000 tons of Chinese production in Malaya would cease.[167] Even if that estimate was somewhat high, £250 still seemed a reasonable level[168] and was confidently repeated.[169] As the price continued to decline, then so did the level at which the imminent shaking out was expected to occur.[170]

There was only one discordant public voice and that came from the United States. In the influential *Engineering and Mining Journal,* Parsons argued that both the upper and lower ranges were set far too high. The estimates of the upper range had not taken into account the stimulus to technical innovation on the consumption side; those of the lower range had neglected the reduction in costs occurring in all sections of the industry. Parsons recognized not only that the Chinese in Malaya were becoming cheaper producers thanks to the adoption of the gravel pump but also that their lack of mobility forced them to produce tin for whatever they could get. He anticipated no reduction in production until the price dropped to at least £180 and predicted a pattern of production and consumption that would suggest a price of £200 for 1929 and 1930.[171] While Parsons' more realistic approach was endorsed by a statistical bureau, International Metal Service, it was felt that he had set the lower range much too low and that production would start to fall off at £220. All producers would find themselves under serious pressure at £200 and that such a price could not remain under this level for 'any prolonged period'.[172]

That criticism of the critic was echoed by Keynes who wrote some of the most careful assessments of the market and the one prepared in September 1928 is particularly revealing. Here Keynes reviewed the growth in dredging capacity and was not sanguine about the immediate effect of a price of £215 in forcing out the Chinese but concluded:

> Looking further ahead, it would seem very probable that consumption will again endeavour to outstrip production and will have to be kept in bounds by a high price. For no new tin-bearing ground is being discovered and many existing sources of supply have a short life. The present weakness of the market may become more accentuated, but it is essentially a passing phase due to new dredger production having matured a little faster than is wise.[173]

Discussion of price levels was not confined to what they currently revealed but extended to the much more speculative question of what they ought to be and how that objective could best be accomplished. The assumption of imminent scarcity lent credibility to Howeson's initial judgement that £300 represented a fair price and control over output provided the more rational means of securing it. In this Howeson had the support of National Lead, which considered £300 to be the right figure to stimulate production without affecting consumption.[174] While many in the industry considered £300 to be 'very high',[175] Pratt objected that stabilization at this level would prove to be much too low since the proper range was between £350 and £400. This claim could only be asserted in the light of the actual price of tin by dismissing the fall by £30 to £265 over the past few weeks as a result of pressure from the bears and of 'little moment'.[176]

Stabilizing price meant both expanding and controlling production. If the famine thesis proved correct then the apparent contradiction would be overcome, otherwise the whole programme appeared quite 'illogical'.[177] Howeson's initial move was met with some scepticism and the idea of tin restriction soon received an unexpected setback as the British government decided to abandon its policy of controlling rubber supplies as a signal failure. If co-operation in rubber was considered impossible then how much more daunting was the task of bringing together tin producers who were far more diversified. Howeson's programme, however, was not abandoned. It simply had to wait for further falls in price to create a more receptive climate.

In addition to floating this trial balloon, Howeson undertook another initiative whose significance was difficult to interpret. Towards the end of 1928 it was clear that he was behind a group acquiring a substantial bull position, estimated at between 5,000 and 8,000 tons.[178] Prices accordingly rose, from £212 in July to £230 in November.

Retrospective assessment has tended to see this initiative as compounding the problem of tin. Such market speculation simply sent the wrong signal, that the long anticipated corner had finally been turned, with the inevitable result that fresh production came on stream. Not only did it stimulate fresh production but it also allowed Howeson to float even more companies which proved to be a major source of the excess capacity revealed in the early 1930s.[179] The initiative is also thought to explain Howeson's role in finding a solution to the problem of tin. Cutting production to raise prices would enable this position to be liquidated without financial loss.[180] None of these claims are easily substantiated and it is possible to place Howeson's intervention in quite different light.

By the late 1920s many were convinced that there was a fundamental irrationality to the way in which the tin market operated on the LME and one opined that:

> the peculiar structure of the tin market is designed to stultify, rather than to assist, the working of economic forces. It would be absurd to maintain that Tin, under the present system, is a free market. It is taken hither and thither, now by blind anticipations of economic tendencies, now by designed manipulation which have their origin in the market itself. None can tell who is the present master, nor who, next week, may arise to command.[181]

As far as these underlying economic tendencies were concerned, there was an equally firm conviction that dredging was only replacing existing production and the production trend was therefore, at best, a steady one. Any suggestion that there was overproduction was therefore widely dismissed as a 'bogey set up by bears'.[182]

The price of £212 was inconsistent with the statistical position. While visible stocks were continuing to rise they were only at the level of December 1925 when the price was £285. It was therefore not unreasonable to suppose that the current price reflected a bear operation which could be beaten by a determined group of bulls.

Estimates of the size of this bull pool are extraordinarily high, since even 5,000 tons was equivalent to the total amount held in warehouses in the main consuming centres and would have required an outlay of over £1 million. Those who have wondered about the source of this finance recognize that it was well beyond Howeson's own resources[183] but have not considered its rationale. If the purpose of this speculation was to give the bears a hiding, then a far lower quantity would be sufficient.[184] As soon as the lesson had been learnt, the stock would be sold off, especially since the threat could always be repeated. This is consistent with the account that Howeson gave Billiton much later when he admitted that the group had held 1,500 tons in November 1928 but that it was then liquidated.[185] Howeson certainly could not afford to be caught in a deception at the very moment when he desperately needed Billiton's support.[186] This episode confirms the danger not only of attributing too much power to Howeson but also of overlooking 'the tendency to exaggerate the unknown'.[187]

Those attempting to monitor the tin market had to deal with many unknowns: technological trends in both the use and production of the metal, demand levels, rates of depletion and discovery, effect of price on production, levels of stocks outside the visible supplies, especially those being held for purely speculative purposes. The market was so finely tuned that even a slight change in any of these variables could easily throw off any assessment and it is, therefore, not surprising that analysts gave up in despair and complained about the perversity of the metal.

Producers, however, had no such luxury. They had to make decisions in light of some assessment, no matter how ill-founded. By mid-1929, at least one feature was clear. Tin was in trouble and solving it would require some fresh initiatives. The extent of that trouble became ever more acute as consumption began to decline and productive capacity continued to rise. The next two chapters will consider the way in which the industry dealt with its predicament and the role played by Howeson in guiding it towards a solution.

5 The depression: initial responses, 1928–1930

The search for a solution to the problem of tin took place on two separate planes: independent decisions by governments and companies and Howeson's efforts to bring about international co-operation in research and production. The next chapter will examine the way in which international control was shaped, while this will explore other responses to the crisis that was beginning to beset the industry.

Since tin is a depleting resource, both governments and producers were forced to decide whether it was in their interests to preserve it in the ground or add it to the swollen supplies which were depressing the market. This was not an easy decision since it involved balancing current and certain but low income against a hypothetically higher one in the future. It was made in different ways by different countries.

Malaya

In early 1929 the Senior Warden of Mines for the FMS expressed his concern about the rate of depletion of its tin resources. Companies tended to 'inflate the output of tin ore for a few years instead of spreading it over a large number of years'. Since this inevitably led 'to overproduction now and a prospect of scarcity in the future', it was in no one's long term interest.[1] During the summer the Mines Department persuaded the government to reduce long term supply by suspending prospecting. In November 1929 the government announced that in future no more land would be alienated for mining, 'except where necessary for the efficient development of existing undertakings'.[2] No new entrants were therefore to be allowed into the industry, unless they were prepared to buy one of these undertakings.

The anticipated self-correction through the falling away of the Chinese failed to materialize. Only around 8 per cent of productive capacity was suspended during 1929–1930 in response to falling prices. Since a high proportion of their costs were in wages, miners passed on the reduction in their own income to their workers. Most had no option but to continue to work under any conditions and many were prepared to do so for their food alone.[3] Chinese mines generally worked on subleases, so that in continuing to produce they were wasting not their own assets but those of the leaseholders. In addition, falling prices had a perverse effect. Owners of mines that remained marginal preferred to turn them over 'to their coolies on tribute with the usual result that the coolies to gain a sufficiency of money worked the harder and produced more ore'.[4]

Siam

Siam eventually followed the Malayan policy on alienation of further mining land. In August 1930 the government announced that no applications for EPLs or leases would be granted except from those operating mines about to be exhausted, 'in order to regulate output of tin ore and thus increase its price, and also to safeguard the supply of this mineral from being prematurely exhausted'.[5] The principle of conservation was reinforced by eliminating the fees charged for cessation of mining operations.[6]

Nigeria

Since Nigerian tin companies considered themselves as high cost producers, they felt entitled to apply for relief from certain government regulations. They had been successful in securing concessions in the slump of 1921 and as tin dropped below £210 most of them were losing money and they sought similar treatment.[7]

As early as 1928 the mining companies started pressing for three concessions: reduction in rents from 5 shillings to 1 shilling per acre, suspension of labour obligations and a moratorium on completion of prospecting under existing EPLs. Unfortunately, the history of the industry continued to cast its shadow and the government expressed little sympathy:

> the tin mining companies consistently forget all that Government has done for them, and will continue asking for more as long as they think there is any chance of getting it. They are not deserving bodies.[8]

Particularly contentious was the demand to stop development by suspending EPLs which was dismissed since it originated in 'the fact that the share pushing industry on the Stock Exchange is what is really present to the mine owners'.[9] Given the overall constraints on labour supply, the government felt that as labour was released from mining it should be redeployed in prospecting and that if the existing holders could not afford to do so then they could always relinquish their EPLs and concede the opportunity for fresh development to others.[10]

The other two issues were eventually conceded since, ultimately, there was a convergence between the points of view of the companies and government which felt that 'it would be better that the land should remain unworked and the potential wealth retained until such a time as an economic price can be got for tin'.[11] As these matters were reviewed in London, that fundamental convergence based on the property rights of the government was reinforced by the new Economic Adviser in the Colonial Office, Sir John Campbell.[12] This marked the first point at which the man who would soon be placed in effective control over the whole industry was forced to think about the problem of tin.

However, there was an important divergence of views between London and Nigeria as to the overall condition of the tin market and it is that which explains the apparent contradiction between expanding prospecting while restricting production. Both the government and many local miners considered:

> that the present slump in the price of tin is not due so much to overproduction, or the working of the laws of supply and demand as to an artificial manipulation of the market by financial interests elsewhere and that the price will not remain for long in its present depressed condition.[13]

While this had been common thinking in 1928, it was clearly obsolete by the end of 1929. However, the deep suspicion of financial interests would often cloud Nigeria's judgements about tin.

Bolivia

Bolivia found the price declines most difficult to deal with. Not only did they affect government revenues but they jeopardized the ability to service the dollar loans that were contracted in 1927–1928. They also revealed the extent to which currency stabilization had resulted in an overvalued exchange rate. With some 77 per cent of minehead expenses incurred domestically, Bolivia had become a high cost producer.

Producers began to put pressure on the government for some relief. It came in the modest form of a reduction of duties on mining imports.[14] A more substantial commitment was made with a decree at the end of 1929, enforcing a reduction in railway tariffs by 50 per cent on both exports and mining imports as long as tin remained under £220.[15]

These gestures proved quite insufficient to protect the small miners and the second largest mine of all, Caracoles, was effectively abandoned and turned over to Aramayo. A large part of Bolivia's productive capacity was suddenly eliminated and whether it could be replaced became one of the sources of contention in Bolivia's always problematic relationship with the ITC.

Bankruptcy involves not only the elimination of obsolete productive capacity but also provides opportunities for reorganization. In the summer of 1929, the old established firm of Bebin Hermanos was unable to meet its obligations[16] and this was the moment to realize a long gestating plan to consolidate several operations on the Cerro de Potosí. Hochschild was able to persuade the Bebins and others to transfer their interests to a new company, Compañía Unificada del Cerro de Potosí in late 1929. With an infusion of fresh capital from the South African mining finance house, Central Mining and its French associates,[17] Hochschild had taken a control of a large mine, which, although currently very marginal, had enormous potential. In 1930 CMO was unable to service its loans from the Banco de Chile and this also provided Hochschild with the opportunity to take control of another large mining complex.[18] The structure of the Bolivian mining industry would be profoundly affected by these two developments since they now created a basis on which a powerful rival to Patiño could emerge.[19]

Overall effect of price declines

From January 1929 to December 1930, tin prices dropped from £223 to £112. A few marginal companies simply went bankrupt and while their properties might eventually be reworked, they would either require fresh infusions of capital or simply be turned over to scavengers. The total amount of such productive capacity which was more or less permanently eliminated is estimated at 8,100 tons, or around 4 per cent of the world total, far less than the fresh capacity that was still coming on stream in 1930–1931.

In most mines, suspension involved low immediate costs to maintain capital equipment and, given the way in which the price decline was read as a temporary phenomenon to be soon corrected, much capacity therefore simply remained idle. In a few cases, companies could have continued to have produced at these low levels but simply preferred to wait until prices recovered before depleting their deposits. Otherwise, suspension was simply a function of their marginal costs which could not be covered. The total amount of productive capacity held in suspension is estimated at another 8,000 tons.

Even if this reduction in capacity had not been vitiated by the continued increase from dredging, it was still far from sufficient to enable the market to correct itself. The amount that was in suspension would come back as soon prices began to recover. But it would take a long period of sustained low prices before the market could be expected to eliminate surplus capacity.

These data also reveal the seriousness of an issue which would continue to plague the ITC, namely the basis for distributing production cuts among the various participants. While Bolivia had not exhausted the options necessary to put her industry on a competitive basis, the fact that 8 per cent of the productive capacity utilized in 1929 had folded by 1930 indicates that she now occupied a very different position in the world market than any of the other major producing countries.

Comparative costs

As prices declined in 1929, the price question took on a new dimension. Whereas two years earlier, the question was one of whether they were high enough to stave off a famine, the question now was one of whether they were dropping too low. A judgement about price level involves the development of an overall model of the tin economy and one of the most interesting and influential was presented in a series of articles in the *Engineering and Mining Journal* by a past President of the Institution of Mining and Metallurgy, Alfred James, with considerable experience in the industry.

James put the price cycle of the 1920s in its historical context and by linking the troughs and the peaks he showed a long term trend towards higher prices. The trend lines reflected the growth in the population and suggested that demand increases at a faster rate than technical innovation could reduce costs. What made his account interesting is the attempt to explain the fluctuation around the trend line by reference to stock holding policy on the part of industrial consumers. An upswing leads first to drawing down on stocks, withdrawal from the market and then to a return to the market so that buying takes place when it is rising, creating the condition of famine prices and speculative flotations. At this point prices begin to decline and consumers draw down stocks. Visible stocks increase and decline turns into slump. This would then explain why the industry could never establish an equilibrium unless 'some responsible body maintained and controlled adequate stocks'.[20] James was sufficiently astute that he recognized the implications of the continued technical development in dredging and hence did not see the current situation as one of simply holding productive capacity until it could be effectively utilized. Instead, he had a more profound conception of the inherent weaknesses of the operation of the tin market, which worked against the interests of both producers and consumers. Stabilization around stock control would become one of the important, though elusive, objectives of the ITC.

With stock control went price control and the question became one of what price level would preserve the current capacity required to meet future demand. Following an extensive analysis of comparative conditions of companies producing around half the world's supply, he concluded: 'If the price of tin does not rise above £240 during the next four years, my view is that consumption will once more exceed supply.'[21] When this was published, the price was £133 and consumers had not had to pay £240 since early 1928 but they were being warned that any continuation would destroy too much of the productive capacity they would rely on to meet their needs. What makes James' analysis of continuing relevance is less the prediction or the timetable over which it would be realized but the methodology whereby he

Table 5.1 Production costs, 1929 (£/ton)

	Concentrate	*Metal*
Banka	54.9	78.4
Burma/Siam/Malaya (low cost)	51.8	74.0
Bolivia (PME)	72.4	116.8
Mean	83.5	119.3
Billiton	90.5	129.3
Burma/Siam/Malaya (high cost)	109.6	156.6
African	142.9	204.1
Cornwall	150.4	242.6

Note: Banka includes depreciation.

came to this conclusion. James first presented the direct working costs which are arranged in Table 5.1.

The data for Burma, Siam and Malaya were compiled on the basis of 64 British companies and no attempt was made to estimate the position of the Chinese. Their significance lies primarily in the demonstration of the wide range of cost conditions which remains for Malaya even when the higher-cost companies in Burma and Siam are excluded. For the Malayan group as a whole their costs were considered to be around, rather than (as they liked to suppose) considerably below the world average.[22]

In order to determine the long term impact of low prices, James added estimates for the realization, overhead, depletion and capital costs, reckoned at a low 8 per cent on the nominal value of issued capital. Once these are included then production of only 24,000 tons of metal was economically viable over the long term at a price of £150. Raising the threshold to £175 would bring in the 21,000 tons of Patiño and a further rise to £200 would cover another 20,000 tons, split between Malaya and Billiton. Even at £200 only a third of the world's current capacity would be safe. Here at least was an objective demonstration that the current price level was getting dangerously low from the perspective of consumers as well as producers. It was one that validated the overall position that Howeson had reached.[23] It was also validated retrospectively since the average level of demand throughout the 1930s was around 50 per cent greater than these producers could meet. Prices would have to be even higher to keep in the more marginal operations and the range that dominated discussions in 1928–1929, £220–£250, therefore seems well justified.

The politics of tin often worked with very simplistic concepts around costs which tended to treat groups and countries as comparatively homogenous: Bolivia, Nigeria and Anglo-Oriental were considered high cost producers, the NEI and Malaya low cost ones. All had a range, at the upper end of which all would have experienced severe pain had an unregulated market persisted. However, the means were quite different. Defining Anglo-Oriental's overall cost position in 1930 is complicated by the fact that its programme for cost reductions in Nigeria had yet to get underway. But if we look more narrowly at the case of Malaya, where the issue of comparative costs created the most concern, and focus even more narrowly on the dredging sector, which was the largest and most dynamic, the following picture emerges. Working costs ranged from a low of £60 to a high of £140 per ton of tin-in-concentrate, with an average of £88. The Redruth and Tronoh groups were by far the cheapest at £66. Following the absorption of Alluvial Tin, Anglo-Oriental was somewhat higher at £81 but still below the average for Malaya as a whole.[24] While most of the companies in the Redruth and Tronoh groups could survive with a price as low as £150, that did not mean that they were indifferent to the current price. However, it did mean that they were under far less pressure to come to an agreement on the terms whereby the market would be regulated. Although Anglo-Oriental could fairly argue that its corporate interest was identical with that of Malaya and that of the industry as a whole, it could never prevent that claim from being challenged.

Expanding the Patiño empire

Smelting

On the smelting side of the industry a major reorganization occurred under the auspices of Patiño. In late 1927 Patiño and National Lead had bought out the one third of Williams, Harvey still held by Cornish interests, to become joint owners of the company and substantial economies were then made. Six months later, Patiño persuaded the Board of PME to buy the whole company and National Lead reluctantly conceded, with the transaction being completed in July 1929.[25]

With the sale of his personal interest in Williams, Harvey, Patiño had £666,000 in cash[26] and this was used to expand his control over the British section of the industry. That included buying shares in ESC, where he became the largest individual shareholder.[27] BMC, through its subsidiary, VYB, served as selling agent not only of ESC but also a Cornish smelter, Cornish Tin Smelting and both were controlled by Budd. Lyttleton now saw an opportunity for amalgamation of all three smelters. The Cornish one could be closed and VYB would become the selling agent for the whole group.[28] Consolidated Tin Smelters (CTS) was then formed which Howeson soon joined, bringing in his Penpoll smelter.[29] Given the size of Williams, Harvey, Patiño, either personally or through PME, now held 55 per cent of the equity of an organization which in turn had around 45 per cent of the world's smelting capacity.[30] That degree of concentration would soon be increased with cross shareholdings and directorates between CTS and Billiton's Arnhem smelter.

The creation of CTS marks an important moment in the reorganization of the tin industry, since it established an alliance between the three most important players, Patiño, Howeson and Billiton.[31] However, its significance should not be overestimated. In spite of its size, CTS had no financial resources with which to exercise market power and it therefore continued to hedge all its purchases of concentrates. Nor was this a platform on which the much more complicated task of production control could be easily completed. Nor was it a particularly firm set of alliances. Patiño explored the prospect of expanding in Malaya in collaboration with Billiton, independently of Howeson, while at the same time Billiton was engaged in negotiations with STC which sought to protect itself against CTS.[32]

Malaya

Another important outlet for Patiño's funds was shares in companies in the Redruth, Tronoh and Malayan Tin groups. As tin prices fell so did the value of the companies' shares and Patiño started collecting them as 'a small boy collects stamps'.[33] By early 1930 he had sufficient holdings to request representation on the Boards of Directors.[34] Patiño was clearly looking towards establishing the basis of co-operation which would put a halt to the deterioration of the condition of the industry and in this he had the active support of Thomas of the Tronoh group.[35] Other Cornish interests, however, reacted with alarm at the implications of this non-British presence. Not only would it undercut their claim to moral leadership of the industry as its pioneers but it could also affect their relationship with STC and hence the overall position of the Malayan industry.[36]

The skirmishing that then ensued prefigured much of the cleavage that would emerge between Malaya and London and between the Cornish and foreign interests. The High Commissioner simply endorsed the Cornish position and proposed legislation which would limit foreign ownership in British companies to a maximum of 25 per cent and deny representation on Boards of Directors on pain of surrendering the mining leases.[37] As this was received in the Colonial Office, Sydney Caine commented:

> No part of duty of the Malayan government to protect the monopoly at present enjoyed by STC. No consideration appears to have been given whether a world rationalization of the industry might not be to the benefit of Malaya in the long run, … inclined to think that part of the complaint of the British interests concerned is simply that there will be no room for them in the world organization which Signor Patiño is attempting to establish … Our claim is that we control Malaya in the interests of the native inhabitants, but no consultation with Malayan rulers or what the Chinese think about it. It is looked at solely from the British point of view.[38]

Although Caine was supported by Campbell and another senior official, Gerard Clauson, the Permanent Under-Secretary, Henry Wilson, overruled them and drafting work on the legislation was then authorized. Details were made public at the very point when Patiño was co-operating in the formation of the ITC and the Colonial Office was asked to reconsider its position.[39] Reconsideration did not directly take into account the implications for the much more important game of international co-operation. Instead, it focussed on the problem of retroactivity and possible retaliation by the Americans in other spheres.

National Lead and Patiño contemplated a more direct form of retaliation, by threatening to build a smelter in the United States and seek protection against imported metal through a tariff.[40] As the Colonial Office became more cautious, Kuala Lumpur became more insistent but after three months of exchanges of views, the proposal was finally blocked.[41] Caine's assessment of the motives of the Cornish proved to be well-founded but they would have drawn a very different lesson. They had won the local battle but could not control the outcome.

Loss on this matter had followed closely on the elimination of the Cornish interest in Williams, Harvey. At the same time, the Cornish position in Malaya was being undermined thanks to Howeson's acquisition of the Alluvial Tin interests. Patiño and Howeson were allied in CTS which was displacing STC as the world's largest smelting company. Almost overnight the world of the 1920s, which had revolved around the Cornish-Malayan tin companies and STC, had vanished. New forces were in control with which the old would find co-operation difficult.

6 The formation of the International Tin Committee, 1927–1931

From the first signs that tin was in trouble to the creation of the machinery that would permit a solution took two difficult years. Getting there required the formation of several coalitions, between the British producers in Malaya, between the British, the Bolivians and the Dutch and finally between the producers and their respective governments. While Howeson was the key figure in shaping most of these coalitions, it was the continued deterioration in the condition of the industry that led others to a rather different assessment of its future which ultimately permitted the emergence of a common front.

Preliminary overtures

For Howeson, trouble began with the decline in price in 1927 since it would delay the introduction of the fresh capacity the world would surely need. Discussions with several British tin interests were held in October and the press was informed that they planned to bring the price back to the £300 reached earlier in the year and stabilize it at that level. The rationale was simple: the world's tin resources needed protection in the interests of consumers and the producers needed protection from market speculators. In some quarters the initiative was heralded as the result of 'statesmanlike public spirit and foresightedness'.[1] While others dismissed it as simple propaganda on behalf of Anglo-Oriental, it proved to be the prelude to discussions with the Dutch. An exchange of correspondence between Nutt and Pan over the winter of 1927–1928 resulted in a meeting in London on 20 March 1928 attended by representatives of Billiton together with Howeson, Temple (also Anglo-Oriental), Byrne (Kamunting) and Waugh (Taiping). The fact that the Howeson team included these latter two interests at this stage is an important indication of the extent to which his initiatives had potentially broad support among established low cost producers.[2]

Byrne took the lead in this meeting and laid out the essential features of the rationalization position: the need to reduce the role of speculators, the irrationality of the decline in prices in 1927 in spite of the decline in visible supplies, the general desirability of high but stable prices in the range of £250–£300 and the possibility of ensuring producer control thanks to limited sources. The Dutch took issue with this assessment, stressing the importance of current marketing arrangements and the role of excessive production increases, especially in Malaya, as the cause of any current difficulty. Byrne then laid out a proposal for a Tin Producers' Association whose members would commit themselves to a common selling price, say £280 and Waugh added a proposal to stop dredges for one month a year. Stabilization was to be accomplished primarily through sales control, backed only by modest production limitations. Such vague suggestions, with no consideration of methods of enforcement, gave a very poor impression and the Dutch politely indicated that they were prepared to at least consider subsequent proposals. After the formal meeting was concluded, the Dutch held a private session with Byrne where they made it quite clear that Malaya had to limit production and the companies operating there had to create a single organization with the ability to enforce such a policy. Getting the Malayan house in order was to be the indispensable condition for any effective international co-operation.[3]

Howeson could review the results of the March meeting with some gratification since he had secured the support of the Dutch for the principle of stabilization, albeit with a profound difference in the level to be aimed at and the means whereby it was to be accomplished. Had he read the meeting carefully, he would have seen that the coalition he was attempting to establish with Billiton would be contingent on the formation of three others: between himself and the Malayan producers; between Billiton and Banka[4] and between himself and Patiño. The difficulty lay in the fact that each set would be dependent on the others.

Howeson next met with Patiño in the summer of 1928 and arranged the first meeting between the three key players. This was held at Don Simón's Paris home on 6 October. Again, Howeson stated his conviction that world tin supplies were very limited and laid out his plan for a common selling organization, aiming at a price of around £275, or a rise of just under 25 per cent. This was predicated on the assumption that the gap between production and consumption was quite narrow, so that controlling small quantities of metal would be sufficient to exert great influence on the price. Howeson may have expected a quick agreement between the major producers which would support his plans to manipulate the market but he was immediately provided with an entirely different assessment. Johannes van den Broek from Billiton pointed out that the statistical picture was quite unclear, especially since deliveries did not reflect actual consumption patterns and stressed that the Dutch had no interest in an artificially high price which would simply encourage further investment. Howeson's assumption of the inherent limitation of supply was simply not valid in light of the extensive deposits in the NEI, so the judgement of an appropriate price could only be made in relation to actual consumption patterns. An agreement to undertake further statistical study was the only concrete accomplishment[5] but certain lines were now drawn. Patiño and Billiton would work together and the latter expressed their reservations about working with Howeson.[6] Howeson had been forced to abandon his initial conception of stabilization, based on maintaining 1927 prices to ensure long term supply, and now had to work on organizing the Malayan producers on terms that would satisfy the Dutch and Bolivians. Although cost considerations were not irrelevant to Dutch thinking, as the cheapest producers with ample reserves they were under considerably less market pressure and the cleavage that emerged was between those countries that were well-organized with dominant companies and the one that was not.

The first proposal for co-operative production control was raised at a meeting of the Billiton Board on 18 April 1929, whereby at least the large producers would bind themselves not to increase production and provide some modest support for Howeson's efforts to hold up the price by buying metal. It was set aside, however, to await a more concrete demonstration of Howeson's ability to organize the Malayans.[7]

Tin Producers' Pssociation

With the uncertainties caused by a General Election in Britain, it was not until mid-summer that a general meeting of all directors of British tin companies was called on the initiative of Sir Edmund Davis.[8] A letter to *The Times* set out the case for international co-operation in terms that still reflected the famine thesis. Tin reserves were 'dwindling rapidly;' while demand 'has expanded persistently'. Since adequate supplies could only be guaranteed with a price well above £200, the industry faced 'chaotic conditions' thanks to its totally uncoordinated character and the signatories called for support for an 'authoritative association to protect the industry'.[9] This optimistic tone was quickly dampened by Stephens. Speaking at the Kramat Pulai AGM, shortly before this meeting, he listed an extensive catalogue of obstacles to effective restriction and concluded in the following ambiguous terms: 'If such a scheme could be formulated, which I regret to say seems to me to be hopeless, it should be supported by all who have the welfare of the industry at heart.'[10]

Three hundred directors representing 167 companies attended this meeting on 11 June. A month later the Tin Producers' Association (TPA) was officially launched[11] with a general council, on which all important groups operating throughout the British Empire were represented and two outsiders, Sir Philip Cunliffe-Lister and Sir William Peat[12] as Chairman and Vice-Chairman. Peat also served as Chairman of the Executive Committee which included not only Howeson and two of his closer allies, Byrne and Temple, but also the two leading Cornish figures, Thomas and Mair.[13]

Perhaps because of the press warnings and Stephens' remarks, production control was considered too controversial to be on the agenda. Instead, the initial focus was on remedies for the weaknesses in statistical information and for further research into uses for the metal.[14] The only immediate result was to initiate negotiations with the British Non-Ferrous Metals Research Association (BNFMRA) and the Department of Scientific and Industrial Research for an extensive research programme.[15] But the combination of the publicity given to the expressed determination to establish better control through the TPA and the anticipated recovery in US demand, confirmed by the impression given by delivery statistics,[16] allowed for some recovery and prices were held at over £200 from July to September.[17]

For the optimists this was but a temporary low before a substantial increase occurred 'within a period which could not be delayed'.[18] A more serious debate about the dimensions of the problem occurred at the Tronoh AGM. Here, Thomas offered both a diagnosis and a solution. The diagnosis had two components: one laid the blame for the current state of affairs on the excessive investment which had expected a price of £300; the second shied away from its implications and focussed on the irrationality of the tin market where 'producers had no voice in price; speculators were in control, not supply and demand'. That led in the direction of 'rationalisation even if it leads to restriction of output'. A solution along those lines was justified since: 'This movement to rationalize tin is part of a great world movement in every sphere of national, industrial and social life to endeavour to substitute co-operation for competition.' A compelling precedent could even be found in the efforts of the Associated Tinners of Cornwall to fix prices at the end of the eighteenth century.[19]

Stephens was also a director of Tronoh and felt sufficiently moved to challenge his Chairman in public. While the TPA was to be supported, restriction was not:

> The industry needs a purge, and unless it takes its medicine and eliminates the unhealthy production it is likely to suffer from chronic illness later on … The sooner the industry is freed from all shackles of artificiality, the better it will be for all who have invested their money in soundly capitalised concerns.[20]

With its Board deeply divided on the question of restriction, Tronoh could only passively respond to initiatives that emerged elsewhere.

The price rise of the late summer proved to be but a temporary respite. Faced with losses in other commodities, speculators had to jettison bull positions in tin. As prices began to fall again late in 1929, the TPA was prodded into taking action.[21] Its first response came at a meeting on 27 November to consider a proposal whereby British-American Tin would finance the withholding of ore at the smelters. That would interrupt the supply of fresh metal rather than rely on restriction of production.[22]

British-American Tin Corporation had been formed by Howeson immediately after the June TPA meeting with a nominal capital of £1 million. The proposal aimed at securing market control by withholding 12,000 tons of tin-in-concentrate at the smelters to be released when the price returned to £225. While British-American Tin would finance the miners, the TPA would compensate the smelters for their loss of revenue.[23] Just what Howeson really expected from this proposal is unclear,[24] since it would simply add to the heavy overhang of stocks, albeit in concentrates rather than metal and it was not well received by either Patiño or the Dutch. Stephens led the British opposition and with the rejection of this scheme, production control moved to the forefront of the agenda.[25]

While Patiño and Billiton offered general support to the TPA, they considered Howeson to be weak. At this stage the differences were quite fundamental. Neither Patiño nor Billiton was anxious to see a speedy return to £225 but rather an orderly liquidation of stocks. Nor were they very enthusiastic about restriction and agreed to develop a common policy. Patiño was anxious to keep his distance from Howeson and refused to accept an invitation to become President of the TPA.[26] In spite of the fusion of their respective smelting interests in the formation of CTS, Patiño was not prepared to support Howeson at the expense of effective collaboration with the Dutch. Antenor Patiño made this distinction quite clear[27] and sought to strengthen the relationship with Billiton by getting its participation in Patiño's policy of acquisition of shares in Malayan companies.[28]

In retrospect, it appears that 1929 was wasted in largely fruitless exercises. Perhaps they were necessary to convince sceptical British producers that all other methods had been exhausted before applying the stringent medicine of production control. Perhaps they simply reflected a misreading of the relation between production and consumption. The most obvious barometer, visible supply, demonstrated a consistent improvement with quantities declining after February and did not begin to rise again until August, so that it is possible that Howeson supposed that the situation was getting under control, albeit at a much lower level than he would have preferred.

Voluntary restriction

The initial moves towards production control were quite modest and were first proposed by Mair.[29] Revealing his inability to work effectively with Howeson, Mair communicated them to the press before consideration by the TPA Executive.[30] TPA members were asked to suspend production from 10 pm on Saturdays to 6 am on Mondays, that is by 32 hours of a 168 hour work week, a theoretical maximum of 20 per cent. Mair came under immediate attack from two extremes. One of the most important shareholders in the Redruth group saw no need for any initiative involving a 'well managed conservative' company, on the grounds that 'What has been good enough for a past generation is surely good enough for the present.'[31] From one of the first advocates of a root and branch approach involving explicit enforceable quotas came a warning: 'Mr Mair's suggestion possesses the inevitable dangers associated with palliatives.'[32] However, Mair's proposals were quickly adopted by the TPA, which added a provision for a suspension of production for a whole week in January and

February and if necessary in March as well.[33] Restriction took the form of time since it could be easily monitored.

Soon after the TPA announced its plans for voluntary restriction, Howeson, Cunliffe-Lister and Waugh went off to The Hague to request Dutch support. The proposal for weekend stoppage was not particularly impressive. Whether it would even cut output was doubted since companies could concentrate all their maintenance work at the weekend. In any case, such a cut had a long way to go to meet the production schedules on Banka which worked ten hours a day with one day off per week and 15 days' holiday per year.[34] While the Dutch conceded nothing at their meeting on 21 December, they began to revise their earlier positions. Howeson was still held in poor regard but at least he had created an organization that was taking production control seriously and one with which they could negotiate.[35] At the Billiton Board meeting on 16 January 1930, it was agreed in principle to support production restriction as long as all important producers did likewise but that, in view of the different rhythms of development, the basis should be an average of 1927–1929.[36] In coming to this decision, Billiton was prepared to upset the general sentiment in Dutch commercial circles which was strongly opposed to any participation in a restriction scheme.[37]

In Bolivia the strongest support for the TPA came from the Aramayo group and its managing director, Malcolm Roberts, expressed his frustration at Patiño's reluctance to declare his position. The inability of the Bolivians to co-operate domestically did not augur well for international co-operation.[38] However, in January 1930 Patiño finally committed himself to participate in the restriction scheme and as the largest single producer he was rewarded with the Honorary Presidency of the TPA.[39]

Billiton and Patiño reviewed their position on 12 February, taking the results of their discussions to the TPA on 18 February 1930. Patiño managed to persuade Billiton to retreat from its proposal to cut back to the average of 1927–1929, since this would be unfair to companies in Malaya which had also not expanded rapidly but were nonetheless expected to restrict. Instead it offered a cut of 5 per cent, from the beginning of the next work-year on 1 June.[40] The TPA also made it clear that they did not expect their own restriction to be sufficient to solve the problem. What they needed from the Dutch was a level of restriction that would sustain the commitment from the British producers;[41] what the Dutch feared was that further concessions might relax the current pressure, stressing that 'every promise on *our* part must secure the objective of *their* restriction'.[42] In addition to the promise of Dutch restriction as a way of encouraging effective restriction in Malaya, Billiton even considered the creation of a weapon to reinforce its urgency. This would take the form of a tin pool of 7,000 tons, assembled in conjunction with Patiño, to which the FMS government would be invited to contribute another 12,000 tons. The threat of a sudden drop in price through sales from this pool was expected to serve as a powerful incentive for the Malayans to co-operate. Although nothing came of this proposal, it serves as an indication of the power that the Dutch considered was ultimately theirs as they sought to establish the terms of international co-operation.[43]

Billiton's modest contribution to restriction was supported with an even more modest gesture by Banka when it was announced in the Volksraad of the NEI on February 15 that it would restrict production in 1930 to 22,000 tonnes, more or less equivalent to its 1929 production.[44] This announcement followed a more general debate in which the Director of Government Industries in the NEI, de Iongh, deflected attempts to lay the entire blame on the expansion in Malaya by stressing that a free market solution was no longer desirable. Prices were now at a level which would drive out not only the marginal producers but those which would be required when consumption returned to normal. Eliminating them would create a

Table 6.1 Tin market, July 1929–December 1930

	Production	Consumption	Production excess		Visible stocks	
	tons	*tons*	*tons*	*%*	*days*	*£/ton*
1929 July–Dec	91,377	84,243	7,134	7.8	52	196
1930 Jan–June	87,699	74,437	13,262	15.1	67	180
1930 July–Dec	77,722	75,819	1,903	2.4	102	124

Source: Economist, 18 July 1931;

Note: Stocks at beginning of period of consumption.

dangerously unstable situation in the future, since consumption would again expand ahead of production repeating the vicious cycle to which the industry had been prone. The Dutch were now publicly committed, not just to the abstract notion of stabilization, but to seeing it implemented in the current situation.[45]

In spite of the fact that the market fundamentals were still poor and there had yet to be any proof of the willingness of the members of the TPA to deliver on their commitments, market response to the announcement of the first restriction scheme was very positive and the price rose from £176/10/– on 6 December to £194/10/– ten days later.[46] Confidence that the tide had now been turned led to some bull accounts which could not be long supported and a large one of 800 tons was eventually liquidated at a loss of £25/ton.[47] Each subsequent move on the part of the TPA would have the same effect of providing reassurances that the problem had been finally solved and that the bottom had therefore been reached. There would be many further costly disappointments before it actually was.

The experience of 1930 is best appreciated against the following picture of the tin market in Table 6.1.

The standard barometer was the position of the visible stocks. At the end of June stocks were already high since they would cover 52 days of consumption. Over the next six months production exceeded consumption by nearly 8 per cent and the visible supply rose by nearly 4,000 tons, sufficient to support 67 days of consumption for the first half of 1930. Production dropped by 4 per cent but consumption dropped at an even faster rate, by 12 per cent. Although the second half of the year saw a slight recovery in consumption, the stocks to support it had risen to 102 days. While production was cut by a further 11 per cent, it was not enough to bring the situation back under control.

While the voluntary agreements would apply to producers in many countries, their success would be largely measured by their effects in Malaya. Assuming that maintenance work normally took around 10 per cent of the work week and that it could all be concentrated into the periods of suspension, full implementation of the first plan by members of the TPA in Malaya would be sufficient to cut supplies by 3,000 tons over the first quarter of 1930. The actual results are presented in Table 6.2.

The results of the first round of cuts were obviously very disappointing. Malaya was unable to reach even half of the 3,000 ton target; shipments from Bolivia and Nigeria actually increased slightly and the most important real, albeit temporary, support came from the reduction in shipments from the NEI.

The TPA put as good a face on these results as it could and announced that of its 141 members, 104 with 47,368 tons of capacity were restricting, while only 37 with 9,038 tons were not. Twenty-one other companies with 68,431 tons were also restricting.[48] Since the main culprit, Malaya, was still producing at a higher rate than for 1929, the Dutch were not impressed and now demanded better control.[49] However, with the announced co-operation

Table 6.2 Tin shipments, October 1929–December 1930 (tons)

	Bolivia	*Banka*	*Straits*	*Nigeria*
1929 4th	9,675	3,371	9,587	2,717
1930 1st	9,798	2,689	8,889	3,039
1930 2nd	9,854	3,279	8,627	2,281
1930 3rd	8,186	2,641	7,301	1,742
1930 4th	8,820	6,222	9,099	1,630
change from 1929 4th quarter				
1930 1st	123	–682	–698	322
1930 2nd	179	–92	–960	–436
1930 3rd	–1,489	–730	–2,286	–975
1930 4th	–855	2,851	–488	-1,087

of Patiño, Billiton and Banka, it was hoped that the programme would eventually see a total cut of 30,000 tons.[50] Reaching this target would require firmer measures and, at its meeting on 16 April, the TPA set specific production targets. Members were asked to cut their 1930 production to 80 per cent of that in 1929 and those commencing after the beginning of 1929 to 80 per cent of capacity. It was not greeted with a great deal of enthusiasm by the market, as one firm commented despondently: 'The immensity of the world's stocks makes even the proposed restriction of 30,000 tons a year pale into insignificance.'[51]

Mair was among the most enthusiastic supporters of this proposal. He too saw the problem as a temporary one of excess capacity in relation to rising demand and responded to the advocates of a market solution in the following terms:

> There is every reason to assume that without regulation, prices will fall even lower and remain at a low level for many months before high cost producers are driven out and then must follow the slow attrition of stocks to an economic level. It has been suggested that the law of supply and demand should control the price. If there were a large number of surplus producers this might be true – that is if the world's demand could be satisfied by two thirds of the producers. But this is not so, and the result has been a definite cycle of high and low peaks in the tin price. I cannot regard such spectacular and violent effects as good for any industry. They must result in making the product unpopular among consumers, and we as producers, low cost and high cost, should not overlook that important fact.

Preservation of capacity and stabilization of prices would become a continuing rationale for restriction. Mair concluded this speech by stressing that:

> Another need is that hearty and loyal co-operation among all producers is essential to bring regulation to a successful issue. I feel that any company producers who stand aloof now and fail to give the scheme every chance of success are shouldering a grave responsibility. They will certainly profit by the sacrifice of the others for a period, but the time may come when they find themselves forced to give reasons for the prodigal exploitation of a vanishing asset at an uneconomic price.[52]

This conclusion captures not only the moral principle on which tin restriction would continue to rest but also the principle which should infuse co-operation. As will be seen, Mair was unable to offer that co-operation to the ITC, not because he did not believe in it but rather because others would establish its terms.

Those others included Billiton and Patiño and they met on 9 and 21 May to review the new TPA proposals. They decided to offer support by increasing the rate of restriction at Billiton from 5 per cent to 10 per cent. When they presented their joint position to the TPA on 27 May, they insisted on a much faster rate of curtailment than implied by the 20 per cent cut and issued an ultimatum:

> If Patiño and Billiton do not feel satisfied with the results of the TPA they intend to follow their own policy [from 1 July] … We greatly desire that this should not occur, as we are convinced that the most sincere cooperation between all producers can eliminate the actual difficulties of the industry. If this is not the case, we shall see lower prices and we should have to be prepared for it.[53]

The immediate result was the formation of a special international committee composed of Billiton, Patiño, Howeson, Mair and Stephens. Now it appeared as though the coalitions were finally falling into place, since this was the first time that British producers, other than those associated with Anglo-Oriental, were invited to participate in the tripartite discussions. At the first meeting of the special committee on 28 May, Mair revealed that he had been in touch with the FMS government to see whether it would help enforce restriction. While the results were not encouraging, at least the question of formal governmental intervention had been raised.[54]

Shifting the explicit basis of restriction from time to output did not generate any immediate results. In the second quarter of 1930, European production in Malaya actually rose over the level of the first. Banka resumed shipments and the only positive sign was a slight reduction in Nigeria. Consumption in fact declined, so stocks rose by an extraordinary 10,000 tons. It was now time for urgent action.

Unimpressed by the efforts of the TPA, Australian groups based in Sydney proposed a total suspension of operations in Malaya and Siam for three full months.[55] As it responded to this bold initiative, the TPA recommended that its members first suspend production for two of the next three months and then cut their production for the remainder of 1930 to 80 per cent of that for 1929. It was this set of proposals that opened up the cleavage that would bedevil the subsequent functioning of the ITC. Thomas provided the support of the Tronoh group but without any confidence that this would be sufficient to solve the problem. Mair publicly rejected them since they failed to address the problem of stocks and stressed that restriction for two months without a similar cut from the NEI was grossly unfair to Malaya.[56] In the FMS, Archibald Glenister, from Osborne & Chappel, managed to shore up support for the TPA and the proposal for a two month stoppage was adopted by Chamber of Mines, with 31 in favour and only two opposed but with 25 abstentions.[57] Although the TPA published a list of 94 companies that were prepared to implement this round of cuts, it did not include any from the Redruth group.[58]

In view of these public divisions, it is hardly surprising that when the international committee reviewed the proposals, the Dutch were still scornful of the 'vague and muddled plans'of the TPA and sceptical of the power of its leaders. At this point, the strongest alliance was between Billiton and Patiño who had developed a 'cordial understanding'.[59] Henceforth, it would fall to Billiton to exercise the leadership required.

As Billiton's financial situation deteriorated, its first move was to indicate its willingness to increase its rate of restriction to 20 per cent if Malaya did the same and that policy was officially announced on June 23. Although this appeared as a concession to facilitate a co-operative game, it was actually the result of an independent decision to suspend work on the deeper lode mines which were no longer financially viable.[60] As before, the most that could be secured from Banka was a commitment not to increase production. This meeting also reviewed Billiton's ideas for a pool and Patiño's even more ambitious plans for a comprehensive selling organization. While the pool was still regarded as a potential weapon to enforce restriction, it became increasingly clear that it would be an essential adjunct to restriction and henceforth the Dutch would always be interested in both.[61]

Coincidentally, a senior member of the FMS Mines Department developed an ingenious valorization scheme.[62] The Lovett scheme addressed the fundamental weakness of the restriction programme that had been initiated in January: it did not have enough bite to lead to a rapid price recovery. Lovett considered that a reduction of 52,000 tons would be sufficient to restore the price to £200 and proposed that it be realized by a 40 per cent cut on 1929 production by the FMS, Bolivia, Siam and Nigeria. What made the scheme ingenious was the incentive system envisaged. The TPA would sponsor a company to offer the minimal immediate finance needed to keep participating companies going on a hand-to-mouth basis, in return for which they would sell tin to be delivered when the target price was reached, with the difference between the current and target prices being split evenly between the producer and finance company. The amount of tin involved would be kept in the ground and the company would then suspend all production.[63]

Lovett was certainly thinking along the right lines about the size of the cut required. However, the practical details were immense and extended not simply to ensuring the level of co-operation required but also to dealing with the governments who would simultaneously have to cover the loss of revenue and the increase in unemployment. Conceptually, the scheme was fundamentally flawed as well. It rested on the assumption that the slump would be of short duration, four to six months, with no contingency plan if that proved invalid. While Lovett hoped that the producers themselves would subscribe to the finance company, he failed to recognize that there was no incentive from that side of the equation. If tin was selling at £140 and expected to rise to £200, there was no point in making an advance of £170. 'Vague and muddled' is, therefore, an apt characterization of more than one section of the Malayan tin mind.[64] But the most serious consequences of this kind of initiative lay in the fact that it defined the parameters of a solution by reference to 1929 production and without any contribution from the Dutch. That could only weaken Malaya's bargaining position.

One exception to this pattern of confused thinking was someone from outside the industry. Gordon published a comprehensive criticism of the TPA which identified several failures, which were summarized: 'It failed to tackle the problem at the root, but was content with half-hearted measures dealing with symptoms rather than the disease.' He also offered a diagnosis of the source of the problem: the absence of 'an executive of some half-dozen men capable of discussing the industry's economics in all its aspects'.[65] In the face of this kind of criticism it was not easy for Howeson to hold together the coalition he had assembled but twelve months earlier.

In early October, Mair again became active. He resigned first from the Council and the Executive of the TPA and then from the international committee. Freed from Howeson, he attempted to establish a coalition directly with Patiño and Billiton. In this he had the support of Thomas and Stephens and while a tripartite meeting took place on 7 and 8 October, this new group of British producers also came with no new ideas, leaving just as poor an impression as Howeson.[66] Howeson's efforts to create a coalition between himself and the other major groups in Malaya around the TPA had now failed. Since Thomas and Mair found they were not able to displace Howeson and insert themselves into the centre of tin politics, they became testy critics of virtually all the moves that would be made thenceforth.

Even before the two month suspension was over, Billiton was already sceptical about the level of voluntary compliance in Malaya and also feared that the weaker producers would soon defect. In addition, there was the fear that Banka would actually increase and then maintain production in the face of low prices given its large reserves and low costs.[67] Van den Broek shared his thinking about Malaya, first in October with Mair, Thomas and Stephens and then in November with Howeson.[68] He had come to the extraordinarily pessimistic conclusion that 'any world wide agreement for restriction seems to be out of the question; that the so-called Lovett or any other plan for general restriction is unpracticable and that the situation must right itself by the survival of the fittest'.[69] There could be no further discussion of any private solution.

Assessment of voluntary restriction

It is evident that the TPA programme of voluntary restriction failed to accomplish its manifest objectives. Whether it is fair to call this a 'fiasco'[70] is an entirely different matter. The results of the third quarter of 1930 were quite promising. Shipments from Banka and Billiton declined and the cuts in Bolivian production were now becoming evident. Nigeria was able to show an especially impressive rate of reduction. The key test was, of course, Malaya. In this quarter, European production was cut by 37 per cent of the level at the end of 1929 but it was not enough to make up for the slow start. When the results for 1930 were in the Europeans could only show a 5 per cent cut on 1929, a far cry from the overall target of 20 per cent.

Without the TPA programme, the market would have fallen faster and, with no prospect of a collective solution, a number of weaker producers would have folded. But the extent to which it sustained redundant marginal production[71] and thereby defeated a smooth market solution depends entirely on the judgement miners were making about the future of the industry. Mining is far from a rational business and the optimism with which it is infected is often incurable. As late as October 1930 many felt that the corner had been turned and some were even prepared to make fresh investments in opening up new mines.[72] However, even if marginal producers had adopted a more realistic view, they could well have decided to exhaust their mines as quickly as possible before abandoning them. Had production in 1930 remained at more or less the same level as 1929, the stock overhang would have prevented any effective recovery for at least another five years.

Three factors prevented the TPA from bringing about the desired level of control over the market. The extent of the production momentum created by the fresh investment of 1928–1929 was clearly underestimated and a similar error of judgement was made about anticipated consumption trends. The most important weakness was the failure to secure sufficient secure compliance from its own members and from other producers.

Although Mair was initially responsible for the proposal to shut down on weekends, the Redruth and Tronoh groups established their own conditions for participation. They were only prepared to support the first proposal if other companies representing 90 per cent of the tonnage of the whole association did likewise and if all the Bolivians and Dutch 'consented to come into line'.[73] Without this assurance, they only implemented a 24 hour suspension.[74] While both groups adopted a common policy in supporting the two-month shut down,[75] Thomas refused to accept the subsequent 20 per cent cut.[76] Similarly, the Stephens group took its time to declare its willingness to co-operate with restriction.[77] Such guarded support from these important groups may have led to less enthusiasm elsewhere.

The only enforcement resource at the disposal of the TPA was a moral one. At the ATMN AGM, Holland laid it out in the strongest of terms:

> A sense of honour, of loyalty, alone would have dictated cooperation. It is not morally permissible for one or two companies to stand out alone and steal an advantage from the sacrifice of others. Nor would the immediate gain – the wage of their treachery – prove other than a loss in the long run, for nothing can be more hurtful than the prodigal exploitation of a vanishing asset at uneconomic prices.[78]

It was not without effect. Immediately afterwards, St John Wynne stated at the Geevor AGM that while 'Cornwall was so negligible that she might be excused from any curtailment whatever, but for the principle of the thing, and so I feel sure you will wish our company to play the game fully.'[79]

The existence of a deep moral template was now being revealed which exposed the flaws in the logic of competition. It was one that was universally shared and derived from a particular game, as Thomas made explicit in his appeal for support for the two month closure: 'It is not playing cricket for any company to do nothing to attain the desired end. Britishers always do desire fair play.'[80]

While implementing the principles of that morality would prove to be elusive, they continued to set benchmarks by which the behaviour of all participants would be judged. Unfortunately, they would soon find that they had different conceptions of the nature of the game.

In Malaya, the TPA established a local committee to which members were expected to put up a case if they felt unable to comply with the policy. The fact that such machinery was used indicates that moral sanctions had some sway but of course it would have been very difficult to turn down such requests. Moral sanctions were reinforced with the publication of the names of those co-operating under such headlines as 'TPA 'Roll of Honour''.[81]

While not members of the TPA, the Chinese miners were asked for support and representatives were included on this local committee. The Selangor Miners Association recommended that the Chinese mines co-operate by reducing the hours of work each day from 24 to 18.[82] While this cut was supported by only four kongsis in January, the numbers rose to the more respectable level of 42 in February and March, which was around half the total number of Chinese mines in Selangor.[83] In the Batu Gajah district of Perak six of the 63 Chinese miners stopped, though all said that they would stop if they found that most of the others did.[84] However, they were subject to other pressures since there was some concern about the danger of social unrest in the Kinta Valley with large numbers of unemployed miners.[85] The extent to which actual restriction took place must have been very modest, since the decline in production from the Chinese sector in Malaya was very similar to that in the overwhelmingly Chinese non-dredging sector in Siam where there was no expectation of co-operation.

The reduction in production cannot be entirely attributed to restriction as part of a co-operative agreement. All producers faced a trade off between absorbing higher operating costs on a lower level of output and preserving their reserves for an anticipated improvement in the market. In many cases restriction was a rational decision at the level of the firm. Yukon Gold, for example, was not a member of the TPA but came to such a conclusion, with the result that it realized one of the largest cuts of all. Its 1930 production was cut by 28 per cent of the level reached in 1929.[86]

The case of Siam reveals several weaknesses in the TPA. In spite of the fact that one of its Directors sat on the TPA council, STS considered that it was not in the interests of its shareholders to become a member of the association, 'considering it preferable *in all circumstances* to maintain complete freedom of action'.[87] It offered a very modest 24 hour restriction. While the more important Australian groups were members, they were not prepared to enforce restriction within all their companies.[88]

It was in Nigeria that the TPA had its greatest support. The Nigerian Chamber of Mines endorsed all the proposals[89] and they were loyally carried out by its main members. Suspension by time was not suitable for Nigeria and instead the companies committed themselves to a straight 20 per cent cut for the first quarter.[90] Anglo-Oriental managed to cut its production in 1930 by 38 per cent in comparison with 1929 and most of the important medium-sized companies also showed significant reductions, Jantar at 33 per cent, Kaduna at 32 per cent, Nigerian Base Metals at 28 per cent. Few Malayan companies could match that accomplishment.

There was always a major gap between the public propaganda on behalf of the TPA and the covert jockeying taking place among the key producers. The programme of voluntary restriction may have had less to do with actually controlling the market than with establishing a series of conditions which would make compulsory restriction a viable solution.

Research

The one encouraging feature of 1930 was the progress made towards putting the plans for a comprehensive research programme onto a secure foundation. The TPA had established a Tin Research and Industrial Applications Committee (TRIAC) and it proposed a five-year programme costing a total of £150,000. Since TRIAC only received an annual budget of £1,910 from the TPA, it began to lobby for governmental contributions.[91]

Several months passed before either Nigeria or Malaya responded. Reflecting its empire-first mentality, Malaya wanted all funds to be spent by the BNFMRA but this restriction was eventually relaxed on the grounds that the BNFMRA was limited in the kind of research it could undertake and had no capacity to develop and promote any of its innovations.[92] Nigeria was initially reluctant to participate at all on the grounds that it had no money but was eventually persuaded on condition that the industry contribute two-thirds of Nigeria's share.[93] This principle was then applied to Malaya where both the FMS Chamber of Mines and the Perak Chinese Mining and Planting Association supported a cess of 5 cents per pikul.[94]

As production restriction looked increasingly ineffective, the expansion of the research scheme appeared to be most attractive way of addressing the problem with tin. Indeed, Campbell considered this 'much preferable to restriction schemes which nearly always encounter unexpected difficulties and break down at the most embarrassing moments'.[95]

However, within a few weeks the limitations of voluntary restriction would force the Colonial Office to address the implications of compulsory restriction and Campbell would then have many opportunities on which to reflect upon the extent to which this principle applied to tin and other commodities.[96]

Towards compulsory restriction

In early November the Dutch and Bolivians developed a formal proposal, presenting it to the TPA Council with a thinly veiled ultimatum: either a quadrilateral plan, enforced by governments, to bring production in line with consumption over the next two years, or no further voluntary co-operation and a return to an unrestricted market. Three moves followed immediately. One was to communicate the principles of the scheme to the governments of Nigeria and Malaya through the Colonial Office, a second was to communicate them to the advisory TPA committees in both countries and the third was to communicate them to the Malayan and Nigerian Chambers of Mines.

With the support of the local TPA committees and the metropolitan Chambers, the colonial governments could be expected to fall into line. In all these communications the TPA stressed that the proposal originated with Billiton. That would prevent it from being seen as irredeemably tarnished by Howeson.

The idea of tin restriction was received with a great deal of sympathy in the Colonial Office, not least because it had become acutely aware of a more general problem with primary commodities in the empire. It had already reviewed the Billiton proposals and a senior official commented: 'Personally I believe that restriction, if it can be applied all around, is the proper policy in the present circumstances for sugar, tea, tin and perhaps rubber.'[97]

While the Colonial Office did not formally endorse the proposal, it made its own support clear as it passed on the TPA message. At this stage, all that was being proposed was the general principle of production control, the general level of the cut aimed at, 20 per cent on 1929, with that year as the basis of the ratios between the four members. It was also expected that once production control was in place then arrangements would be made to deal with the question of excess stocks.

In orchestrating the necessary support, Howeson had intended that two separate messages would go to the governments of Nigeria and Malaya. Since the Cornish had made their position clear on the international committee, Malaya was to be told that it was expected to 'make equitable allowance for production initiated in 1929 and 1930'. That would give them grounds on which to claim quotas for the extra production for which they were responsible. He did not want the same explicit message to go Nigeria, since he expected that the primary beneficiary would be his own ATMN. Instead, the government was to be given full discretion in the allocation. That would allow ATMN to make a claim but whether it was granted or not would clearly be the responsibility of the government rather than the TPA. He was not anxious to stir up more trouble than necessary. Unfortunately, while the messages sent by the TPA to the Chambers and the local committees reflected the distinction, those that went to governments did not, since the Colonial Office simply copied the same cable to both.[98] That crucial oversight would create serious difficulties in Nigeria.

The issue of compulsory restriction became public as soon as the TPA became the broker for Billiton and Patiño and only one significant figure raised a voice in opposition. That was Hochschild who was always optimistic about the prospect of large scale production making low-grade ore bodies viable.[99] Otherwise, the prospect of broad industry support was good and with a nihil obstat from the Colonial Office, the way was now clear to set up a meeting at which the basic principles of a restriction scheme could be established.

Until this point, plans for compulsory restriction had been an industry initiative but the meeting at Cunliffe-Lister's home on 26 November marked an important transition. Fortuitously, de Iongh and Sir Cecil Clementi, High Commissioner to the FMS, were both in Europe so there could be direct discussions between representatives of the two most important governments and the three major industry players, Billiton, represented by van den Broek and Houwert, Patiño, represented by Antenor Patiño and Ricardo Martinez Vargas, and the TPA, represented by Howeson, Peat and Cunliffe-Lister.

Consensus was easily reached around the overall objective, the establishment of a balance between production and consumption with a concomitant reduction in price oscillation. Although there was some discussion of the level of restriction that would be actually required, at 22.25 per cent it was not much higher than the rate being sought by the most recent voluntary scheme. Debate then focussed on how this would be distributed among each of the members.

Two contradictory perspectives were at work and both were dressed in the language of equity. One started with a diagnosis of the origins of the current problem for which Malaya bore a disproportionate responsibility and for which she should, therefore, bear a heavier burden. The second started with existing productive capacity and supposed that, since all producers would benefit from the price recovery, all should share equally in its costs. It was not simply that the Dutch and the Malayans were adopting expedient arguments; they were deeply felt positions which reflected the quite different character of their industries. Between them there could be no consensus, only a compromise.

Clementi found himself at something of a disadvantage in arguing the Malayan position and the balance of power in these negotiations remained with the Dutch, especially with Banka. All along, they had consistently argued that the British had to come up with serious proposals before they would commit themselves to full support. If the proposals on the table failed to meet their objective, it could be some time before they were prepared to meet with the British again. In addition, of course, it is extraordinarily difficult to operationalize the concept of capacity. Indeed, estimates for Malaya alone range from 80,000 tons to 93,000 tons.[100] In spite of the weakness of his negotiating position, Clementi did extremely well since he was able to deflect the pressure from Banka on to the other partners.

The terms of the compromise involved three sets of concessions. Malaya conceded its claim for current capacity in return for 1929 production; the Dutch conceded their claim for ratios based on an average of 1922–1929 in return for 1929 plus concessions to Banka made by Bolivia, Nigeria and Billiton. The concessions by Bolivia and Nigeria effectively preserved the ratio between Malaya and the NEI established in 1929; those by Billiton granted most of Banka's claim for 1922–1929. That such a compromise could be reached in one short day was due to the flexibility of both Clementi, 'whose tact and skill in his discussion with Mr de Jongh[101] were beyond all praise', and 'the Billiton representatives. Although they had both might and right on their side, they were prepared to make the greatest concessions of all.'[102] Any concession from the two extremes would inevitably benefit Bolivia, so that the rivalry between two strongest participants redounded to the advantage of the weakest.

While discussion focussed on initial production quotas, these were not important in themselves. What was crucial was the ratios they established. Table 6.3 indicates the concessions that were made from the three negotiating positions.

Table 6.3 Negotiations on standard tonnages

	Preferred positions			*Outcome*				
	NEI	*Bolivia*	*Malaya*	*1929*		*Final*		
	1922–9	*1929*	*1930*	*Agreed*	*Transfer*	*ST*		*cf 1929*
	%	%	%	*tons*	*tons*	*tons*	%	%
Malaya	40.9	42.9	48.9	69,366		69,366	42.9	0
Bolivia	27.6	28.6	24.7	46,338	–2,274	44,064	27.2	–4.9
Nigeria	6.1	6.4	7.4	10,412	–444	9,968	6.2	–4.3
NEI	25.5	22.1	18.9	35,730	2,718	38,448	23.8	7.6
	100	100	100				100	
Banka				21,122	3,392	24,514		16.1
Billiton				13,480	–914	12,566		–6.8
Singkep				1,128	239	1,367		21.2

Notes: 1930 capacity estimates, Nigeria 13,000, Malaya 90,000, Bolivia 43,000 and NEI 36,000 tons. Nigeria's actual production proved to be higher and her standard was increased to 10,734 tons.

Although the agreement reached here was still subject to review and ratification, the actual negotiating process inevitably fixed these ratios of restriction. Those not present could always claim that a better deal could have been struck but the prospects of actually bringing it off were considerably weakened by the fact that the concessions had already been made. The informal character to this critical meeting is particularly striking. While the Dutch had a delegation that could speak with some authority from both an industry and government perspective, neither Howeson nor Patiño had any negotiating brief on behalf of Nigeria or Bolivia; indeed, neither had been in direct communication with their respective governments. While Bolivia certainly could not have done any better, Nigeria, in fact, gave up a great deal. Clementi was in an even more difficult position. He was an outstanding Sinologist and had only recently been transferred from Hong Kong to Singapore to deal with the anticolonial agitation there. A 'patrician scholar who found relaxation in Latin poets and Mandarin love-songs',[103] he had little grasp of the complexities of the Malayan mining industry and may well have naively supposed that the concession would be quite palatable domestically.[104]

At the request of Clementi, the communiqué announcing the results deliberately hid these compromises:

> That the quotas of production permitted to be exported during 1931 and 1932 should be based upon the actual output in 1929 in the following ratios, namely – Federated Malay States 35.9%, the Dutch East Indies, Bolivia, and Nigeria 49.6%, the rest of the world having produced in 1929, 14.5%.[105]

The reference to 1929 as a base was ambiguous. The immediate impression was that the standard tonnages simply reflected the ratios of production established in that year, the last one of unrestricted production.[106] In fact, it was only a base in the sense of being a point of departure. Combining the NEI, Bolivia and Nigeria together made it appear as though consensus had been reached and disguised the particular position of the Dutch. Although this would be revealed as soon as the actual production quotas were published, it still served to hide the privileged position of Banka.[107]

The other issues were quite straightforward. The scheme would run for two years commencing on 1 January 1931. Decisions would be made by unanimous agreement, failing which a member had the right to withdraw on six months notice. Siam would be approached as soon as Malaya had formally endorsed the agreement, with the full anticipation that she would follow the Malayan lead on this as on other mining matters. The level of restriction required would be just over 22 per cent to meet a world demand of 145,000 tons, assuming Siam's participation and a certain falling out of producers in all other countries.[108] Cunliffe-Lister summarized the position:

> All over the world, governments, businesses, and economists are trying to bring about the rational production of primary products. In many instances the problem seems insoluble. In tin we have found a solution. It seems to me the adoption of that solution is peculiarly in the interests of Malaya … I shall take an impartial view, as I came into this business as an outsider, and my whole work has been that of an independent chairman and conciliator trying to bring the parties together.[109]

Two interpretations of the origin of the ITC have become current. One focusses on Patiño and sees it as a result of the internationalization of his empire, especially through the formation of CTS and his acquisition of shares in Malayan companies. This is naturally popular in the Bolivian literature[110] and among those who see wish to see the ITC as an attempt on the part of high cost producers to save themselves rather than the industry. The other, of course, focusses on Howeson and his very visible efforts in agitating for tin control and organizing the TPA. It is something of a tribute to the brilliance of the Dutch strategy that their crucial role has been generally concealed.[111] While all three were indispensable, Howeson was the weakest. He was the newcomer whose financial empire had been but recently constructed and under boom conditions and was therefore more vulnerable to the slump; he was also the smallest of the three major players. Howeson's importance derived from his ability to shape a coalition of Malayan interests. Such a coalition was needed to persuade the Dutch of the viability of international co-operation and the reason Patiño played such a passive role is that there was little he could have done to facilitate Howeson's work with the Malayans on the one hand and to broker his relationship with the Dutch on the other. The fundamental features of the political process that would shape the operation of the cartel were revealed in its origins. The Malayans would continue to experience considerable difficulty in formulating a coherent and united position, the Dutch would continue to exercise leadership behind the scenes and Patiño would generally take a back seat.

Once a framework agreement was in place, the next stage was to secure approval from the respective governments. This would entail three steps: (1) commitment to the principles of the Lygon Place Conference, so that a meeting of formal government delegations could be struck to work out details; (2) endorsement of their proposals; (3) passage of appropriate domestic legislation.

In the case of the NEI, these matters were quite straightforward and no legislation was needed, given the government's control over Banka and Billiton and the latter's willingness to formulate a joint policy with Singkep. Similarly, the Bolivian government needed no time to review the proposals and committed itself to participate in the scheme on 12 December 1930.[112] It was the British who found it difficult to make the next moves with any degree of urgency.

Clementi also wanted a prompt resolution, by getting the Colonial Secretary to settle the question of principle and simply asking the governments of Nigeria and Malaya to prepare the necessary legislation. The Colonial Office, however, insisted on a process that involved considerable delay. While the two governments would be encouraged to support the scheme, they would have to come to an independent decision in light of local circumstances.[113] In addition, the Colonial Office also insisted that since the British Cabinet had been responsible for the implementation of rubber restriction, the tin agreement would require the same level of authorization.[114] Cabinet would not act independently of the governments and governments would not act until they were convinced that their domestic mining communities were overwhelmingly in favour.

In the case of Nigeria, these matters were easily resolved. The Nigerian Chamber of Mines met to discuss the proposal as presented by the TPA[115] and endorsed it without qualification.[116] Since the Chamber spoke on behalf of 95 per cent of Nigerian production, the government could fairly claim that 'our miners themselves have asked for it'.[117] Unfortunately, the 'it' that the government thought was being requested included the Malayan provision for accommodation for fresh capacity, while the Chamber thought it was endorsing a straight 1929 principle of allocation. No one recognized the confusion when the Nigerian Governor, Sir Graeme Thomson, visited London and in the general discussion, 'he confessed himself entirely in favour of the principle'.[118] The government in Lagos quickly gave its formal endorsement, adding only: 'subject to quotas being in your opinion equitably fixed'. In its reply the Colonial Office naturally provided the assurance requested, though omitted to point out the concession from Nigeria's straight 1929 tonnage.[119]

It was Malaya that was troublesome. Clementi's commitment in London carried no particular weight in Kuala Lumpur. The local government had three problems on its hands. In spite of explicit declarations that it would remain strictly neutral, STC was conducting an informal campaign against the principle of restriction.[120] Complaints about Malaya's share were being raised by the Cornish.[121] Finally, it would have to mount a complicated administrative operation with virtually no lead time. The response was typically bureaucratic: it sought to make administration more straightforward by seeking deferral of the implementation date for two months until March 1 and insisted that the quota be fixed for a full twelve months.[122] While there could be no commitment to a specific quota in such a rapidly changing context, the Dutch were prepared to concede the request for a deferral of implementation in order to secure a quick overall agreement.[123]

Given the differentiated nature of the Malayan industry, the FMS government could not adopt the easy position of supposing that it spoke with anything like a united voice through any particular organization. Nor would it follow Clementi and adopt a position of its own:

> Government is quite satisfied that the responsibility for accepting or refusing principle of restriction must lie with the mining industry and that it is no part of the Government's business either to express its own views or to give a lead to the mining community on that point.[124]

Instead what it did was to send a circular to all miners, laying out the basic terms of the agreement and the qualifications that the government saw as necessary, together with an outline of the way in which it would work in Malaya.[125] It is unclear whether the deception was deliberate but the circular was extremely misleading since it stated that each party would restrict in the same proportion. In laying out the position of the government, it stressed the importance of deferring implementation until at least 1 March, with full disclosure of hidden stocks. Consistent with the understanding reached in London, provision was to be made for fresh capacity introduced in 1929 and 1930 and miners were warned, though in rather imprecise terms, that once total capacity had been assessed, their individual quotas would be lowered below the standard 22 per cent cut.

The Chinese were sufficiently desperate that they would support restriction on almost any terms and these proposals received enthusiastic support from both the Selangor and the Perak Chinese Chambers of Mines.[126] When the Europeans met under the auspices of the FMS Chamber of Mines on 7 January, the Cornish had indicated their reservations. While quite prepared to support the principle of restriction, 'provided we think it is based on sound and equitable lines towards Malaya as an economic and progressive producer',[127] the proposed scheme signally failed to meet this criterion. In addition, they presented a quite different reading of the market, arguing that the 22 per cent would be quite insufficient to accomplish the desired objective and that one of 30–40 per cent over one year was what was really required.[128] In effect, what they wanted to do was renegotiate the agreement so that it was, in their terms, sharper, shorter and fairer[129] and in this they received considerable support from the British financial and mining press.[130] Anticipating that the meeting would be difficult to control, Glenister indicated that he would only allow a question and answer period with the Senior Warden of Mines, Greig, and that no vote as such would be taken. While support through a vote was pre-empted, the meeting demonstrated its enthusiasm when virtually all those present stood up physically to indicate where they stood on the matter.[131] With these manifestations of support, together with the response to the circular, the FMS government felt as though the conditions for its own support had now been met.[132]

Efforts were made by unofficial members in the Legislative Council to persuade the government to resist coming to such a conclusion and they indicate something of the deep suspicion with which the whole project of international co-operation was regarded. It touched an imperial nerve, as Rae worried about 'going into partnership not with people of our own nationality ... with the Dutch who have always been quite capable of looking after themselves and ... with a somewhat irresponsible Latin Republic in South America'.[133]

This lack of trust in the prospective partners made it impossible for Rae to support the principle of restriction without knowing precise details of its implementation, especially of the price level to be aimed at and the nature of subsequent plans should the 22 per cent cut prove to be insufficient. The absence of such details made the votes recently taken quite worthless. But no scheme could possibly operate if such conditions were to be established in advance.

Malaya therefore went into the restriction agreement deeply divided, less about the principle than the crucial details. These divisions are hardly surprising in light of the experience of the previous two years. Perhaps the marginalization of the Cornish in the key preparatory negotiations[134] meant that they were unable to appreciate the force of the Dutch position but as a group they were also in a very strong position and with low costs and high reserves they were not under any immediate pressure to come to an agreement. Rejecting the terms of this agreement meant an invitation to renegotiate and, while the Dutch would not have conceded the fundamental point of the basis of restriction, the results of those negotiations could well have seen a more stringent and effective scheme. In any case, rejection staked out a position to which they would frequently return not only during the life of this agreement but all its successors.

By insisting on a commencement date of 1 March, Malaya was also responsible for weakening the agreement. All producers were now in the position of being entitled to produce without restriction for two months, knowing full well that severe restriction would immediately follow. Exports for January and February 1931 from the FMS were back at the level of 1929. Ostensibly this concession was requested on the grounds that it would make it easier for the Chinese to clear up their accounts at the end of their financial year, Chinese New Year, but it was also an issue for the Cornish.[135] One factor prompting their resentment of the Dutch was the decision of Banka to ship 3,000 tons of metal from its stocks in the NEI to the Netherlands, prior to the commencement of restriction. While it clearly stated that such stocks would be frozen, they would inevitably add to the perceived stock overhang. Some felt that the request for a delay was seen as giving Malaya a parallel opportunity to organize itself in readiness for restriction.[136]

Once formal support had been received from the colonial governments, the Colonial Office could present the case for tin restriction to the British Cabinet. In spite of support from the Board of Trade, the Cabinet initially expressed some scepticism and the matter was referred to a special committee. J. H. Thomas was concerned that consumers in the United States would be particularly upset, seeing tin control as a return to the experience with rubber and was not reassured by the observation of the Colonial Secretary, Lord Passfield, that the level of stocks would prevent prices from rising unduly. Clementi was then invited to meet with the committee and he provided the arguments that were needed to permit a unanimous report:

> The committee has been impressed by the representations made to them by the Gov of SS that it is essential from the point of view of Gvt of M that the scheme should be brought

into operation and that if this is not done the financial and economic consequences will be very serious.[137]

The Cabinet could then ratify the principle of restriction,[138] followed by an official announcement in the House of Commons which stressed that because the 'restriction scheme is controlled by a committee of representatives of the four governments, the interests of consumers will be protected against any excessive increase in price'.[139] Such bland reassurances were not sufficient to allay all Parliamentary opposition. One Labour MP, George Strauss, was a metal broker and agitating against any solution he considered was solely designed to rescue Howeson and his friends from their speculative excesses.[140] He would continue to use his position to raise awkward questions about the operation of the ITC once it began its work.

By the end of February it was possible to work out the specific details of the scheme. Since these belong more to the machinery of restriction than its conceptual foundation, they will be reviewed in the next chapter. Once they were shaped, they could then be communicated to the participating governments, with the expectation that they be ratified so that the scheme could actually commence. Pressure to provide a public announcement which would allow for a market rally was resisted by Campbell, who insisted that both Malaya and Nigeria be given a full opportunity to review and assess them:

> I do not like to do anything which may appear to be rushing Malaya or Nigeria, because that may produce undesireable reactions. I feel that rather strongly. They have by force of circumstances been left out of this up to date to a considerable extent ... They are thousands of miles away and not getatable and that is always liable to produce an inferiority complex and make them a little more susceptible than they otherwise would be.[141]

This proved to be an abortive gesture. It took some time for the details to reach Malaya and it was not until the middle of April that the FMS Legislative Council met to pass the necessary legislation. Unofficial Members only received the precise terms as they arrived and naturally resented being expected to pass it without much scrutiny. Glenister seized on this as indicative of the overall problem:

> This is only too typical of the way in which affairs in connection with Tin Restriction have been conducted, and it is yet one more example of the policy of silence, the policy of asking the people of Malaya to leap first and look afterwards, which has caused so much trouble in the industry and has led the country into accepting ... an insufficient quota allowance.[142]

Either out of ignorance or malice, he went on to criticize Nigeria for having conceded some of its quota to the Dutch, which reflected both disloyalty to the empire and the ineffective bargaining ability by Malaya's representatives. That was not a challenge directed to the government presenting the legislation but to the one at home. Unfortunately, it required a stronger local government to establish its primary responsibility and authority to speak on behalf of the country.[143] Unqualified support, however, came from the Chinese member, who was more interested in the principles of internal distribution since the smaller mines needed preferential treatment in the allocation of quotas to ensure their survival as producing units.[144]

Since the procedural problem derived from the apparent marginalization of the government, the unofficial members modified their objections and decided to vote for the bill. However, the point had been registered. Excluding an established European miner from the negotiations meant that the results were never considered satisfactory in spite of all the representations to the contrary. There remained, therefore, a lingering resentment in Malaya that her interests had not been well served and this resentment would continue to eat away at effective co-operation. Such resentment annoyed the Dutch since it failed to recognize the validity of their own claims and the government attempted to reduce it by issuing a communiqué stressing that it considered that: 'Malaya had received equitable treatment and nothing that has since transpired has shaken its belief in the equity of the terms or cause it to question its decision to participate in the quota scheme.'[145]

Although this gesture was appreciated by both the Dutch and the Colonial Office,[146] it was not sufficient to overcome the apprehension that the FMS government was too weak to resist the pressure from the Cornish and Campbell worried:

> one fears that if Malaya is, pressed by fear as to the attitude of their unofficial members, to approach each communication sent them in this attitude of offended dignity and unjustified suspicion, the prospects of the intl restriction scheme are not bright … Malaya seems not to visualise the position at all … It is not an easy matter to ride four horses; and it is certainly not going to be made easier if Malaya persists in this present attitude. All of us have quite honestly done our best to keep Malaya informed of everything impt; and the Gvt there could be convinced of that – and use their own method for passing on that infmn – it might facilitate things materially. It is so important to remove this feeling of offended dignity that every effort shd be made to do so.[147]

Far from removing that feeling, the Colonial Office would soon find that the administration of tin control would increase it substantially with all the consequences that Campbell had predicted.

The origins of the ITC reveal several tensions that would shape the way in which it could address the problems of tin. They are less about the principle of political control than about who would exercise it and to what precise end. Such control granted the industry something of a reprieve from the constraints of the market but, in so doing, it opened a fresh set of conflicts which set their own constraints on the exercise of that newfound power.

7 Constructing the machinery of control

The agreements that had been reached over the winter of 1930–1931 simply set a framework which needed fleshing out in several different ways. The most pressing were to establish the principles whereby decisions at both the international and the domestic levels would be made. Once those parts of the machinery were in place it was possible to reinforce them, first through a pool to control stocks and then through the recruitment of the most important producer still outside the agreement, Siam.

Principles of international control

Operational details of the control agreement were settled at a meeting held on 27 February 1931. Government delegations were joined by representatives from the TPA (Cunliffe-Lister, Howeson, Stephens and Waugh) and a metal broker, Lazarus. The first detail to be addressed was a statement of the objectives of the scheme for communication to the public. It could not directly address the question of price since it would have to publicly justify the price level being aimed at[1] so it settled on the following rather bland statement: 'The scheme is intended to secure a fair and reasonable equilibrium between production and consumption with the view of preventing rapid and severe oscillations of price.'[2]

The statement of objectives has been seen as a rather disingenuous formula for a group with an overriding interest in income maximization, designed merely to 'convince industrial consumers of the benefit, or at least innocuousness, of the scheme'.[3] Perhaps there was some anticipation that this formula would defuse consumer criticism but it genuinely reflected current thinking, especially on the part of the Dutch. They had always stressed the importance of eliminating the violent oscillations in price which served neither producer nor consumer, together with the need to end the features of a classic commodity cycle. The objective was not simply to protect the producers but to do so in a way that also served the interests of consumers. Much later, the question of a potential conflict between the two would arise but it was far from anyone's mind in the deflationary days of 1931.

Production was to be controlled so that exports corresponded to quotas at least on a quarterly basis, though at this stage there was no provision for reducing the level of minehead stocks. On the advice of Lazarus, it was agreed that changes in quotas should be made as seldom as possible and with at least a three month interval between changes. That was to provide some degree of predictability, both to the market and to producers. The final important points of agreement were that the scheme should run for at least two years to 1 March 1933 and the conditions under which withdrawal was permitted was confirmed. The 22 per cent cut had been built into the first quota but thereafter unanimity among all governments would have to prevail for any alteration. Delegates had no formal authority to bind their respective governments so that all decisions were subject to ratification.

Just as important as the results was the process, as Campbell commented:

> The meeting went very well ... Our anticipation that the Malayan and Nigerian delegations would find it easy to work with the Dutch seems likely to be fulfilled. The Bolivians were also accommodating, sensible, and pleasant to deal with. ... I think we can continue to count on common sense, a reasonable spirit of compromise, and a strong sense of realities on the part of all the delegations.

Campbell also provided some reassurance to those who were sceptical of Howeson:

> Mr. Howeson took a very minor part in the discussions. Sir P Cunliffe-Lister was most helpful, sound, and tactful throughout ... with him in the saddle, we ought not to experience any difficulty with the T P A. The impression I got was that Mr. Howeson was very much under control![4]

One objective was dropped. The Colonial Office at first wanted to dissociate itself from playing too prominent a part in the scheme by nominating a Dutchman as the first Chairman but was persuaded to allow Campbell to take it on. In light of the role that he and the Colonial Office subsequently played in moderating tin politics, that proved to be very wise. However, some semblance of parity of participants was secured by agreeing to rotate meetings between London, Paris (for Patiño) and The Hague.

Given the way in which the ITC has been regarded as a creature of the TPA,[5] it is worth noting how the Colonial Office defined the relationship. At first it was proposed that the agenda of ITC meetings be divided and that the technical advisers be excluded from the point at which actual decisions were made but Campbell commented:

> We shall have to treat the TPAssn people rather better than is proposed. After all they originated the scheme, they will be in the pool and possess much information which we shall doubtless on occasion be glad to have. If any purely Committee matter arises, where it seems desirable cttee shd sit alone, it will be easy to get a suggestion to that effect from the Dutch. I think it unwise to do anything to antagonise the TPA reps. They will in any case have no weight – except such weight as may attach to sound advice coming from them, and accepted by the Comm: as a whole. And, so far, we have no reason for rubbing in the fact that it is a purely governmental committee.[6]

Apart from a modest propaganda role on behalf of the ITC, the TPA would eventually become completely irrelevant to the shaping of its policies. But in the early stages, Howeson remained an important source of information and advice, without at any point moving the ITC away from its formation as an intergovernmental institution.

Composition of delegations

While the ITC may have defined itself as quite independent of the TPA, the delegations made a variety of different arrangements for including representation from the industry. The NEI delegation was exclusively Billiton, with van den Broek and Houwert from BM and Groothoof from GMB.[7] Formally, the membership of the Bolivian delegation was nominated by the Ministerio de Hacienda on the recommendation of the AIM but the main members, Antenor Patiño and Martinez Vargas, were drawn from PME.[8]

Only the British delegations confined industry representatives to the marginal position of technical advisers. The Colonial Office would not allow the Nigerian and Malayan governments 'an entirely free hand [in selecting their delegates] especially in view of Secretary of State's undertaking that the interests of consumers would be protected'.[9] Instead, it kept both delegations firmly under its control by making Campbell chairman of each. Additional members were appointed and were also granted the right to communicate to their respective governments. In the case of Nigeria this was straightforward and the position went to Sir Frank Baddeley, a retired civil servant with experience in both Nigeria and Malaya.

Malaya proved difficult. The Colonial Office suggested that the second member be the head of the Malayan Information Agency in London, H. W. Thompson. Clementi responded with two controversial nominations: a miner, George Simms (Thomas' son-in-law) and another retired civil servant, Sir George Maxwell, ex-Chief Secretary of the FMS. Neither was acceptable to the Colonial Office[10] and Clementi grudgingly accepted Thompson.[11]

The delegations were therefore quite asymmetrical in their composition. Bolivia was purely producer, Nigeria and Malaya largely metropolitan government, the NEI largely producer. This would have profound consequences for the development of the ITC. As long as Howeson was intimately, albeit informally, involved with the ITC, there was no need for a strong voice on behalf of the Nigerian industry voice to match those of Bolivia and the NEI. But in Malaya, the failure to secure direct representation by either colonial government or industry simply compounded the sense of resentment already festering in many quarters as a result of the way in which the overall agreement had been negotiated. Since real power in the British delegations was held by the Colonial Office, it also meant that the core set of relationships was between the Colonial Office, the Dutch and Patiño. Subsequent events would make Patiño marginal, leaving the Colonial Office-Dutch axis as the fundamental basis on which the agreement would be implemented.

Negotiations with outsiders

The quadrilateral agreement was seen simply as a prelude to a more comprehensive arrangement which would include other producers, especially Siam and Burma, since in both these countries a number of companies had supported the TPA programme. The resources available to the ITC to deal with outsiders would be clearly revealed in the negotiations with each.

Burma

Decisions on Burma involved four separate departments: the provincial government of Burma, the central government in India, its representative in London, the India High Commission and the India Office. The initial moves were made through the Minerals Advisor to the High Commission. Anglo-Oriental and CTMB exercised as much pressure as they could to secure Burma's adherence[12] but they were opposed by Anglo-Burma. It grasped at several arguments, including its own inability to afford restriction, future difficulties in replacing labour, the particular problems of hydraulic mining in Burma with its five-month season and the fact that much restriction was occurring anyway.[13]

When Campbell approached the India Office he proposed the same terms on which the four signatories had joined: 1929 production, estimated at 2,445 tons and the same level of restriction, 22.3 per cent.[14] The standard tonnage corresponded to Lower Burma and did not take into account the Mawchi mine in the Shan States where the government had no administrative authority. In Lower Burma production was indeed falling off rapidly so the whole issue appeared moot.[15] However, the ITC did not let the matter drop and a formal approach was made to the government of India which passed it on to the government of Burma.

By June 1931, the process of consultation in Burma itself had elicited a generally negative reaction. Anglo-Oriental continued to press for restriction[16] but most of the others who could not work year-round were strongly opposed. In addition to arguments which identified specific problems in Burma, two issues raised were troubling. Considerable development had taken place since 1929 and since Burma was small, her participation would make no difference to the price.[17] In other words, she preferred to play the role of the freerider without constraint. This was a serious failure for Howeson, especially given the comparative importance of Anglo-Oriental in the Burmese industry.

South Africa

Negotiations with South Africa foundered on a different problem. Her 1929 production was 1,218 tons and she was offered the same terms as the signatories, so that the initial quota would be 950 tons.[18] This was considered 'not unreasonable' by the government mining engineer who observed that although South African production was hardly likely to affect the market, 'the moral effect might be desirable'.[19]

The four producers were then consulted and they generally expressed support in principle. But since current production was running at 408 tons, far below the proposed quota, the government was not prepared to proceed with a scheme which would have no practical effect, especially since it would raise awkward legal questions concerning the right to enforce restriction on minerals won from private land.[20]

Siam

Siam, of course, was far more important. It had always been expected that Siam would be prepared to follow Malaya's lead on this as on other mining matters and the first proposal simply put the same set of terms, 22 per cent cut on 1929 production[21] but this seems to have been simply ignored. The Lygon Place Conference had badly underestimated the extent of her anticipated production for 1931 and the 7,728 tons offered was quite unacceptable.

As soon as the ITC appreciated the Siamese position it became clear that some limits had to be placed on production, otherwise the plans for the pool would be jeopardized. When serious negotiations commenced, the ITC dropped any attempt to define the extent to which she was responsible for the degree of overproduction but simply focussed on current productive capacity, now running at around 13,900 tons.[22] Implementation of the same cut as the others on the basis of her 1929 production would involve a real cut of 50 per cent. Since this was somewhat in excess of the level expected of any other member, it was clearly unreasonable. Instead, what was proposed was a fixed flat rate of 10,000 tons.[23]

These proposals were received sympathetically by the Ministry of Foreign Affairs which recognized a moral obligation to support the ITC but not by the Ministry of Agriculture which had primary responsibility for the administration of the mining industry. It was reluctant to give an immediate response and in any case felt that the figure should be 11,000 tons, a slight reduction from the 1930 output of 11,175 tons.

The Siamese government was under considerable pressure from the mining industry not to participate in restriction at all. The Siam Chamber of Mines passed a resolution to that effect and it did so without much thought as to the implications for the restriction scheme and the price level that it, too, was anxious to restore.[24] This Chamber had no metropolitan counterpart, and spoke without the specific authorization of the directors of the companies it represented. Opposition was not to the principle of restriction but many companies were fearful that the level of restriction aimed at would be far too severe and not give adequate provision for the fresh dredging capacity that had emerged over the previous two years. Since much of this fresh capacity had been financed through debentures, restriction would prevent the meeting of such obligations, forcing companies into bankruptcy.[25]

The Siamese government had found it difficult to deal with the Chamber and the TPA then sent Edwards of Anglo-Oriental to meet with the Ministry of Commerce in Bangkok to provide another industry voice. In addition to 'doing the proper thing', there were two material advantages. Smuggling was costing the government royalties on around 1,500–2,000 tons and it would be much easier to control this if Siam were a member of the ITC. Edwards was also able to stress that with production uncontrolled it was most unlikely that the 30,000 ton pool then being planned could be formed.[26] Given this mutual interest in the outcome, there were no actual negotiations about the terms.[27]

The internal debates within the Siamese government were resolved when the Ministry of Commerce refused to support the claim for 11,000 tons on the grounds that it was too much and that it would look as though Siam was not making the kind of sacrifice expected of other countries. The flat rate figure of 10,000 was therefore agreed but Siam asked for two concessions which were impossible to refuse. One was to abstain from contributing to the research scheme[28] and the second was to ask for the same kind of administrative delay in implementing restriction that had been granted to Malaya. Restriction would therefore not begin until 1 September.[29]

As these terms were formulated in the Colonial Office it was recognized that they were likely to cause some resentment in Malaya. The very principle which related restriction to productive capacity which had been firmly denied to Malaya was being easily conceded to Siam. On the other hand, Siam was not prepared to recognize this as a concession but rather as an entitlement and would be very upset by criticism from Malaya.[30] Although attempts were made to get the FMS government to put a positive gloss on these results, it was difficult to get the Malayan industry to recognize that they involved genuine restriction. One obstacle was the fixation on 1929 as the basis of the agreement. Since Siam's production for 1929 was recorded by the ITC as 9,939 tons,[31] the figure of 10,000 looked extraordinarily generous. It has therefore been difficult to look beyond this to the actual level of restriction that was implemented.[32] As representatives of the Malayan industry tried to digest the implications of this agreement, they turned to a familiar moral template:

> If this were cricket, it would appear that Siam is being permitted to play with 15 men. I do not think that Siam would wish to have this advantage, and I hope, Sir, that you will indicate that Siam does not intend to place all her men in the field.[33]

With little appreciation of the difference between the artificial conventions governing the very British game of cricket and the real world of international commodity politics, it was difficult for Malayans to feel anything other than profoundly aggrieved at the apparent success of both her neighbours.[34] The higher the levels of actual restriction, the worse this problem would become.

In addition to the more extended negotiations with Burma, South Africa and Siam, overtures were made to China, Portugal, Japan, Belgian Congo and Australia but none expressed any real interest.[35] Portugal certainly took the request seriously but was not prepared to accept the 450 tons of 1929 as its standard tonnage, arguing instead for a figure nearly three times as large, based on current potential and past performance.[36] As one of the smallest producers being approached, she could also claim that her production was really too insignificant to make any difference.

Domestic control

The structure of the industry determined the overall features of the system of domestic control. In the case of the NEI this was quite straightforward. The original pattern of negotiations had already settled the distribution between Banka and the other producers and it was unnecessary to enforce restriction through legislation. Elsewhere, legislation was required in order to ensure that domestic production was consistent with the international quota. In addition, the other members had to establish a specialized machinery to resolve three issues: (1) entitlement to a production quota; (2) the size of the quota; (3) the precise bundle of new rights that was granted with a quota, especially the right to transfer production entitlements.

Federated Malay States

Given the large number of independent mines, the process of establishing an effective administrative machinery was most difficult in the FMS.[37] The industry was divided into two main categories: the dulang washers who were given specific production targets but with no transfer rights and the miners who were first granted an assessment on which a specific production target was calculated on the basis of the ruling international quota. These assessments were the domestic analogue of the international standard tonnage.

Eligibility for an assessment was extended to all mines that had actually produced in 1929 and 1930 and to those who were in a position to produce in 1931 for which the required investment had been made by 1930.[38] Each mine had to make an independent application, backed by data about production in 1929 and 1930 and current capacity, to an assessment committee established for each state, with the right of appeal to a central committee. At both levels the committees were made up of officials from the Mines Department and miners.

These assessments were mainly based on 1929 production, with modifications to reflect fresh investment but there was no attempt to introduce any other principle of equity. When the assessment process was complete, it was not subject any major modification, so that fresh investment did not provide the basis for an increase; however, they were monitored to ensure that the mine could always produce its assessment. Although these assessments were initially based on a specific production unit, they could be transferred in order to allow the miner to move from one parcel of land to another. All holders of active assessments were entitled to a production quota.

The legal form of the production quota was the Certificate of Production (CoP), issued by the Mines Department to each producer for each specific quota period. All concentrates, apart from those produced by dulang washers, had to be connected to a specific certificate. Any that were not were illegal and subject to seizure. This was the main enforcement device to ensure there were no unauthorized exports in excess of Malaya's international quota.

Although CoPs were issued to specific mines, the system allowed for two kinds of flexibilities. In order to ensure that the quota issued was actually produced, amounts could be transferred should a mine experience a temporary interruption. More important was the right granted to holders of CoPs to group them so that the aggregate production entitlement could be exploited in the most efficient way. Where the various CoPs were held by the same owner, grouping was a simple matter. Grouping among separate mines took the form of a semi-permanent co-operative agreement, subject to a detailed set of rules which determined who could participate and under what conditions.[39] Applying and monitoring the implementation of these rules generated a great deal of work for the Mines Department but, since it was also free to change them in light of actual experience, it had effective control over the way in which the Chinese sector of the industry would evolve. One fear was that the Chinese kongsis would transfer their quotas to the more efficient dredging companies and reduce total employment; grouping between them was therefore prohibited.[40]

Although Malaya had insisted on deferring the start of the scheme, she had not taken advantage of this delay to put the proper machinery in place and it was not until May that appointments were made to the central and state committees which then drew up the general administrative principles.[41] Given the amount of time required to review some 1,200 applications, the Mines Department only started to issue CoPs in June, before the process of assessment was complete. In doing so it made two assumptions that proved to be fatally flawed. The first was to assume that the initial cut would run for the six months that had been part of the informal understanding of the agreement and the second was to estimate the total capacity of the FMS at 31,500 tons for the six months. To give itself a margin of safety to allow for adjustments once the assessment process was complete, the Mines Department imposed a cut of 25 per cent rather than the final amount that was estimated at 23.8 per cent.[42] The Department admitted that 'the thing had not been worked out with any attempt at closeness', but 'felt that it would have to be settled by some kind of trial and error basis'.[43] While such a principle could work domestically, the error would have serious international consequences.

The estimate of 31,500 tons is quite remarkable since at an annual rate of 63,000 it is below the standard tonnage allocated to the FMS which in turn, of course, was considered in many quarters to be much below her legitimate claim. Insofar as it had a basis in fact, it reflected the reduction in actual exports on the part of the Chinese but it proved to be under the final total by 6,500 tons.[44] The resulting overexports would set the ITC off to a bad start.

Although assessment was a zero-sum exercise, there was a tendency for assessment committees to err on the side of generosity. That would reduce the prospect of appeals and the increased advantage to one producer would only have infinitesimal consequences for each of the others. When the consequences of a myriad of independent decisions were added up, the total assessment was greater than the actual productive capacity it supposedly represented. The higher the assessment, the greater the gap between the domestic and international quota rates, which naturally fuelled the overall resentment in Malaya about its international standard tonnage.[45]

Siam

Since Siam had a similar structure to her industry as that in Malaya and implemented restriction five months later, she could draw on the Malayan experience as the basis of formulating her own system of administration. It followed the Malayan model in detail, with three important exceptions. While grouping was allowed, it was generally confined to established corporate networks. Entitlement to an assessment was not limited to existing producers, so productive capacity could continue to grow. Administration was in the hands of the Department of Mines, with no joint consultative committee on which the industry was directly represented.

Nigeria

Nigeria adopted the Malayan system of assessments (admitted claims) and certificates (permits) with a similar bundle of attached rights and obligations. The Department of Mines established an administrative committee, misleadingly called the Quota Committee, which dealt with all matters relating to the domestic administration of restriction.[46] At first, industry representation was confined to the three large companies and one of the medium. Eligibility for an assessment was more narrowly defined to those actually mining before 1 January 1931.[47] The major departure from the Malayan model was in the adoption of a much more flexible set of principles in determining the assessments, ones that would both recognize and reconcile different kinds of claims made by different categories of producers.

One group of producers, the small independent miners, had been excluded from all the discussions about restriction and when they learnt that Nigeria had committed itself to join the ITC, they protested vehemently.[48] Three major objections were raised. The first took issue with the very principle of restriction as a contravention of 'the first principles of economics' and as impractical in the case of tin, doomed to 'lead to disaster'. The second argued that the Nigerian tin industry was largely a creation of the small miner and that the recent pattern of concentration of capital had been based on an unrealistic assessment of the economics of the Nigerian minefield. While the small miner could continue to produce profitably even at current prices, the large could not and such an 'attempt to bolster up the undesirables at the expense of sound propositions',[49] would therefore be against the long term interests of the country. The problem with tin was seen as a result of excess production by the large and the small miners could not be expected to bear the burden of solving something not of their own creation. Finally, they protested against the fait accompli which had secured Nigeria's consent on the basis of representations made by London-based organizations, such as the TPA and the Nigerian Chamber of Mines which served only the interests of the large without even consulting the small, local producers.[50] Right from the outset, therefore, the government had to deal with a politics that derived from cleavages within the industry and it was anxious to deflect this initial attack by granting the small miners generous terms. Each miner was granted an allotment, or 'basic allowance' of 3 tons of concentrates per quarter, designed to ensure an 'irreducible minimum or living wage'.[51]

Whatever obligation was contained in the original expectation that Nigeria also make equitable allowance for new capacity was reinforced by the Chief Inspector of Mines who felt that the industry should not be frozen at a particular point in time but rather permitted to continue its rhythm of development.[52] However, operationalizing this new capacity was much more complicated in Nigeria. Whereas in Malaya it had taken the form of new dredges, whose capacity could be easily measured, in Nigeria much of it had taken the form of expansion of existing operations as a result of the new source of cheap power from Kurra Falls. Since that had only come on stream in July 1930 it was too early to measure its overall effect with any degree of precision, so that a rough estimate had to be made of this new capacity. It was hoped that the actual production experience would provide the basis on which these estimates could be revised. To permit that revision, these estimates were defined as 'special claims' and kept separate from claims based on a straight 1929 production, termed 'ordinary claims'.

Actual production quotas were therefore made up of three elements: (1) a basic allowance granted to every miner, regardless of size; (2) a quota based on ordinary claims; (3) a quota based on special claims. Complicated though this system proved to be, it rested on a conceptually simple foundation. The government considered that restriction was something the industry had requested, both internationally and domestically; its own responsibility was, therefore, to be confined to effective administration.[53] The system created by the Chief Inspector of Mines was simply one which would place 'upon the mining community the responsibility for carrying restrictions into effect in their own way, the Resident and myself only acting as referees to see that no section of the community is unduly penalised'.[54]

At first it appeared as though the system would work well. Of the 90 cases dealt with, only seven were the subject of a difference of opinion within the Quota Committee and of these only two involved important amounts.[55] Unfortunately, it was not possible to gauge the political implications of the adoption of these principles until they were actually applied and the Chief Inspector would soon find his sense of equity under attack. Table 7.1 indicates the distribution of special and ordinary claims by different categories of producer and the impact of the first quota allocation.

The small miners had benefited from the application of two quite different principles of equity: preservation and development. There were also more pragmatic considerations supporting their position. Bankruptcy on the part of small miners posed two potential problems: they might remain in Nigeria, becoming a group of poor whites and upsetting the established link between class and race or they might demand the costs of their repatriation.[56] If the number of white miners had to be reduced, far better that they come from the companies who had the responsibility of ensuring that dismissed employees be shipped home.

While the full implications of this system were not made public for some time, as soon as miners received their allotments it was clear that a major breach had been made in the understanding that they would be based on 1929 production. Those who had been expecting a cut of 25 per cent were shocked to discover that it was over 40 per cent. ATMN, with its well known extensive development, became the inevitable target and a group of companies began to try and change the administrative system. They felt they had a good argument since they had assented to restriction in November on the assumption that 1929 would be at least

Table 7.1 Nigeria quota allocation, 1931

		Admitted claims				*First quota*			
		A	*B*	*C*	*B/C*			*D*	*D/A*
		Ordinary	*Special*	*Total*		*Basic*	*AC*	*Total*	
	no.	*tons*	*tons*	*tons*	*%*	*tons*	*tons*	*tons*	*%*
Small	51	687	808	1,495	54	612	807	1,420	207
Medium									
Objectors	15	4,072	320	4,392	7	180	2,370	2,550	63
Other	13	2,207	847	3,054	28	156	1,648	1,804	82
Large									
Latilla	1	1,032		1,032	0	12	557	569	55
AO–ATMN	4	3,791	1,800	5,591	32	48	3,018	3,066	81
AO–LNTM	1	3,431	292	3,723	8	12	2,010	2,022	59
Total	85	15,220	4,066	19,286	21	1,020	10,410	11,430	75

Source: Maxwell Report, Appendix V.

Notes: Quarterly quota calculated on annualized basis. AC: quota based on admitted claim. Tons are of concentrate the primary basis of allocation.[57] As they grasped for moral principles with which to bolster their case, they also pointed to the dangers of supporting further monopolization and the desserts owed to the pioneers of the industry.[58]

Their case was supported by the Nigerian Chamber in London and the Chairman pointed to what he felt to be a fundamental moral contradiction:

> How could that [development] be when the companies that answered to that description were already under an honourable agreement to restrict production. The idea was to restrict ... and if they chose to develop, they did it at their own risk ... The scheme could not possibly succeed except in an atmosphere of complete good-will. The last thing to produce that atmosphere was a sense of injustice.[59]

Such sentiments were lost on the Chief Inspector of Mines who continued to insist on the equity of the principle of special claims.[60] The disaffected minority, therefore, bypassed the newly established administrative system and secured the intervention of the Governor in Lagos, who proved sympathetic to their case.[61] However, as he took up the issue, he was met with the argument that the scheme had been designed in the best interests of the mining community as a whole. Before any alternative could be taken seriously it had to address both the particular problem of the small miners and the equity claims of the Kurra Falls investment.[62] The Governor was therefore caught between his sympathy with the objectors and the difficulty of formulating a coherent alternative.

Consistent with this strategy, the objectors were able to use the local council to attack the Quota Committee, claiming that it could not represent the industry as such, since members were in a position to advance their particular interests. In effect, they wanted to turn the Quota Committee into a government rather than an industry body, thereby creating a structure which would permit precisely the kind of overt politics that the government had tried to avoid. As the administrative machinery tried to defend itself, it dismissed the substance of the issues raised, as merely the result of 'frayed and sore nerves'.[63]

Some immediate flexibility was found as a result of a small increase in the international standard tonnage awarded to Nigeria, as a result of the belated discovery of a mistake in the statistical basis of the original calculation. This windfall could be awarded as relief to those who were particularly hard pressed, so that a new moral claim was tacitly recognized without any regard to the overall principles which formed the basis of the scheme.[64]

The issues in contention were joined at a general meeting when the Governor visited Jos in early August 1931. By this time it was clear that the overall level of restriction was far more severe than that originally anticipated, on the basis of which the complex formula for internal allocation had been designed. This meeting therefore provided an opportunity for review of the entire scheme.

The large companies tried to identify a fundamental contradiction in the position of their opponents. If 1929 was taken as the base year, then it would reflect the increase in productive capacity initiated in 1928. In which case, equity demanded that the increase in capacity that resulted from Kurra Falls, also started before the slump but which did not materialize until 1930, should also be recognized. Concessions from the government had enabled them to proceed with the scheme and these did not allow for a slowing down in its rate of completion. They also claimed the moral high ground of being the greatest source of the basic allowances which operated in favour of the small miner. However, in a willingness to secure a compromise, they proposed that any further increase in the international quota be assigned entirely to ordinary claims.[65]

For the medium companies, the issues were equally clear cut. Increase in capacity in 1930 was not justified, so the principle of special claims simply rewarded those who were responsible for the problem for which restriction had to be imposed. The issues had not been properly discussed when restriction was initiated and as result the companies found themselves bound by a principle to which they had not assented. They were forced to restrict at a much higher rate than anticipated, for the benefit of undeserving interests! Both the process and its results had placed them in an unacceptable position. For their part, they were prepared to offer a compromise which would recognize the principle of special claims but cut them in half!

The small miners had three compelling arguments. They had been granted special assurances as the price of suspending their opposition to the principle of restriction and these special assurances had to recognize the cost and good faith with which they had invested in fresh development in 1929 and 1930. There were also the moral claims that came with their responsibility for the overall development of the minefield.[66]

Resolving these conflicting positions required some willingness to compromise but that was confined to the large owners. Otherwise, the Chief Inspector of Mines commented:

> Every group … who thought that by making trouble might secure better terms for themselves, did so on every possible opportunity. Settlement after settlement, including generous concessions by the large owners were agreed to as final, as soon as made, the dissatisfied owners would find some hair to split in order to go back on it.[67]

The government was just as divided as the industry. While the Governor appreciated the case for special claims, his fundamental sympathy remained with the medium companies; the Chief Inspector of Mines still supported the large and the small. In these deliberations, the government could not formulate any coherent policy of its own; it was confined to supporting the principle of restriction and to seeing its own role as administering that principle on behalf of the industry. What could not be settled in Jos was then referred to London and the positions argued by the large and the medium groups were turned over to the Nigerian Chamber of Mines for a final decision. Since at this stage the Chamber was controlled by the large companies, it simply endorsed their position.[68]

In attempting to develop a comprehensive system in which all principles of equity would be recognized, the Department of Mines found that it had opened up a set of fundamental cleavages within the industry and ones that were now reflected in a division within the government itself. Those tensions would obviously be exacerbated as restriction bit harder and the way in which they were resolved will be considered in relation to the subsequent cuts. Some important lessons might well have been learnt from this episode. Much of the difficulty derived from divergent conceptions of the principles of restriction and the failure to engage in comprehensive consultation with all the interested parties. More important was the trap set by trying to establish an overarching principle of equity; it was bound to generate moral resentment on the part of those who held different but equally firm views of equity. In this respect, the internal politics within Nigeria reflected the external politics conducted by the Cornish in Malaya.

Bolivia

Of all the members of the ITC, Bolivia faced the most difficult problem of domestic administration. In part, this was technical. Lode mining did not lend itself to the comparatively easy determination of production capacity that was possible in relation to specific dredges, draglines and gravel pumps working particular deposits. A further technical problem arose from the fact that restriction came at a point when the industry was undergoing a substantial change, with important mines closing and others developing, so that no uniformly acceptable base date could be established. In part, this was administrative. Without a separate Department of Mines, the government had no comprehensive production statistics. Above all, it was political. Unlike her partners, Bolivia was an independent state where her miners had real political rights and ones that were easily articulated through conventional representative institutions. That combination of a weak administrative structure, organized political mobilization and absence of any uniform body of technical data would make it extraordinarily difficult to translate the principles governing the international scheme into their domestic equivalents.

Instead of developing a system of assessments, the Bolivian government allocated actual production entitlements on a monthly basis. It was therefore always open to political pressure on each such occasion for a specific increase. Production entitlements fell into two categories. The large mining groups of Patiño, Aramayo and Hochschild were granted production rights for the whole complex, which meant that quota transfer between their individual mines was permitted freely. The medium and small miners were granted rights linked to their mines without any possibility of transfer. However, without a system which could tie each parcel of concentrates to the authorized quota, it was impossible to prevent such transfers, or even prevent excessive accumulation of stocks. Since responsibility for enforcement rested not with a Department of Mines but the Ministerio de Hacienda,[69] it was easier to secure exemptions which would allow for excess exports.

As a result of the closure of several mines in 1930 and continuing low prices, effective capacity in 1931 was not much greater than Bolivia's initial international quota. Although an early start was made in developing a system of quota allocation, following meetings between representatives of the miners and the Ministerio de Hacienda,[70] the government found itself subject to a great deal of pressure from Hochschild who threatened to suspend operations at Unificada unless the mine were granted a large increase over any entitlement based on performance in 1929.[71] As is evident from Table 7.2, he managed to get his way at the expense of the both the small and medium miners. They were cut by nearly five percentage points and Hochschild increased by about the same.

In fact, the actual experience of 1931 diverged considerably from the initial allocation. Hochschild, Aramayo and the medium mines were unable to use their full entitlement. While the small miners exceeded theirs by the considerable margin of 40 per cent, there

Table 7.2 Bolivia quota allocation, 1931

	1929		*1931 quota*		*1931 actual*	
	tonnes	*%*	*tonnes*	*%*	*tonnes*	*%*
Patiño	27,790	58.8	15,400	58.7	16,334	62.0
Aramayo	4,719	10.0	2,460	9.4	1,773	6.7
Hochschild	4,596	9.7	4,142	15.8	3,763	14.3
Medium	5,594	11.8	2,211	8.4	1,639	6.2
Small	4,528	9.6	2,029	7.7	2,839	10.8
Medium and small	10,122	21.4	4,240	16.2	4,478	17.0
Total	47,227		26,242		26,348	

Source: Revista de Economía y Finanzas, February 1932.

Note: 1929 based on 1931 ownership.

was still a deficiency which was made up by Patiño under a 'gentlemen's agreement' with the government.[72] The Hochschild problem would eventually resurface but that of the small miners would become more acute as production levels were cut further.

International Tin Pool

As soon as the question of production restriction was finally settled, the problem of the large overhang of stocks had to be addressed. Not until these were down to normal trade levels would any balance between production and consumption be reflected in a substantial price recovery. The Bandoeng Pool provided something of a model and two of its principles were followed in the first pool formally linked to the ITC. The total quantity of stocks frozen were published and included in the visible supplies, as were details of the price at which the market could expect them to be released. However, this pool differed in two important respects: the stocks were held privately rather than by governments and those private commitments could only be secured against specific agreements on restriction. While Howeson was being gradually excluded from the way in which the intergovernmental agreement on production operated, he would play a critical role in the formation and operation of the pool.

Pools carry considerable risks. One is defection of participants and this was solved by appointing McKenna as an independent trustee to hold the warrants, the legal title to the metal. A second is independent action on the part of those otherwise able to affect the market. This would be best resolved by securing the participation of all the relevant players, pre-empting any options that might otherwise work against the interests of the pool.

Initial thinking about the size of the pool established an objective of 30,000 tons to be distributed in three equal syndicates, Dutch, British and Bolivian. It soon became clear that Bolivia could not match the others and it was revised to 20,000 tons: Dutch and British at 7,500 tons each and Bolivian at 5,000. Howeson had already secured commitments from Patiño for 2,500 and Aramayo for 600 and now wished the Bolivian government to arrange for the completion of the Bolivian Syndicate with a further 1,900 tons.[73]

Since the NEI government would be part of the Dutch Syndicate, the UK government was approached to make sure it had no objection, otherwise the Dutch could be exposed to a further round of criticism in Malaya. Where Campbell saw the pool as a corollary of restriction, Passfield thought otherwise. His objections were three-fold. The losses on the Bandoeng pool, estimated here at £3 million, set a discouraging precedent. Since the pool proposed to acquire part of its stocks in the market, it rested on a gamble. Most important for Passfield was the moral position that would subsequently emerge as a general basis of criticism of the pool: 'It would be repugnant to me to be concerned in a gamble in which I could myself influence the factors determining the profit or loss at the expense of others outside the combination.'[74] Although Clementi was also interested in participating,[75] he was never given the opportunity to address Passfield's objections before they were endorsed by the Cabinet and became official policy. The Dutch were therefore told that the Colonial Secretary was 'in favour of declining to participate but saying to promoters that HMG prefers to express no opinion as to the action of private owners of tin'.[76]

That was only a green light for a further round of negotiations, since the pool could only be formed if there could be some assurance that the restriction agreement itself would not be terminated before the pool was liquidated. While an open-ended commitment was not feasible, three years from the formation of the pool seemed a reasonable period. Until that date, or the prior liquidation of the pool, members of the ITC would have to waive their existing rights to withdraw. Malaya gave its consent without difficulty; Nigeria only did so under pressure.

The Bolivian government was not interested in participating itself. It had received an extraordinarily frank and pessimistic report on the tin situation from its Chargé d'Affaires in London, who considered that the market had far from bottomed out and the price could easily drop to £80–£90. He thought it was unlikely that further restriction would be implemented so that it would be years before the stock overhang could be eliminated. He was also sceptical that financial resources could be found to support the pool and warned that the opposition in Malaya was determined to break the scheme. Such pessimism was reinforced by a criticism of the economic rationale for restriction.[77] However, the government was prepared to give the commitment required to allow others to form the pool and agreed to:

> adhere to restriction agreement until end February 1933 or in case pool not liquidated August 12, 1934. Undertakes that until then will accept and enforce any decisions arrived at by the ITC. This undertaking given on the understanding that the ITC will examine all suggestions for alterations in quotas in an objective manner, without taking into account the special interests of the ITP; and that the Committee's decisions will be based upon an impartial examination of all the existing data, on its estimate of the trend as regards visible supplies, production and consumption.[78]

It had sensed the same problem as Passfield, though it remained to be seen whether this proved to be a solution.

The Nigerian government, however, considered Passfield's objection as fatal. It needed no prompting in the light of its memory of the distortions wrought by financial interests. All it could see was 'miners prohibited from stocking, yet pool holders make large profit', and felt that this would introduce a fundamental conflict between the restriction levels aimed at by the pool holders and the working miners, with the latter being sacrificed to the benefit of financiers.[79] Since the Colonial Office could not bring any pressure to bear, it needed Howeson, working through the Nigerian Chamber of Mines, to persuade the government to withdraw its objection.[80]

Separating governments from the actual discussions and the thinking underlying these pool proposals was clearly making the administration of restriction difficult and Campbell mused:

> One feels that given plenipotentiaries who know the facts and trusted each other and could take immediate decisions and had in fact the confidence of the Governments the thing would work fairly well. But one never gets such conditions. And the more I see of this relatively simple case of restriction, in actual practical operation, the less inclined I am to have anything to do with restriction schemes![81]

The subsequent experience of the ITC would provide reassurance for nervous governments, eliminating some of the distrust. But the formation of the pool encountered another source of distrust, as it tried to make Patiño deliver on his commitments.

The size and shape of the pool underwent some modification in the early summer of 1931. Restriction was not yet having much bite, largely as a result of administrative weaknesses in the FMS. As the level of visible supplies continued to increase so was the target, which was now set at a minimum of 30,000 tons to be distributed as follows: British Syndicate 10,000, Dutch 12,500, Bolivian 7,500, with the amounts being supplied from current stocks, market purchases and future production.[82] However, the NEI government was not prepared to allow its syndicate to have a higher level of exposure than the British. It was still smarting from the Malayan criticism of the results of the distribution of quotas and was not willing to be vulnerable to any allegation that it had an unfair advantage.[83] Since the limits of the British position proved to be 10,000 tons, the target was reduced to 27,500 tons.

As soon as the pool moved towards actual implementation, Patiño disappeared. As Campbell commented:

> There is a strong reason to think that M Simon Patiño is trying every dodge to avoid signing the provisional agreement as to the 27,500 ton tin pool. He has left Paris: his whereabouts are unknown: he is said to be returning on dates when he does not in fact return: and so on …

He then raised the question of alternative plans in case the 27,500 tons could not be assembled:

> Howeson and the Dutch felt that they should go on with the pool at 22,500 tons. I pointed out that they had emphasised that 30,000 was really the minimum: was 22,500 any use? … They felt that the market wanted support very badly; and they would rather risk the 22,500 not being enough than give no support. The 22,500 comes from British-10,000: Dutch – 10,000: Bolivian – 2500, on the original promise, *unconditional*, made by M. Simon Patiño.

But, of course, since Patiño was refusing to meet on the 27,500 ton pool there was no guarantee that he would complete the agreement for the reduced one of 22,500 tons:

> In that case, the whole thing would drop; the Dutch, particularly, are now so suspicious of the Bolivians that they would not come into any pool unless Bolivian tin-actual tin in a warehouse – was there to guarantee their good faith. They even have some doubts now whether Bolivia means to remain in the control scheme-though they cannot understand why it should wish to withdraw. Howeson agrees – and is equally unable to supply any reasonable motive for Bolivia's attitude … If he [Patiño] goes back on his promise to put up 2500 tons in any case – a promise renewed before me twice at Paris –the whole thing drops absolutely … My own guess is that the probabilities are strongly in favour of it breaking down. That was the opinion of Mr Howeson and the Dutch also – though they still hope for the best.[84]

Patiño was not prepared to participate in the absence of either the producers or the government of Malaya[85] and may have simply preferred to lie low to prevent this decision from jeopardizing the plans of the British and the Dutch. However, it came at some cost:

> The Bolivian attitude – and their conduct of negotiations – have produced a most unsatisfactory impression on the Dutch, and on the British syndicate. The Dutch now feel that they cannot trust the Bolivians: they have therefore begun to have doubts as to their continued adherence to the control scheme. They feel, also, that the pledged word of the Bolivian Govt: is something which will have to be scrutinised, and somewhat askance at that.[86]

In spite of Patiño's refusal, the British and Dutch went ahead, with the amounts shared equally between them but without much confidence that the now much reduced pool would be adequate to affect the market.[87] However, given the dismal production results, especially as a result of Malaya's overproduction, some such gesture was certainly required. On 24 August a communiqué was issued announcing the formation of the International Tin Pool (ITP) and its schedule of release prices was established. If the spot price averaged £150 over a calendar month, 5 per cent would be released, rising by another five percentage points for each £10 increase. While no target was specified, the market was periodically informed of the total amount of metal held.

While the formation of the ITP did not reshape the nature of the overall agreement, it did create some formal anomalies. The ITP was a consortium of very different interests, only one of which was simultaneously involved with both the ITP and the ITC. Some had interests in tin mining, while others were simply speculators. In spite of Passfield's desire to keep the ITP quite separate from the ITC, the two were closely linked by making Campbell Chairman of both.

The two syndicates were constructed in a quite different manner. The Dutch Syndicate was simply formed by Billiton (BM 1,000 tons, GMB 3,167 tons) and Banka (8,333 tons).[88] BM transferred the 1,000 tons it had bought in 1930, replacing this with a further 1,000 tons.[89] Banka could draw on the 3,000 tons already shipped to Amsterdam and had another 630 tons in storage in Singapore but the remainder had to be supplied from current production.

The British Syndicate was made up of several disparate interests. The bulk came from those who already held metal. British-American Tin and BMC supplied the lion's share at 7,557 tons, followed by CGF (501 tons) and individual speculators, Sir Hugo Cunliffe-Owen (1,001 tons), Yvonne Le Baron and Leon Cotnareanu (1,000 tons). Some came from diversion of current production and included Central Mining (500 tons), Hochschild (500 tons) and Aramayo (601 tons). The remaining 2,939 tons were then bought on the LME, of which 2,670 came at an average price of £118/12/–, with funds supplied by Dean Finance (Cunliffe-Owen), Prudential Assurance, Whitehall Trust and Lazard Frères.[90] The coalition represented here largely reflects the range of contacts that Howeson had assembled over the previous few years.

The crucial role played by Howeson in establishing the ITP provided a good opportunity to test the extent of his integrity and as far as Campbell was concerned he passed it in an exemplary fashion:

> The Dutch trust him now – they told me so, though they did not originally … He brought that telegram [advance information concerning Clementi's proposals for participation in the pool] at once to me, and asked what he should do with it. I told him to keep it entirely to himself till I had official infmn. He did so – disclosing it to no one. The Dutch did not know – I verified that later. It was to his interest to disclose it. I have watched Mr. Howeson – starting with a fairly strong bias against him – and I have found absolutely nothing to complain of.[91]

Four years later, Campbell would be forced to qualify that verdict but in the interim Howeson had established a good working relationship which would help the ITC respond to additional challenges.

Research

Once the essential elements of the restriction scheme were in place, the ITC turned its attention to the research programme. The relationship between TRIAC and BNFMRA deteriorated as the latter became increasingly 'suspicious of the bona fides of the TPA' and felt that interest in research was likely to 'fade away' once the price had recovered.[92] However, the commitment to a strong research programme was soon strengthened as both Bolivia and the NEI followed the British lead and decided to participate, though in the former case the contribution was borne entirely by the industry.

By the end of 1931 TRIAC was formally wound up[93] and a new organization was created, the International Tin Research and Development Council (ITRDC). In addition to assuming the responsibilities of TRIAC, the ITRDC also began the task of compiling and publishing comprehensive statistics on production, stocks and consumption. Given the need for some symmetry between the British and the Dutch, the technical functions of the ITRDC were largely concentrated in the UK, while Billiton managed the statistical functions from The Hague. Meetings of the ITRDC would be held at the same place as meetings of the ITC. While there was considerable overlap in the delegations to both organizations, the ITRDC became an independent body and one whose fate would not be tied to the ITC itself.[94]

In spite of the pressure created by the deterioration of the tin market, it had taken well over a year between the point at which it was recognized that an effective solution to the problem of tin would require co-operation at the intergovernmental level and the point at which such co-operation was in place. But it was now operating in a much more comprehensive fashion than was originally envisaged. Five governments covering over 90 per cent of the world's capacity had been recruited. They had all agreed to give up their right to break away from the cartel to allow a solution to the problem of excess stocks. Four of them were prepared to make an investment in research to secure the industry's future over the long term. It now remains to be seen just how successful these measures were in actually solving the problem of tin.

8 Rescuing the tin industry, 1931–1933

From the Spring of 1931, it was evident that the ITC had an extraordinary challenge ahead. Stabilizing the industry would involve three stages: getting production to the level of consumption, getting it below consumption and, finally, getting it sufficiently below consumption that the accumulated stocks could be liquidated. As consumption continued to fall throughout 1931 and 1932, the magnitude of the problem simply increased. However, as the ITC began to take effective control over the tin market, it attracted more and more attention as not only setting a model which could be fruitfully applied to other commodities but also as an important weapon in the arsenal of counter-depression measures. Getting to that point required several difficult decisions.

The second cut – 20,000 tons (34.5 per cent)

Although it was clear from the outset that the 22.3 per cent cut would be quite insufficient,[1] the ITC was bound by the framework agreement that this initial level would be sustained for at least three months. It was therefore not until May that the ITC addressed the question of the kind of cut that was really needed. Lazarus took the initiative in the discussion and estimated consumption at 137,000 tons. With current levels of production, stocks would continue to rise by a further 14,000 tons by year end. He proposed increasing the cut from 22.3 per cent to at least 42 per cent, an annual increase by 32,000 tons which would at least begin the process of reducing the accumulated stocks. This position was slightly modified on the basis of comments from Pearce and Howeson who offered a lower consumption estimate of 130,000 tons but queried by Houwert on the basis of the Dutch sales experience. The meeting now had to address the implications of asking governments to impose a further cut of between 18,000 and 25,000 tons for 1931, before the scheme itself was working at all smoothly. To make the recommendation more palatable to governments, Campbell excluded Howeson and the other representatives of the TPA from any further participation and the meeting ended with an endorsement of the more cautious Dutch position, recommending a cut of 20,000 tons to be implemented as soon as possible.[2]

Since the second cut started on 1 June and Malaya was working on a six month basis, the level of overproduction, already evident in May, would become even more serious. When that first six month period expired in August, the total excess exports for the FMS was 4,902 tons, or 18 per cent of her quota, 8 per cent of the total for all four signatories.[3]

Since the whole point of the restriction scheme had been to establish a satisfactory level of control over Malaya, its failure could not have been more serious. It came at the worst possible moment since it coincided with the political crisis of late August which saw the end of the Labour government. Nervousness about the future of a sterling commodity was already leading to liquidation of speculative accounts in Paris and market sentiment had turned very strongly against tin. In such an uncertain context it was perhaps inevitable that Malaya's failure was not simply regarded as one of the teething pains of a new and complex scheme but as an indication of her complete inability to carry out her obligations. Such a judgement could easily lead to a wave of speculative selling and the ensuing maelstrom would catch those who had been planning to provide the necessary finance to complete the British syndicate's share of the ITP.[4]

In spite of the urgency of the problem, dealing with the Malayan excess ran into several obstacles. As Clementi was brought into the picture, he expressed his determination to take drastic measures and get legislative authority to hold up exports and prevent the excess production from spilling over into the market.[5] Unfortunately, there was no space for the storage of concentrates at ports within the FMS and Clementi made rather vague plans to hold up the excess at the Straits smelters and to liquidate by the end of February.[6] In doing so, he appeared to be indifferent to the harm created by the continued fall in price caused by this excess production and the fact that the ITC had, so far, signally failed to impose an effective level of restriction. As High Commissioner, he was as much concerned with the political implications of the severe levels of restriction that would be required by a more rapid rate of liquidation. He had been appointed in order to deal with the Chinese question and was particularly concerned that a sudden drop in authorized production would bankrupt many marginal Chinese mines simultaneously. The prospect of mass repatriation of Chinese workers was not easy to contemplate, either politically or financially. To provide some kind of cushion for the industry, Clementi was not prepared to authorize a local cut greater than 60 per cent on the most recent assessment.

As far as the Colonial Office and the Dutch were concerned, there was only one solution and that was for the FMS government to buy metal and liquidate it as domestic production was brought down. Given the financial constraints of the government, it had to be done as cheaply as possible, so arrangements were made with the ITP to sell and repurchase the metal. This was by no means easy given the way in which the government had been unable to enforce its earlier obligations. Nor was it easy at the Malayan end, since Clementi baulked at the prospect of reopening a debate on restriction in the Legislative Council to secure the authorization necessary to finance this arrangement and had to be overruled by the Colonial Office:

> While appreciating the embarrassment it will cause you, I have decided that the purchase scheme must be adopted, I am now taking measures accordingly … The official majority must be used to sanction the expenditure required if necessary.

This led to a testy exchange of cables, as Clementi initially refused:

> I must with all respect remind you that Federated Malay States are not a Crown Colony and to commit Federal Council without consultation to the purchase of 4,000 tons of tin is unconstitutional. Such action is moreover most impolitic when active discussion of decentralisation proposals is in progress … There is no obligation on Federated Malay States to buy this Tin under International Agreement. I submit that Bolivians and Dutch cannot dictate in this manner to Federated Malay States … Difficulties created by unconstitutional action will far exceed any advantages which may accrue.

Whereupon, the Colonial Office responded in the strongest possible terms:

> Political difficulties in Malay States have not been overlooked ... Central fact in situation is that Malaya by failure to carry out her obligations has put herself in the wrong and it is absolutely essential to extricate her from a position in which she stands charged with breach of faith. It is unnecessary to remind you that this charge inevitably reflects not only on Malaya but on also on His Majesty's Government. We must at all costs put ourselves right with the world.[7]

The tin world looked very different in London than in Malaya. As far as London was concerned what was at stake was not simply the future of the tin agreement but the whole moral fabric that enabled such agreements to be made in the first place. It was not insensitive to the prospect of social dislocation and political chaos but these were risks that simply had to be run for the sake of a more important game. Such risks were naturally weighed more heavily by those who would actually have to deal with them but the fact that Malaya was several steps removed from the main centre of the tin market made it difficult for her to appreciate its particular fragility. One potential link was STC. However, this was most unreliable not simply because of Bagnall's expressed antipathy to restriction but also because he was quite prepared to use information he received from Clementi for his own purposes.[8]

Clementi was in fact caught in a very difficult situation. The other mainstay of the Malayan economy, rubber, was also in serious trouble and it was even more imperative to raise the price of tin. The only immediate device to hand was to increase the size of the ITP and Clementi again sought permission to participate. It would provide some justification for the guarantees Malaya had given which allowed the formation of the ITP but above all it would help re-establish Malaya's prestige vis-à-vis the Dutch, so sorely tarnished as a result of the overexports.[9] But the Colonial Office gave no indication that it would encourage Passfield to revisit his earlier position in the light of these practical arguments. Groothoof even tried to get McKenna to intervene but in late August 1931 the Cabinet was in no position to reconsider its earlier decision.

Some time before the problem of Malaya's excess became particularly acute, pressure was building from several quarters, including Howeson and Mair, for yet another cut, this time of at least 10,000 tons.[10] However, it was not one that the Dutch were prepared to contemplate, at least not while Malaya was still unable to demonstrate that it could deal effectively with the problem it had created for them.[11] That in fact came quickly. In September, Malaya wiped 1,220 tons off her excess and was restored to the good graces of her partners.[12] It was then possible to consider the next cut.

The third cut – 15,000 tons (44 per cent)

As was so often the case, the ITC found itself trying to catch up with a deteriorating situation. Unfortunately, the constraints imposed as a result of its governmental character prevented any speedy solution. Howeson tried to bring pressure by identifying the widespread support among Malayan producers for the new cut and although Campbell readily conceded that there was a strong statistical case, the situation was far more complicated and, in a rapidly changing situation, he disliked 'taking decisions further ahead than need be'.[13] It was not until October that the issues could be joined at the next ITC meeting. Calder was obligated to present the position of Nigeria and Malaya, both of whose governments had been consulted prior to the meeting, and neither of which was prepared to support a further cut. In arguing for the status quo, Calder stressed that the rise in the visible supplies was attributable to the abnormal factors associated with the delay in implementation and the Malayan excess and that at the current level of production and consumption, stocks would be liquidated at the rate of around 2,000 tons a month.

The Bolivians and the Dutch had a very different perspective. Devaluation of sterling in September had created an important cleavage within the ITC. Since the bulk of world demand came from countries remaining on gold, the immediate effect of the devaluation was to raise the sterling price of tin but the gold price continued to decline. The Dutch remained on gold and while Bolivia was far too dependent on tin exports to the UK not to follow the devaluation of sterling, half her imports came from gold countries. It is therefore not surprising that the Bolivian delegation boldly declared that the objective of restriction had to shift from establishing an equilibrium between production and consumption to one of raising the price.

Billiton provided a detailed statistical assessment which challenged the naivety of the assumption that consumption would remain at its current level; indeed it predicted that it would drop from 12,000 to 10,000 tons a month. At that rate the reduction in stocks would be closer to 4,000 tons per year and, with surplus stocks at around 30,000 tons, a further cut of at least 15,000 was required if any real improvement in price were to be realized. In this it was supported by both Pearce from CTS and by Howeson and the TPA.

Governments had to deal with the social implications of any further cut. While Malaya would simply prolong the domestic cut already imposed to deal with its excess, it was fearful about the rate at which this would force out the Chinese. Nigeria was more concerned about the criticism from producers who would experience a disproportionate increase in the cost of production. While this was an argument easily refuted by the producers present, Baddeley could not get beyond the simple formula that the cut would cause great hardship.

The ITC was now running into some constitutional difficulties. Delegates were speaking on behalf of governments. In the case of Bolivia and the NEI, these were associated with dominant companies who knew what support they could count on from their respective governments. In the case of Malaya and Nigeria, these were civil servants, governed by bureaucratic conventions of consultation with governments which had to take into account a variety of considerations. Since their principals had already spoken, delegates felt that even endorsing a recommendation would compromise their relationship.

However, since the mandate of the ITC was to regulate the industry on the basis of the kind of analysis presented by Billiton, the issues raised were presented to both of the recalcitrant governments.[14] Nigeria was prepared to concede, though it insisted on one proviso, namely that it be for only eight months, after which quotas would revert to their current levels.[15] Malaya remained unco-operative but capitulated once Nigeria had conceded.[16] This bureaucratic inflexibility had cost the ITC a full month and the new cut could only be endorsed at another formal meeting, to be implemented from 1 January 1932.

Although governments had exercised their ultimate power on this occasion, it had become clear that for such power to be effective it had to be based on a sound grasp of the issues that would be thrashed out at the level of the ITC. Delegates could then work with a clear negotiating brief from a position that would command some respect. Nigeria was particularly deficient in this regard, as Campbell commented:

> Their reply is almost as cryptic as a Delphic oracle. They give no reasons: attempt no solution … they merely instruct [their delegates] 'not to support a cut for an indefinite period' and leave it at that. … All the delegates resent this sort of treatment: it is unfair

> both to them, and to the other participating Governments. And this kind of thing must provoke resentment on the part of these other Govts: and must militate most seriously against the smooth working of a scheme which has substantial enough difficulties to encounter, without adding to them by this indefensible method.[17]

With an understaffed Mines Department, which was chronically overworked in the simple administration of tin restriction, formulating negotiating briefs was beyond the capacity of either colonial government but that would not restrain the way in which they continued to try and exercise the power that was formally theirs.

1931 – an assessment

Just as the TPA had dismally failed in its efforts to correct the tin market in 1930, the first ten months of the ITC showed that it was even less successful. Table 8.1 compares the actual experience of 1931 against the objective established when the scheme was first formulated in November 1930.

In retrospect it appeared that three estimates were too optimistic: the extent of outside production, the terms acceptable to Siam and the level of consumption.[18] Thanks to its own internal politics the ITC had been unable to cut its own production to compensate for more than the underestimates on the production side. The overestimate on the consumption side spilt over into the visible supplies, which now stood at a truly daunting figure of 51,000 tons or 136 days supply. At least the year had been a fruitful learning experience and as the tin market continued to deteriorate throughout 1932, the ITC had a much better sense of what was involved in dealing with it. Both production control and stock control would be pressed to their limits.

Just as the best that could be said about the TPA policy for 1930 is that it prevented an even more severe drop in price, so the same could be said about the ITC. The price history is presented in Figure 8.1.

March was more buoyant thanks to the expectations held about the impact of the new ITC; the period from April to June reflects the recognition that the first cut was too low. A boost was given with the announcement of the second cut, the participation of Siam and the formation of the ITP but from then on the course was inexorably downwards, with only the

Table 8.1 ITC performance, 1931

	Plan	*Actual*	*Difference*
	tons	*tons*	*tons*
Malaya	53,925	52,642	
Bolivia	34,260	30,397	
NEI	29,910	27,143	
Nigeria	8,001	7,222	
Total Core ITC	126,096	117,404	–8,692
Siam	7,685	12,430	4,745
Other	11,219	15,636	4,417
Total production	145,000	145,470	470
Consumption	145,000	136,395	–8,605
Visible stocks	42,140	50,954	8,814
Days consumption	106	136	

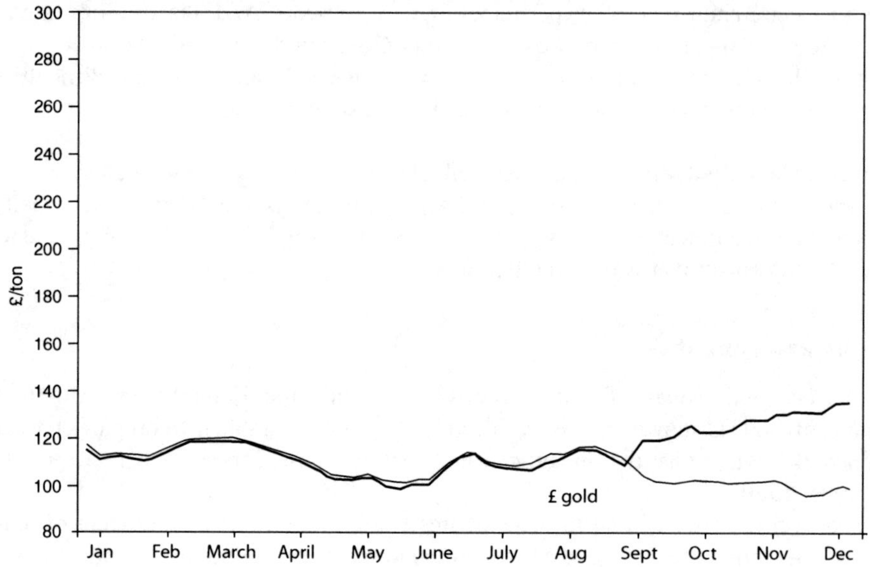

Figure 8.1 Tin prices, 1931

appearance of an improvement thanks to the devaluation of sterling. At least it did not fall to the level of £50 that was otherwise anticipated.[19]

1931 political crisis

The financial-political crisis that beset Britain in the late summer of 1931 had an extremely important outcome. The new National Government brought Cunliffe-Lister back into office, first at his old post at the Board of Trade and then in the reshuffle that followed the General Election of October as Colonial Secretary.[20]

With the change in personnel came a change in economic policy. It was imperative to ease the pressure on Britain's balance of payments and while tariff policy was the most important weapon in that struggle, higher prices for commodities sold outside the sterling area was another. The ITC could therefore confront the difficult task ahead confident that it could count on full support from the highest levels of the British government.

Mair's challenge

Confronting that task was complicated by the fact that by the end of 1931 Mair had found new grounds to challenge restriction. The fundamental asymmetry in the standard tonnages was now exacerbated by two factors: (1) devaluation of sterling had changed the comparative cost conditions, making the Bolivians and the Dutch more dependent on high prices for their survival; (2) consumer preference for particular brands was not matched by the supply, so stocks of Banka and standard were piling up but not Straits. Not only could Malaya produce economically at current price levels but there was no shortage of demand for her particular brand of metal.[21] If there were no shortage of demand then a further cut would mean a shortage of supply, forcing US consumers to turn to other brands and ending the privileged position of Straits tin.

Behind this lay the fact that the Straits premium in New York had generally declined since devaluation. While this could be interpreted as a result of the comparative stock position, it could equally be a result of technical improvements in the quality of standard tin which had encouraged the Welsh tinplate manufacturers to switch from Straits. Stock movements are also the result of price changes and, with a low premium for Straits in New York, American demand for standard was reduced, so that standard stocks accumulated in Liverpool. While reading the stock and price statistics was not at all easy, it was evident that Mair had overstated what could only be a parochial case.[22]

The obsession by Mair and the Cornish with the preservation of the comparative position of Malaya in the world tin market had made them not only insensitive to the need to preserve the comparative position of the market as a whole but also to the implications of their carping for the attempts made by others to save that market. Already fixated on the 'unfairness' of the allocation of standard tonnages, they now added another about the future of Straits tin and the premium it enjoyed. When the pool was added to this mix, Mair's imperial conscience led to the following explosion:

> The industry without consent has been handed over to a group of unknown financiers ... Absolutely wrong to allow countries – countries moreover without the British Empire – overproducing unwanted tin to harass and interfere with countries within the Empire which are producing tin that can be absorbed.[23]

Mair's very public statement simply confirmed the impression current in many financial and mining circles that the ITC had not yet learnt how to exercise its responsibilities:

> The ITC has shown itself singularly unable to appreciate the complexities of the tin trade thus far. We still maintain that an essential portion of their task is to create and develop an entente with the merchanting and consuming sides of the industry and to win their confidence.[24]

Whether this was a reasonable expectation in the light of the actual situation facing the five delegations that made up the ITC is another matter. It was not one that was easily addressed. Van den Broek raised the question as to how the ITC should meet the challenge that had been flung down and Campbell advised:

> It was always undesirable for an official body, such as this Committee, to attempt to controvert such statements, or to enter into discussions as to the merits or demerits of the action they had taken. That only provided fuel for further misrepresentations and discussions. The Committee could not carry the arguments beyond a certain point, and their irresponsible critics would then assert that they had reduced the Committee to silence.

It was then agreed that the best way for such rebuttals to be made was through the means Mair had used, addresses by sympathetic Chairmen, such as Stephens, at the annual meetings of their companies.[25] As a result, the public debate among the various factions was conducted in the rather bizarre form of Chairmen's speeches which were then widely reprinted in the mining and financial press.

Mair pressed yet another issue, namely the absence of any direct connection between the ITC and the Malayan mining industry and offered the following olive branch:

> I am anxious to strain every way I can, so that if possible we may end any quarrels … and establish peace. There is in several quarters a considerable sense of injustice: In anything where men are concerned if that sense of injustice is not removed it may at any time burst into flame. It would be so simple a thing to stop this. I am not aware of anyone either in England or the F.M.S. connected with either of the Chambers that wants to wreck the Quota Scheme, but having regard to what has been done … there is a strong feeling of injustice that there is nobody on the International Tin Committee representing Malayan tin miners. I cannot imagine what reason there *can* be against what is to me an entirely reasonable request that on the International Tin Committee there should be somebody there who is a Malayan miner … seeing that Bolivia and the Dutch have such representatives.[26]

Howeson and Calder felt the proposal had some merit but Campbell thought otherwise; in part because bureaucratic protocol meant that such matters had to be settled through Clementi but he also added:

> To my mind, it would be merely stupid to put on one of a party which has hitherto criticised the whole scheme with the utmost venom … The scheme is a fait accompli: it is there – clear, precise, obligatory. The thing has moved into the governmental sphere now; and the experience of the 'Malayan miner' does not seem likely to be of any particular help … I'm always prepared to build a bridge of gold, if need be, for the powerful enemy who wants to retreat; but I see no point in building a bridge to let a declared enemy … walk into my stronghold, and quite probably cause endless trouble there, without being able to do anything more than cause such trouble.[27]

Recognition of the real issues that had to be addressed was simply lost in this personal invective.

Completion of the International Tin Pool

Weakness in the overall position of the ITC was not easily compensated by strengthening that of the ITP. With the third cut came a revision of the release schedule of the ITP.[28] The Bolivians attempted to get the pool to work with the old schedule but on a gold basis, since that was likely to prolong the time required before liquidation and hence the agreement itself. The old gold release price of £150 was a new sterling £190. But the Dutch were prepared to make a major concession and raise the new release price to only £165. Not having participated in the formation of the pool, the Bolivians found they had no leverage with which to press their position.[29]

The change in this schedule made the ITP less attractive to outside speculators and the extent of the financial support available to the British Syndicate was reduced by some £250,000.[30] That in turn limited the amount that was actually bought to 2,670 tons and most purchases were concentrated in November. They were not sufficient to provide any real boost to the price.[31] Over the 88 days on which the LME was open between the end of August and the end of December, the pool bought on 31 and of these the market rose on only 17. Since the market actually fell on 12 days on which the pool was buying, the average rise over the previous day's price was just 0.5 per cent, only slightly better than the 0.1 per cent average rise for the days on which the pool was inactive.[32]

By the end of 1931 the British Syndicate was able to meet the minimal commitment that it had made to the Dutch of 10,500 tons and a formal announcement could be made that the ITP held 21,000 tons. However, 25,000 remained the operative target and with the new cut from January 1 the Dutch needed more time to make their contribution from their now reduced supply.[33] Patiño was again approached to see whether he was now prepared to make the earlier promised contribution.[34] Since Cunliffe-Lister could get the Cabinet to change its earlier policy, he had a 'good exchange' with Clementi[35] who agreed to take 2,000 tons, matched by the Dutch.[36] Unfortunately, his local advisors, apart from the Senior Warden of Mines and Hake of the TPA, were not prepared to support him and the matter had to be dropped.[37]

It proved to be impossible even to retain the 1,150 tons of the Malayan excess that the ITP was still holding and there was no basis on which the Colonial Office could exert any further pressure on Malaya. Campbell considered that there was no moral obligation to participate and that it was simply a business proposition in which the high probability of the gain in royalties had to be balanced against the costs of carrying the metal and the slight possibility of a failure of the ITP.[38] On this occasion, the Malayan assessment of that balance prevailed over that made in London. Had Malaya been recruited, then Patiño would have followed and Howeson could well have found the further funds needed to complete the pool. But 21,000 tons proved to be its limit.

In fact, it turned out to be even lower. Hochschild proposed that his contribution take the form of slags of 8 per cent tin, which the other members thought quite unsaleable.[39] Hochschild continued to stall and his movements made it difficult to pin him down. In spite of the fact that the slags were later considered marketable, the pool members insisted on actual metal and Hochschild never made good on his commitment.[40] Since the British Syndicate actually had warrants for only 10,000 tons and the Dutch had the right to reduce to a similar level, the actual amount firmly frozen was 20,000 tons.

The ITP came in for some criticism, on account of the secrecy that 'shrouded' its membership and operations.[41] While some members like CGF and Aramayo were quite prepared to publish the size of their participation,[42] it is not clear that others could have been recruited if their identities were to be revealed. As far as its market operations were concerned, these were purely technical. Given the difficulties in securing Hochschild's commitment, it is probably wise that the market was kept completely innocent of yet a further complication in completing the machinery for effective tin control.

The fourth cut – 20,000 tons (56 per cent)

With the ITP stuck at an inadequate level, all further proposals had to address both production and stock control. Although the third cut had yet to show any results, 1932 began with a recognition that it was still too low. As consumption declined in the United States, deliveries declined at an even faster rate as users drew down on their existing stocks. From an average of just under 6,000 tons a month from July to October deliveries had dropped to just over 3,000 from November to January. As a result, all the new cut would accomplish was a simple balance between production and consumption, with no effect on the stock overhang.

On this occasion it was Patiño who took the initiative. Production costs were rising in relation to the rate of restriction; reserves were being depleted at a loss so the situation now needed much more radical measures. A short suspension rather than a longer reduction would not only give the market its needed boost but would be more economical for most producers. Suspending production for a month would take 9,000 tons from the market and with another increase of 9,000 in the holdings of the ITP, the remaining visible stocks would be brought down to a normal level, at which point producers could expect 'an increase of certain importance, if not appreciable in the price of tin'.[43]

These measures were far too radical for any of the other members. However, there was no contesting the diagnosis and the two month target of around 18,000 tons was accepted by the other delegations. That meant discarding the rule that had been adopted but two months earlier when the endorsement of the third cut was made conditional on there being no further cut for at least eight months. Discussion focussed on the way in which it would be implemented. Increasing the size of the ITP was contingent on an increase in its scale of release prices.[44] Since that proved to be unfeasible, it was dropped. The suspension part was not supported by any of the other members and this was then converted into a further reduction on the quota to take effect from June 1. The date had to be delayed until then since Nigeria, in particular, was not prepared to implement any further cut until Malaya had wiped off her excess. The Nigerian government was particularly incensed at the failure of the Malayan government to implement restriction effectively; after all, it had very similar problems associated with imposing high levels of restriction in a differentiated industry but had not capitulated to any internal pressure by allowing excess exports.

Patiño replaced the ITP part of the proposal with one for a month's withholding of production, to be released at the rate of 5 per cent over the following 20 months. This, too, met with insufficient support. In Malaya, the Senior Warden of Mines argued for it but he was overruled by the Finance Committee of the FMS Federal Council. It was sceptical of the very concept, since withholding might only produce a temporary fillip followed by an even lower price level and would not make a permanent dent in the stocks. But behind this line of argument lay a sense that the limits of control had already been reached in Malaya. Clementi was not prepared to accept a cut beyond the level already being imposed and in any case wanted to see how other participants would cope with a similar level at current prices, 'in the belief that such countries will not produce full quota'.[45] This was but a restatement of the old concern with the inequitable principles of the distribution of restriction and could hardly be more irrelevant to the current situation. In rejecting this proposal, no fresh thinking emerged from Malaya; indeed, Clementi indicated that he was not really open to any new ideas: 'If present proposals were adopted precedent would be constituted for further withholding and any psychological effect on markets of adoption of present proposals would be based on inference that government would come to rescue again in future.'[46]

The Dutch responded more constructively with a proposal to put this month's production into a government pool, independent of the ITP but with the same release schedule. It met with strong support from the Colonial Office which saw advantages not simply through taking charge of a large portion of the excess stocks but it would also 'bring the participating Governments together in a co-operative endeavour which would be direct and public … and the psychological effect of this on price is likely to be considerable'.[47]

Government commitment to stock control would provide greater reassurance than its less dependable commitment to production control. But none of the other partners were persuaded. Not even the offer of a bank loan was sufficient for Malaya to set aside its nervousness[48] and that at least confirmed the premise on which the earlier recommendation had been made.

Patiño objected to the proposal because it would require government financing and would deprive the producers of their freedom of action.[49] But he returned with another plan which met with full support from the other delegations. Governments were to withhold one month's production for four months and then release it at the rate of 5 per cent per month, with provision for a higher rate of liquidation on the same release schedule as the ITP. To avoid the mistake of the ITP, any metal bought would have to come from the visible stocks.[50]

In this long drawn out process of determining the shape of what was hoped to be the final cut, one problem was revealed. Someone in the Ministerio de Hacienda leaked the contents of Patiño's cabled report on the February ITC meeting and it appeared in the *Daily Metal Reporter*.[51] Fortunately the terseness of the cable prevented a recognition of the difficulty the ITC was having in formulating a comprehensive solution. The fact that the problem did not arise again is indicative of the ability of the members of the ITC to sustain a complex pattern of international consultation in confidence. A price was to be paid in the reputation of the ITC as being closed and secretive but without it the prospect of reconciling very different perspectives would have been remote.

Malaya alone stood against this new plan and all that was left was for the ITC to recommend that the two components be collapsed together and become a simple reduction in the quota by 20,000 tons to take effect from 1 June. That recommendation was made public without proper consultation with Malaya. On learning of the decision, Clementi responded to the Colonial Office's explanation in the strongest terms he could find:

> The fact that, as stated in your telegram, 'the condition of the tin industry has deteriorated since July 1931', had not escaped my own observation. But the price of tin and the condition of the tin industry is only one of many facts which the Malayan Governments have to consider; and in my telegram of 24th July 1931 I had informed you that a local cut of 65% would be both economically and politically disastrous. The local cut now to be imposed … will be 67% and the Malayan administrations have been committed to it in great haste, without adequate opportunity for consulting the local tin industry, and without consideration being given to any except statistical arguments.

Not only did the position taken the previous July still stand but Clementi had repeated his insistence that no further cut could be authorized without his approval on two recent occasions.

He then went on to point out that in response to earlier suggestions for a further cut, he had arranged meetings with various miners' associations but that the process was pre-empted by the announcement of the ITC decision:

> I may state here in the opinion of the Senior Warden of Mines that had the subject been been allowed to proceed on consultative lines, a majority recommendation in favour of the cut would have shortly emerged owing to the lead by the Selangor Chinese. The odium that has since accrued to the FMS Government, by reason of what the *Malay Mail* has characterised as 'a hypocritical show of consulting those who would have to bear the brunt of the new cut', would have been avoided and any political repercussions averted.[52]

What made the situation particularly intolerable was the way in which the guarantees that Malaya had given to the ITP in August were now being used to constrain her own freedom of action.

While Clementi recognized the fait accompli, he tried to secure a promise that never again would the Malayan delegation be allowed to act as plenipotentiaries. Cunliffe-Lister had clearly done no better than his predecessor in establishing an effective administrative structure which linked the Colonial Office, its representative in Kuala Lumpur and the Malayan mining industry. One obstacle to this was the absence of a radio-telephone connection.[53] Whereas the Dutch routinely picked up the phone to speak to Batavia, the British were constrained by the terseness of the telegram and the formalism of the dispatch. Without a more direct form of communication each party could rest content with a stereotype of the insensitivity of the other.

The final cut and the Byrne scheme (66.6 per cent)

While governments had failed to find a point at which they could unite around a rapid solution, at least the issues that had to be faced had been given a thorough airing. They were now reshaped into a proposal that emerged from the industry itself. It was presented by Byrne but with the support of Thomas, Mair and Glenister and hence could be considered as a united Malayan proposal.[54] It had three features: (1) an immediate two month suspension; (2) resumption with a quota of 40 per cent for ten months (equivalent to 33 per cent spread over the 12 months); (3) a plan for the liquidation of the ITP. The first would bring the level of visible stocks outside the ITP down to normal levels. The second would cut production below consumption. The third would restore a balance between the two met, in various proportions, from increases in quota and releases from the ITP tied to a pivotal price of £200–£220.[55]

The Byrne scheme addressed the major weakness with the current governmental character of the ITC. The pace at which governments were prepared to move was simply too slow to provide the companies with the benefits they expected in relation to the costs of restriction. The experience of the past 12 months had exhausted the patience of many and Inchbald fulminated:

> From first to last, instead of anticipating events, the ITC has been content to follow in their wake, always six months behind the Fair, always too late ... those in control incorrigible optimists ... not too willing to take advice from more experienced men who have been connected with tin all their lives.[56]

Now Malayan producers were forcing the pace. It was one that was required if the fundamental problem were to be solved before the agreement expired in August 1934 and before the uncertainties about renewal loomed large.

The third feature of the original Byrne scheme created two problems. It constrained releases from the ITP until quota levels were raised to 60 per cent and the price sustained at £200 but without a reciprocal commitment to continue restriction until the ITP was liquidated. It also envisaged a fluctuating quota triggered not by changes in consumption but simply by price.[57]

The Byrne scheme was soon adopted by the ITC but with several important modifications. The two month suspension would take effect from July and would only apply to the marketing, rather than the production, of concentrates.[58] On Malaya's insistence, while the effective 33.3 per cent quota could be raised after six months, there could be no discussion of a further cut for at least twelve.[59] Similarly, she forced a redefinition of the link with the ITP.[60] The existing release price of £165 was maintained but was now subject to the quota being raised to at least 40 per cent during this twelve month period. While these modifications reduced the bite of the original scheme, it was now clear that the industry was finally on the road to recovery. It came at a particularly critical moment.

In June one of the largest and most respected metal brokers failed. Two months earlier Lewis Lazarus had made the following prediction in a market newsletter: 'No doubt that tin cannot decline to lower level in US. The rest of the year will bring more serene days, and the pool will soon liquidate stocks.'[61]

With the price just over £100 it looked as though the trough had indeed been reached and a group of Paris clients then developed a large bull position, with Lazarus selling the metal forward. When the market turned against them, they were unable to cover their position, forcing the temporary suspension of the firm in June.[62] Fortunately the LME was able to make arrangements to cover the 6,000 tons that would otherwise have been released, averting a total collapse of the market but it remained quite depressed until the Byrne scheme was announced. One interesting detail emerged. Samuel Hoare was among Lazarus' list of creditors. Why a senior cabinet member was speculating in tin itself became the subject of some speculation and it revealed a close link between the political and financial networks that served as the background support for the ITC.[63]

With the Byrne scheme, the asymmetry between the levels of restriction in Malaya and Siam became acute.[64] What was an unthinkable cut for Siam in 1931 had been far surpassed in Malaya 12 months later. Overtures were made to Siam to accept her share and impose a further cut and were received sympathetically by the Siamese Legation in London and Prince Damaras. But there were two overriding constraints. Chinese miners in Siam were local born and could not therefore be repatriated, making it difficult to address the ensuing problem of unemployment. In July a revolution had occurred against the absolute monarchy and it was therefore considered 'essential for them to be ultra-nationalistic in their outlook and avoid any action which could be represented as damaging to any important local interest'.[65]

Compliance with the terms of the scheme, which required suspension for July and August, was complicated by the inherited over and under exports. Only the NEI suspended exports of its entitlement for those months but exported most of that which had been carried forward. Bolivia hardly reduced its exports, while Nigeria exported at 46 per cent and Malaya at 72 per cent of their respective entitlements.[66] When the exports against the inherited entitlements are included, exports for two months in which they were supposed to be suspended ran at 45 per cent of the average of the previous six months. This was a significant reduction but hardly the boost that the sponsors had hoped for.

In spite of this initial setback, the Byrne scheme accomplished its objective of getting production below consumption and by the Spring of 1933, the ITC had to assess the likely course of the market for the second half of the year. Keeping the quota at its current level had two obvious advantages. The most pressing was hastening the process of liquidation of the tin pools and other surplus stocks. The sooner the ITP was liquidated, the sooner negotiations for a new agreement could be completed. It was clear that a return to unrestricted production could not be contemplated for several years and renewing the agreement would provide a further element of market security. Keeping quotas low was the least risky strategy. If production were raised, there was always the danger of it again exceeding consumption and a subsequent correction would be difficult to absorb.

Getting to the point where a formal decision could be made involved a certain amount of arm-twisting, especially with Nigeria. In 1931 it had managed to impose a time limit on a cut and attempted to do so again. Such conditions were not only difficult to enforce in practice but they violated the assumption on which the ITC operated, namely that it made an objective analysis of the statistical situation as the basis on which quota recommendations were made. That meant that local considerations could be noted but were ultimately irrelevant. If the

Governor could establish his own analysis of the condition of the world market, then that would be taken into account but, of course, he was hardly in a position to do that.[67]

The second tin pool

Just as the discussions within the ITC finally culminated in the Byrne scheme, so the discussions around the completion of the ITP finally culminated in the formation of a second tin pool. Unlike the ITP, it was not formally connected to the ITC but it was also designed to complement the new cut.

However, the second tin pool had a very different character from the first. Whereas the first, once completed, was held entirely in frozen metal, passively awaiting a rise in the price, the second was formed in both cash and metal and was designed to stimulate the market by jobbing in and out. Howeson's proposal was based on the assumption that there remained 22,000 tons of surplus stocks, which would be reduced by 10,000 by the end of September as a result of the Byrne scheme.[68] The Dutch were not easily persuaded, especially in light of Patiño's refusal to participate in the first pool but agreed once they were convinced that the objective of this syndicate was to simply encourage market activity rather than take control over it. As usual, the argument was made that their commitment was required to secure the commitment of BMC, which in turn was required to secure the £1.5 million needed in cash.[69] The tin would come from the Dutch at 2,000 tons and BMC at 3,000 tons, to be frozen for a year, while Howeson and Patiño would raise the cash to buy 10,000 tons.[70]

Howeson then set about raising the £1.5 million required to secure control over these 10,000 tons. £900,000 came in the form of a credit from Lloyds Bank guaranteed by BMC and the remaining £600,000 was supplied by Cunliffe-Owen, CGF and Patiño.[71] The nature of the second pool was seriously misrepresented by Patiño in a letter to President Salamanca:

> I have been able, as the result of prolonged efforts, to interest a powerful London metal firm, and to form a syndicate with the object of defending … the position of tin. The syndicate has a capital of £1,500,000 and its first objective will be to increase the price of tin to about £160. Once this objective is attained … we shall consider what further measures will be needed to reach £180 a ton. You know full well that this kind of business is … an extremely uncertain one and the syndicate might well sustain heavy losses. Nevertheless I have not hesitated one moment in sacrificing … my remaining cash assets and in giving my guarantee to English banks. I am doing this to defend my mining interests and also as the major contribution I can offer my country to save its economic situation.[72]

This letter should be compared with a later one apparently addressed to the AIM:

> I organised a syndicate with Consolidated Gold Fields, with an effective capital of £600,000 and a bank credit of £900,000, making a total of £1,500,000. The syndicate signed a contract with British Metal Corporation to operate pool number two, with the immediate result that the price rose from £110 to £150 …My participation in the syndicate is two thirds and that of Gold Fields one third. I have taken a great risk in the service of the tin industry …[73]

It was this kind of self-promotion that has left the impression that Patiño played a far larger role in tin politics than was the case. Given the fact that production was now substantially below consumption, a price rise was inevitable, so the prospect of capital loss was extremely remote. Whether Patiño was responsible for securing the support of BMC is unclear but this was not the independent initiative that it appeared. Patiño was just a participant, albeit a crucial one, in a scheme orchestrated by Howeson, intimately linked to the existing pool and restriction schemes.

On Patiño's own instructions to the other participants, the objective was not in fact to raise the price substantially[74] but rather to prime the pump of a stagnant market at the point when the Byrne scheme was beginning to bite. In this regard it was very successful, at least as judged by the course of prices. By the end of August it had acquired some 7,000 tons[75] and the price rose from £127 at the end of July to £156 at the beginning of September. Although production continued well below consumption, it was primarily due to the work of the pool that the price remained at around that level for another seven months until April 1933.[76]

In order to further speed the rate of recovery, Howeson developed yet another pool proposal. This was to be the mother of all pools since it would involve a consolidation and an extension of the existing ones to a total of some 46,000 tons. He hoped to bring in an additional 5,000 tons from Malaya and Nigeria and to arrange financing for a further 5,500 tons to be bought in the market at £170.[77] Although Patiño indicated his willingness to participate,[78] uncertainties concerning the future of the overall restriction agreement appear to have prevented this new effort.

Corporate consolidation

At the same time as the ITC began to take control over the market, the core alliances within the industry were being reshaped. Three interrelated proposals were afoot. In March 1932, British-American Tin was divided into two new companies, one to hold the metal, Tin Holdings and the other to hold the equity investments, British Tin Investment Corporation (BTIC). Patiño acquired a 50 per cent stake in BTIC on the transfer of his investments in Malayan companies.[79] This consolidated the relationship between Howeson, Lyttelton and Patiño that had begun with the formation of CTS and since Lyttleton became Chairman he soon emerged as the leading figure in British tin finance.[80]

A second development was a strengthening of the relations among the smelters. Billiton had decided to build a second smelter at Arnhem to treat concentrates from the NEI and invited both CTS and STC to take a share of the capital required. STC declined, even though it was losing the largest single source of concentrates for its smelter in Singapore but CTS accepted, with a 40 per cent stake. A third initiative was designed to bring Banka, Billiton, CTS and STC together in a single selling organization. While this came to nothing, the very suggestion was an important indication of the way in which the Dutch were now committing themselves to making tin control more or less permanent. The shape of the core network and some of its affiliates is presented in the Figure 8.2.

1932–1933, an assessment

With two tin pools and production levels well under consumption, the ITC had largely reached the objective of market control that had been sought for over 18 months. There were two limitations. It could not control the level of outside production, nor could it fully control output from its own members.

The original hope, that much productive capacity on the part of outsiders would be expelled by the normal operation of market forces, was not realized. Instead, it grew by over

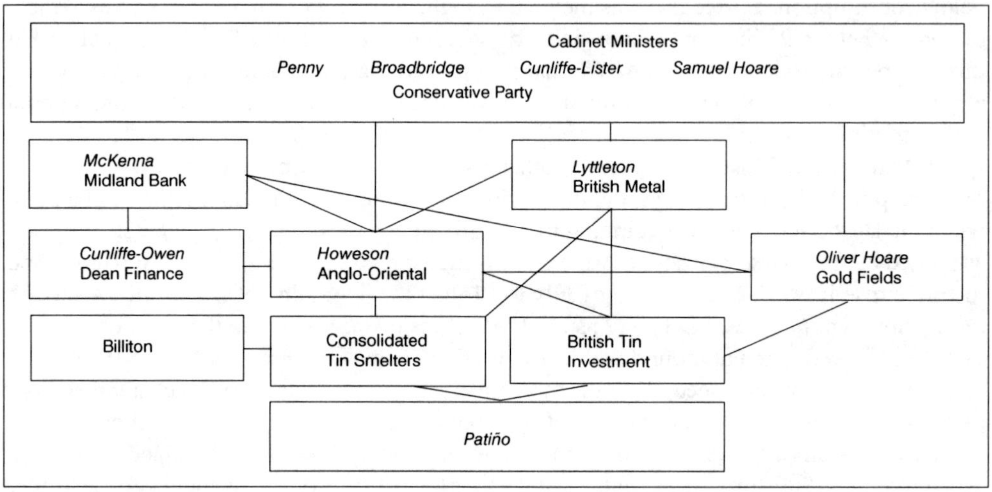

Figure 8.2 Core political-financial network

20 per cent from 1931 to 1932 and then by a further 20 per cent from 1932 to 1933. As a proportion of the world market the figures were even more ominous. From 11 per cent in 1931, by 1933 it was over 25 per cent. The extent to which the ITC itself was responsible for this development will be considered in Chapter 13 but it raised a new issue that would have to be addressed in any renewal of the agreement.

For reasons that will be discussed shortly, Bolivia had found it difficult to impose the full extent of the cut required from the beginning of 1932 and by the end of May she was 1,200 tons in excess. Although virtually all the new production cuts were passed on, she was never in a position to work off the accumulated excess. That naturally caused considerable embarrassment for the Bolivian delegation on the ITC but the Bolivian government simply ignored repeated requests for its regularization.[81]

In the case of Siam, Malaya and Nigeria restriction regulations had left open one loophole. Since these countries imposed export taxes on the basis of a nominal metal content in the concentrate, 72 per cent in Siam and Malaya, 70 per cent in Nigeria, the same principle was also applied in the allocation of domestic quotas. It had always been possible for miners to work their concentrates up to a higher level and avoid paying any royalty on the difference but that had to outweigh the additional costs involved. With restriction they not only avoided the extra royalties but also gained the extra income when the concentrates were smelted. Since it was concentrates and not metal that was actually controlled, this increase would not be officially recorded. Evidence of this practice came from CTS which showed that in the case of Nigeria the average assay value had risen from 70.97 per cent in 1929 to 72.49 per cent in 1932, in Malaya from 74.65 per cent to 75.71 per cent and in Siam from 73.68 per cent to 73.93 per cent.[82] Over the life of the first agreement this meant that 4,000 tons of metal were being produced in excess of the authorized quotas. The Bolivians naturally seized on the way in which this loophole had been exploited to deflect some of the pressure on them for their much more blatant excess exports and the final solution to the whole problem of all these excesses had to be deferred until the next agreement.

Under the first agreement there was only one overriding objective and that was to bring stocks closer to a normal range. While there was some debate about the size of that range,

Howeson and Lazarus both agreed that it was between five and seven weeks of actual consumption. Until the adoption of the Byrne scheme, the gap between consumption and production was being more than met by running down invisible stocks. Some of this was a function of the flushing out of some speculative holdings but most was simply the anticipated result of the decision of consumers to bring their existing stockholdings down to a new normal level set by a lower rate of consumption.[83] The visible stocks actually increased and the new rate of consumption meant that the excess stocks outside the ITP rose from 33 to 76 days. For the first six months of the Byrne scheme, that pattern was modified only slightly and a small reduction in the visible occurred; but by the end of 1932 the excess was still as high as 54 days.

By the beginning of 1933 invisible stocks had been run down closer to normal levels, so the gap had to be met mainly from the visible. But the ITC now enjoyed a fortunate but unanticipated development on the consumption side. Devaluation of the dollar in March signalled an easing of the depression and American demand began to pick up. This was particularly striking in May when American tinplate mills found themselves operating at over 80 per cent of their capacity, by contrast with just 50 per cent in April. With the ending of Prohibition came an interesting new market, as the can became an important form of packaging beer. Expansion of consumption meant a restocking of consumer inventories[84] and the prospect of a further devaluation led to some speculative buying.[85] That increase in demand meant that at long last visible free stocks were getting to the normal range. As stocks declined and demand recovered, the conditions now existed for an increase in price.

The Patiño syndicate had been keeping prices steady but doing so under these conditions forced it to sell off its holdings and by the end of April its resources were largely exhausted. Meanwhile, the ITP remained constrained by the terms of the modified Byrne scheme, so that for May and June there was no stabilizing force in the market. The price response was therefore very dramatic. From £164/ton at the end of April it jumped to £225 by the end of June. In New York it rose at an even faster rate, from 30 to 45c/lb.

A certain element of misjudgement of the market may have been the cause of a shortage of spot tin, reflected in a slight backwardation in London throughout July and a momentary but large premium of £21 for Straits tin in New York. This naturally revived Malayan anxieties about the incentive being created for US consumers to make a permanent shift from Straits to other brands.[86] They would be reinforced by a superficial reading of the changing nature of the American tin market. Whereas in 1929 over 75 per cent came from the Straits, in 1933 that proportion had shrunk to 50 per cent, with the difference being made up by metal from Banka, China and Britain.[87] How much of this should be attributed to the artificial character of the premium for Straits deriving from tin control is another matter. The development of a more efficient tin smelter in Yunnan had resulted in a higher quality brand available for the American market and Banka was also interested in exploring its potential to supplement traditional markets in Europe.

The issue of the distinctive market for Straits tin that had been raised by Mair in 1931 continued to fester. In early 1933 he took it up again at the annual meeting of Gopeng Consolidated, claiming that since the stocks of Straits tin had actually declined from 1930 to 1932 while those of standard tin had increased, there was in fact no shortage of demand for the metal produced by Malaya.[88] The ITC was now better prepared to respond and Campbell asked Houwert to look into the matter. Mair had simply misread the published stock statistics and aggregated several different brands, including some standard afloat to the USA, as Straits, a misreading that could hardly be accidental:

> It is evident that an inaccurate statement of the quantity of a definite brand occasions a dual error in the comparison of the 'increased tonnage' and this wrong statement, not only exist, but – what is still worse – Mr. Mair knew, could know that his statement is inaccurate. ... Viewing the matter therefore in this light we cannot very well speak of 'Malayan's goodwill' which point Mr. Mair has thought it expedient to accentuate with so great an enthusiasm.[89]

Mair's obsession with a purely regional perspective on the tin world had led to a serious distortion of Malaya's role within it.

The sudden rise in price had another important result, since it revealed the existence of considerable stocks in the Straits. In April the smelters produced 485 tons more than they received in shipments in the previous month and by July this had risen to 1,939 for a total of 4,430 tons over the four months. Not only did this indicate that the statistical sources available to the ITC were much more limited than had been supposed but raised the worrisome possibility that even more would be flushed out in the months to come. At first, Howeson accused the STC of holding this for its own account but the company denied this[90] and strongly hinted that of the 3,700 tons of invisible stocks held at the beginning of restriction, 3,000 belonged to GMB.[91] If so, this indicates some of the limits to effective co-operation between the key participants. One immediate result was a request to the smelters for full stock disclosure, on a confidential basis, to the ITC.

The terms of the ITP meant that the first agreement came to an end before stocks were reduced to a normal level. By December 1933 they were still at twice that point, similar to the position in March 1930. The problem of the industry was far from being solved and the overall verdict must remain that all the ITC had been able to accomplish was to prevent a difficult situation from turning into a disaster. Even that judgement must be qualified by recognizing that had the American recovery not come in the Spring of 1933, the members of the ITC may well have lost their stomach for continued restriction beyond August 1934. Each stage of the deterioration of the market could be attributed to Malaya. The problem of 1929 was a function of her expansion, that of 1930 a function of her limited commitment to the voluntary programme, that of 1931 a function of the deferral until March and her excess, that of 1932 a function of her inability to contribute to the ITP. At the same time, there was an escalating chorus of complaints from Malayan interests. It would be expecting a great deal of any international agreement to continue long in the face of such tension.

The price history for 1932–1933 together with the relationship between production, consumption and stocks is presented in Figures 8.3, 8.4.

The significance of the critical barometer of the visible stocks should be assessed against the background of the total stock position. Invisible stocks include those at the beginning of the pipeline, as concentrates are shipped to smelters and those at the very end in the form of consumers' stocks. For purposes of illustrating the trend in these stocks, it has been assumed that a normal level of such stocks should support 200 days of consumption. While the level of normal visible stocks remained the subject of some discussion, within the ITC it was assumed to be 50 days. Both trend lines indicate the extent to which the actual level of stocks was over or under the normal level.

It is evident that the visible is extraordinarily sensitive to changes in production and consumption. From 1931 to mid-1932, the level of abnormally high stocks continued to rise, even though production was more or less in balance with consumption, simply because consumption levels were falling. When they dropped off very seriously in the third quarter of 1932, that level of abnormality skyrocketed. What made the situation particularly serious

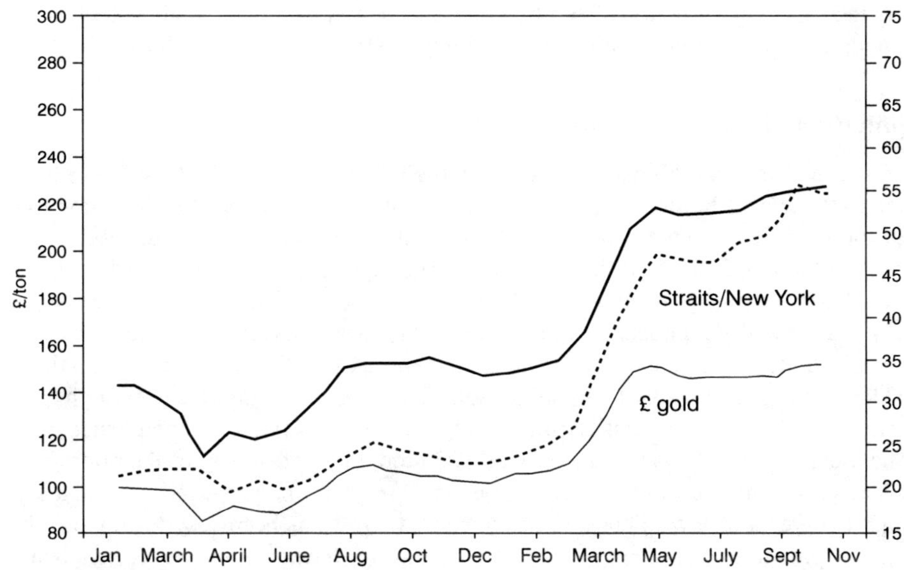

Figure 8.3 Tin prices, 1932–1933

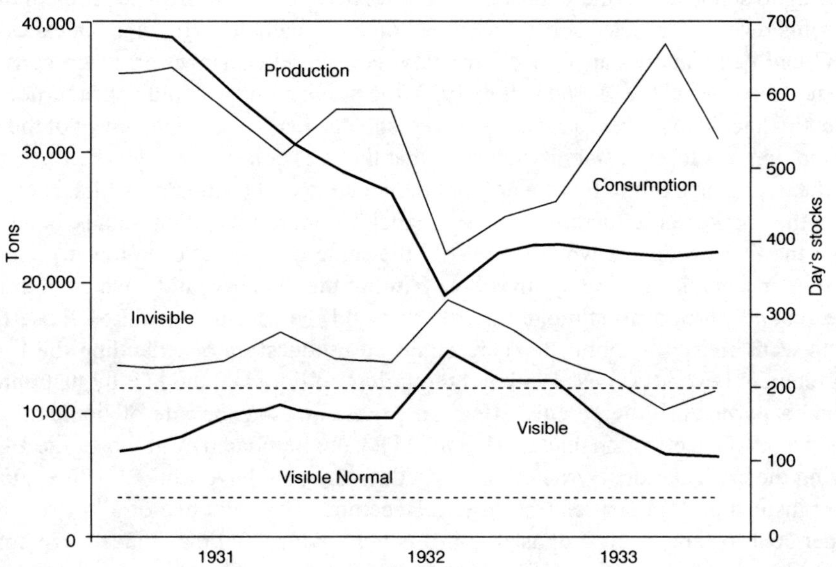

Figure 8.4 Production, consumption, stocks, 1931–1933

was the fact that the level of abnormality in the invisible supplies also rose. It was the dependence on the consumption level that allowed these ratios to rise even though total stocks were falling. As consumption began to rise in early 1933 in the face of a stable level of production, the level of abnormality of both forms of stocks dropped dramatically. In the case of the invisible stocks, they had approximated their normal level, so that the problem facing the ITC on the expiry of the first agreement was excessive visible stocks, most of which were in the process of being sold off by the ITP.

Liquidation of the International Tin Pool

The Byrne scheme had originally constrained the ITP by preventing releases until the quota reached 40 per cent but although the 33.3 per cent quota was kept for the second half of 1933, the modified scheme meant that that condition no longer obtained.[92] Releases were now limited to a maximum of 20 per cent per month, triggered by increases in price. From July to October rising demand was fed by sales from the pool and by the end of November the ITP was formally liquidated. Liquidation simply meant that metal was released from the restrictions under which it had been held; selling was quite a different matter. With the end of the ITP there were some 8,000 tons still unsold. The market was not strong enough to absorb this amount[93] and it was not until April 1934 before that was finally accomplished.

The metal in the ITP was sold at around £226/ton, higher than any of the prices at which it was acquired and the resulting profits inevitably rankled the producers who had suffered through two years of low prices and heavy restriction. It has been easy to suppose that the interests of producers were being subordinated to those of the pool holders.[94] Resentment at the apparent benefits accruing to financial interests, while the producers had barely survived, would complicate later plans for stock control. Although it was certainly disingenuous for Campbell to claim that the interests of the pool and those of the producers were identical, it is equally misleading to suppose that there must be a conflict between the two.[95] The problem that these high profits revealed reflected something much more fundamental. To wipe out excess stocks, the ITC had deliberately cut production below consumption. However, when consumption levels suddenly rose in the second half of 1933, the production tap could not be turned back on sufficiently quickly to prevent a runaway price[96] and one major accomplishment of the ITP was the restoration of a level of stability to prices that the market had never seen before.

As the experience of the Bandoeng Pool demonstrated, the extent to which excess stocks depress the market is a function of the financial commitment that stands behind them. Without the ITP, producers would have sold the same quantity at even lower prices but all such counter-factuals are largely invisible. Without the ITP, it would have been impossible for the second tin pool to stimulate the market and the sacrifice made by those financial interests went entirely unnoticed. There could be no question of extending the life of the first agreement beyond August 1934 and the failure to liquidate the ITP by that date would have had a particularly depressing effect on price. Slowing the rate of liquidation would have had a perverse effect on quotas. Had the ITP been liquidated evenly over the 14 months following the expiry of the Byrne scheme, producers would have worked with a quota of 46 per cent instead of 33.3 per cent for July to December 1933 and one of 37 per cent instead of 46 per cent for January to August 1934. It is not obvious that many faced cost conditions that would have left them better off as a result.

Without a comprehensive valuation of the metal held, estimates of the actual level of profits can be very misleading. One method was to compare the market price for metal during the formation of the ITP (*c.* £130) with that during its realization (*c.* £226) which shows a 'considerable net return'.[97] In the case of the British Syndicate this can be calculated more precisely. The 2,939 tons bought at a total cost of £350,000 were sold for a net gain of £310,000, or an annualized rate of return of some 30 per cent. However, most of the metal brought in had been acquired much earlier at higher prices and the return on that would have been considerably less.[98]

Table 8.2 Tin Holdings, return to shareholders

	Investment £	*Return* £	*Annual profit rate* %
BMC	34,911	60,643	15
CGF	9,504	41,058	80
Other ordinary and preference	27,899	42,256	10
Ordinary only	45,199	34,841	–9
Total	117,513	178,798	10

Source: BT31/33325/263565.

The experience of Tin Holdings reveals a more complicated picture. When the company was formed out of British-American Tin in March 1932, the 1,948 tons of metal transferred were valued at cost, £168, rather than at the current market of £130. In addition to the metal, the company also inherited the liability of over £200,000 which had been borrowed to buy it in the first place. Within a month, the price dropped to £109 and the assets were now less than the liability. To meet the margin call of £5,000, 6 per cent cumulative preference shares were created and offered to all shareholders, pro rata. Most considered that there was little point in throwing good money after bad and the bulk of these new shares were taken up by BMC and CGF. When the final distribution was made the whole of the net profit of £62,000 went to those who risked the additional £5,000 and the ordinary shareholders had to take comfort in the fact that their losses were reduced to only £27,000 on their original investment of £112,500.[99] Table 8.2 indicates the actual return to shareholders with different combinations of ordinary and preference shares.

That returns on the metal contributed by Tin Holdings were even as high as 10 per cent was a result of the fact that it was highly leveraged. When that feature is taken into account, the experience of this quantity of metal suggests a very modest annual return, closer to 2 per cent. The fact that a high proportion of the ordinary shareholders were prepared to abandon their investment is an indication of the level of risk[100] that was associated with a pool, even one as firmly held as the ITP. That, too, was largely overlooked once the liquidation was complete.

Domestic tensions

While the ITC had gone a long way to rescuing the tin industry at the international level, the severity of the cuts required engendered considerable tension in two cases, Bolivia and Nigeria, both of which jeopardized the prospect of renewal.

Bolivia

With the more severe cuts of 1932, the problem of administration had to be taken more seriously and the politics intensified. A Tin Rationing Committee was established which attempted to final a comprehensive formula covering four sets of claims: (1) from established miners; (2) from those developing new mines; (3) small miners; (4) workers living in urban centres. The two most important new operations were those initiated by for the reinvigoration of Unificada following consolidation in 1929 and the dredge launched by Bolivian International Mining Corporation in 1930. Small miners could make a similar claim, namely that their mines were often the basis on which future large scale developments

Table 8.3 Bolivia, quota distribution, 1932–1933

	1929	*1932 Jan–June*		*1932 July–Dec*		*1932 Total*		*1933*	
	%	*tonnes*	*%*	*tonnes*	*%*	*tonnes*	*%*	*tonnes*	*%*
Patiño	58.8	6,674	56.4	4,195	56.2	10,869	56.3	8,211	55.2
Aramayo	10.0	1,044	8.8	377	5.1	1,421	7.4	763	5.1
Hochschild	9.7	2,039	17.2	1,421	19.0	3,460	17.9	2,575	17.3
Medium	11.8					1,451	7.5	1,828	12.3
Small	9.6					2,102	10.9	1,491	10.0
Medium and small	21.4	2,085	17.6	1,468	19.7	3,553	18.4	3,319	22.3
Total		11,842		7,461		19,303		14,868	

Source: Estaño en Bolivia, La Paz, 1935.

occurred and could add two powerful arguments, their comparative poverty and the fact that less of their income leaked abroad. Their position was taken up in several influential quarters.[101] Prefects and Deputies representing Oruro and Potosí supported the claims of workers resident there.[102] They were concerned about the problem of public order as result of extensive unemployment, arguing that workers in the remoter mining camps could generally be dispersed back to their original agricultural regions.

The initial results of this new allocation show that the claims of the large established mining groups of Patiño and Aramayo were both scaled back in order to meet the other claims, all of which benefited Hochschild. Not only was he developing Unificada in Potosí but his other large mine was also in an urban centre, Oruro, and he continued to serve as the primary dealer serving the small miners. Table 8.3 shows the allocations for 1932 and 1933.

Hochschild had continued to improve his position but now this was at the expense of all other groups apart from the small miners. The distribution for 1933 shows a considerable improvement in the position of the medium miners, met by concessions from all other groups.

Bolivia did not tighten her administrative machinery with the commencement of the third cut in January 1932. At first overexports continued at a modest rate but in May a sudden increase meant that she was over by 1,221 tons or 60 per cent of her monthly allotment. Table 8.4 shows that the primary source of this problem was in the small miners.

In spite of lower quotas the small miners had actually increased their production in 1932 over that in 1931 and now there was far less flexibility on Patiño's part to absorb their excess.

Continued excesses were reported to the ITC, accompanied by promises for rectification. Another rise in June made Patiño's position on the ITC intolerable and he pressed for firm measures: 'We find ourselves in a difficult situation, especially since only country with

Table 8.4 Bolivia, over–exports, 1932–1933

	1932			*1933*			*Total*
	Exports tonnes	*cf quota tonnes*	*% quota*	*Exports tonnes*	*cf quota tonnes*	*% quota*	*tonnes*
Patiño	10,824	–45	–0.4	8,203	–8	–0.1	–53
Aramayo	1,430	9	0.6	784	21	2.7	30
Hochschild	3,550	90	2.5	2.574	–1	–0.0	89
Medium	1,702	251	14.7	1,445	–383	–26.5	–132
Small	3,350	1,248	37.3	1,874	383	20.4	1,631
Total	20,856	1,553	7.4	14,880	12	0.1	1,704

Source: Estaño en Bolivia, La Paz, 1935.

excess. Need measures to maintain credit of the country within the agreement. Need details for next meeting; this hurting our country's prestige and responsibility.'[103]

Moral suasion had some effect and the first set of comprehensive regulations were promulgated in July. Customs was instructed to grant export authorizations only in relation to previously approved quotas.[104] This was followed in November with a law that entirely changed the position of the small miners. Associations were created in each Department which would be allocated a block quota. Small miners were required to join their respective association and negotiate for their share of the quota.

Just as Bolivia began to get production under effective control, the severe cuts of the Byrne scheme commenced. While the new machinery ensured that she lived within her new quota, she never found the flexibility required to impose a further cut to wipe off the excess. Much to the frustration of her partners on the ITC, Bolivia simply carried the excess through the remainder of the first agreement. In explaining the source of Bolivia's excess, Antenor Patiño commented:

> With universal suffrage, and with tin amounting to 80% of the whole economic activities of the country, the small producers dominated the situation electorally. They knew nothing about the connection between price and output, and so on; they saw tin 'high'; they were prevented from producing or exporting freely; they felt the whole thing was a ramp, worked in the interests of the large men – of Patiño and Aramayo, in fact.[105]

While this may account for the pressure, it did not explain why the government had not been able to meet it at the expense of Patiño and Hochschild. After all, Hochschild was working with a quota designed to allow for growth and Patiño was working with a more generous quota than was another large lode mine, Pahang, and had far more flexibility with which to meet it. The other members of the ITC were not impressed and Bolivia's reputation for unreliability was confirmed.

The overall political weakness of the Bolivian government was frankly admitted by President Salamanca in a letter to Patiño:

> I am beginning to be seriously preoccupied with this matter [tin restriction] because of recent developments. As soon as the price of tin began to go up, an almost violent struggle broke out to participate in the production quota assigned to Bolivia. I presume that you know all about this as you, to quieten the agitation of the small Potosí producers, ceded them several tons annually from your own quota … This agitation will keep on increasing as the tin price goes up because, in brief, it is inspired by an appetite for profit. As it will be impossible to satisfy everybody, while keeping within the quota fixed for Bolivia, they will finish up by viciously attacking the agreements which limit production. I believe it will be extremely difficult to prevent these agreements being rescinded.[106]

As will be seen, these internal politics not only came with a moral cost but they jeopardized the renewal of the very agreement that sustained these small miners. Unfortunately, that was not the only occasion on which internal conflict in Bolivia would spill over into the international arena.

Nigeria

As Nigeria had to find a way of getting its industry to live within even lower production levels, it moved first to rectify two of the obvious initial mistakes. The first was to reconstitute the Quota Committee so that it was more fully representative, with two members from each section of the industry, large, medium and small. The second was to turn the basic allowances into compensating or compassionate allowances, that is of a minimal entitlement of 12 tons.

With low levels of production there could no longer be any justification for the requirement that quotas based on special claims be won exclusively from the development that had supported them. Special and ordinary claims were therefore consolidated and free transfer among producers was then permitted.[107]

While the rate of restriction on the small miners had increased considerably, it was at a slower rate than that of every other sector which strengthened their comparative position.[108] The Resident of Plateau Province took the opportunity to inform the mining community that he now wished to review the principles of restriction allocation. It was clear that the large companies could no longer afford the concessions made to the small and the Chief Inspector of Mines now considered them 'pure sentiment' to the detriment of the 'real backbone of the industry', the large company.[109] Unable now to depend on the original assurances, the small miners organized themselves as the Private Mine Owners Association to defend their interests.

The medium miners also wanted to use the opportunity to reopen their case against the special claims and were prepared to challenge the principle of the compensating allowances.[110] The small miners now began to counter attack, questioning both the statistics and their interpretation and introducing a new criterion, comparative profitability.[111] It was clear to the government that any reopening would involve endless politics, especially since there was no staff available to undertake a complete assessment of the only basis that stood a chance of adoption, current productive capacity.[112]

However, the medium miners could not be simply ignored. Since one of the concessions made to them the previous summer had now expired, it was felt that a further concession would be justified and for this the small miners might have to be sacrificed:

> The small miner is of no particular use to the industry and if Government has to choose between annihilation of the small owners and the bankrupting of large concerns … then clearly the latter are more worthy of consideration and Government in its own interest must protect them.[113]

This position was articulated more provocatively and publicly in the press, where it was argued that the small miner had no 'moral rights … as no such thing is recognised by big business', and the anonymous author looked forward to the day when the Nigerian Chamber would be 'completely controlled by one group which will eliminate all disputes and enable united front being presented to Government'.[114]

As the small miners responded, they reopened the question of the principle of restriction and the interests that it served.[115] In this they were joined by some of the companies who were growing impatient at the delay in price recovery. With a much wider range of issues being raised, there was no attempt to call a general meeting to search for a new basis of agreement. Instead, everything was referred to the Nigerian Chamber of Mines in London.

At a special meeting of the Chamber a committee was struck under the Chairmanship of Sir George Maxwell, Chairman of the TPA, and included two representatives of the original group of objectors.[116] Its report went to considerable lengths to explain the source of the original misunderstandings and hence validated one important element in the grievance nursed by the objectors. Its statistical analysis undermined the other element, namely that there was a substantial distortion created as a result of the special claims made by the large companies. While this benefited ATMN it did so at the expense of the other Anglo-Oriental company, LNTM. This also confirmed that the major problem lay in the special treatment accorded the small miners.

The committee approached this crucial problem very gingerly and its major recommendation was to encourage the government to give the small miners special terms to entice them out of tin and into gold. Otherwise, it turned the issue back to the government to reopen the whole issue through a special committee.[117] On the question of special claims, it criticized the Mines Department for allowing transfer of quotas between special and ordinary claims since this prevented an evaluation of the statistical basis of the special claims and encouraged a systematic review of all such claims with a view to reducing them.[118]

An additional set of issues addressed by the Maxwell committee concerned the composition of the Chamber itself. It stressed that members of the council should not consider themselves representatives of their companies, encouraged the Chamber to recruit individual members and to change the voting structure to reduce the power of the large companies.[119] In short, it wanted the Chamber to become a body which could resolve disputes rather than become part of them. With the elimination of a number of petty grievances and the introduction of a process whereby others could be aired, it is not surprising that earlier contention within the Chamber was replaced by an air of 'harmony'.[120]

However, when the Department of Mines inherited the relevant recommendations there was no willingness to rethink the premise of the original conception. To do so would involve a substantial commitment of time and resources to determine current productive capacity at all accurately.[121] The problem of determining a satisfactory basis for the allocation of production quotas remained intractable.

By mid-1933, the price recovery began to diffuse much of the pressure of quota allocation and a number of small miners took the offer to make the transition from tin to gold.[122] However, for the government, the most important lesson learnt from the entire experience is that the politics of tin were unnecessarily troublesome. As will be seen, it had no enthusiasm for the renewal of the ITC and was certainly not interested in lending support to any strengthening of it through a buffer pool.

The Nigerian difficulty with internal allocation simply replicated the problem that existed at the international level. All parties worked with a moralistic discourse, attempting to establish universal criteria of fairness and equity, in terms of which they could couch their own interests and challenge those of their opponents. Treating the question of distribution as though it can be reduced to the administration of consensually established principles was largely responsible for the escalation of the actual politics. But those politics never challenged the assumption that the process should be inherently apolitical. The historic distance between government and mining industry had prevented the development of an overall policy for the most efficient management of Nigeria's tin resources according to criteria that the government itself established. Had that been done, the government would have escaped from the conceptual trap it initially set by defining itself as the implementer of a policy formulated by a superficially homogeneous industry.

Malaya

Malaya had a far simpler administrative system which avoided the kind of difficulties that emerged in Bolivia or Nigeria. However, it was not entirely free of controversy. Frisby, Chairman of Pahang Consolidated, decided to contest the legal basis of restriction, at least as it applied in its case. Central to the argument was that the company operated a concession rather than a lease and that the terms of the concession precluded limitation of production. In addition, the concession had been renewed in 1888, prior to the date at which Pahang became part of the FMS. While this would not affect the overall viability of restriction in the remainder of the FMS it would tighten the bite on the other producers and deal a blow to the legitimacy of restriction. Indeed, it may have been that last consideration that prompted Frisby, having lost at both the initial and appellate levels in Kuala Lumpur, to take the case to the Privy Council in London where he received no consolation, apart from the apparent satisfaction of his honour.[123] However, this legal failure would prove to be but a prelude to a sustained campaign against the very principle of restriction.

As restriction intensified, the value of quotas increased. In December 1931, with a 60 per cent cut, they were selling at S$15 per pikul, or 20 per cent of the market value of the metal; ten months later, with a cut of 75 per cent, they were selling at S$45, or 58 per cent of the value of the metal. While the marginal cost of production may have made this economical for the miner who bought, the bulk of the marginal increment must have been transferred to the miner who sold. With the quotas at such a high value the extent of the parasitism grew and with it some concern that the primary beneficiaries of restriction were not working miners but those in a position to live off the sale of their quotas. While that indicated a weakness in the administrative system, short of fixing and enforcing a 'fair' price, there was little that could be done to remedy it. It would add to the catalogue of complaints about the unfairness of restriction in Malaya.

The International Tin Committee and its reputation

While there may have been much criticism within tin circles of the specific structure of the ITC, the fact that an intergovernmental agreement had actually found the resources with which to tackle the problem of a major commodity was generally received with much sympathy. In the absence of many other counter-depression measures, the ITC was naturally considered an important pioneer. The first accolade came in an adjournment debate in the House of Commons when the Chancellor of the Exchequer commented: 'There is one scheme, that for the control of tin, which has gone to the root and which has decreased production. That is the one scheme that has been successful. Others are just keeping supplies off the market.'[124]

More enduring was the reputation established at the World Monetary and Economic Conference in July 1933. This conference was designed to consider forms of concerted action to counter the depression. The central issues were monetary and thanks to the lack of support from the USA they could not be successfully addressed.[125] However, the conference did give considerable attention to the cartelization of primary commodities as an effective counter-depression measure.[126] Cunliffe-Lister was anxious to see the conference lend strong support to additional commodity agreements and laid out the general principles that should govern them, including:

1 comprehensiveness. They should command a general measure of assent amongst producing countries and within each country a substantial majority of the producers themselves.

2 fairness to all parties, both producers and consumers, and worked with the willing co-operation of the users themselves.
3 of lasting value, with a sufficient period to give an adequate sense of security to all concerned.
4 consistent with efficiency. Restrictions that are enforced should have due regard to the factor of relative efficiency.[127]

In addition to laying out these general principles, the British delegation addressed the specific problem affecting the ITC. Maintaining a price at the current level of £200 would induce such a large expansion of production outside the control of the ITC that its members could not be expected to continue to shoulder the burden of restriction, unless they were joined by others.[128]

Campbell therefore had two tasks. One was to start negotiations with the relevant delegations but no one had the authority to do so and the 'net result was that the proceedings terminated rather in the air'.[129] The other was to ensure that the report of the subcommittee on tin to the Sub-Commission on Coordination of Production and Marketing supported the ITC. Since Cunliffe-Lister served as the Chairman and Campbell as the Vice-Chairman of the subcommittee that was considerably easier and the final report explicitly connected these two issues:

> The Sub-Committee has examined the existing international scheme for the control of tin, to which it is suggested that such countries as now produce the metal in appreciable quantities, and are not at present participants in the scheme should adhere. Discussion did not elicit any substantive criticisms of that scheme; no suggestions were made for its amendment; no alternative methods of control were proposed.
>
> The Sub-Committee consider that the existing scheme of control is framed upon sound lines; that it is in accord with the principles which have been adopted by this Conference as those which should govern the framing of plans for the co-ordination of production and consumption; that it has worked smoothly in actual practice for a period of over two years; and that it has been largely successful in achieving its main objectives.

It then went on to stress that the ITC 'had been consistently mindful of the interests of the consuming countries', and explained that 'the scheme does not attempt to regulate or control prices, except indirectly by adjusting production to demand', unless there is the prospect of an 'undesirable price rise', independent of the balance between production and consumption. The report concluded with the spectre of an end to the agreement as a result of excessive outside production and laid out the detailed terms on which the ITC could expect to negotiate its inclusion within a new one.[130]

The American Delegation felt as though something stronger than assurances concerning the importance of securing consumer interests was required and the Secretary of State, Cordell Hull, responded to the report:

> The American Government realizes the urgency of the conditions which led to the formation of the International Tin Pool, which is stimulating this further effort to develop a more inclusive agreement for the regulation of production and international trade in that metal. It does not wish to be understood as being in opposition to that effort. However, it wishes to take this occasion to express its natural concern that any agreement that is elaborated should be equitable to the consuming countries.

> The tentative form of agreement which has been under consideration would not appear to include any explicit automatic provisions for varying production or the release of stocks in accordance with price changes, such as would appear to be required in order to properly safeguard consumers. The American Government, therefore, observes with great interest the assurances expressed in the draft report submitted by the Sub-Committee ... to the effect that the International Tin Pool has been and will continue to be consistently mindful of the interests of the consuming countries. It will follow attentively the operation of any agreement which is elaborated.[131]

Although Hull refrained from mentioning price, it would soon come to the forefront of the attention that he promised. The price issue was now particularly acute in the United States thanks to the continued devaluation of the dollar throughout 1933 which meant that prices in New York were still rising at twice the rate of those in London.[132] Indeed, by the end of 1933 consumers were paying 53c/lb, a price not seen since January 1928. As the price issue became more pressing, the American government would find stronger terms with which to express its assessment of the ITC.

Whatever the difficulties the ITC had experienced in trying to formulate a coherent policy during the first agreement, at least the committee itself found that it was able to work in a more or less harmonious fashion. In Campbell it had found a Chairman with considerable resourcefulness. In March 1933 he was awarded a presentation hammer and Groothoof paid him the following tribute:

> You have shown us how to do business in an international committee. We have learned from you to do teamwork and to appreciate each others' special viewpoints. You have made most excellent minutes by dictating what we have said and also what some us have not said but intended to say at several occasions. And last but not least you have shown us how to handle our Governments in very delicate positions.[133]

Those skills would be put to a particularly severe test as the negotiations for the renewal of the agreement proceeded.

9 Renewing the second agreement, 1933–1934

The sense of crisis that had dominated the formation and operation of the first agreement had dissipated by 1933. While the industry had finally moved towards an equilibrium between production, consumption and stocks, productive capacity remained far in excess of demand. As long as this condition remained, there was little question about the necessity of maintaining the ITC. Negotiations for a second agreement took place at three levels: among the core members over the question of standard tonnages; between the ITC and outside producers; and within the ITC over the modifications that should be included in a new text. Only the last of these was at all straightforward.

The new text

Since so much of the work during the first agreement had focussed on the problem of stocks, it is not surprising that the new text gave much more attention to their significance. The objective of the scheme was modified: 'The scheme is intended to secure a fair and reasonable relation between production and consumption with a view to prevent rapid and severe oscillations of price and to ensure the absorption of surplus stock.'[1] Not only did this include a reference to stocks but it also replaced the ambitious term 'equilibrium' with the more modest 'relation' in defining the fundamental relationship between production and consumption.

The conditions under which members could withdraw were similarly modified. At the beginning of the first agreement members had the right to withdraw if the others simply did not endorse their proposals for a modification in the quota within six months. Now that right would be tied to a change in the visible supply. Clause 21 identified a variation of 5,000 tons over two quota periods, as triggering consideration of a change in quota level to compensate. Should any such proposal not be endorsed within three months, then any of the signatory governments had the right to withdraw and the scheme would collapse de facto. This clause was perhaps more controversial since Malaya originally proposed tying the right to a much narrower variation in visible supply, 10 per cent and introducing a similar link to changes in price. Given the asymmetry between Siam and the others, the term 'signatory' was introduced to identify the four core members.

The most important modification was the introduction of a new clause, 20, which specified that:

> If all the signatory Governments and Siam are satisfied that the estimated production of all other countries has, over a period of six consecutive months, exceeded 25% of estimated world production during that period, or fifteen thousand tons of metallic tin,

> whichever is the less amount, it shall be competent for any signatory Government or Siam to give six months' notice of its intention to withdraw from the scheme.[2]

The 25 per cent reflected the level of outside production during 1933; the 15,000 tons over six months allowed for some further growth. The clause simply defined the limits of outside expansion; with what success will be considered later.

The new text attempted to address the problem of over-exports by specifying that any over-export in one period had to be liquidated within the next. At least a timetable for liquidation was established, though there was no mechanism for its enforcement.

Standard tonnages

At the core of the first agreement were three ratios in the standard tonnages: between Malaya and the NEI, between the NEI and Bolivia/Nigeria and between the four signatories and Siam, all of which involved compromises. In early 1933 the Dutch raised the question of the renewal of the agreement. Without the time pressure that had led to these compromises in November 1930, they hoped to recover their concession. Just as the Dutch wished to recover their concessions so too did the other partners and the process of renewal would therefore not be at all straightforward.

The Dutch position was laid out by Houwert. Its principles did not differ materially from the position taken in 1930, namely that the problem to be addressed was one of overproduction and that the burden of restriction should therefore fall disproportionately on those who had created it. Production over 1924 to 1926 was defined as normal and that over 1927 to 1929 was excessive. Recognizing the need to grant something to those who had actually invested in this surplus capacity, Houwert proposed as a compromise the average of 1924 to 1929. Applying this formula to the actual standard tonnages would result in an increase to Bolivia of 325 tons and to the NEI of 2,238 tons at the expense, of course, of Malaya of 1,792 tons and Nigeria 771 tons. One formal inference from these statistics was that the NEI 'whose position was repeatedly envied have strictly speaking made the greatest sacrifice'.[3]

An opportunity to test the substance of this argument was provided by Clementi who invited the Dutch and Malayans, together with an observer from Siam, to a meeting in Singapore. The Malayans were not at all interested in a discussion of abstract principles of equity based on the experience of the previous decade. Instead, they wished to correct the apparent imbalances that had emerged during the course of the agreement. One had been identified by Mair and focussed on the threat to the role of Straits tin in the American market. The second was the way in which sacrifices by low cost producers redounded primarily to the benefit of the high cost ones. Both arguments pointed towards a renewal of the agreement but one based much more on an assessment of current market conditions. It could only be accomplished by an alliance between the NEI and Malaya against Bolivia.

The prospects of such an alliance were not at all promising. While both the Dutch and the Malayans wanted an increase in their standard tonnages, it soon became evident that it would be extremely difficult to renew on any other basis than those already agreed. Clementi, in particular, felt that regardless of the merits of Malaya's case it was 'too late to go back on the bargain made in 1929' and added 'If you wish to reopen the question of what the Standard Tonnage should be there is every chance of the whole restriction scheme being wrecked.'[4] That effectively ruled out the possibility of an early return to a free market, even as a device against which a better agreement could be secured.

At this stage the Dutch were even more sceptical about the desirability of a free market and van den Broek sketched out its implications:

> If today restriction is not going to be continued, everybody has to protect himself against what is going to happen. It means there will naturally be overproduction and we shall all then realise that we will get very poor prices. And then it comes to this: Who will come out as the winner? Personally I don't think anybody will be the winner. Just as in the War, we shall all be killed. But I don't believe a country like Bolivia, which has always been the highest cost producer, will ever go out of production. I cannot believe that they will close down their mines. They will work on any wages. It is the only thing on which they can rely. As long as they can sell tin at any price, they will produce.

Even if were possible to squeeze out Bolivia, it was not desirable in light of the long term consumption needs. Otherwise a consumption recovery would produce a runaway price with permanent damage. In any case, the market had not produced a solution to excess production in rubber or copper and van den Broek feared that:

> For years and years to come, with the enormous machinery that has already been installed in different mines, every rise in price, the slightest, will increase production and the lowest cost producer will have to be content with a very small profit.[5]

The timetable for renewal allowed for several opportunities for far more thorough consultation in Malaya than had been possible in the formation of the first agreement. Some sections of the press argued for a prompt return to a free market:

> Some very well-informed people believe that without the so-called international tin control project, Bolivia's mining industry would have collapsed and therefore Malaya's chief rival would have suspended production. In other words, we would have obtained an automatic restriction through this suspension of the world's second largest tin producer … We would add that whilst the mines belonging to the Chinese in Malaya could have continued to produce tin at this very low price [£109], we very much doubt if Bolivia could have done the same.[6]

But in the FMS Legislative Council, Rae was quite prepared to put whatever criticisms he had made in1931 behind him and noted:

> I do not suppose there is any person in a responsible position in Malaya who would suggest for one moment that tin restriction should cease in August next year. It is quite obvious that it must continue for a longer period … we are now committed to this policy and we have got to see it through in the light of experience.[7]

For him what was at issue was simply the policy of the ITC. Not only was it undermining the position of Straits tin but it aimed at far too high a price which simply stimulated outside production.

The Cornish, however, sent rather ambiguous messages. Thomas stated that it was:

> not in the best interest of the industry to contemplate, at this stage, the abandonment of restriction but it might be better to face even this issue than to enter into an agreement which did less than justice to the claims of all the parties concerned.[8]

A few days later Mair supported Thomas but was more explicit about the need to revise the standard tonnages and added two more ominous notes: one flagged the danger of a government sponsored pool and the other the threat to the share of Straits tin.[9] Simms was even more explicit. He was opposed to restriction in any form, especially since no scheme could protect Malaya's special interests.[10] Unfortunately, the others could never declare their position should the modifications they sought prove to be elusive.[11]

The subcommittee of the FMS Chamber of Mines charged with monitoring negotiations attempted to address the Dutch memorandum on the standard tonnages. This recognized that each country could argue that restriction had 'operated with greatest harshness in its particular case', and then went on to propose 1913 as a better base year than 1929.[12] The only case for 1913 was the quite specious claim that Malaya's development had been adversely affected by the war but it allowed the growth in the late 1920s to be cast in a quite different light. That growth was simply a matter of 'making up the leeway lost in the war years', in which race she still had some way to go! Having disposed, at least to its own satisfaction, of the Dutch argument about who bore the greatest responsibility for excess production, the subcommittee then went on to sketch out the case for current productive capacity. While it endorsed the figures of current domestic assessment as reflecting that capacity, its own data fell far short of that amount. Nor was there any attempt to address the key questions that would arise in any negotiations: how could expansion in Malaya during 1930 and early 1931 be rewarded when the NEI was preventing any further development; how far should Malaya press her position? Without any reflection on them, the Malayan argument remained a vacuous demand for greater fairness in the full knowledge that virtually every one of her partners could construct a no less compelling case. While that may not have constrained the FMS Chamber of Mines, it certainly affected the government.[13] In response to a letter from Lowinger,[14] Clementi issued the following instructions:

> You say … you will be surprised if can find we have a case for anything relatively better than 1929 and I agree with you it will be very difficult to put up a case for a higher tonnage for Malaya. Instructions: press Malayan claim to larger tonnage, not jeopardise the scheme by insistence on a larger tonnage, cannot agree to lower. Strong opposition should be put up against any suggestion that Bolivia and Nigeria should surrender any part of their tonnage to the Dutch.[15]

At least Clementi was prepared to back the industry in this regional rivalry.

While Nigeria has been regarded as a high cost producer and therefore with an overriding interest in restriction, this was not the view held by her Governor who alone argued for ending the ITC:

> I am of the opinion that further consideration should be given to the question of terminating restriction in August 1934, or at the most continuing it for a very limited period for the sole purpose, publicly expressed, of easing the break from drastic restriction to unrestricted production. In August 1934 producers will have had 3½ years in which to reduce their costs and put their affairs in order. Moreover they will have been carried over the worst period of the depression with lowered stocks.

> If artificial conditions and prices are maintained for a prolonged period there is clearly a danger that undesirable financial interests will acquire control of the mining industry. That danger will be lessened, if not avoided, if the industry is allowed to return to the more healthy conditions of free competition.[16]

This fear about undesirable financial interests was, of course, long standing, given the role they had played in the formation of the industry. The difficulties in finding a satisfactory formula for the domestic administration of tin restriction simply confirmed the desire to curtail this experiment as quickly as possible. While Cameron could express his views, he ultimately had to submit to the combined position of the Nigerian Chamber of Mines and the Colonial Office who informed him that he stood virtually alone in holding them.

Bolivia did not have to debate the principle of restriction but rather its basis. If restriction were to be extended then the standard tonnages for all participants would have to directly reflect their 1929 production. That meant that the concessions made to the NEI would be terminated but the ability of its delegation to press this position on the ITC was severely compromised by the lack of resolution of its problem of overproduction.[17]

Pressure to come to a decision on renewal continued to build up throughout 1933. At first it appeared that the slow rate of price recovery would prevent the liquidation of the pool before August 1934. While the other members were prepared to extend the existing agreement to permit liquidation, Bolivia was not. The sooner the agreement came to an end, the sooner Bolivia would recover the concession made in 1930 and alleviate the domestic pressure. With no progress towards a new agreement, the Dutch now sought to revise the pool agreement, tying liquidation not to price but to the decline in visible supplies which would have the effect of reinstating the August 1934 termination date. Once again the Bolivians refused.[18] The Dutch commitment to restriction that had been so forcefully expressed in Singapore in April would now be severely tested, since Bolivia was not prepared to renew on anything other than the strict 1929 basis.[19]

It is not that 1929 had any particular significance since Bolivia would have received somewhat more had the Dutch managed to secure their preferred position. It was much more a matter of the moral principle that should govern the new agreement. The old one had pointed the finger at those who had been responsible for what was defined as a crisis of overproduction;[20] the new one had to address a problem of underconsumption for which no producer could be held responsible. The Dutch position was therefore obsolete and 1929 was simply the cleanest solution to the problem. Not only was the Dutch position obsolete but it was also in practice incoherent. It required a further sacrifice from Malaya which was quite impossible and without that it could only be accomplished by insisting on a continuation of the sacrifices from Bolivia and Nigeria. Since the strict logic of the position gave Bolivia more rather than less, she could not accept the need to grant a concession that was not asked of Malaya.[21] The stage was therefore set for a major conflict between Bolivia and the NEI.

Although economically weaker than the NEI, Bolivia, in fact, held several cards. One was the position of the other partners. Nigerian support could be taken for granted, since any concessions Bolivia secured would automatically apply to her. Persistence of regional rivalry in the East effectively neutralized the Malayans. The Dutch position was more narrowly a Banka position. While the Billiton representatives loyally pressed it in the negotiations with Bolivia, their ability to do so was compromised by the fact that a concession to Bolivia could reopen the internal distribution within the NEI.[22]

Both Bolivia and Banka placed themselves in a position in which ideological considerations outweighed the results of any assessment of the benefits of continued co-operation. In principle, war was preferable to capitulation. Campbell summarized the cleavage:

> The Dutch said their Govt. was irrevocably opposed to granting the 1929 'straight' tonnage as the quota basis. They would not, and could not accept this; they would rather, having counted the cost, have economic 'war'. The Bolivians put it that whatever the merits might be (and they felt passionately convinced that the merits were with them) their President had wired them saying that their domestic situation was such that he could not get anything through which did not concede, fully, the 1929 'straight' quota basis … The intense local discussions had put the whole business on a 'national' basis; it had become a matter of national pride and prestige.[23]

Although both positions were quite rigid, there was far more flexibility on the Dutch side than on the Bolivian. Bolivia was constrained by the politics of the small miners who had never accepted the principle of restriction. In addition, her position of principle was tied to a particular number, whereas the Dutch position was not. The Dutch therefore sought a compromise and wanted to extend the current agreement to provide the time necessary to reach it.[24]

The Patiños badly miscalculated the position of the Dutch and thought they could force a settlement. Campbell tried to stop them but in vain:

> Bolivia absolutely refused to consider any extension of the existing agreements while negotiations were pending. I refused to accept this attitude; and said they must put the facts properly to their Govt. It means that the existing agreement will almost certainly end on the 1st November 1933 … that the pool cannot take an effective measure to protect itself – unless they now force tin on the market and depress the price very materially … . The Bolivian view is that with heavy unsold stocks of pool tin, the N.E.I. must agree to their terms. The N.E.I. feels this very strongly and bitterly resents it. That is one reason why they have got so angry that they would prefer 'war' … and are now deeply distrustful of Bolivia as partner and colleague.[25]

The level of distrust soon reached the point where serious discussions became impossible. During one negotiating session the Dutch presented a table which, unwittingly, would have given Bolivia virtually all her demands but it was 'refused, hotly'.[26]

As the agreement was blocked at one level, new processes emerged. One was orchestrated by Campbell who developed his own compromise position. It was ingenious since it addressed the principles of both parties. Banka would continue to receive the whole concession at the current level of restriction but would gradually return it as the quota levels rose. It would be therefore be exempted from the full burden of restriction when quotas were low but would not enjoy the full benefit when they were high. However, such an initiative would only bear fruit if other means were brought to bear on the two intransigents and that required the intervention of both Cunliffe-Lister and the Dutch Prime Minister, Hendrikus Colijn.[27]

Cunliffe-Lister's first thought was to inquire as to whether there was any way in which counter-pressure could be put on Bolivia, that is by blackmailing the blackmailer. But Bolivia had no other international games apart from tin and hence there was no leverage against her policy.[28] He then allowed Campbell to speak directly to Colijn who confessed that he was:

> ready to throw away millions for the sake of a few thousands … but, there was in all such cases a point where national prestige and national feeling became so strong that the arithmetic of the thing must be disregarded.[29]

However, Campbell was able to get a discussion of the basic arithmetic of his compromise and eventually secured Colijn's endorsement. He then attempted to get Bolivia to make the reciprocal gesture and presented the position to Patiño in the direst of terms:

> I have been so anxious regarding the tin situation, and so convinced of the appalling disaster which failure to conclude the new agreement would result in that I – entirely on my own initiative – obtained the sanction of my Secretary of State … to ask Dr. Colijn to accord me an interview. I urged him in the most insistent manner possible, to do everything that lay in his power to secure the conclusion of the new agreement. With the utmost difficulty and obviously with the greatest hesitation and dislike, Dr. Colijn finally said that he would be prepared to accept the compromise, despite the attitude of the Netherlands East Indies Government which is resolutely opposed to that compromise, and has definitely rejected it as wholly unacceptable.[30]

Howeson now became involved in adding support to Campbell's pleading but to no avail. Colijn's acceptance of the compromise had reduced the gap considerably and Campbell now approached van den Broek to close it completely:

> There are too many millions of pounds at stake to make any other course possible. The thing has now been narrowed down to something which just does not come into the picture, as compared with the appalling loss and disorganisation which failure to secure a renewal of the agreement would undoubtedly lead to. Nor would the thing end with "tin" clearly. There would be an inevitable repercussion on "rubber". And the effect of a breakdown would almost certainly have further consequences on the various "world" control schemes.[31]

As the failure of tin control loomed large, the stakes had become very high indeed.

Howeson's approach to Patiño resulted in the basis of a solution.[32] He agreed to the lowest concession that was generated by Campbell's formula, 300 tons but only for one year, 1934. Nigeria was expected to contribute on the same terms, 55 tons. Van den Broek advised the Colonial Ministry to endorse this proposal which at least conceded the principle to which the Dutch were so firmly attached and then discussed it with Banka.

De Iongh's immediate reaction was to dismiss the offer as of no interest, nothing more than a 'beggar's tip'.[33] But he was not prepared to issue a declaration of war. Colijn had softened his own position on learning that there was a strong possibility that Bolivia could come to a direct arrangement with the United States to feed a protected smelter[34] and there was some feeling that he might intervene again. In any case, the amounts at stake were now too small to warrant throwing away all that had been accomplished during the first agreement. As de Iongh conceded, one lesson was clear, the NEI had to equip itself to be in a stronger position during the next round of negotiations when the prospect of economic warfare might have to be taken more seriously.[35]

In these protracted negotiations one important issue fell away. At the preliminary conference in Singapore in April 1933, the Dutch raised the question of a buffer stock with which to stabilize the market and secured an agreement from Clementi that Malaya would participate on equal terms with the NEI and Bolivia.[36] No progress could be made on this matter[37] until the question of the standard tonnages was settled and by that point the first agreement was about to expire. As soon as the second agreement became operational it remerged and the surrounding controversy might have been obviated had the issue been addressed as part of conventional negotiations.

Although Patiño may have felt pleased with the outcome of the negotiations, Bolivia had violated the norms of negotiation, especially by insisting that the Dutch speak from an apparent position of comparative disadvantage as a result of their unsold pool stocks. Their inevitable resentment would continue to weaken the ITC. Not only had Bolivia paid a heavy moral price but the material benefits in fact proved to be transitory as war with Paraguay over the Gran Chaco prevented her from using her increased entitlement.

Siam

The flat rate of 10,000 tons that Siam enjoyed under the first agreement was clearly a stopgap measure and she was expected to move to a position more consistent with that of the signatories. The first offer to Siam was therefore of a minimum of 7,000 tons rising when the international quota reached 65 per cent.[38] As this proposal was reviewed by the Siamese government it received some support, since the country's reputation as a genuine supporter of tin restriction was at stake. But there were too many practical problems in imposing such a severe cut, especially in relation to continued expansion in production capacity. As Siam reflected on the moral implications of holding fast, she took comfort in several irrelevant arguments: the ITC had been formed without her participation; it should be looking to capture the 17,500 tons produced by outsiders; low levels of production were only imposed to allow the liquidation of an unnecessary pool.[39]

As long as the signatories remained at logger-heads they were unable to formulate a policy to deal with Siam. In June, the country went through a political upheaval[40] and several months passed without any response. Once the negotiations with Bolivia had been completed, the ITC was ready to endorse a new agreement but, of course, would not again make the mistake of doing so without Siam's participation.

A special meeting of the ITC was called at which the following strong resolution was passed:

> The Committee desires to emphasise, with all possible insistence, that unless the new agreement is signed by Siam as a participating country the present control agreement will lapse in the very near future ... The Committee are convinced that the consequences will be disastrous for Siam, which is and has been profiting greatly by the heavy sacrifices made by the participating governments ... The delegations of Bolivia, Nigeria, the N.E.I. and Malaya feel that the situation warrants the strongest and most urgent diplomatic representations .... [41]

Although the threat of termination was serious, it was not tied to specific terms. The Siamese government could therefore respond with a counter offer: the existing flat rate not as a maximum but as a minimum, rising by 1 per cent for each percentage point whereby the international quota was increased over 65 per cent. The only qualification that was made was to put the 10,000 tons on a true assay basis. A true assay basis was complicated to administer, since it involved minor retroactive adjustments once smelters' returns were received and Siam claimed that it was too difficult to implement. She was therefore allowed to control exports by recording concentrates as though they had a nominal 72 per cent metal content of 9,800 tons but which, at 73.47 per cent, really contained 10,000 tons.[42]

When Siam's assent to this minor stipulation was received, the agreement was ready for formal ratification. There remained but one detail to settle and that was the carryover of the excess production from the old agreement to the new. Each signatory member was granted a special quota of 4 per cent from which excesses due to deliberate overproduction, as in the case of Bolivia and the inadvertent overproduction through the rise in assay values, as in the case of Nigeria and Malaya, would be deducted.

Placing every member on a true assay basis involved a modification in the statistical basis of the new agreement. Since control was exercised over concentrates, modifications were applied according to the following formula. The figures in the first agreement were grossed up by the notional assay value applicable in 1931 to get a total figure for concentrates to which the true assay value established in 1933 was applied.[43] That meant that all figures were raised slightly and, since the rate at which assay values changed varied between the members, the ratios changed as well. The results of this new distribution are presented in Table 9.1.

One result of this process of readjustment was that, in spite of the aggressive way in which she had attempted to recover the full extent of the concession made to the Dutch in 1930, Bolivia had failed to establish her strict 1929 ratio. On 27 October the ITC could finally announce that the new agreement was in place.

A public confession of the difficulties in reaching this point soon came at the CGF AGM, when the Chairman commented:

> The agreement was accomplished in the face of the most bitter hostility on the part of several producers, prompted in some instances, not so much by genuine conviction that the scheme was unsound but by motives of a less worthy character. In addition, these negotiations for some time were carried on in an atmosphere of wire pulling and intrigue that would have daunted many men with less strength of character than those responsible for the conception of the scheme and bringing it to fruition, magnificent and unselfish part played by the leaders of this movement, the very first of whom we count our esteemed colleague, John Howeson.[44]

This was not a universal sentiment. The metal traders, Brandeis, Goldschmidt, considered that 'abandonment of restriction would not only have helped industry and created more employment right through the tin trade but would also save the Tin Syndicate from a collapse, which is inevitable if the present policy is pursued'.[45] This was but a warning shot in what would become an open battle between the LME and the ITC.

In its rush to make the announcement, the ITC placed the Colonial Office in an embarrassing position. Not only had it violated the norms of courtesy, since the Foreign

Table 9.1 Standard tonnages, 1934–1936

	First Agreement		*1934*		*1935–36*	
	tons	*%*	*tons*	*%*	*tons*	*%*
Malaya	69,366	42.8	71,940	43.4	71,940	43.4
Bolivia	44,338	27.3	46,190	27.9	46,490	28.1
NEI	38,090	23.5	36,685	22.2	36,330	21.9
Nigeria	10,384	6.4	10,835	6.5	10,890	6.6
Total	162,178		165,650		165,650	

Office only learnt of this through the press, but the much more serious norms of diplomatic protocol. An international agreement had been signed without being properly vetted and in any case such agreements should only be signed by metropolitan not colonial governments.[46] It was an issue that would recur on other such occasions.

Recruiting outsiders

The ITC had three arguments at its disposal in its attempts to expand the scope of the agreement. The World Monetary Conference had laid out the case for commodity agreements as an effective counter-depression measure and if tin were to fail, then the prospects of success in more complicated cases were remote. There was also the moral obligation on each producer to do the right thing by its fellows. However, the only real backing for these positions was the threat of a return to a free market. Had the ITC been able to negotiate with all outsiders as a group, as Campbell had hoped at the Conference, then a coherent strategy might have been developed. Instead, the ITC was forced to negotiate with each producer separately.

Negotiations proceeded under two serious constraints. The first was the logistical problem of communication. Not only were there several links in the chain from London to the local government to the local industry but they were connected by letters and memoranda, resulting in considerable time delays. The second was that, with one exception, while tin was often of regional importance, it was not of national significance in any of the outside countries. Governments therefore had no well-formulated policies for the industry and often the industry itself was not well-organized.

The issues of principle that had been so contentious in internal discussions were set to one side in the discussions with outsiders since they were irrelevant to the issue of who bore responsibility for the historic problem of overproduction. All the ITC could realistically hope to accomplish was a restriction not of actual production but the extent to which it could grow. Since new members would still be considered as outsiders for the purposes of article 20, its implicit threat might well be strengthened. The terms settled with Siam provided the framework for all other discussions, so that there was consideration of both a minimum flat rate and a standard tonnage.

Cornwall

The approach to Cornwall was initiated through the Department of Mines. The ITC secretary, Blenkinsop, stressed the importance of curbing the production of outsiders as a condition of renewal of the agreement. While this threat was taken seriously, the Department felt under two constraints. One was the fact that an official committee had but recently recommended an increase in metalliferous mining in Britain[47] and the Department did not want to be under any obligation to offer a quid pro quo in the form of government assistance.[48] However, the Department was prepared to serve as an intermediary and passed on Campbell's proposal to the Chairman of the Cornish Chamber of Mines, Arthur Thomas, brother to Charles who was expressing reservations at the annual meetings of Tronoh and Sungei Besi.

This overture was made in the midst of the World Monetary Conference. Campbell was not only manipulating the tin subcommittee but he also managed to get Arthur Thomas to call a special meeting of the Council of the Cornish Chamber of Mines and pass a resolution indicating the intention of the Chamber to engage in negotiations with the ITC. This was then announced at the end of the last meeting of the tin subcommittee, providing some momentum for Campbell's other initiatives and 'its first fruits came at once in the shape of a similar promise by Portugal – not a very exciting recruit, but still a recruit'.[49] What was not made public were the caveats attached to the resolution which not only included references to equity and the special condition of the Cornish industry but also to the participation of other countries. Cornwall actually gave away nothing but Campbell got a useful gesture.

Actual negotiations between Campbell and the Cornish Chamber took place in September. While Campbell thought that 'Cornwall will probably prove to be a particularly easy case',[50] he could not have been more mistaken. Negotiations started with a proposal for a flat rate of 1,800 tons which was already generous in the light of the current production of 1,500 tons but Thomas managed to get this raised to 2,200 and which was then endorsed by the Chamber.[51] Even this was not straightforward as Thomas reported to the Department of Mines:

> I am greatly relieved I got this decision; there was the usual opposition. One greatly wonders why some people cannot appreciate the position the industry would find itself in if restriction was not in force and how much more secure if it becomes more general.[52]

The small miners of Bolivia were not alone in their inability to grasp the fundamental logic of cartelization.

The decision of the Chamber was simply a recommendation which had to be approved by the directors of the companies affected. It came under attack in the press as the *Daily Express* opined that 2,200 was much under what Cornwall could do and that in any case there should be no limitation on 'home production of a natural commodity'.[53] In the end unanimity could only be secured around a flat rate of 2,600 tons, a level quite unacceptable to the ITC.[54]

Matters were therefore at an impasse. Prospects of successful negotiations were substantially reduced in March 1934 when Charles Thomas declared that while he had agreed with the position taken in July, it was now time to move towards decontrol of the industry and that if Cornwall came into the ITC it should be on the same terms secured by Siam.[55] With negotiations at an effective end, the three principal companies, South Crofty, East Pool and Geevor, decided to approach the ITC independently and struck what was defined as a gentlemen's agreement. They agreed that they would operate with a flat rate of 1,700 tons, increasing with a rise in the international quota over 65 per cent. While this was actually higher than their share under the 2,200 tons proposed for the whole county, it was accepted by the ITC without further discussion. It was a gesture that left Arthur Thomas, who had worked for so many months to secure a more inclusive arrangement, very bitter, since real gentlemen would have at least kept him properly informed.[56]

Portugal

The tentative commitment made at the World Monetary Conference was followed by a discussion between Campbell and the Portuguese Ambassador in London. These discussions proceeded on a quite different basis from those in 1931. There was no question about a cut in production, nor was the industry involved in any significant way. A simple exchange of views resulted in the following recommendation: a flat rate of 650 tons rising when the international quota was above 81 per cent. This was equivalent to a standard of 800 tons and bore some relation to the past being just under Portugal's peak output of 1926. The flat rate of 650 tons was just about her current level, so that while it involved some limitation it was certainly 'no sacrifice'.[57]

French Indo-China

The World Monetary Conference provided an opportunity to make considerable progress with the French about the participation of Indo-China.[58] The metropolitan government was entirely supportive of the conference's overall position on commodity agreements and appears to have been quite prepared to make a decision without local consultation.[59] Since the estimate of potential annual productive capacity remained at 5,000 tons, it was far too large to remain outside the scope of the agreement. Negotiations simply focussed on the rate at which French Indo-China would be permitted to expand. They were completed by the end of October, with the result that French Indo-China was granted a flat rate of 1,700 tons for 1934, 2,500 for 1935 and 3,000 for 1936. This was a pure flat rate with no provision for increase with the international quota.

Belgian Congo

At the World Monetary Conference the Belgians expressed no interest on the grounds that their production was far too small. The first formal overture was rejected by the Ministry of Colonies, which argued that the whole matter should be deferred until productive capacity had been established.[60] However, once the French had decided to participate, it was felt that the Congo could not afford to continue to be excluded.[61]

Negotiations with the ITC were not easy, since the Belgians demanded very generous terms. Although output in 1933 had been 1,957 tons, they claimed a total of 17,000 tons covering 1934 and 1935 and no limit at all for 1936.[62] Even these terms were insufficient to secure the assent of the second largest producer, Symétain.[63] In addition to arguing that productive capacity was already increasing, the Belgians also claimed that they were prepared to run the mines at considerable losses, provided that they generated high levels of employment.[64] Such a position undermined whatever threat to freeriders might be contained by article 20. If employment, not profits, were the object of mining policy, then low prices were not to be feared. Indeed, they might actually increase production.

The ITC, however, was not prepared to concede the full extent of the Belgian demands. Initially it offered some 9,000 tons which led to forceful condemnation in the Belgian press and the breakdown of negotiations. By February 1934 they were resumed with an ITC offer of a standard tonnage of 10,500 tons which provided the basis of a solution. The Congo would be granted an increasing flat rate – 4,500 tons for 1934, 6,000 for 1935 and 7,000 for 1936, with this last figure being subject to an increase of 25 tons for each percentage point by which the international quota exceeded 65 per cent.[65]

Such terms were very controversial and criticism was particularly severe in Malaya.[66] But Clementi was not prepared to oppose them and within the ITC the most serious reservations were expressed by Baddeley on behalf of Nigeria. In part, this was a function of regional rivalry since the Congo would now overtake Nigeria as the largest producer in Africa but it also expressed annoyance at the fact that the Belgians had lied but a few months earlier during the World Monetary Conference when they claimed their output was inconsequential.[67] However, the ITC had little choice but to grant the substance of the Belgians' case. Since their plans for expansion had been shaped long before the ITC had demonstrated its ability to control the tin market, they could not be accused of being conventional freeriders; the trade-off between production and price for the Congo was quite different from that of the established producers. It was the magnitude of the levels of production that might come from this new source that struck alarm, so any restriction would be desirable and Campbell

Table 9.2 Distribution of production quotas, 1934–1936

	1934			*1935*			*1936*		
		Quota 45.0%			*Quota 58.8%*			*Quota 92.5%*	
	tons	*tons*	*%*	*tons*	*tons*	*%*	*tons*	*tons*	*%*
Standard tonnage									
Malaya	71,940	34,220	35.4	71,940	42,265	35.6	71,940	66,544	37.8
Bolivia	46,190	20,988	21.7	46,490	27,313	23.0	46,490	39,516	22.4
NEI	36,685	18,189	18.8	36,330	21,344	18.0	36,330	33,605	19.1
Nigeria	10,835	5,015	5.2	10,890	6,398	5.4	10,890	10,073	5.7
			81.0			82.0			85.0
Flat rate									
Siam	9,800	9,800	10.1	9,800	10,290	8.7	9,800	12,495	7.1
BC	4,500	4,500	4.7	6,000	6,125	5.2	7,000	7,687	4.4
FIC	1,700	1,700	1.8	2,500	2,500	2.1	3,000	3,000	1.7
Cornwall	1,700	1,700	1.8	1,700	1,830	1.5	1,700	2,418	1.4
Portugal	650	650	0.7	650	684	0.6	650	741	0.4
			19.0			18.0			15.0

comforted himself with the thought that 'at least we gain definite limits, and we get the Belgians where we can argue with them'.[68]

The results of all these negotiations are presented in Table 9.2 which demonstrates the way in which the entitlement of those on a flat rate basis changed in relation to the overall international quota.

Although these four new members were recruited at quite different stages, it was not until the agreement with the Belgian Congo was in place that the terms of their participation was officially announced.[69] By then, all the other efforts of the ITC to secure universal membership had come to nought and the experience demonstrates the limitations of the moral appeal issued at the World Monetary Conference.

Union of South Africa and Swaziland

The South African government received the ITC's request for support with far less sympathy than in 1931. It found a convenient obstacle in the fact that base mineral rights lay with the surface holder and legislation would therefore raise some complicated jurisdictional questions. However it had no objection to negotiations being undertaken with the producers, following the Cornish model. While the producers were supportive, negotiations languished.

The neighbouring protectorate of Swaziland offered more promise. In spite of its size, Swaziland's peak production in recent years was only 270 tons; Campbell was keen to get its support for its 'psychological' value. Bringing in all the areas under the control of the British government would help strengthen its hand in dealing with the others. Campbell, therefore, proposed terms which would simply curb potential rather than affect current output, a flat rate of 150 tons with a standard tonnage of 270.[70] The Resident Commissioner endorsed the principle, suggesting a slight increase in the flat rate to 180 tons, to stimulate the only industry that provided an alternative to agriculture but without encouraging new production.[71] Campbell naturally agreed and it seemed as though the matter was settled.[72]

It came unstuck because South Africa was dragging her heels. Swaziland felt uncomfortable about being asked to make any kind of sacrifice that was not being imposed on her neighbour. It could also have political ramifications if the European population seized on this as a means of supporting the South African government's overture for a complete absorption of the protectorate.[73] Deferring a definitive agreement had provided an opportunity for a modest expansion on the part of the local industry which now pressed for a much larger flat rate of 300 tons.[74] Since there were limits to Campbell's definition of 'reasonable', the matter was dropped and he drew the following conclusion:

> The local producers and the local administration take what is I am afraid the obvious and the general attitude in such cases – "let us ride on the back of the others, the maximum extent possible. Let them make the sacrifices: let us take the benefit. It is quite a safe policy for us in the circumstances". And that I am afraid is that! I don't see that we can do anything further about it: except quote the Bab Ballads: "If this is human nature, Oh, Then isn't human nature low."[75]

What Campbell failed to realize was that the strategy of one-on-one negotiations with the small producers set up a situation in which general moral arguments were almost bound to founder on local practical ones. In any case, the longer such negotiations proceeded, the shorter the time before the expiry of the agreement and the weaker those moral arguments became. As is evident here, failure had a cascading effect. Other strategies, backed by the real threat of invoking article 20, might have been more fruitful and there were other prices that the ITC could have exacted in return for issuing tickets for free rides, such as contribution to the costs of the research scheme.[76] While token, they would have at least been consistent with the moral universe within which Campbell operated.

In the case of South Africa, the companies eventually proved 'willing, almost eager' to participate and drew up a detailed operational scheme.[77] However, such support was not sufficient to lead them to accept Campbell's proposal of a standard tonnage of 1,300 and a flat rate of 800 tons.[78] By mid-1934, Anglo-French had broken ranks and all that remained was the prospect of a Cornish style gentlemen's agreement among the others.[79] None felt sufficiently moved to initiate one.

China

China was the big prize but the prospects of securing an agreement were remote. After all, there were no organic links between any part of the Chinese industry, from mining to smelting, with the rest of the tin world. Discussions were opened through diplomatic channels in both Yunnanfu and Peking but without any expectation of success.[80] Yunnan was offered a flat rate of 8,000 tons and a standard tonnage of 11,000 and although these produced an acceptable counter-offer of a flat rate of 10,000, negotiations 'faded into oblivion'.[81] The British Consul-General in Yunnanfu warned that the warlords of Yunnan were hardly likely to keep to any agreement[82] and the ITC would certainly have been far worse off if they had formally agreed and then manipulated the statistics as to the actual quantities being produced.[83]

Burma

The threat of non-renewal unless the scope of restriction was widened that Campbell had issued at the World Monetary Conference was taken very seriously by the Indian delegation, which advised Burma to 'come into the scheme'.[84] The Tavoy Chamber of Mines was divided on the question, in part because of potential complications in imposing tin restriction on mines producing both tin and wolfram and, while the local government was also reluctant, it was prepared to accept a suitable flat rate 'if abstention of Burma is likely to affect adversely attitude present participating countries'.[85] Campbell and Howeson represented the ITC in the negotiations with the India Trade Commission, acting on behalf of Burma and quickly produced an agreement on a standard tonnage of 3,850 and a flat rate of 3,080.[86]

Without the principals at the table, there was some scope for revision of these terms. The Tavoy Chamber pressed for an increase to allow for new productive capacity and a counter offer was made accepting the 3,850 not as a standard tonnage but as a flat rate.[87] Since this was a level that implied virtually no restriction, Campbell expressed doubts that it would be acceptable, especially since it would adversely affect negotiations with others, adding his disappointment 'that the Government of India and the Government of Burma did not take a more international view of the scheme'.[88] In spite of this, the ITC accepted the increase but tied it to an explicit commitment to use it only for new production.[89] Then the Tavoy Chamber found further grounds for prevarication and, as the matter dragged on, the support of the Burmese government waned. The international agreement was no longer contingent on Burma's adherence and the government would have to incur substantial administrative costs to ensure compliance with terms that produced only nominal restriction.[90] All the ITC got from Burma was a suspension of the issuance of new EPLs.[91]

These negotiations reveal several difficulties. One is the implementation of the condition laid down by Cunliffe-Lister, namely that there should be broad support within the local industry. That eliminated any leverage against exploiting the freerider option. The second is the existence of too many links in the chain connecting the tin producers to the ITC. Negotiation at the governmental level involved five intermediary links and that dissipated the resources of explanation, persuasion and compromise.

Australia

At the World Monetary Conference, the Australian delegate expressed little sympathy with the general principle of restriction as a means of addressing the problem of primary commodities, though he indicated willingness to consider particular cases.[92] The government followed up the ITC's formal request in September 1933 but the industry was in no position to provide any assistance. By 1933 only one company was operating and that was responsible for but 10 per cent of Australian production.[93] With only 1,000 tons spilling over onto the international market, no apparent prospect of any increase, the Australians could justifiably claim that there was little point in making even a token gesture.

Other initiatives

The ITC also approached Mexico and Japan. In Mexico new deposits were being exploited but they were on such a small scale that it is difficult to see what form participation could have taken. Similarly, the approach to Japan was doomed to failure, given its position as a net tin importer and one which imposed a heavy import duty on tin metal.

It was a pity that the attempt to recruit new members was not made more thoughtfully and selectively. Only with an industry of a certain size could participation be considered to have any real impact and that should have ruled out South Africa, Swaziland, Mexico and Japan. Similarly, the industry needed a rudimentary administrative structure and that should have ruled out Australia. In its search for universal adherence, the ITC was collecting more rejections than acceptances and that weakened its ability to maintain pressure on the only two countries that really mattered, Burma and China.

10 Stabilizing the tin market, 1934–1936

As a temporary measure to solve a short-term problem, the ITC could count on a great deal of support but as it became a more permanent feature of the industry, its policies proved much more controversial. Working towards equilibrium where none existed was comparatively easy, especially when all agreed that stocks were too high and prices too low. Trying to define an equilibrium and then maintain it was much more difficult, especially in the face of new destabilizing forces. Prices became controversial. Metal brokers fought against stock control. Uncontrolled production emerged as a potential threat. Even production under control became unpredictable as the Chaco war began to affect Bolivia. However, the ITC now had some experience on which to draw, as it tried to bring stability to the industry in the face of a deteriorating external political environment.

Regulating production

The task of setting production quotas was approached in the first instance as though it were a purely technical one, based on estimates of physical indicators. Consumption trends were reviewed to provide an estimate of the anticipated demand over the forthcoming quarter. Stock levels were reviewed to see whether they were too high, in which case consumption would have to draw on them; if they were too low, then production would have to exceed immediate consumption in order to build them up. The level of production not subject to quota was then assessed, together with any deficit accumulated by members of the ITC. That left a production target to be met from the quota.

This conceptually simple process was complicated by two factors. One was the need to take into account external reactions which could come from several quarters: governments, producers, consumers and traders. The other was the reliability of the various estimates. By 1933 the statistical section of the ITRDC was in a position to devote some effort to producing and publishing estimates of the crucial indicators.[1] Production estimates were reasonably sound but estimates of real consumption, even when retrospective, were inevitably guesswork.[2] The critical variable, of course, was the level of invisible stocks, especially those in the hands of consumers and that was also a matter of informed guesswork. With uncertainty about the level of consumption and stocks, the process of setting quotas inevitably moved from the purely technical realm into the political one of who should bear the risks of misjudgement.

Setting the quota for the first quarter of 1934 illustrated many of the issues which the ITC would continue to address. It started with the technical arithmetic presented by Lowinger which involved the following steps: (1) estimate anticipated consumption for the following quarter, 29,400 tons; (2) deduct estimate of extent to which this should drawn from existing stocks, 2,500 tons, leaving a production target of 26,900 tons; (3) from that target deduct amounts that would come from fixed sources, outsiders 4,700, Siam 2,500 and the entitlement of core members from the 1933 adjustment, 1,075, for a total of 8,275 tons; (4) the residual of 18,625 tons, spread over the new standards of 165,650 tons, produced a quota of 45%. In this presentation, the only issue was the rate at which stocks should be brought down and Lowinger wished to move cautiously, reducing them to normal levels over the course of the next 12 months. The more conservative the stock figure, the higher the residual to be met by the quota. A mistake on the generous side would be corrected by cutting the quota for the following quarter.[3]

Such an approach forced the producers to bear the risk of misjudgement and it was one that was not easily supported in light of the experience of the first agreement. Houwert reminded the meeting of the difficulties in getting production down fast enough and this was echoed by Patiño who pointed to the particular problem Bolivia had encountered with the small miners. A quota of 45 per cent, coupled with the entitlement from adjustments at the end of 1933, meant an increase of 32 per cent in actual production on the part of the signatories. Cutting it again soon after such a substantial rise would not be easy and caution therefore dictated a lower rather than a higher quota. That was then reinforced by van den Broek's desire to see stock levels reduced at a much faster rate, not least to allow the ITP to sell its remaining holdings. As a result, the meeting settled on a 40 per cent rather than a 45 per cent quota.[4]

This was the first point at which the ITC had actually increased a quota and it was preceded by an editorial in *The Economist* which opined: 'in view of the substantial reduction in the costs of production, the price of £225 is certainly not justified'.[5] The ITC was caught in something of a dilemma: whether to pay attention to the price implications of the way in which it wanted to align production, consumption and stocks, or should it disregard price altogether. As usual, Campbell experienced that dilemma more acutely than anyone else. The editorial provoked an internal discussion within the Colonial Office in which he commented: 'I much prefer £200 to £230. I have time and again expressed the view to Mr H and to the Dutch that we must guard against putting or allowing price to get too high.'

Yet at the same time he noted that the stock position was such as to justify the 40 per cent quota that was adopted. With prices on the high side, visible and invisible stocks too large, the quota could only be an uncertain step towards stability and he gave up trying to work out an optimal path with: 'Strong probability that the producers know their business best. It is their life and livelihood. We officials don't suffer from the mistakes.'[6]

One direction towards stability was offered by the Dutch who, with the support of the Bolivians, pressed for the introduction of a buffer stock. This proved to be the most internally divisive issue of all.

Buffer stock

The conception and role of the buffer stock differed quite fundamentally from that of the pools. The pools were designed to give support to a sagging market by temporarily freezing metal that was already in its commodity form. The buffer stock, however, would be used simply to fine tune the relationship between supply and consumption and would be made up of metal produced specially for the purpose. Most important, it would not comprise excess stocks but rather be part of the normal level of stocks. As a result, the buffer stock could be used to stabilize prices, at least in the short term. As soon as it was formed, it could sell in order to prevent prices from rising and subsequently the cash received would allow it to buy to stop them from falling.

The argument for a buffer stock was quite simple. The lag between fresh production from the mines and its appearance as metal in the market was simply too long to allow for any close relation between production and consumption to be established. That interval was over two months, to which had to be added the time required for the ITC to make a decision authorizing a change in the quota and the time required for the miners to produce it. Such a gap provided an opportunity for speculators to drive the price of the metal up or down, preventing any stability in the market.

The buffer stock would be formed by a special quota of 5 per cent of the standard tonnage of the four signatory governments and each producer would have the right to participate so that, at least in principle, the interests of stock holders and producers would be identical. Since Siam was on a flat rate and hence exempt from changes in quotas, at least at the current level, she was not invited to participate. The buffer stock would be held in addition to normal commercial stock levels, so that it would serve as a cushion against an unanticipated rise in demand. In order to obviate a repetition of the problem experienced with the ITP, the buffer stock would be liquidated a full year before the expiry of the overall restriction agreement itself. Price policy was not part of the formal proposals but informally the Dutch made it clear that they wanted prices stabilized within a range of £140 to £160 in gold. That would make the pivotal price in sterling £240, just above the current market level, so the buffer stock was considered as a means of trying to bring stability to the existing structure of the industry.[7]

In view of the way in which postwar thinking about commodity agreements placed great stress on the indispensability of such a buffer, the controversy about the formation of this buffer stock is somewhat surprising.[8] However, it served as a lightning rod with which to express deep-seated reservations about the very conception of restriction. One source of difficulty was the resentment generated by the lingering sense that the ITP had profited at the expense of the producers.[9] More substantial was a concern about the relation between normal commercial stocks and the buffer stock.

Since there was no accurate information about either the levels of commercial stocks, or the anticipated demand they were designed to serve, there was no precise signal of the point at which the buffer stock should be released to compensate for a decline in normal stocks. Still less was there was any obligation on those who controlled those stocks in fact to release them. The buffer stock therefore conferred a substantial power in the market. It was a power that would inevitably be exercised against the metal brokers whose entire livelihood depended on providing an alternative form of buffering. It was also one that could be exercised against consumers by keeping stocks so tight as to sustain prices at an artificially high level. Considerable trust would therefore have to be placed in the good judgement of those responsible for managing the stock. Since the stock was to serve as a buffer, details of its composition in cash and metal would not be made public; the ITC would have a monopoly of all the relevant statistical information and this in spite of its professed desire to make tin statistics more reliable. There would therefore be no independent basis for an effective commentary on the judgement of the members of the buffer stock committee who could determine at least short term prices.

As the Colonial Office presented the case for the buffer stock to Nigeria and Malaya it also frankly reviewed some of the difficulties, especially with respect to the implications for renewal or termination of the agreement.[10] Two conditions were laid down. Just as with the overall agreement itself, a substantial majority of producers had to be in favour for the scheme to go forward. In order to ensure that all producers were in a position to exercise their right to participate, governments would have to be prepared to extend a loan to cover the cost of producing the additional quota. Since the metal in the buffer stock was not yet in its commodity form, governments would run no risk in providing such finance as the value of the metal when realized was bound to be considerably higher than its marginal cost of production.

As soon as the process of negotiating the buffer stock commenced, a vigorous campaign was launched against it, especially in Malaya. The inherited 'personal antagonism, cleavages and mistrust'[11] of the previous round of tin politics made it impossible for many to examine the proposal on its own merits and come to an objective assessment. The Cornish interests, in particular,[12] defined it as yet another Howeson-Dutch maneouvre, designed to extend the scope of tin control and further marginalize Malaya and Straits tin. In this, they were naturally supported by STC and most trading associations. Formal resolutions opposing the scheme were passed by the Council of the London Metal Exchange,[13] British Association of Straits Merchants, Association of British Malaya, FMS Chamber of Commerce, the Straits Settlements (Singapore) Association and the Singapore Chamber of Commerce. The buffer stock proposal simply reproduced the same kind of division that had emerged on other aspects of restriction. Clementi was strongly in favour[14] and he had some support from Hake in the debate in the Federal Council who had sharp words for those who were harassing the ITC based on half-knowledge.[15] But other representatives of the mining industry spoke against;[16] they could do little else given the fact the Council of the FMS Chamber of Mines had voted 9/4 against the scheme, as had the Perak and Selangor Associations of Chinese miners.[17]

Opposition to the buffer stock received considerable support from a fresh quarter. The Chairman of the TPA, Sir George Maxwell, resigned over the issue.[18] He had two fundamental objections. The level of normal stocks should be closer to 25,000 rather than the 15–18,000 proposed, so the scheme presaged a severe shortage. In addition, it detracted from the fundamental objective of the second agreement which should be to move towards decontrol and it therefore implied the perpetuation of restriction. Howeson had secured the support of the TPA Executive Council without properly informing him and Maxwell expressed his opposition very publicly and forcefully.[19] As intended, Maxwell's resignation served to destroy whatever lukewarm support there might have been. It certainly jeopardized the referendum Clementi had promised to Malaya since that would have resulted in a substantial majority against the scheme.[20] It also provoked Campbell to one of his more intemperate outbursts:

> nothing the ITC – or the Angel Gabriel for that matter – can do will gain the confidence of the Malayan producers as a whole. There is an irreconcilable minority which does not want to be placated and will always refuse to be pleased. (One wonders if they are of Irish origin?) They want a grievance to hug and will always find one. They did nothing and said nothing which could hinder the renewal of the agreement; when it was renewed the chorus of disapproval burst forth afresh. That is not cricket. But I am under no delusion that these tactics will not be repeated.[21]

In fact there was an element of rationality to the opposition. They did propose an alternative, which was to increase the level of minehead stocks of concentrates ready for shipping. This would avoid the problem of direct control over the market and preserve an independent role for Straits tin within it. But it could not serve as a stabilizing device since it would only be a one-way stock, absorbing increases in consumption but not cushioning declines. Nor could it avoid the inevitable price fluctuations between the shipment of concentrate and its appearance as metal.[22] It is therefore hardly surprising that this approach was not taken very seriously.

Nor was there any enthusiasm for the buffer stock proposals in Nigeria. Government participation was too unorthodox for Baddeley to support the Colonial Office on the issue.[23] Profits made by the ITP still left a bad taste[24] and local officials agreed with the Malayans on the alternative of raising the level of minehead stocks and expressed indifference to the implications for price fluctuations. In any case, they were not prepared to see government provide financial assistance.[25] A meeting with local representatives found the industry divided along existing lines, with the large companies in favour, the small strongly opposed and the medium companies supporting but wanting assurances that this would not be a step towards permanent control.[26]

In communicating the reservations of the Mines Department back to the Colonial Office, the Governor made one important modification, namely that he would be prepared to support the adoption of the buffer stock and provide government finance, as long as this was requested by a substantial majority of the Nigerian Chamber of Mines in London.[27] The Nigerian Chamber duly met and, after much discussion, passed a motion supporting the buffer stock, 'subject to the understanding that the acceptance of this proposal does not commit the industry to any further prolongation of the tin restriction scheme beyond that fixed in the agreement recently concluded'.[28] Such a declaration was, of course, quite empty but it did reflect the persistent distrust of the TPA and its future plans and went some way to mollifying the opposition expressed on behalf of the small miners.[29]

The question of the buffer stock dominated all ITC meetings at the beginning of the second agreement since the Dutch were under explicit instructions not to agree to any increase in the regular quota without it. Bolivia's position was just the reverse. Faced with the need to finance the Chaco war, she wanted to maximize current revenue and then form the buffer stock.[30] Before this impasse in the ITC could be resolved, Cunliffe-Lister decided to disregard Campbell's advice[31] and broke the log-jam in Malaya by announcing that both Nigeria and Malaya would support the scheme.[32] In explaining the decision to an embarrassed Clementi, Cunliffe-Lister stressed the responsibility of the ITC to consider consumers' interests. March had provided a dramatic demonstration of the value of such a buffer stock. Deliveries to both the USA and the rest of the world jumped by 30 per cent over February. Although the quota had been increased from 1 January, fresh supplies had yet to come on stream and it was only sales from the remainder of the ITP that prevented prices from rising to unacceptably high levels.[33]

It was easy to extrapolate the consumption trend[34] and as long as increases in production were expected to lag behind those in consumption, consumers would expand their stocks. With the invisible stocks accumulated during the previous years of low prices still at a high level, such buying would be 'insensate' but it could drive the price up to as high as £260.[35] The simple announcement of the formation of a buffer stock with its commitment to provide adequate supplies could be sufficient to defuse any further buying movement. With sales from the ITP exhausted, there was no other way of preventing a rise in price. Mair, unfortunately, remained unconvinced and saw this experience as not the reason but the excuse to force his group 'in a straitjacket designed not by those who have to wear it but by theorists who think they have a genius for planning other people's clothes'.[36]

As prices continued to rise throughout April, Campbell called a special meeting to settle both the quota for the second quarter and the buffer stock. In addition to removing the obstacles in the way of forming the buffer stock, the ITC also recommended that the quota be raised from 40 per cent to 50 per cent but with an important caveat, it would operate for six months only.[37]

Getting to this point had not been easy since there remained considerable mutual bitterness and suspicion between the Bolivians and the Dutch as a result of the negotiations over the renewal of the agreement.[38] As Campbell put pressure on the Bolivians to overcome their reticence, they insisted on several conditions that maximized their control and minimized their exposure. The agreement would automatically expire at the end of December 1935, with a provision for possible termination as from the end of June. Since they trusted neither Howeson nor the Dutch, they wanted Campbell to take full responsibility 'for the whole thing'. Responsibility for day-to-day market decisions was far too much to ask and Patiño therefore stipulated that no decision could be made without the unanimous agreement of the buffer stock committee. Campbell naturally considered it all 'rather unsatisfactory', and worried about the role he would have to play:

> I rather think the whole thing will work out on this basis:
>
> 1 Howeson and the Dutch will agree on something. H will discuss with me. He'll phone our joint decision to Paris and Bolivia will agree.
> 2 If they don't agree, they'll state their objections and not improbably leave it to me to decide.
> 3 If they don't agree and don't leave it to me, we'll have to start again.[39]

This was a clear confirmation of the fundamental structure of the power relations that underpinned the whole ITC.[40] Malaya and Nigeria had no independent voice; Bolivia did but it was one that was almost exclusively reactive.

It took Bolivia two months to provide the necessary confirmation of her participation and it was July before the ITC could issue a formal communiqué announcing the formation of the buffer stock scheme.[41] The stock would be managed by a committee composed of representatives of the participating countries with Howeson representing Nigeria and Campbell representing Malaya and serving as its Chairman. To give it maximum flexibility in the market, the stock could be held in any recognized brand, without any special role for Straits tin. Price policy would be determined by the committee without any reference to the ITC. A reasonable level of overall stocks was to be 18 per cent of the deliveries over the preceding 12 months.[42] These stocks were defined as the visible supplies, including the buffer stock,[43] together with the smelters' carryover. Howeson secured McKenna's co-operation who agreed to serve as trustee and who would therefore hold the actual buffer stock warrants. All transactions would be undertaken through the LME with the business being equally distributed among all the dealers.

The formation of the buffer stock confirmed the widely-held scepticism about tin control in Malaya. Producers had been led to expect that they would have the final say, at least de facto,[44] but they found that the Colonial Secretary was prepared to overrule both them and his own his advisers. There could be no trust in any of the assurances that this was simply a temporary device rather than one designed to make tin control permanent and perpetuate its perceived inequities.[45] The tension that had always characterized the relationship between the ITC and the Cornish interests in Malaya reached a new height, creating 'far reaching acrimony' which would be 'so prejudicial to the smooth working of the scheme'.[46] Henceforth, their criticism would be aligned with that coming from several other quarters.

Since producers were granted until the end of December with which to fill the special buffer stock quota, a private pool was formed in June 1934 by LTC, BMC and BM with upwards of 6,000 tons.[47] This pool provided the basis for a brilliant piece of market stabilization and operated until the buffer stock came into effect in February 1935. Over these eight months, tin prices fluctuated by no more than 3.7 per cent; by contrast, the figure for copper was 21.3 per cent.[48] Such success, however, came at considerable cost to the sponsors, with a loss of some £12,800.[49]

A further cost was borne by the dealers on the LME. The announcement of the formation of the buffer stock in June had led to a weakening of the market as consumers realized that they did not need to maintain their traditional level of stocks. In combating a bear raid the buffer stock bought three month contracts but as these came due in September a shortage of spot metal developed. The inevitable consequence was the emergence of backwardation, at one time as high as £3/5/–. Backwardation increases the risk of hedging, so the dealers found themselves caught in a double bind. Price stabilization had reduced the need for hedging but its mechanism increased the cost. The higher proportion of spot metal contained within the buffer stock, the greater that risk would become. While the managers were sensitive to the needs of the market and reduced the backwardation by buying forward and selling or lending spot,[50] that simply confirmed the fact that effective market power had passed into their hands.

In the propaganda war that broke out around the formation of the buffer stock, each side claimed the support of consumers. Cunliffe-Lister went public with his admission that he had not consulted the producers because of the overriding interests of consumers and the need to eliminate speculators.[51] Market operators, however, pointed to anonymous consumers who were strongly opposed,[52] while other sources reported their endorsement.[53] But, without the inclusion of consumers' representatives in the deliberations of the ITC, it was difficult to demonstrate how the buffer stock served everyone's interests.

Smelters' stocks

One uncertainty facing the ITC was the level of invisible stocks held at the smelters but not under warrant and hence not included in the publicly declared carryover. Since the unanticipated release of these invisible stocks had created a problem during the first agreement, the ITC was particularly anxious to receive this information from the smelters.

CTS and Billiton were quite happy to co-operate, at least as long as such information remained confidential within the ITC. STC was not and Bagnall used this issue as a way of continuing his campaign against tin restriction, much to the annoyance of the Colonial Office.[54] For Campbell this was confirmation of the 'intense conservatism' and general 'stupidity' of the company and commented that:

> Ever since control started, we have had nothing but bitter, factious opposition from the Straits Trading Co. Nor has it been at all a "straight" fight. One can respect and admire a strong and clean fighting opponent; the S T Co. has conducted its campaign in a manner which does not appeal to me at all.[55]

The fight with STC could only be resolved through legislation. In the debate on 16 April 1934, Bagnall, as the Senior Unofficial Member of the Straits Legislative Council, expressed the fear that information would be used to the advantage of ESC and worse, that it was part of the 'adroit efforts being made to manoeuvre the Malayan tin industry and the Straits tin market into a position where they will be compelled to surrender their last liberties'. He concluded that the use of the official majority to pass the legislation would be 'a travesty of justice and the odium of forcing this partisan legislation on Malaya will discredit the Home Government as being anti-Malayan in its policy and practice'.[56]

Coincidentally, the manager of ESC also served on the Council and made the case for disclosure on sober technical grounds, challenging Bagnall's irrationality by stating: 'In my professional career I have been associated with four metals, lead, copper, gold, and tin. Never till I came to be associated with the tin industry did I find this astounding preference for secrecy.'[57] The Acting Governor went even further:

> Professor James in one of his manuals on psychology said that suspicion will never yield to explanation or argument because it is of the very nature of suspicion to impugn both, and I am afraid that Mr Bagnall is in the position of those men of whom we read in the Psalms who were brought into great fear where none exists.[58]

Only a minority of the unofficial members were prepared to support Bagnall and the legislation passed without resort to the official majority. Unfortunately, Bagnall was not prepared to see this as forcing any modification in his overall position on tin control.

1934

The surge in American consumption that had emerged in mid-1933 and which had allowed for final sales from the ITP was not sustained through 1934. While other industrial countries increased by 3 per cent and the developing world by 19 per cent, the decline by 27 per cent in the USA meant an overall decline in consumption as measured by purchases by 7 per cent. This recorded decline does not take account of a substantial drawing down by American consumers on their stocks but, even on generous estimates of these stocks, the decline would have been at least 15 per cent bringing the overall decline in final use consumption by 2 per cent.[59] Fluctuations in consumption were broadly matched by changes in production and the following figures demonstrate the extent to which the ITC had been able to bring stocks to their normal levels. With stock normalcy went a remarkable degree of price stability.

What now became controversial was whether the ITC was wise in ensuring stability around £230. By the end of 1933 tin prices had recovered to the point where they were at 85 per cent of their level in 1928 but prices for other major non-ferrous metals all remained at between 50 and 60 per cent of 1928. The very success of the ITC was creating a new problem since tin was now clearly out of line with other metals which were potential sources of competition. Measured in terms of gold, the price was still low; in terms of sterling, it was reasonable; in terms of the metals market, it was high. While the latter would not be a problem in the short run, if it stimulated the search for substitutes at a faster rate than the ITRDC stimulated fresh consumption, the long term future of the industry would be jeopardized. The search for an equilibrium point appropriate for all sectors would therefore be elusive.

Many Malayan producers now argued that the price was far too high not simply in relation to other benchmarks but in relation to their production costs. Frisby of Pahang went so far as to claim that the producers in the East could meet the whole world's supply at £100 and still make a 'good profit'.[60] A year later he lowered his estimate of costs to £50–£60/ton metal, so that the ruling price of £230 was denounced as nothing less than 'commercial robbery'.[61] Given the dollar price levels as a result of devaluation, the argument was naturally well received in the USA. The State Department now adopted an aggressive posture, claiming that the 'increase in price has, in the judgement of many, already been inordinate'.[62] This was reinforced by the

Economist which editorialized: 'The crux of the situation is the Committee's conception of a reasonable price. In our view it is certainly below £200 per ton.'[63]

It was against this background of increasing pressure from Malaya and the USA that the ITC met to consider the quota for the fourth quarter. The decision to revert back to the 40 per cent of the first quarter has been interpreted as marking an important turning point in the history of the ITC, as a shift from simple stabilization of production and consumption to protection of the high cost producers through stabilization but at a high price and at the expense of the low cost producers.[64]

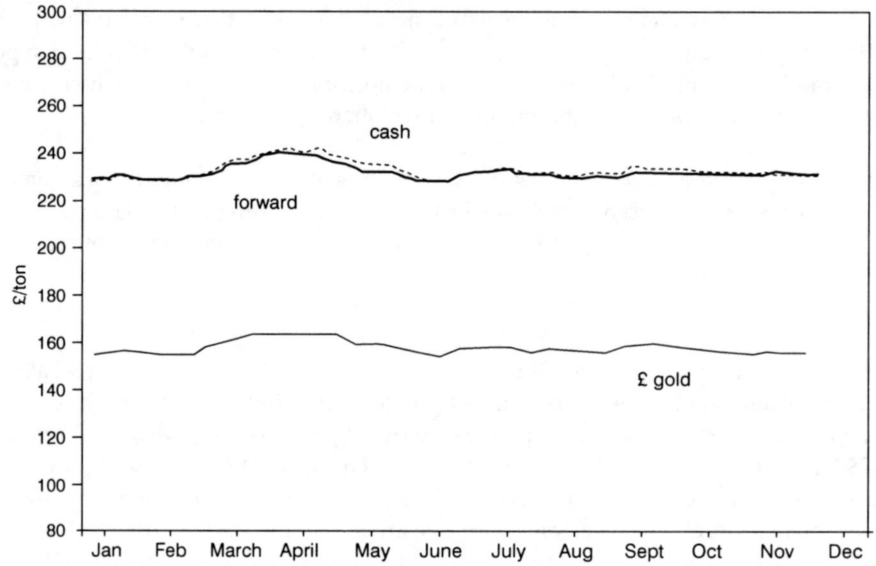

Figure 10.1 Tin prices, 1934

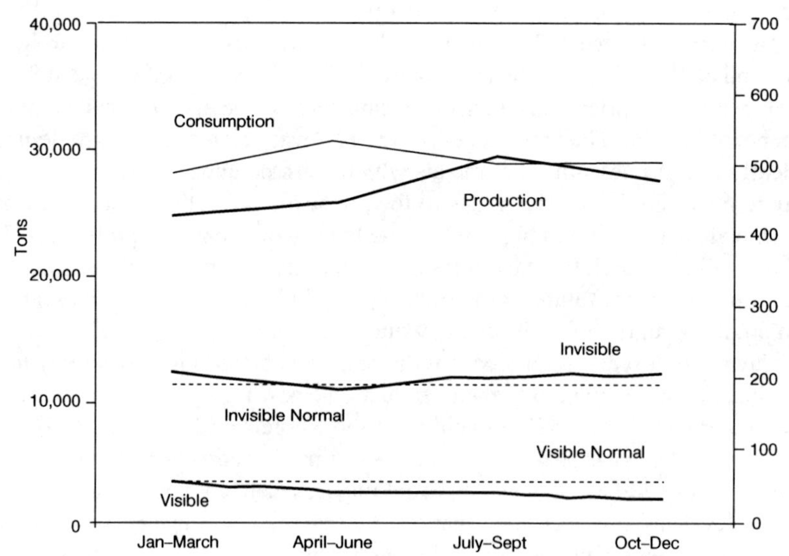

Figure 10.2 Production, consumption, stocks, 1934

When the ITC met in August to determine the quota commencing in October it found several sharp differences of opinion but none were grounded in cost conditions. On the production side, the key unknown was the extent to which both Bolivia and Nigeria could make up their current arrears during the coming quarter. On the consumption side, the key unknown was the course of American demand. Extrapolating from the data available for July as to the condition three months later was simply a matter of guesswork. There were certainly good arguments for caution on the demand side given that deliveries in the US had shown consistent declines from their peak in April. The Bolivian problem could also be discounted as one simply deriving from the inherent difficulty in expanding production from lode mines rapidly and the Nigerian one could be related to the seasonality of production there. These were then reinforced by anecdotal evidence about consumer stocks. A decision to drop back to a 40 per cent quota could therefore be easily justified on technical grounds and this was anticipated by the financial press.[65]

However, on this occasion the ITC paid less attention to the technical parameters than the political ones. Cunliffe-Lister was particularly concerned about the implications of a cut in quota for the reputation of the ITC in both the USA and Malaya; in spite of his earlier position, Campbell dutifully pressed the case. Resistance came from the Dutch.[66] Their thinking about price was shaped by the fact that they, together with the rest of their major European customers, were still on gold. More important was their commitment to the buffer stock. The higher the quota, the slower the rate at which it would be assembled and meeting unanticipated demand from the metal in the buffer stock would generate the cash needed to support the market if that demand proved to be transitory. The meeting, therefore, divided five ways: Calder arguing for 50 per cent on technical grounds; Howeson combating this with a pessimistic technical assessment; Campbell also arguing for 50 per cent but on political grounds; the Dutch arguing for 40 per cent because of the buffer stock; and the Bolivians arguing that it should be settled without any reference to the buffer stock at all.[67] The Bolivians agreed with Malaya and Nigeria on 50 per cent but offered the compromise of 45 per cent.[68] It was not one that the Dutch were prepared to settle for, at least not at that meeting and Campbell ruled that in the absence of any decision, the quota must revert to the 40 per cent. Given the political interest in the question, it was an important decision, since it meant that the Dutch were prepared to force the ITC to take a position clearly opposed to the wishes of the United States and one which 'may seriously prejudice the whole scheme'.[69]

While the position presented by the Malayan delegation is consistent with the notion that their interest lay in a higher quota and lower prices, it derived much less from an economic basis than a political one. Indeed, in his report on the meeting to Kuala Lumpur, Lowinger expressed considerable sympathy with the Dutch position concerning the buffer stock and about levels of demand.[70] The Dutch and Bolivian positions are directly inconsistent with the assumption that simple cost conditions are the primary determinant of quota policy. The high cost Bolivians had argued for higher quotas, while the low cost Dutch argued for lower ones.

The conflicts that emerged at the August meeting revealed the limits of trying to settle the question of quotas on technical grounds. Perhaps it was in order to introduce a more technical perspective that the ITC decided to invite representatives of consumers to offer their advice on consumption trends.[71] Picard raised the issue and Campbell proposed the model already adopted by the IRRC. Their role would be purely advisory[72] and the agenda could be divided into two parts to allow full members of the ITC to discuss and settle matters in their absence.[73] John Hughes of the United States Steel Corporation was nominated by the American Iron and Steel Institute (AISI) and attended the November meeting. Nine months later, Sir William Firth, Chairman of the steel company, Richard Thomas, and of the Welsh Tinplate Manufacturers' Association,[74] became the second member of the advisory panel.[75]

1935

While 1934 was a very successful year for the ITC,[76] it ended on a difficult note as Congress became a vehicle for the expression of American opposition to tin control; 1935 began with a much more serious attack in the House of Commons following the exposure of Howeson as a corrupt, speculative capitalist. By mid-summer the buffer stock was creating far more problems that it solved and the LME was in open revolt. From this point the ITC was under siege.

United States Congress and tin

In 1934 Congress decided to take an active interest in the position of the United States vis-à-vis the world tin industry. This was prompted by three independent initiatives. One was a question raised by Cordell Hull of acquiring tin as a credit against British war debt.[77] A second was a concern expressed by Roy Veatch on behalf of the State Department about the potential problem of supplies during war and about the price being imposed by the ITC.[78] But the most pressing issue was triggered by a surge of exports of fresh tinplate scrap to Japan which was able to bid up the price to a level that jeopardized the position of the two detinning companies. Japan's competitive advantage was based on several factors. Labour costs were a tenth of the level in the USA. Shipping rates across the Pacific were lower than domestic rail rates and there was a strong demand for the tin residues from silk producers. The detinning companies turned to Congress for protection and the House Committee on Military Affairs held hearings from which the Faddis Bill emerged which proposed an embargo on virtually all exports of tinplate scrap. Since detinning had a modest but important contribution to make in ensuring tin supplies during war, this could be justified on strategic grounds.[79]

Under pressure from Millard Caldwell, the House of Representatives gave a subcommittee of its Committee on Foreign Affairs much wider terms of reference 'to investigate the extent to which the United States is dependent on foreign nations for its supply of tin'.[80] Over the winter of 1934–1935, the subcommittee, chaired by Sam McReynolds, held hearings which addressed five basic issues: potential sources of domestic supply of ore, the viability of a domestic smelter, development of substitutes, export of tin-bearing scrap and the creation of a government stockpile.

The stockpile proposal was pursued by Bernard Baruch who, as a previous Chairman of the War Industries Board, had some experience with the strategic problems posed by tin[81] but it was dropped by the committee on the grounds of cost and its failure to address what in its estimation was the fundamental problem of tin.[82] The concept of dependence had some substance when trying to think about the problem of securing wartime supplies but it slipped over too easily into an unreflective diagnosis of the structure of the tin market.

American dependency on external sources rankled on several counts. Control appeared to lie in the hands of the British, in large part because of the pattern of smelting, and it was one that was fundamentally intolerable on several counts. It enabled the creation of the ITC which excluded the United States from any participation in its pricing policy. As a country with outstanding war debts, Britain was hardly entitled to exercise such power against its creditor, especially since it protected an even less worthy country, Bolivia, which had defaulted on its debt. The smelter issue was used as a way of defining an economic abnormality in which ore was produced in one country, smelted in a second and sold to a third, a situation considered 'fundamentally uneconomical and fraught with the continuing danger of violent readjustments in the industry'.[83] The McReynolds committee then went on to propose a new world order in tin:

> In order to perfect a harmonious and continuing control of production, it will be found necessary to enlist the United States as a participant in the [ITC] scheme in the role, not only of consumer but of producer, effectuated by the active participation of American capital in ore production and by the establishment of an American tin smelting industry in the United States on a scale large enough to supply her entire needs for metallic tin. With this readjustment, international control would become more truly representative of the interests of the industry as a whole, and ... it might continue to function indefinitely ... The value of a smeltery lies in its ability to control. The transfer to the United States of approximately half the tin smelting of the world would not mean any redistribution of the profits now being realized by the tin ore producers. Because of the opportunity it would afford for the perfection of the organization of the industry, the development of tin smelting in the United States would in reality represent a net gain to the great bulk of the capital at present invested in tin production.[84]

Tin politics had generated its share of utopian speculation among the producers but nothing matched this heady brew of aspirations and contradictions. What gave this vision some currency was the undoubted fact that the ITC had effective control over the market and could therefore be denounced as a 'monopoly'. Whether the McReynolds model would solve more problems than it created was less important than the fact that it represented a challenge to the existing structure.[85]

This was not a challenge that received any support from the tin interests that were invited to testify. Both the major tin consumers, National Lead and US Steel, indicated they were quite satisfied with the current system but the absence of any real American interests that would be served by a domestic smelter would not deter the committee from pursuing that will o' the wisp.[86] Nor was there any particular enthusiasm for reducing dependence through substitutes. The military representatives stressed the importance of ensuring that such substitutes were commercially viable, since war hardly provided an opportunity for the development of new techniques. Research would therefore have to proceed on the assumption that there would be no subsidies and controls available for the implementation of any substitutes. In any case, the scope for modifying the structure of demand was so limited that it could make no important contribution to solving either the economic or the strategic problem of tin.

On the question of prospecting for domestic sources of ore, the committee heard contradictory testimony and chose to set aside the conventional geological wisdom represented by the Bureau of Mines that there were no commercially viable deposits to be found in the continental USA. Such an assessment had led to only modest private prospecting efforts and that could be used to justify the expenditure of public funds to support a more comprehensive programme.[87]

It was only on the question of tinplate scrap that the committee received the evidence that it needed to support its position and those arguments had already been well rehearsed before the Faddis Committee. As this issue was taken up in the final report, the weaknesses in the committee's thinking became all too transparent. In discussing the economics of detinning several witnesses had made it clear that the actual quantity of tin in the scrap being exported to Japan was infinitesimal, about 500 tons per year. Instead of representing this accurately as 0.7 per cent of primary tin imports, the decimal point was dropped and the figure became a much more impressive 7 per cent.[88]

The committee completed its work by proposing legislation to create a Board of Strategic Minerals and laid out the objectives of the policy it should pursue: impose a tariff of 6c/lb to protect a domestic smelting industry, fund prospecting and research and control tin scrap exports.[89] When this appeared the American tin trade began to mobilize[90] and a metal dealer, C. S. Trench, circulated a pamphlet warning about the disruption caused by a subsidy for substitutes.[91] One of the partners wrote to his cousin, President Roosevelt, identifying several weaknesses in the legislation and commented that it was nothing more than the 'political aggrandizement of certain members of Congress'.[92]

Although the McReynolds investigation produced no immediate results on the main issues on its agenda,[93] it established two principles that would shape American thinking about tin: the current structure of the industry which allowed for its control by the ITC was unacceptable and that a central element of changing it was a domestic smelter. As long as Bolivia was committed to that structure, it could remain intact but should she ever be tempted to strike a separate deal with the USA and provide the feed for her smelter,[94] it would collapse. A year later one of the committee's concerns was taken up and Congress passed legislation prohibiting the export of tin scrap except under licence.[95] While the committee discouraged further thinking about a stockpile on the grounds that it would be interpreted as a belligerent move,[96] that did not prevent Hull from continuing to work in that direction.[97]

Poorly thought out as the McReynolds investigation may have been,[98] it was extraordinarily well-timed. Just as the committee was ready to release its report a major scandal erupted which, in the eyes of many, confirmed its assumption that there was something fundamentally wrong with tin control.

Shellac, pepper and tin

Whenever commodity prices dip below what are considered to be historic norms the incentive to form a bull pool is irresistible. In spite of the lessons that should have been learnt from the experience of tin in 1929, pools were created in shellac and pepper in 1933–1934 but with far more serious results.[99] Since they drew in Howeson and his associates, these failures provided tin control with much unwelcome publicity.

Shellac is a resin collected from trees by Indian peasants and was primarily used in gramophone records. In early 1933 the price was only 54 shillings per hundredweight, nearly half its 'reasonable' level of 80–100 shillings. By May 1934 a pool formed by Cunliffe-Owen and McKenna[100] had forced the price up to 115 shillings prompting the release of large hidden stocks in India and by early February 1935 the price was dropping to 68 shillings and stocks were over ten times their normal level.[101] The instability in shellac coincided with a similar problem in white pepper.

In February 1934 the initial success in shellac led to the formation of a pool in pepper. Initial purchases were made in1934 at 9d/lb but as the pool pushed the price to 1/5, hidden supplies emerged, mainly through the conversion of black pepper into white. By early 1935 the pool had exhausted the financial resources with which to cover its position. Again, the level of stocks that were about to be released was enormous, the equivalent of two years' world consumption. In February 1935 both markets crashed with a total loss of around £2 million. While shellac had been largely financed,[102] pepper was not. Dealers had undertaken forward sales commitments solely on the basis of their trust in the reputation of the purchasers and at least two were forced into bankruptcy.

The central figure in the pepper and shellac scandal was a long-standing business associate of Howeson, Garabed Bishirgian.[103] He had his own business, Bishirgian & Co., and he and

Howeson were joint owners of a commodities brokerage firm, Williams, Henry.[104] One of Howeson's associates more closely linked to tin, Hardy, served as managing director of Williams and Howeson was also indirectly involved in commodity trading through another family firm, Jayandee.[105] Throughout the first half of 1934 Bishirgian was engaged in speculative trading in pepper and shellac on behalf of Williams and in April he decided to acquire a firm with ring dealing privileges on the LME to save brokerage commissions. James & Shakspeare was chosen as a suitable candidate and arrangements were made to buy and transform it into a public company which would then acquire Bishirgian & Co. and Williams, Henry.

James & Shakspeare was floated in September 1934 and the prospectus implied that the funds being raised, £412,500, were required to simply consolidate and expand established brokerage businesses. In so doing, it misrepresented the fact that Williams was engaged in highly speculative trading which at that time had an open position in both shellac and pepper for a total of £1,366,000 supporting which required additional funds. Even without this embarrassing item, only 17 per cent of the issue was taken up by the public and the rest was therefore left with the underwriters, which included McKenna, Cunliffe-Owen and Dean Finance.[106] There were therefore two sets of City victims: those who lost their investment in James and Shakspeare[107] and those bankrupted because of the way that investment had been used to create an unsustainable position. Heads had to roll. Bishirgian was charged under the Larceny Act with issuing a false prospectus and Howeson and Hardy were charged with aiding and abetting him.

There were four fundamental issues at stake in the trial. First, were the stocks of pepper acquired by Williams designed to corner the market? Second, was the liability on the open pepper and shellac contracts held by Williams at the time of the flotation of James & Shakspeare reasonable in light of normal commodity trading practice or was it exposed to undue risk? Third, were the funds raised through the flotation inappropriately used? Fourth, what responsibility did Howeson bear?

Bishirgian's defence was that these were normal market transactions. They therefore did not need to be disclosed in the James & Shakspeare prospectus and that the funds used were consistent with such a policy. It was not easily sustained in the light of several facts: (1) details of the pepper contracts had not been properly recorded; (2) Bailie, one of the directors of Williams, had resigned because he had not been kept properly informed; (3) with the contracts coming due in February, Williams had control over two years' world supply of white pepper. Bishirgian was therefore found guilty and sentenced to 12 months in prison, while Hardy received nine.

Howeson's responsibility was much less clear. He had certainly participated in pepper discussions at the time when these were more obviously a standard bull operation. Evidence was presented at the trial which suggested that Howeson had not in fact learnt of the extent of the exposure until very late and that is consistent with the impression left by his attempt to try and rescue the situation in early 1935. It was only in January that he and van den Broek started to try and persuade the NEI government, as the largest producer, to restrict production.[108] Such a guarantee would allow the financing of the February contracts, preventing the inevitable crash and the NEI government would take effective control over the stocks. If Howeson had been a party to a deliberate deception of McKenna and the others in September,[109] he would not have been exploring the prospects of financing additional pepper in November and the kind of rescue attempt he was launching with van den Broek would have occurred much earlier.

But the question of what Howeson actually knew was legally less relevant than the question of what he should have known and what questions he should have been asking.[110]

The extent of his connections with Bishirgian were far too extensive to allow him to claim complete innocence. After all, he lost £300,000 of his own money. Too little is known about his character to make a judgement as to whether he was so absorbed in dealing with tin that he could easily neglect to pay any attention to the fate of such a sum at risk in pepper. The jury was not prepared to grant him the benefit of the doubt and he was too received 12 months. This scandal forced him out of tin. Unlike some figures guilty of a similar offence,[111] Howeson was not rehabilitated. His outsider background and his earlier reputation as a shady financier had finally caught up with him.

The collapse of the pepper and shellac markets provided the occasion for a rambling debate in the House of Commons in which most attention was turned on tin. Howeson's involvement in pepper was used to raise serious questions not only about his role in both the private and the official buffer pools but about the whole edifice of tin control. While Cunliffe-Lister tried to shore up the reputation of the ITC by quoting extensively from the endorsement it had received at the World Economic Conference, it was difficult to dispel the damage done to the principle of stock control as an indispensable adjunct to production control.[112] The scandal broke at precisely the same moment that the United States was beginning to formulate its view of tin. Restoring the reputation of the ITC would not be easy.

Setting quotas, 1935–1936

When the ITC met in October 1934 to set the quota for the first quarter of 1935 the statistical indicators provided confirmation of the wisdom of the pessimists of August. Although Bolivia again proposed a 50 per cent quota, this was simply in deference to the government's desire to maximize short term revenues and she met with no support. The existing nominal quota of 40 per cent was confirmed without much discussion.[113]

The overall policy of the ITC remained one of driving down stocks to what it considered to be normal levels and then taking control over the market through its own buffer stock. Quota levels had to be on the low side in order to feed the market from the stock and thereby provide it with some cash. However, a fundamental weakness was soon exposed. Defining normality for the market as a whole was difficult enough but it also involved defining it for each specific region. Such definitions needed constant revision in order to deal with the dynamic character of the market.

To further complicate matters, an important shift was occurring in the way in which the tin market operated. Consumers could acquire metal either directly from the smelters or from public warehouses. As metal was shipped to these warehouses it was recorded as a delivery and the monthly statistics of deliveries provided the only reliable guide to the demand side of the market. Their usefulness was now beginning to decline, as a higher proportion of metal was shipped direct from the smelters.[114] Relying on delivery statistics prevented early monitoring of important shifts in demand and made the definition of the normal level of stocks increasingly obsolete.

When the ITC met in February to settle the second quarter's quota the statistical position appeared good. Consumption was exceeding production and stocks, outside the buffer pool, were being reduced and there was a tentative agreement to raise the quota to 50 per cent. However, it was decided to wait a month since there were some signs of market weakness. But March found the ITC meeting in the shadow of the pepper scandal.

The Dutch came to this meeting following a frank internal debate which revealed fundamentally conflicting positions. Van Buuren, the Director of Transport and Waterworks which now had responsibility for Banka, wanted a quota of at least 50 per cent in light of his assessment of both consumption and stocks. Houwert provided a more refined view of the stock position and pointed to the decline in US deliveries which argued for maintaining the 40 per cent quota. The Parliamentary debate and the open criticism in the mining and financial press simply stiffened his position. It was all the 'work of opponents who will cheer if we put up the quota'.[115] He had not been prepared to back away from a fight with the USA in the fall and there was no reason to capitulate to critics now. Nor was Houwert willing to concede van Buuren's position that the 'current price [£216] was too high for a sound basis of the restriction scheme', which had led him to advocate 45 per cent. In view of the uncertainty of the future of restriction, Houwert adopted a short term perspective and wanted high prices to compensate for past losses. Groothoof concluded this internal debate with a proposal to defer a decision with the option of a retroactive increase. Holding the current quota and selling off half the buffer stock would be the safest course for the moment.

The British and Bolivian delegations had a quite different view. For Bolivia, the overriding issue was short term revenue maximization to deal with the financial implications of the Chaco war and it wanted a quota of at least 50 per cent.[116] Cunliffe-Lister was anxious to defuse the growing criticism and wanted 55 per cent and again the Malayan delegation could provide the optimistic reading of the consumption trend required to justify it. Of course, if it was wrong and the buffer stock remained all in metal, the ITC could find itself very constrained. Under pressure from Campbell, the Dutch eventually dropped their opening position of 40 per cent and the meeting therefore settled on 45 per cent.[117]

The experience of the second quarter confirmed the expectations of the ITC. Consumption proved to be higher than anticipated and this was met by the running down of stocks, including sales from the buffer stock. By June the overall stock objective had finally been reached. However, when the ITC met to settle the quota for the third quarter it found itself in a difficult position, especially since Bolivia had made it clear that it would oppose any proposal to continue the buffer stock. Her interest in its liquidation was based on two considerations, both of which derived from her overall financial situation, ever more precarious as a result of the Chaco war. She could not afford to run the risk of a non-renewal of the agreement and wanted to mollify Malaya by withdrawing this irritant. Nor could she afford to forego whatever cash that liquidation would generate.[118]

The June ITC meeting began with a tribute to Howeson, now under indictment for his role in the pepper scandal, and Campbell placed on record:

> The deep obligation under which they were to Mr Howeson for the work he had done … their sense of the very great assistance which Mr H had throughout rendered to the cttee; his encyclopedic knowledge of the subject and his sound judgement had been of the utmost value to them. I very much regret that this association is now ending.[119]

His departure would pose a major challenge to Campbell's ability to continue to administer tin control.

This was Firth's first appearance and he tried to change the terms within which the ITC set the quota. For him, the key was price. Increases since 1931 had placed producers in a very favourable position at the expense of tinplate manufacturers and an appropriate price target was £175. In the background was the competitive pressure from the USA who enjoyed the advantages of a depreciated dollar in some of Britain's traditional export markets and Firth was clearly looking for some relief.[120] Houwert challenged the details of Firth's argument and Campbell had to explain that the ITC simply linked production to consumption and left the price to be settled by the free play of market forces.

Matters were rarely that simple and certainly not on this occasion when the liquidation of the buffer stock had to be factored in. The regular members of the British delegations again argued for raising the quota to 55 per cent but Campbell desisted in the light of what he considered to be the obvious weakness in current demand. Unless the buffer stock were fully liquidated by year end, the metal would be returned to producers creating a risk of market disorganization as they disposed of it. The safest strategy was therefore to raise the quota but only to 50 per cent.[121] Firth left with a quite different impression, convinced that keeping prices high was the ITC's primary objective.[122]

The warning signals sent by Firth were reinforced from other sources. Viscount Elibank, Chairman of Malaysiam Tin, who was deeply committed to restriction, expressed his concern about the implications of a tight stock policy for American consumers. The high level of backwardation, around £9/10/– or 4 per cent of the spot price, thanks largely to forward selling by the buffer stock, and a high price were making Americans very irritated, though he expected they would be calmed if the price were kept within the range of £200–£230.[123] Unfortunately, Hughes had stopped attending ITC meetings and since the Colonial Office had little other evidence of irritation, Elibank's concern failed to register.[124]

The decision taken in June proved to be a serious mistake[125] and one that would prove difficult to correct. Liquidating the buffer stock on a rising market meant that visible stocks soon became very tight and they would remain so throughout the rest of the second agreement. The mistake was clear by the August meeting and there was no dissent from Campbell's recommendation that the third quarter quota be raised retroactively to 65 per cent.[126] No fresh information was available for the September meeting which extended the 65 per cent for the final quarter.[127]

Again, the ITC found itself in a volatile situation which made its statistical projections irrelevant. In September, Italian war preparations resulted in a depletion of the visible stocks to a crisis point and by the end of the month those in the United Kingdom had dropped to 555 tons. The ITC was now forced to confront the contradictions of its earlier policy of first taking control over the market through a buffer stock and then abandoning it. Immediately following the invasion of Ethiopia, an emergency meeting was called to head off a strong upward price movement and the quota was raised to 70 per cent, as the probable maximum production immediately available.[128]

At the end of October the ITC met yet again to settle the quota for the fourth quarter. Without a buffer stock, the Dutch argued for an increase to 80 per cent and this had to meet both the real consumptive demand and rebuild stocks. Discussion found a new alignment of arguments. Such an increase, especially to rebuild stocks, was likely to be temporary and many producers had already reached the limit of their immediate productive capacity. In Malaya, this was because so many Chinese had been repatriated during the previous years of low production and there was an obvious reluctance to bring them back for only a short period. In the case of Bolivia, miners had been conscripted to die in the Chaco and Patiño frankly admitted that the country could not produce at more than 75 per cent. Calder, who had always argued for larger quotas, was now the one to be cautious and agreed with Patiño on 75 per cent. Campbell pressed the case for 80 per cent primarily on the grounds that the statistical projections were now so unreliable that the risks of misjudgement on the consumption side were sufficiently large that they should be absorbed by the producers, even at the cost of a later downward adjustment. The decision to recommend a quota of 80 per cent therefore attempted to correct the policy that had been pursued for the previous four months. Van den Broek took the opportunity to point out that, had the buffer stock remained intact, the issues would have been far easier to address.[129]

The first buffer stock – an assessment

The experience of the buffer stock revealed several difficulties. First was the simple problem of assembling it. Bolivia was the only country that could not meet the December deadline; indeed, it took until the following July before her contribution was considered complete.[130] Second, was the fact that making decisions proved to be far more complicated than Campbell had anticipated. Without Howeson's assistance and experience, Campbell was forced to shoulder the primary responsibility for the administration of the buffer stock. While the buffer stock subcommittee developed a general price policy which aimed at stabilizing around £230 within the limits of £215–£245,[131] it provided no authority for McKenna or Campbell to make concrete decisions. They could only be made by the unanimous agreement of all members of the subcommittee. Campbell found dealing with Antenor Patiño particularly difficult since 'he won't attend meetings, funks responsibility', and 'does not understand the issues in the least'.[132]

As a higher proportion of the visible supplies moved into the buffer stock, free spot metal became increasingly short and by May, 90 per cent of the visible stocks in the UK were held by the buffer stock. Thanks to the rigidity of the price policy, Campbell did not have the authority to sell physical metal and the inevitable consequence was a serious backwardation which remained a persistent feature of the market from May 1935 to December 1936. The extent of the problem is demonstrated in Figure 10.3.

Not only did this destroy the basis for hedging but it adversely affected the interests of some consumers. The tin dealers staged an unprecedented demonstration by walking out of the LME on 8 May.[133] The LME followed this up with an official protest to the Colonial Office and a delegation met with Campbell. It did not go well. The LME delegation was not interested in a discussion of the technicalities of the interface between the buffer stock and the tin market, but rather in presenting its overall view of tin restriction. Since this came soon after the exposure of Howeson, it naturally focussed on his speculative dealings and flotations and saw in tin control an unworthy project designed solely to drive up prices. Breaking the commitment of

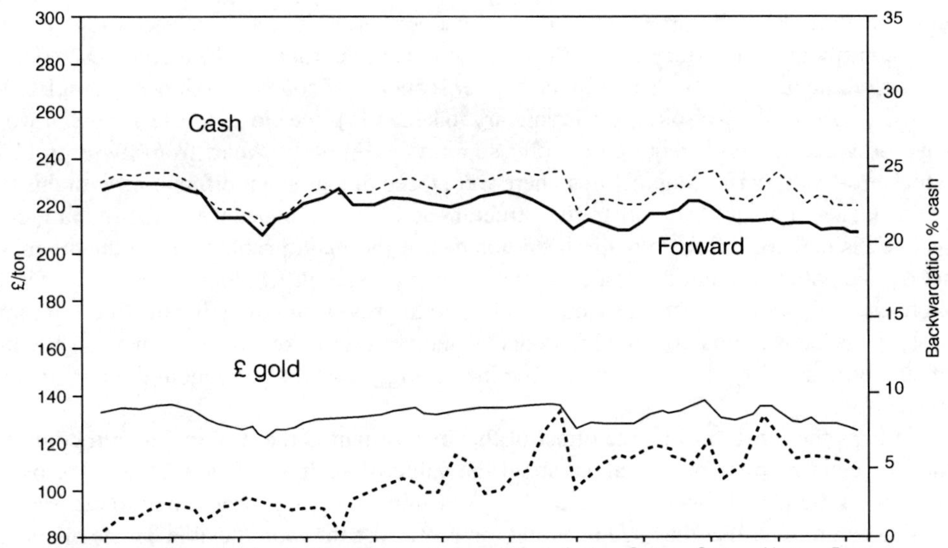

Figure 10.3 Tin prices,1935

the Colonial Office to the very principle of restriction was obviously impossible and the LME remained very frustrated at the lack of much sympathy with its concerns.[134]

Campbell did concede one point and that was to recognize that there were small consumers who did not buy forward and who therefore needed spot metal to cover their immediate requirements.[135] Short supplies not only raised the price against them but also encouraged them to switch brands. Since prices were well under the target of £230 neither the Dutch nor Patiño were prepared to be co-operative but eventually Campbell persuaded both to allow him to sell or lend spot metal.[136] The precise course of prices shows a slight reduction in the backwardation towards the end of Malaya and then its reopening throughout the summer. On 19 June the backwardation reached the highest level hitherto recorded, with a premium of over 5 per cent for spot metal. A month later, 22 July, the system collapsed, as McKenna failed to make available any warrants and spot trading had to be suspended for the complete absence of any physical tin. This caused considerable embarrassment since such trades were the basis of the LME's published settlement price which was a benchmark price for many contracts. The LME was therefore issuing as a market price one which had not been actually determined in the market.[137] Following this fiasco, the buffer stock committee agreed to make available whatever supplies were required to meet the market, even going so far as to borrow common tin for the purpose.[138] As a result of the decision to wind up the stock at the end of the year there could be no dealings in the forward market after the end of June since that would saddle the committee with physical tin whose disposal might be difficult. Market dealings were therefore very confined.

Short physical supplies raised yet another problem. Severe shortages in one region could be relieved by shipments from another. In July the buffer stock committee decided to move 1,260 tons of Straits tin from New York to London, causing some anxiety among American consumers and derisory comments among metal traders.[139] Not only did this incur unnecessary shipping charges but in a few weeks another boat would be bringing standard tin back to New York from Liverpool. Substantial quantities were also shipped back and forth between the Netherlands and London for the same purpose of providing temporary relief.

This buffer stock experiment must be regarded as a complete failure. It was not designed to make a profit for the producers and it almost certainly did not.[140] Its very existence may have contributed to the steepness of the price decline in February as the market reacted to the allegations about Howeson's role in the pepper scandal. Of course, had it been completed on time, it might well have sold some in January to keep the price closer to £230 and that would have provided the cash required to offer support in February. Apart from these problems which arose at specific conjunctures, there were three fundamental difficulties with this type of stock. The inflexible administrative structure and the consequences of the running down of free visible stocks have already been noted. But the main problem lay in the attempt to set a price policy without the resources with which to back it. At around 4 per cent of world production, the buffer stock was simply too small to provide anything but the finest of tuning around a price it ultimately could not control and the tin market was still far from any kind of structural stability. That would become increasingly evident throughout the remainder of the decade.

Perhaps the most unfortunate effect of this experiment is the way in which it confirmed the long held scepticism in Malaya about the value of such a buffer stock and the overall wisdom of the ITC. Lowinger visited Malaya in June and publicly admitted to the Malayan mining community that they had been right in preferring the more flexible device of keeping such reserve stocks in the hands of the miners.[141] That reputation would make it difficult to return to the question of a buffer stock when again it became indispensable.

1936

Much of the price stability the ITC had provided to the market during the first two years of the second agreement was largely lost in its final year. A general decline in price occured for the first three quarters of 1936 and then a sharp recovery in the last and is illustrated in Figure 10.4.

This pattern followed the general trend of commodity prices and by comparison with the overall average of other commodities, tin prices were actually somewhat more stable. Two other factors shaped the particular rhythm of the tin market. Throughout the year, tin politics were dominated by the question of the renewal of the agreement. As will be evident in the next chapter, this was far less straightforward and far more public than had been the case in 1933 and uncertainty had inevitable repercussions on the market.[142] The year also provided a demonstration of the limitations on the ability of the ITC to predict the changing pattern of both supply and demand.

On the demand side, consumption in the USA continued to grow, by some 17 per cent, but elsewhere it largely stagnated. Important declines were registered in Germany but one encouraging development was the emergence of the USSR as a major purchaser. Overall, real consumption increased by some 7 per cent.

Production policy was shaped by two considerations: the need to raise the level of stocks in relation to the new levels of demand and the continued inability of Bolivia to produce her quota. This problem had first emerged in the last quarter of 1935 and was due primarily to a shortage of experienced miners thanks to their conscription for the Chaco war. A partial and temporary solution was found as the NEI responded to Campbell's request to draw down its stocks, so the total supply remained more or less as anticipated. However, the situation deteriorated throughout 1936, as the Bolivian mining industry found itself under increasing fiscal pressure from a more nationalist regime and was unwilling to deplete its mineral reserves without any prospect of profits.[143]

Following the experience of the last quarter of 1935, the ITC adopted a more flexible quota policy. While Bolivia would be entitled to the nominal quota that applied to all of the

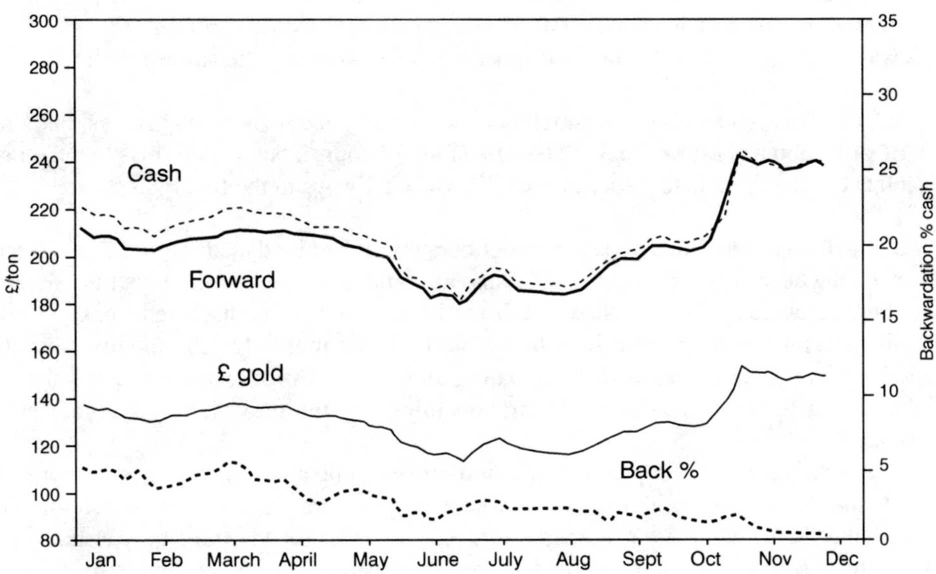

Figure 10.4 Tin prices,1936

signatories, production estimates were based on an assessment of her effective capacity. That had a rather perverse effect since it meant that the more Bolivia underproduced, the higher quotas had to be for the other members to meet the needs of the market and the higher the nominal quota, the higher the rate of underproduction. Until this was resolved with the surrender of the accumulated tonnage, the fact that Bolivia could suddenly bring a large quantity of metal on to the market before adjustment of the quotas for the other members was feasible had its own depressing effect. In many respects, therefore, the position was quite unreal since the published statistics of underproduction disguised the fact that this was an entitlement that was not expected to be realized. It took some time for this more flexible policy to reflect the actual situation in Bolivia and it was not until the last three quarters of 1936 that there was a reasonable balance between expectations and performance.

When the ITC met in February to set the quota for the second quarter, the full extent of this continued underproduction was not yet evident. The quota had been set for the first quarter at 90 per cent in order to build up stocks[144] and the key question was whether it was now time to reduce it to prevent stocks from getting too large. That in turn rested on a judgement about Bolivia, since the more she could produce, the lower the actual quota needed to be. Divisions now emerged for the first time within both the Nigerian and Malayan delegations, which split as to whether the reduction should be to 85 per cent or 80 per cent.[145] Bolivia naturally supported 80 per cent, since the lower the quota, the less her underproduction would be exposed and any increase in price would contribute to the rebuilding of her productive capacity.[146] The Dutch were also divided, along the same fault line that had often erupted in Malaya.[147] Batavia, reflecting the views of van Buuren for Banka, remained anxious to see an 80 per cent quota since the marginal cost of increased production for just one quarter was too high.[148] By contrast, The Hague, especially since it was responsible for maintaining a comprehensive statistical base, was focussed on the overall market position. That suggested that stocks should be raised from 16,000 to at least 23,000 tons. Physical metal in the visible supplies in Britain was particularly short at 800 tons instead of the more appropriate 4,500 tons.[149] Keeping the quota at 90 per cent would show the ITC's determination to deal with the immediate problem. As usual, Campbell proposed a compromise of 85 per cent. But the stock situation was far too desperate for such a decision to be received with anything but severe criticism.[150]

It was not a criticism that Campbell was prepared to accept as he commented:

> Why the fuss about short tin supplies … not a word about short lead stocks, but a hell of a fuss about short tin stocks. They *are* short, of course, but it is obvious that the main object of the thing is to discredit the ITC. We are always in the limelight.[151]

Campbell considered this agitation as deliberately stoked by the editor of *The Economist* who had long been very critical of the ITC and he could rest more content about the situation since he was aware of the existence of substantial quantities of undisclosed stocks.[152] What he failed to appreciate was that in defining the ITC as a model for commodity control, it could be held to a high standard in regulating an otherwise irrational market and that one of the crucial features of eliminating irrationalities was the provision of accurate public statistics.

Quite fortuitously, the 85 per cent decision proved to be about right, since consumption fell off slightly and the visible supply showed a marked improvement. The Dutch therefore came to the June ITC meeting prepared to keep this level for the third quarter and found support from a fresh source in Firth. However, they were met with strong opposition from both the Nigeria and Malayan delegations who were under severe pressure from the British government to raise the quota to 95 per cent to cope with any military contingencies as a result of the uncertain political situation in Europe.[153] This would be far in excess of anticipated consumption and would double the level of visible supplies. Campbell dropped his Chairman's neutrality and reinforced this talk of the dangers of war, trying to get an agreement at least around 90 per cent. An objective assessment of the condition of the market was clearly impossible given the British position and the Dutch were prepared to concede in order to preserve a united front for the more important game, which was renewing the agreement.[154] Bolivia reserved her government's position.

The market was caught by surprise by the decision to raise the quota to 90 per cent and the price immediately dropped from £183 to £174. It was not a price that Bolivia could accept and an emergency meeting was called to deal with one depressing element, the overhang as a result of her accumulated entitlement. By the end of May, this had reached 10,288 tons and Martinez Vargas proposed that just over half be simply dropped and the remainder, or 5,043 tons, be distributed among the other members.[155] The formula for distribution reduced everyone to a 75 per cent quota with a 15 per cent increase to Malaya, Nigeria and the NEI bringing them back to 90 per cent. The other four members received the residue of 576 tons. This meant that production would be slightly more on a nominal 75 per cent quota than on a nominal 90 per cent, an outcome that would not be easy to explain to the tin market which, while 'not perhaps a highly intelligent organisation; it was, however, highly sensitive'.[156]

That contrast between intelligence and sensitivity made it particularly difficult to interpret market trends, leading Campbell to simply throw up his hands: 'Hunches are much more useful than reason when dealing with the tin market. The whole thing is wildly illogical as experience has time, time and again shown … It rises and falls in defiance of all reason.'[157]

However, the long term trend proved to be positive as a sustained, gradual rise took place in consumption. Continued underexports on Bolivia's part ensured an equally gradual price recovery from July to October and there was no serious discussion on revising the quota for the fourth quarter.

November saw an unanticipated breakthrough in the negotiations with Siam which permitted the renewal of the agreement. Manufacturers had been drawing down their stocks in anticipation of the lower prices that would follow its dissolution and were now forced to rebuild them. Overnight tin jumped from £211 to £233 and joined the rising tide of commodity prices. To prevent the price from completely running away, Campbell called around the four main delegations and secured their agreement for a retroactive increase in the quota to 105 per cent.

The second agreement ended on a very encouraging note – production was virtually unrestricted, prices were good, backwardation had disappeared and the stock situation was finally under control. However, as soon as the agreement was renegotiated, the ITC would find that the boom underway simply posed new challenges and reopened old conflicts.

Decision-making

The decision-making process operated quite differently in the second agreement from that in the first. Formally, the quota decisions reached at the level of the ITC were still subject to endorsement by the governments affected but none raised any objection to the recommendations. The ITC was now being granted much more autonomy in the way in which it controlled the industry. Whereas quota discussions in the first agreement were entirely focussed on the urgent question of how far and how long to cut production to squeeze out excess stocks, discussion now allowed for a much wider range of views. Decisions had to be

Table 10.1 ITC quota discussions, Second Agreement

	Final Decision	*Other*	*Total*
Unanimous	8	7	15
Division	5	2	7
Total	13	9	22
High Position			
Malaya	4	2	6
Nigeria	5	2	7
Bolivia	1	1	2
NEI	3		3
Low Position			
Malaya	1		1
Nigeria	1		1
Bolivia	4	1	5
NEI	2	2	4

reached by consensus and the recovery of the market meant that they could be more easily deferred from one meeting to another to permit a more informed assessment.

During the course of the second agreement the ITC considered production quotas on 22 occasions and the opening positions that the delegations brought to the discussion are presented in Table 10.1.

Unanimity prevailed more frequently than division, especially after the buffer stock was settled and when that occurred the Dutch generally preferred to shift the risks of misjudgement to the consumers by opting for the low position, while the British delegations preferred to shift those risks to the producers. Such a difference reflects not a simple variation in cost conditions but rather the differential location in a complex world of international commodity politics.

This pattern of consensual decision-making only emerged in relation to quota decisions, since any error for one quarter could be corrected for the next. It would break down entirely when the agreement itself had to be renegotiated and the members would have to live with any mistakes for its duration.

11 Renewing the third agreement, 1935–1936

The uncertainty associated with the future of the agreement encouraged the parties to start negotiations early in 1935. There was no question yet of whether it was now time to end this experiment but simply one of the form it would continue to take. The ITC had learnt some lessons from previous experience but it would find itself in a particularly difficult situation as it tried to resolve the old question of the basis of the distribution of the standard tonnages. Renewing the agreement would take almost as much energy as administering it.

The ITC was determined to end the anomaly that had characterized the previous agreement. That had defined several categories of producers: (1) those whose current production was being controlled; (2) those whose future production was controlled; (3) those outside the agreement and not subject to any control at all. The first group was in turn divided between the four core members with standard tonnages and Siam on a flat rate. The second and third groups were combined as outsiders whose overall production would be monitored since it could trigger an end to the agreement. Henceforth, all members were to be on the same basis, that is, granted a standard tonnage to which the same quota would apply. As will be seen, there were variations around these tonnages as a result of the particular pattern of bargaining, which made this change more technical than substantial but it placed the third agreement on a quite different basis from its predecessors.

Since the focus of the negotiations around renewal was on the conversion of those on a flat rate to a standard tonnage, there was no interest in reopening the basis on which the standard tonnages of the four core members was determined. Houwert and Antenor Patiño agreed that they provided the only basis for renewal[1] and they were endorsed by Malaya and Nigeria.

Malaya

Malaya entered the process of negotiation following a full debate. A new High Commissioner, Sir Shenton Thomas,[2] had recently replaced Clementi and he was far more guarded in his position on tin control. However, his overall posture was revealed at the outset of the process. In an interview with the *Straits Times,* he was not prepared to give an opinion pro or con on the issue of restriction but said he would listen to the FMS Chamber of Mines and then act 'having regard to the interests of Malaya as a whole'. The absence of any reference to the Chinese was symptomatic but even more instructive was the way in which he evaded the issue:

> I suppose I ought to say something about restriction but over one shoulder I have the shadow of John Howeson, and over the other the shadow of John Bagnall, and as the shadow of John Bagnall is bigger and nearer, I think I had better say nothing at all.[3]

Thomas would become ever more beholden to the local European mining community[4] and his weakness would exacerbate the already strained relationship between Malaya and the Colonial Office.

Avoiding some of the earlier difficulties in determining Malaya's position was ensured by organizing a referendum which provided overwhelming support. Voting was by domestic assessment. Of the total of 2,033,709 eligible votes, 1,678,853 were cast in favour, 83,013 opposed and 213,714 ruled invalid since they added qualifications to their decision. The only important opposition came from Pahang Consolidated, with 63,185 votes explicitly opposed and the Tronoh group, included among the 128,234 votes conditionally opposed. The Redruth group, however, voted conditionally in favour.[5] Immediately following the referendum, the FMS Chamber of Mines met to consider a three part resolution:

> The ratios of tonnages of signatory countries shall not be altered; the industry shall be consulted and express opinion on agreement before ratification; there should be no deterioration of the position of Malaya vis à vis all other participating countries.

Here the opposition had somewhat greater influence but the resolution passed with 451 in favour and 65 against.[6] The outcome marked an important turning point. Malaya had dropped its claim to base these tonnages on current capacity and the Cornish interests were not only now divided but they had failed to carry much wider support. Their claim to speak on behalf of the real interests of Malaya would inevitably lose much authority. Soon afterwards, the Nigerian Chamber met and passed an identical resolution, *mutatis mutandis*. Consultation in this form was a mistake. The only force behind the otherwise pious 'should' in the third part of the resolution was the willingness to walk away from an unsatisfactory agreement and a belief in that willingness was undoubtedly diminished as a result of making such commitments a full 12 months before the new agreement would come into force.

This resolution was endorsed by the FMS government but, since it failed to recognize that the context for these negotiations had fundamentally changed as a result of the growth of both the Congo and Siam, it provoked some consternation in London:

> I am appalled at the stupidity of this letter and of the Malay attitude generally. There is something which makes for a static mind (climate; fortunes made too easily and too fast?) in everyone in Malaya. If Malaya is going to mess up the industry because she does not realise that the Congo exists and that Siam is important, the Secretary of State will have to educate them.[7]

In fact, Thomas soon showed that he had no commitment to this position and was quite prepared to make whatever concessions were asked by Siam and the Congo.[8] That would 'mess up' negotiations and Thomas would provide the Colonial Secretary with many opportunities for instruction.[9]

Peripheral producers

Shifting French Indo-China to a standard tonnage was quite straightforward. It had been unable to produce its minimum flat rate and Picard was persuaded that making a modest concession would strengthen the ITC's hand in dealing with Siam and the Congo.[10] He was therefore prepared to take the current flat rate of 3,000 tons and define that as the standard tonnage but with a new minimum of 1,800 tons.[11] French Indo-China would therefore have a lower entitlement in the third agreement than it had at the end of the second, though it was not one that it would ever fully use.

Portugal and Cornwall were less co-operative. Portugal wanted to increase its flat rate from 650 to 1,000 tons and only one of the Cornish producers, South Crofty, was prepared to renew on the existing terms. When they had been first recruited, it was hoped that membership would provide an opportunity for them to learn more about the operations and rationale of the ITC and then offer real co-operation.[12] In this regard, the ITC had clearly failed and there were no serious discussions about the conditions under which their continued adherence would be acceptable. Without the prospect of real restriction, it was 'on the whole better to have them out than in'.[13]

Renewal provided an opportunity to review the position of the two major outsiders, Burma and China. Unfortunately, the Burmese government was not even prepared to extend its moratorium on the issuance of EPLs, though it was considerate enough of the delicacy of negotiations not to make any formal announcement until they were completed.[14] Overtures to China were more promising and although 'Yunnan was brought to the verge of an agreement, it proved impossible to get the Central Government of China to take a decision.'[15]

The major difficulty in renewing the third agreement came from the two partners whose productive capacity had substantially grown under the umbrella of the second, Belgian Congo and Siam.

Belgian Congo

The negotiating position of the Belgians largely repeated the basic claims made but two years before. As a young industry, it was still entitled to room for further growth and negotiations focussed on the way in which that demand could be reconciled with the ITC's insistence on some level of effective restriction.

Continued exploitation of the minefield gave the Belgian position fresh strength, since its performance was now the subject of serious attention in the mining press, which cited some very impressive figures. One large company was working alluvial deposits of 3 kilos/tonne, over six times as high as a good deposit in Malaya and overall reserves were established with as much as 160,000 tons of metal.[16] Houwert paid an extensive visit to the Congo in the spring of 1936 and came back with even higher figures, especially concerning the largest producer of all, Géomines, which claimed some 140,000 tons in its reserves and could deliver metal in Europe at under £60.[17] The financial resources behind most of the Congo companies eliminated the prospect of using the threat of a protracted struggle. Indeed, should restriction be abandoned, any subsequent negotiations for renewal would find the Congo in an even stronger comparative position.[18]

However, the Belgians could not simply dictate their own terms since the results of these negotiations would set a precedent for those with Siam. The greater the concession to the Congo, the more would be demanded by Siam and the greater those demands, the less the prospect of a renewal of the overall agreement. The Congo had now become too large to explore the limits of the freerider position without restraint. This explains why earlier resistance on the part of certain producers to the principle of participation was now overcome.[19]

In the final year of the second agreement, the Congo had been granted a flat rate of 7,000 tons. Since this rose when the international quota reached 65 per cent, it was considered as equivalent to a standard tonnage of 10,770. Negotiations proceeded on the basis of this figure and continued to concede the claim that productive capacity was still expanding, so the notional standard tonnage of 10,770 for 1936 was raised to 11,200 for 1937, increasing at an annual rate of 1,000 tons over the life of the agreement.[20] In return, the ITC secured the elimination of the flat rate and should the quota again be reduced to low levels, the Congo would have to develop the machinery for effective restriction.

The terms on which the Congo joined the new agreement naturally caused some concern in Nigeria which was now permanently overtaken as the largest producer in Africa but they also rankled in Batavia which had never been happy with the standard tonnage of the NEI. Van Buuren's frustration was taken out on van den Broek who was accused of capitulating to the Belgians because of the financial support Billiton was providing for some of this fresh investment.[21]

Siam

Placing Siam on an equal footing with the other members of the ITC involved several considerations. Not only was there the question of the standard tonnage which would obviously dominate negotiations but the ITC was also anxious to end two anomalies. One was the way in which Siam's quota was calculated, not on the basis of the true assay value but on its notional metal content at 72 per cent. A second was the exemption of support for the research scheme. It had always been recognized that negotiations with Siam would be complicated but no one anticipated just how tortuous they would be and the extent to which they would jeopardize the prospects of a renewal of the agreement.

The political upheaval in 1933 had installed a new regime in Bangkok which was anxious to break with the pattern of subservience to neighbouring European colonial powers and assert a new form of nationalism. Constitutional expression for this was to be found in placing ultimate sovereignty in the People's Assembly, half of whose members were nominated by the government and the remainder subject to popular election. A factional struggle within the government[22] created a complex and unpredictable relationship with the Assembly. One faction was headed by Luang Pradit (Pridi Phanomyong), of Chinese extraction and educated in France. Considered the 'idol of the Siamese intelligentsia', he was the leader of the civilian 'liberals' whose nationalism was not explicitly anti-Western.[23] A cabinet reshuffle at the beginning of 1936 considerably strengthened the Phibun faction which now controlled not only the military but also two portfolios with a direct interest in tin, Finance and Agriculture. In this reshuffle Pradit was moved to the margins of political influence as Minister of Foreign Affairs.[24] As Phibun increased his power, the nationalist current that was prepared to look to Japan for support against the West became more pronounced. Negotiations with the ITC would inevitably play into the tension between these two factions.[25]

For entirely different reasons, the ITC recognized that negotiations would be particularly difficult, since it feared a repetition of the pattern that had emerged during 1933–1934 when Siam was included in the negotiations that led to the first Rubber Restriction Agreement. Rubber was just as remote from Bangkok as was tin and during the initial round of negotiations the government had insufficient data on the extent of its cultivation. The IRRC offer was considered generous and accepted but when the terms were discussed in the Assembly representatives of the rubber region considered them quite inadequate. That forced the resignation of the government and when negotiations were renewed Siam received a considerable increase in the basic quota.[26] Siam was then authorized to produce at ten times the level reached during the peak of the rubber boom in 1926 and the ITC considered that success as a result of this 'outrageous black-mail' would stiffen her position on tin.[27] Since the Assembly had already taken an active position on rubber, the new government could never be sure that it would not suffer the same fate as its predecessor if it failed to deliver on tin.

At the heart of the conflict between Siam and the ITC lay a division about the appropriate moral basis for the allocation of standard tonnages. For the ITC, especially for Malaya and the DEI, Siam was an old established industry much like theirs. It was, therefore, expected to accept a standard tonnage broadly consistent with the terms they had accepted. For the first two agreements Siam had been granted a flat rate which was considered generous, especially during the period of severe restriction of 1932–1933. Under the existing arrangement Siam would only get 13,400 tons when the international quota was 100 per cent and she was now offered a standard of 15,000 tons.[28] In placing Siam on the same basis as the other members, she was being asked to give up the protection of the flat rate in periods of low quotas in return for increases when quotas were high. Since the standard of 15,000 tons was considerably greater than even productive capacity in 1931, let alone actual production in 1929, the ITC regarded it as close to the limit of any reasonable position.

The Siamese government was not prepared to take the existing arrangement as a starting point for negotiations.[29] The 1931 agreement had been signed as a stopgap measure and had been renewed in 1933 'when Siam was still passing through a very critical stage of her political career and the Government was too fully occupied with internal affairs to spare serious thought for outside consideration'.[30]

As the depression took hold in Siam, the policy of turning off the production tap through limiting new prospecting and mining leases had been reversed in 1933.[31] The Chinese miners were able to press for new concessions and assessed claims continued to rise.[32] In using the tin industry as a way of absorbing unemployment and reducing the extent of potential Communist agitation,[33] the government was redistributing the burden of restriction, so that the established European companies found themselves restricting at a higher rate than even their Malayan neighbours.[34] In addition to the symbolic significance of a revision of the standard tonnage as the basis of a new relationship with her partners in the ITC, Siam faced immediate pressure from the industry. The new STS Chairman went to Bangkok and 'pressed for Government to obtain larger tonnage'.[35] Some Australian firms went even further and insisted that Siam could be safely parasitic.[36] In the background was a large new lode mine being developed by British-American Tin Mines which would soon claim a generous assessment.

All these factors committed Siam to obtaining a radical revision in its standard tonnage, to a level which reflected its new productive capacity. They were reinforced by several arguments. The basis of the first agreement was considered to have been productive capacity, so the Siamese considered they were making the same claim. On that false assumption, the equity argument received further support, since Siam was restricting by about 50 per cent while the others were doing so only to the extent of 10 per cent. Since the Belgian Congo had been granted room for further growth in its tin industry, simple equity required the same concession to Siam. Nor could the Siamese understand why there could be no simple transfer from Bolivia which remained a chronic underproducer.[37]

Current capacity was estimated at 18,600 tons but a 'reasonable' allowance for the future would bring the total to at least 25,000 tons. This figure first became public in *Sri Krung*, a paper that articulated the views of many members of the Assembly and was widely endorsed.[38] It was one that caused considerable alarm and split the ITC into two camps.

In London, Lowinger shaped the hard position. Fifteen thousand tons represented the outer limit, beyond which anything would be a 'complete surrender'. Not only was this morally wrong since it would be very unfair to Nigeria, which was prevented from playing a similar game and it would complicate negotiations with the Congo.[39] In Kuala Lumpur a very different position emerged and Thomas replied to Lowinger:

> At our meeting [of the local Tin Advisory Committee] we were all unanimous that the offer of 15,000 tons to Siam should be made for the purposes of negotiation and should not be regarded as the last word. We all agreed that it would be better to get Siam to join with a tonnage of 18,000 or even 20,000 tons rather than a refusal of the offer of 15,000 tons should break the scheme. One unofficial went still further and said he would give Siam all she wanted … Our view is one of cash. It will pay us to get Siam in on a 20,000 basis if the alternative is abandonment of restriction.[40]

Thomas was not prepared to contemplate the abandonment of restriction under any conditions. For him, a fall in price would have unfortunate political and administrative consequences since it would cut government revenues and raise unemployment. That made the resolutions passed by the Malayan industry empty sentiments and the ability to put some force behind them was further vitiated by the position of many miners, especially those with interests in both Malaya and Siam. One frankly admitted: 'As an FMS miner I think we might well throw Siam a sop of a few 000 tons over 15,000 and as a Siamese miner I would be glad to receive it.'[41]

Capitulation to Siam was excused on grounds that reflected the prevailing colonialist mentality in Malaya, as Thomas explained to his London critics:

> You will ask why we suggest such generosity to Siam and not to the other partners. The answer is because we see no other way of getting her in. She is neither sensible nor reasonable: she is a young, small country and has the obstinacy of an inferiority complex. The facts must be faced. One cannot expect from her what one would expect from negotiations with highly organised European nations … the grant to Siam [of a tonnage higher than 15,000 tons] must not be followed by concessions to any other participator. We should all be unfairly treated vis-à-vis Siam but half a loaf is better than none. We must suffer for being sensible.[42]

The advice that Thomas offered to try and change Siam's position was consistent with that mentality:

> There ought also to be a vigorous campaign of propaganda in the vernacular Press by means of articles which would no doubt be translated and published at a fee. Much can also be done by broadcasting; it is stated that wireless sets in Siamese towns and villages are very popular. It has to be remembered that the Assembly has to be persuaded as well as the Government, and the mental capacity of many members of the Assembly is said to be low indeed.[43]

What gave these otherwise outrageous remarks some substance was the way in which bombastic statements in the Siamese press reflected a lack of understanding of some of the basic features of the operation of the tin market. *Sri Krung* even advocated leaving the cartel altogether, thinking that it could continue without Siam and still sustain prices. It supposed that a deal could be struck with Japan, ignoring the fact that Japan imposed an import duty on tin and could absorb less than half Siam's capacity. It even accused the Straits smelters of pocketing the 2.5 per cent difference between the actual metal content of the concentrates they treated and its nominal content used for royalty purposes.[44] Nor did the opening negotiating position of the Siamese government suggest any degree of reasonableness, since it claimed room for further growth. It wanted a continuation of the flat rate principle but increasing the current level of 10,000 to 18,000 tons, with a further 1,000 tons for each of the following two years.[45] That would then be equivalent to a standard of over 33,000 tons![46]

The Siamese received mixed signals in response to their initiative. One, of course, was that this was so far outside the thinking of the ITC that it was 'wholly unacceptable' and that it would be better to abandon restriction altogether. The other was presented by the British Minister in Bangkok, Sir Josiah Crosby, who, in complete innocence of the position of the ITC, attempted to broker a deal along the following lines:

> The Ministry of Foreign Affairs tells me that 15,000 standard tonnage is unacceptable to the People's Assembly. Government minimum is fixed flat rate of 16,000. ITC should offer 15,000 on understanding that under pressure from Siam Government they will subsequently consent to 16,000. Government can then satisfy the Assembly that the best possible bargain has been struck. I do not believe that the Committee will get better terms.[47]

A most unfortunate prelude to serious negotiations had been set, since the soft messages coming from the region, from Crosby and several sections of the Malayan mining community, would only stiffen the hard position within the Siamese government and overcoming that would take many frustrating months.

The extent of hardness is simply a function of the willingness to walk away from an unsatisfactory agreement. On this occasion all parties found that option easier to contemplate thanks to the increasingly buoyant tin market throughout 1936. Overall, the Siamese took a harder position than did the ITC. On their side, the hard position was taken by the Ministry of Agriculture, which had direct responsibility for mining policy, thanks primarily to its Mining Advisor, Barry Connell. The soft position was taken by the Ministry of Foreign Affairs and that placed Crosby in a strategic role.

Crosby was anxious to support Pradit in the struggle within the Siamese government, fearing that his failure in the politics of tin would strengthen the Phibun faction.[48] Crosby's very involvement tended to define the ITC as a British organization and that raised the further spectre of a backlash against other British interests that he felt obligated to protect. Crosby therefore worked to distance himself from the ITC, defining it as 'obdurate', insensitive and 'shabby'.[49] Just as the ITC tended to consider the Siamese as fundamentally unreasonable, so Crosby provided the basis on which they could develop a similar view of the ITC.

Among the members of the ITC the soft position was broadly taken by the Malayans and the hard by the Dutch. Neither Nigeria nor Bolivia played any independent role. Nigeria was too small and Bolivia was too preoccupied with protecting its own standard tonnage. Dutch thinking about tin had always been shaped by long term considerations and these were reinforced by a decision to make substantial fresh investments in new dredges and machinery.[50] That both increased capacity and cut costs. With 50 years of ore reserves, they could produce perhaps 60,000 tons a year at around £60 a ton.[51] The fortuitous decision of the Dutch government to abandon gold in September 1936 meant a 24 per cent devaluation in relation to sterling, which further strengthened their position.[52] In addition, they were still sore about the way in which they had been compelled to accept their reduced standard for the second agreement.[53] Further concessions to Siam were quite intolerable given the many concessions they had already made.

Within the Dutch delegation, the hardest position of all was taken by van den Broek, perhaps as a result of the personal rebuke he had received over the negotiations with the Congo, while Houwert was much more flexible. Before any final agreement could be reached, two sets of internal politics had to be resolved: among the Dutch and among the Siamese. Anything that was acceptable to the Dutch would certainly be acceptable to the other members of the ITC. Since any sign of resolution and hence softening of one position would harden the other, these internal processes operated at the same time as negotiations among the principals. That made their outcome both tortuous and uncertain.

In spite of the poor start in March, negotiations in July were conducted in a very cordial atmosphere. An ITC delegation composed of Lowinger and van den Broek went to Bangkok and met with officials of the Ministry of Agriculture. Discussions began to shape the terms of an agreement and the ITC had two elements of flexibility. It could concede a minimum flat rate in return for a lower standard and it could grant a higher standard on condition that the agreement ran for five years. That would give all parties an opportunity to develop their productive capacity so they would confront the question of renewal in 1941 on an equal basis without some being constrained by the agreement in 1930. These principles were operationalized with the following offers: (1) a three year term with 15,000 standard and 9,000 minimum, or (2) a five year term with either a simple standard of 18,000 or a standard of 16,700 with a minimum of 10,500 tons. Siam responded with a request for a standard of 20,000 and a minimum of 15,000 tons.[54] As the delegation left Bangkok, these offers were withdrawn but it was clear that if any agreement were to be reached it would be on the basis of five years and at least 18,000 tons.

The breakdown of negotiations at this point provoked some panic on the part of both Aramayo and Hochschild who proposed that the AIM endorse a resolution offering Siam part of Bolivia's tonnage. Antenor Patiño quickly moved to prevent such a 'grave error which would make the situation worse'.[55] Indeed, the very offer could have led to a complete impasse, with Siam more confident about her claim and the Dutch less willing to grant it.

In September negotiations resumed and moved to London, though the Siamese delegates brought no real authority with which to secure a final settlement. Their opening position tried to make the 20,000 standard more acceptable by cutting the minimum from 15,000 to 14,000 tons; the ITC reciprocated with acceptance of the principle of the minimum but at 10,500, to make the 18,000 acceptable. Siam's position was then modified to a standard tonnage of 19,000 and a minimum of 12,500 but was rejected by the ITC[56] though not by Malaya. In spite of Campbell's advice that Siam would come down further, Thomas cabled his immediate acceptance.[57] Fortunately, the Dutch held firm.[58]

Siam then countered with 18,500 and a minimum of 11,100 but made it clear that they could not move to a true assay value basis for its calculation. With assay values running at 74.5 per cent the actual metal content of an 18,500 standard on a notional 72 per cent basis was 19,142 tons. The Siamese delegation could provide no assurance that even this amount would be acceptable to the Assembly and since the parties were still over a thousand tons apart negotiations had reached an impasse.

The Dutch had drawn the line at 18,000 tons. The cost to them of taking their share of the gap of 1,142 tons was trivial, under 0.6 per cent of their own tonnage, or 2 tons for each percentage point of the international quota. But there was an overriding judgement that a line had to be drawn somewhere, regardless of the short term consequences. However, the

Dutch were divided on the question of whether to impose such a line on others. Van den Broek remained 'bitterly hostile' to any concession but he was overruled by the government appointees to the Dutch delegation.[59] Others were, therefore, free to find a solution.

It was the French and Belgians who panicked first and Campbell agreed to arrange dinner with them and the Bolivians to see what could be 'put in the hat for Siam'. But he had come to the end of his patience and was sceptical about the outcome:

> There was not time enough; the Assembly had risen; the Siamese here could not negotiate. Nor did I believe that the Assembly would in fact ratify any reasonable terms which might finally be offered to Siam ... I said that they had all played poker so long that the gambling hell was now closing for the season and there was not in fact time for any further game.[60]

The French offer to contribute was rejected since it was considered to be over assessed and hence the contribution would not involve a real reallocation of restriction. Instead, the Belgians offered 300 tons or 2.6 per cent of their standard and the Bolivians the remainder of 842 or 1.8 per cent of theirs and the Siamese demands were finally met.[61] Campbell had done well to prevent Malaya from being pressed for a contribution, which it would have been quite willing to provide. The People's Assembly was convened in a special session and, contrary to many expectations, ratified the agreement without dissension on 25 December.[62] While there was considerable relief in many quarters, it was not shared by Campbell:

> All things considered, I personally would prefer to see control end, and I think ... that it would be to the advantage, in the long run, of control schemes generally. One cannot go on for ever on this black-mail basis – with every country refusing to give any measure of real cooperation. Tin would supply a most useful warning.[63]

In the haste to secure an agreement, two important symbolic details were overlooked. One was the question of Siam's contribution to the research scheme. The Siamese delegation had promised that they would now participate on an equal basis in the existing scheme which was due to expire in 1938 but this item was not included in the proposals presented to the Assembly, with the result that the offer was withdrawn. Since this violated the only remaining principle of symmetry among the members of the ITC, it left the Dutch feeling particularly bruised.[64] It was a quite gratuitous slight since, at the first meeting of the ITC under the new agreement, Siam indicated her willingness to join.[65]

The second was the question of voting arrangements under the new agreement. The Siamese had wanted one vote per delegation to reflect their sense of equality with their partners and their delegation had not kept Pradit properly informed that equity could only be established in relation to standard tonnages. When the actual text of the new agreement was received in Bangkok, the new proposals caused considerable concern and again Crosby supported Pradit's threat to terminate the negotiations unless his position was granted.[66] As will be seen, the new voting arrangements had considerable, albeit symbolic, significance and the tension this issue generated is a good indication of the fundamental gap between two conceptions of equity.

One individual emerged from these negotiations with considerable credit, albeit of limited recognition, and that was Antenor Patiño. Given the chronic underproduction in Bolivia throughout 1936, Mair proposed she should cede as much as10,000 tons from her standard tonnage to be distributed among the other three original signatories, with a little over for Siam.[67] Thanks in large part to the relationship Patiño had built up over the years with Campbell, these proposals were never formally presented.[68] In spite of the mounting evidence that Bolivia's problems were structural, Patiño was able to spin them in a more positive fashion, always deflecting questions by pointing to the transitory effects of the Chaco war.[69] He had also foreseen the nature of the end game and had secured authorization to make the crucial concession long before it was actually required.[70] At the same time, he managed to conceal this weakness by waiting for others to make the initial concessionary moves, reducing the sacrifice Bolivia would have to make. While his father received considerable and largely unmerited, credit for his role in securing a favourable tonnage for Bolivia in 1930, the contribution of the son in preserving it in 1936 went unnoticed.

Crosby's reputation, on the other hand, was seriously tarnished. The bitterness on the part of the Dutch led to them to make some serious accusations, especially that he told their legation in Bangkok that he considered they were bluffing and let that be known to Pradit.[71] Such a gross breach of diplomatic protocol resulted in the final impasse and the Dutch complained to Campbell. Since it fitted with Campbell's perception of the role played by Crosby, the Colonial Office sent an official complaint to the Foreign Office.[72]

Siam's ability to secure a dramatic increase in her standard tonnage has generally been regarded as a manifestation of the inevitable way in which these forms of collective action must tolerate the free rider.[73] The formal statement of the free rider problem is very simple. The large producers face three options: (1) restriction on a symmetrical basis; (2) restriction on an asymmetrical basis that is with a free rider; (3) no restriction. While the first is obviously optimal, as long as the second is preferable to the third, it is difficult to achieve. The only formal conditions under which it is possible arise when producers experience different costs, in which case the threat of the third option can be used to secure symmetrical compliance. That turns the co-operative game whereby all are winners but some win more than others, into the inverse one whereby all are hurt but some more than others. Before that competitive game can be played, the costs have to be assessed in relation to the ultimate benefits expected.

These formal features fit certain elements of the tin case. Bolivia clearly could not contemplate the third option on purely economic grounds. The reluctance of the Malayan industry to consider it in view of earlier commitments to lowering price and expanding the market for Straits tin is more puzzling but that may be simply a function of the assessment of the comparative costs and benefits. It clearly fits the problem faced by the FMS government which operated within a more severe set of economic constraints than a private corporation. Unemployment is not a cost to a corporation and a government may be less prepared to forego immediate revenues than a corporation is willing to forego immediate profits.

This model makes one crucial assumption, namely that the symmetrical solution against which a free rider can be established rests upon a shared moral basis. The moral dimension is evident from the frequent use of the term 'blackmail' with which the ITC referred to the position of Siam. But the metaphor was only relevant if Siam's position in the competitive game was stronger than that of her opponents, ruling out the threat of unrestricted production as the means of forcing a symmetrical solution. And that was certainly not the case.

If Siam had violated any moral rules, it was to have allowed her productive capacity to grow. It was the same rule that Malaya was considered to have broken at the end of the 1920s and if she had to pay a price then clearly so should Siam. Such a price would have involved some concession from the outer limit of her productive capacity in the early 1930s. The 15,000 tons of the original offer already conceded to Siam what was denied to Malaya. Anything further represented a tacit acquiescence in the different moral assumption that underlay the Siamese position. Once that threshold had been crossed it was impossible to decide just where a new one should be established.

The Siamese clearly considered that they were trying to establish a new basis for their relationship with the ITC. Intelligent exploration of the free rider position entails preserving the original agreement, that is ensuring that its members do not consider the benefits foregone through the free ride higher than the benefits of restriction. Siam had pushed some members beyond that point and there were only contingent factors restraining them from acting on that basis. Why Siam had allowed that danger to arise cannot be explained on simple economic grounds.

The core network that lay behind the ITC had only weak connections to the Siamese tin industry and none to the Siamese government. With no organizations representing the small miners and little communication between the Siam Chamber of Mines and the government,[74] policy could not be informed by a fine sense of comparative cost conditions. This economic insularity was reinforced by the absence of any connection at all with the day to day operation of the tin market. The ITC was run by those who knew the market and enjoyed the confidence, albeit in varying degrees, of their respective industries and governments and that made Siam the marginal member. The processes of negotiation reflected that asymmetry since, with one exception, they did not take place between principals. These structural features allowed political and ideological, rather than economic, considerations to become paramount.

Tin provided an ideal target for a nationalist regime anxious to assert its own independence. It could be used as a way of distancing itself from its predecessor which had rather passively accepted the terms dictated by the established colonial powers. Tin would also demonstrate the capacity of the regime to serve local interests. Siamese assertiveness resonated with other political forces.[75] In the background was always Japan and the prospect of developing a new smelter either in Japan or in Siam itself.[76] The Americans also watched these developments with interest, since the larger the Siamese production, the greater the prospect of building a domestic smelter which would also draw on the only other source of concentrates not already firmly tied to particular smelters, Bolivia. Much more was at stake than simply securing the optimal position within an economically circumscribed pattern of bargaining.

That was also the case for many connected with the ITC. The Colonial Offices of both Britain and the Netherlands had backed the formation of the ITC since it provided an opportunity to see how governmental commodity control could serve as a counter-depression measure and its success had been widely recognized. Capitulation to Siam would now dent this reputation and hence undermine the viability of the overall policy of cartelization. Those bureaucracies had other games than just tin, in rubber, tea and sugar, and that led Campbell to conclude that tin control was finished.

However, as tin negotiations reached their impasse, other forces came into play. Several quarters in the City approached the Governor of the Bank of England to intervene, threatening to embarrass the Colonial Secretary in the House of Commons unless he took a more active role in settling the matter.[77] Had the other members of the ITC shown the firmness advocated by Campbell and van den Broek, it is quite possible that they would have been overridden at a higher level. Politics is always an unpredictable business and there was certainly nothing inevitable about the outcome of these negotiations for the renewal of the agreement. They could have gone either way and in each case it would be as much extraneous factors as those inherent in the economics of collective action that would have been responsible for the outcome.[78]

Text of the third agreement

The text of the Third Agreement introduced a number of modifications, of which the most important was to change the basis of decision-making. No longer would decisions of the ITC on quotas be formally defined as recommendations to governments, reached by unanimous agreement. Instead, they would stand on their own authority with any differences resolvable by vote.[79] The formal allocation of votes represented a compromise between equity based on standard tonnages and one based on membership. Table 11.1 shows the distribution of votes established under Article 13.

A strict standard tonnage basis could have generated complications by granting the British delegations slightly more influence and accentuating the regional rivalry between Malaya and the NEI. Formally, it was designed to ensure that no two members could either carry or block a proposal from the other five.[80] The introduction of the principle of voting did not in fact make much difference, since Campbell insisted on continuing to try and find consensus[81] and only one vote was ever taken. But it did reduce two sources of tension. Quota decisions could no longer be used as leverage for the introduction of supplementary features, as the Dutch had done in 1934 over the buffer stock. Nor could governments continue to impose their own will on the ITC through their delegations.[82] While the NEI government tried to protect the particular interests of Banka by shaping the policy of its delegation,[83] the friction that had often characterized the relationship between the Nigerian and Malayan governments and their respective delegations began to dissipate.

A second significant change was to address the stock question along the lines proposed by Malaya. Previous agreements had required governments to regulate production in relation to exports, so that stocks could be no higher than those carried at the beginning of restriction. Malaya had proposed minehead stocks of four months' production[84] and Banka, which had to carry stocks for both mine and smelter, made a similar request. Article 18 met this need by allowing stocks within any territory to be as high as 25 per cent of standard tonnage.

Otherwise, changes were more of a routine character, reflecting the experience under the previous agreements. Article 1 redefined the objective: 'The Scheme is intended to regulate the production in, and export from, producing territories, with the object of adjusting production to consumption, preventing rapid and severe oscillations of price, and maintaining reasonable stocks.' This language dropped the qualifier 'fair and reasonable' in relating production and consumption, perhaps as generating unnecessary controversy as to just what such terms meant and the earlier commitment to 'ensure the absorption of surplus

Table 11.1 Distribution of votes

	Actual Votes	*Votes based on standard tonnages*
Malaya	5	7
Netherlands East Indies	4	4
Bolivia	4	5
Nigeria	2	1
Siam	2	2
Belgian Congo	2	1
French Indo-China	1	1
Total	20	20

stocks' was obviously obsolete in light of the stock problem that had plagued the tin market throughout 1935 and 1936.

On the question of consumer representation, existing practice was formalized. Under article 6, two representatives of tin consumers were to be invited to attend meetings and their role was spelled out as being 'to tender advice to the Committee regarding world stocks and consumption'. This formal limitation has been seen as confirmation of the supposition that the ITC simply operated as a producers' cartel[85] but estimates of stocks and consumption trends remained the critical variables which determined production decisions.

One Bolivian problem was solved by controlling the carry over of any deficiency at year-end. Article 17 limited this right to 8.33 per cent (one month) of the total permissible exportable amount for the year.[86] Article 24 addressed another. In the event of hostilities, countries were granted the right to temporary over-exports.

With the Belgian Congo and French Indo-China now full members, the definition of outside production had to be modified. Article 22 established new limits for outside production at the lesser of 12,500 tons or 15 per cent of world production over six consecutive months. At that point members could threaten the termination of the scheme. It was a token signal to free riders but of limited effect since it would only become operative during periods of low consumption when the need for restriction was at its greatest.

Finally, the agreement tried to establish a timetable for its own renewal. Article 3 required the ITC to formulate proposals for renewal at least nine months before the date of expiry with delegations being granted but three months in which to endorse them on behalf of their governments. The market would therefore be given six months' notice before the likely expiry of the agreement.

The form of the third agreement followed that of the second and its announcement provoked the same rebuke on the part of the Foreign Office. As before, it felt that an international agreement on behalf of colonies, rather than metropolitan governments, was most irregular, especially since it bypassed the standard procedure of laying such an agreement before Parliament. In these negotiations, the Colonial Office had not paid any attention to that position. It had lost the letter sent by the Foreign Office in 1934 and in any case negotiations had hardly followed the timetable that would allow such a delay.[87]

The Foreign Office position was defined as a 'ceremonial dance', and it was seen as creating some real problems. If the United Kingdom were the signatory, it would be difficult to justify the omission of Cornwall but the legislation required to impose restriction there would be very controversial. In addition, the experience of the IRRC showed that such a 'hard and tight' formal agreement lacked the flexibility necessary to deal with unforeseen issues.[88] However, in spite of these arguments and the strong preference of the other delegations for the current form, the Foreign Office was not prepared to concede and negotiations for the next renewal would be adversely affected as a result.

Campbell and the British delegations

One further detail remained to be settled before the ITC could proceed to administer the new agreement and that was the question of the status of its Chairman. It was only an issue of concern to Malaya where local European members of the mining community had long felt that Campbell's joint position as head of the delegation and as Chairman of the ITC had compromised his ability to secure their best interests. In Lowinger's mind, this tension had come to a head during the debate about the buffer stock in 1934, to which he was opposed as contrary to Malaya's interests, while Campbell had settled the matter with Cunliffe-Lister without even consulting him. When he went out to Malaya in the summer of 1935, he not only reinforced local scepticism about the buffer stock but also shared his scepticism about Campbell. On his return to London his position mellowed considerably and he saw the advantages to the dual role, 'in view of Sir John Campbell's special position and influence with the foreign delegates'.[89] Unfortunately, the seed he had sown soon sprouted. The Redruth group passed a resolution asking that the mining industry nominate two delegates and that the Chairman be independent of any delegation. In spite of the fact that nomination as such should lie with the government, Thomas passed this on as a 'reasonable request'.[90]

Thomas' subordination to the local mining industry was confirmed when actual decisions had to be made. Again, he supported their efforts to block Campbell's appointment to the new Malayan delegation, adding that there were grounds on which to suppose that Campbell was actually hostile to Malaya.[91] The proposal at first came as some relief to Campbell personally as he would have been quite happy to rid himself of the burden of trying to manage tin politics but it caused considerable alarm in the Colonial Office,[92] which cabled back to Thomas:

> The prospect of Campbell no longer serving as Chairman fills me with apprehension. ITC most delicate piece of international machinery operation of which calls for highest qualities of tact and judgement. Campbell has performed his difficult task with distinction and success. You may not always have seen eye to eye with him on points of detail but you will agree that in its general results policy of tin control has under his guidance conferred wide benefits upon the tin industry as a whole. Satisfied that belief [Campbell as unsympathetic with Malaya] is quite without foundation.[93]

It was not just the personal qualities of Campbell that the Colonial Office was anxious to retain; there was a more general principle at stake. The ITC was, after all, a body of governmental delegations which would now be endowed with plenipotentiary powers on major issues. It had no mechanism for the selection of a Chairman who was not a member of a delegation. The Malayan solution to this problem would have seen Campbell remain as Chairman but through his leadership of the Nigerian delegation. Since Nigeria had now fallen to sixth place among the members of the ITC, this was quite unacceptable to the Colonial Office which considered that the Chairman needed 'the status and authority attaching to representation of one of the leading parties to the agreement'.[94]

While this was sufficient to convince Thomas to change his position, not all members of his advisory committee were prepared to accept it and substantial sections of the mining community remained very sceptical about the extent to which their interests would be effectively represented in the new agreement. One concession, however, was made to Malaya and that was to appoint one miner as a full member of the delegation and Glenister was promoted from his role as technical adviser. Wilcoxson, the Managing Director of STC's new British smelter, replaced him in that position. Now the Malayan opposition would have two voices on the ITC.

Changes also occurred in the Nigerian delegation. Calder replaced Campbell as its head. Baddeley was dropped, in part because he was 'very deaf' and in part because he could be 'peculiarly narrow and rigid in his outlook'.[95] The Nigerian Chamber of Mines also asked to be included in the nomination of a member of the delegation. Since Campbell attached 'little or no practical importance to the representation of producers who normally know nothing of the questions the ITC is concerned with',[96] it would have come as some relief that the Chamber proposed Lyttelton.[97] As will be seen, Lyttleton had replaced Howeson as the key figure in British tin circles and his participation made the role of the TPA as advisors to the ITC redundant. While the organization would continue to play a propaganda role on behalf of the ITC, it was not only excluded from its deliberations but even from the process of consultation.

12 Riding the commodity roller-coaster, 1937–1939

While the first two agreements had a singular focus, rescuing the industry and then stabilizing it, the third faced a multiplicity of challenges posed by a rapid boom-slump cycle. The magnitude of these swings was far greater than anything the ITC had to confront in the past, as is evident from Figure 12.1.

January 1937–June 1938

The boom that was underway at the end of 1936 gathered considerable strength throughout early 1937. This was part of a general recovery in commodity prices driven both by increased civilian consumption in the United States and the anticipation of a heavy demand for armaments by the United Kingdom, accentuated by restocking.[1] This boom was far more dramatic in all other non-ferrous metals than in the case of tin. Copper advanced by 40 per cent over the first two months of 1937, while tin was up by just 7 per cent and that only because of the 'most irrational of all factors – sympathy with the other base metals'.[2]

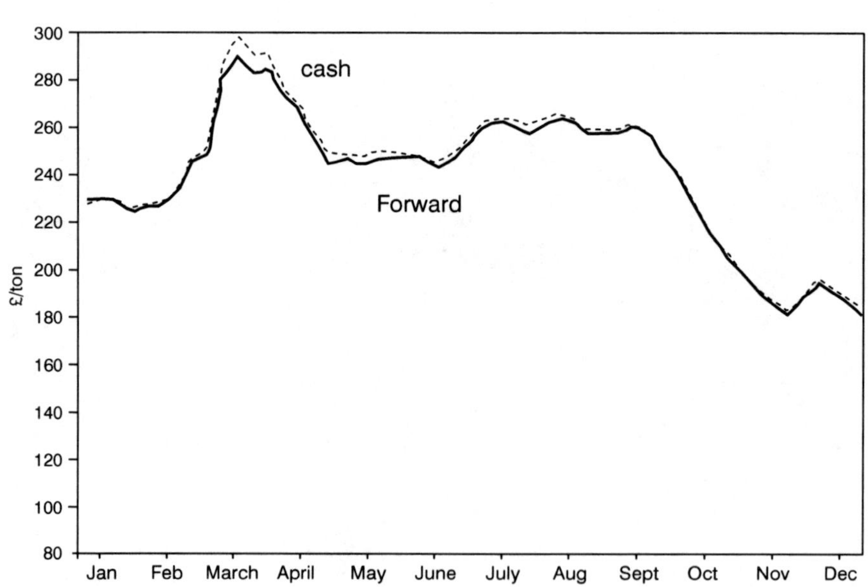

Figure 12.1 Tin prices, 1937

Tin was the only metal in which it was thought that there was little possibility of a shortage developing and prices were generally considered very reasonable. Even those who had been very critical of the ITC over the previous two years were prepared to concede that it had now earned the considerable respect of the tin market.[3] At the beginning of the year, concern was being expressed in some market quarters that stocks were in fact too large and that the ITC should consider reducing the quota to prevent a future problem from building up.[4]

This confidence was soon shattered. The fear that the ITC would actually reduce the quota attracted the attention of speculators. Trench warned of the dangers unless immediate action were taken and advised the ITC to follow the example of copper and allow unlimited exports for two months.[5] Campbell considered that any such move was liable to be construed as a panic measure but did want to offer a public reassurance that the quota for the second quarter would remain at the 100 per cent level. It was not forthcoming since the Dutch preferred to wait until the regularly scheduled meeting of the ITC ten days later.

When the ITC met in March it was anxious to prevent the kind of runaway market that was continuing in copper, especially in light of a forthcoming conference in Geneva which would review all commodity agreements.[6] Bolivia pressed for an increase in production to deter speculation and was supported by Firth who wanted a more explicit price policy to be formulated so that the price would be cut back from its current £256 to a range of £200–£225 but these arguments failed to carry wider support and the quota was confirmed at 100 per cent.[7]

In light of the overall stock position, the 100 per cent quota should have solved the problem. It meant that stocks would continue to rise[8] but it soon proved to be a serious error of judgement.[9] The market had read the situation quite differently. It saw the modest decline in visible supplies, not as a result of transitory factors but as confirming continued growth in demand. Since none of the ITC members was yet in a position to produce at a quota of 100 per cent, the market also saw an uncertain level of supply.[10] When the LME opened immediately following news of the ITC decision, the spot price jumped by over 7 per cent to £269 and by the end of the week it was over £301. The following week it started at £311, the highest level since March 1927, and Campbell feared that the ITC was unable to face up to the political problem that these prices presented:

> I hope the ITC will agree to 'take the roof off'. I doubt if they will … The solid core of opposition will come from the NEI Govt. The Dutch delegation cannot at present emancipate themselves from Batavian control and van Buuren has for a long time been insisting most strongly that the tin quotas are much too high … If they dont agree, the price will bound up again, I fear. The USA Gvt. is being particularly troublesome at the moment as regards rubber; and I think it will not be long before they turn their attention to tin also. 'Politics', in fact, is the dominant practical factor, in both control schemes, at the moment; and I have been rubbing that primary fact into both committees for some time now … it seems to me imperative that we should make every effort we can to prevent prices from getting wholly out of control, whether we succeed or fail.[11]

While Campbell received no support for this extreme position, the ITC meeting at least agreed to raise the quota to 110 per cent. Quotas at the 110 per cent level could only be produced with a fresh infusion of capital, either to rehabilitate idle dredges or to develop new sources. Such investment could only be justified with a longer period of effectively unrestricted production than just three months and Malaya proposed that 110 per cent run for the rest of the year. That simply raised the spectre of excessive stocks once again. With the support of the Dutch, Patiño proposed that they be absorbed by a buffer stock but there was no willingness to think on anything other than a short term basis, so the quota was simply set for the second quarter.[12] With less than three months into the new agreement, the ITC found itself in a situation which it was quite ill-equipped to control.

These two decisions coming within a week of each other upset the reputation of the ITC. Hitherto, it had always based its decisions on an assessment of the stock position in relation to consumption, an assessment that those outside the committee were also in a position to make so that announcements rarely came as a surprise.[13] Now, it was also forced to respond to price and this added a fresh element of uncertainty. Furthermore, it was a price dictated solely by operators with a 'brutal contempt for the ITC and for consumers'.[14] Even those sympathetic to the ITC expressed 'their disgust at the practical operation and results of the International Tin Control Scheme'.[15]

Fortunately, the problem the ITC would eventually have to confront could be postponed. Actual consumption was better than anticipated and by the end of the second quarter it was just about balanced by production. When the ITC met in June to set the quota for the third quarter it found that only the NEI had been able to meet its 110 per cent quota in either April or May and that everyone was behind in meeting their total entitlements since the beginning of the year. With the total shortfall around 23 per cent, prices remained too high. Malaya therefore proposed to raise the quota to 125 per cent. Bolivia responded by arguing for a cut, back to 100 per cent. She could not benefit from the increased production at high prices and a general cut would reduce the extent to which she would share in the low prices to come when the excess stocks finally made themselves felt.[16] The others were anxious to get more time in which to show whether they could in fact produce at 110 per cent, so the current quota was simply confirmed.[17] However, some progress was made at this point about the possible structure of a buffer stock so that a definite scheme would be ready when serious consideration to its implementation could be given.

For the third quarter the production side of the problem began to ease, thanks largely to the effect of the new fiscal regime in Bolivia which again made the prospect of fresh investment profitable. Otherwise the position remained as before with no one consistently meeting their 110 per cent target, so it was naturally kept at this level for the final quarter.[18] No sooner had the ITC announced its decision than a stock market crash, first on Wall Street and then on London, destroyed all the confidence in the recovery that had been building throughout the earlier part of the year. Just as the boom in other non-ferrous metals had pulled tin up quite independently of the underlying statistical position in the early part of the year, so the slump pulled it down.

In light of the new financial crisis, Antenor Patiño called a special meeting of the ITC to revisit the question of the final quota. It was now evident that quarterly quota changes were too crude an instrument with which to handle a rapidly changing situation. Actual levels of production would only be known in four months' time and since they were now well above actual consumption, the stock situation could become very serious indeed. All countries, including Malaya, were seriously in arrears at that point, so the solution lay in a combination of the elimination of these arrears, a retroactive cut in quota and a buffer stock.

Patiño's initiative provoked some acerbic comments from Firth in his response to Campbell's request for his views:

> He felt some difficulty in the matter because no one paid any particular attention to what he said ... World prices were generally falling and he supposed the committee wanted a price of £250–£260. When he had discussed price previously he had been told, quite

politely, that it was rather outside his ambit and that the committee was not interested in price but in equilibrium between production and consumption. He noticed, however, that when the price of tin fell a special meeting was a called as a matter of urgency.[19]

While it was an unfair comment, especially in light of the way in which the ITC had acted in March to head off a runaway price, it did reveal the gap between the perspectives of producers and consumers. Lyttelton gently rebuked him by indicating that the moves were designed to prevent a further fall rather than to regain the earlier prices.

If the Bolivian diagnosis proved wrong, the ITC would again find itself in a situation much like that in 1935 when stocks were short and there was a general outcry against its policies. Van den Broek had visited the United States during 1937 and met both McReynolds and Roosevelt who stressed the need to restore price stability.[20] Fear of upsetting the United States appears to have been a primary factor in the cautious attitude adopted by the committee in leaving the quota at 110 per cent. However, Patiño managed to get at least a discussion of the idea of an 85 per cent quota for the first quarter of 1938 and this prompted the first formal proposal for a redistribution of the standard tonnages.[21]

Unfortunately, the Bolivian diagnosis proved quite sound and showed the decision to keep the 110 per cent quota to have been a serious mistake. When the initial quota for the year was set, it was recognized that it would result in excess stocks but the total increase amounted to around 30,000 tons, of which only 3,000 was recorded within the visible supplies.[22] Consumers had built up their invisible stocks and producers would soon have to pay for their generosity.

The experience of 1937 was instructive in several other respects. It demonstrated the extent to which price volatility in one market is a function of often irrational forces in others. That makes it impossible to assess the performance of the ITC with respect to its objective of curbing these oscillations, except in relation to other markets. It also demonstrated that the problem of excess capacity that had prompted the formation of the ITC was largely eliminated by the end of 1936. Without the existence of some reserve capacity, especially in the form of companies capable of undertaking fresh investment, tin would have seen far greater volatility. The experience of real shortages and extraordinary prices would have reinforced the incentive on the part of manufacturers to reduce their exposure to such a fickle metal.

1937 also exposed a fundamental asymmetry in the allocation of the standard tonnages and jeopardized the whole future of tin control. The dimensions of the problem are presented in Table 12.1.

Table 12.1 Tin production, 1937

	Quota	*Actual*	*Difference*		*Change From 90%*	
	tons	*tons*	*tons*	*%*	*tons*	*%*
Malaya	77,331	77,542	211	0	12,095	18
NEI	39,052	39,779	727	2	6,728	20
Nigeria	1,177	10,447	–1,260	–11	540	5
Siam	20,136	16,385	–3,751	–19	–656	–4
BC	11,620	9,286	–2,334	–20	–729	–7
FIC	3,225	1,531	–1,694	–53	–1,198	–44
Bolivia	49,407	25,024	–24,383	–49	–16,780	–40
Total	212,478	179,994				

Over the course of the year only Malaya was actually close to meeting its entitlement and that generated a widespread sentiment that in all previous years she had been carrying the rest of the industry on her back. The NEI suffered a major setback in her production schedule when a new giant dredge, Kantoeng, sank in the English Channel but put on a spurt in the last quarter when it exported at the rate of 146 per cent of its standard in 'response to the annoying propaganda in the Malayan press that she was not able to produce the 110 per cent'.[23] By year end she had slightly exceeded her entitlement and would be exempt from further Malayan criticism.

The quota had been set as high as 110 per cent in the full knowledge that Bolivia would not be able to produce its entitlement. Since clause 17 established a firm limit to any carry over, it was quite safe to give her a quota that could not be produced. The 110 per cent therefore effectively transferred entitlement from Bolivia to Malaya and the NEI. The extent of that transfer is revealed by comparing the actual results with those that should have obtained under a 90 per cent quota. This is the level at which the same total output would have been reached had all members been able to produce their standard tonnages. Bolivia lost nearly 17,000 tons of which Malaya gained 12,000. On that basis, Malaya had nothing to complain about, since she produced 18 per cent more than she otherwise would and at generally remunerative prices.

While the Bolivia problem could still be defined as a temporary one, the results provided some disconcerting evidence about the position of both Siam and the Belgian Congo.[24] Both had a standard tonnage which reflected potential production, so their inability to meet a 110 per cent quota came as no real surprise. However, the extent of the shortfall indicates that they could not even meet a 90 per cent quota.

In Malaya these asymmetries were defined as profound and intolerable inequities. The extent and significance of the problem was easily overstated.[25] While Malaya had exported her full quota, she had not in fact produced it. The difference, perhaps as much as 8 per cent, had come from running down minehead stocks. One prominent miner pointed this out in an effort to 'stop all the controversy about Malaya being badly treated'.[26] It went unheeded.

The ITC had responded to the boomlet of 1937 by opening the production tap as wide as possible. This was designed to reassure consumers and to give Malaya the necessary incentive with which to rebuild her capacity. Continuing the 110 per cent quota for the fourth quarter in the face of a decline in consumption left the ITC in a position it had not seen before. Over the course of the year total stocks had risen by some 32,000 tons but none of this showed up in the visible statistics. Stocks in the hands of consumers had gone up by at least 10,000 tons and unless consumption recovered that could only mean a decline in immediate demand. The only comfort was that at least neither the Belgian Congo nor Siam was in a position to follow the NEI example and put on a last minute spurt.

A new dimension into the politics of the ITC was introduced during 1937. It had lost control over the market, not because its assessment of the physical parameters was incorrect but because of the unwillingness of the market to trust that assessment. Trust in such an assessment was more easily granted when demand was weak and potential supplies excessive. Under such conditions any mistakes could be corrected. The ITC was caught in an unfortunate squeeze as demand rose at the point when productive capacity was being reduced. The risks of mistakes were therefore increased and the rise in price was the inevitable result. Raising production in these circumstances demonstrated that the producers were unable to impose their will on the market but rather had to absorb whatever risks that market defined. The rest of the peacetime history of the third agreement tried to work out the implications of this painful lesson.

In setting the quota for the first quarter of 1938, the ITC first had to address the problem of the asymmetry in real productive capacity among its members. While it was obvious that the quota had to be cut, it could only be determined after several clauses in the current agreement were suspended. Otherwise the four members who were overassessed would be carrying forward large quantities of unused tonnage and those who were not would be carrying forward slight deficits. That would simply accentuate the inequality since the former would be restricting at a far less severe rate than the latter. No solution could be found which did not involve transfers from Bolivia, Belgian Congo and French Indo-China.

Making matters more complicated was the fact that by the end of 1937 outside production was slightly in excess of the amount specified in clause 22. Malaya was now free to withdraw from the agreement altogether. While the clause was designed to give the ITC some negotiating power against complete outsiders, there was nothing to prevent it from reopening internal agreements. It was not a threat that could be used easily. Formal notice of withdrawal would simply create a period of uncertainty, with even lower prices and this was too much to pay for rectifying admitted inequities.[27] Nor was it evident that Malaya would be allowed to do so. Confirming that bigger issues were at stake in tin than just its internal politics, Clauson noted:

> My own feeling is that for reasons of major policy going far beyond the particular case, HMG would not want to destroy this agreement even if the majority of tin mines in Malaya and the Government wished to do so.[28]

However, since none of this was communicated to Bolivia the threat forced her to adopt a flexible position.[29]

The contentious question of the principle of the basis of a new allocation of standard tonnages was avoided by developing a complicated formula of transfers[30] and the results are presented in Table 12.2.

These new standards were only operative for the first half of 1938 and were conceived as establishing a temporary *ex ante* basis for the kind of *ex post* transfers that had occurred in 1937. But they marked a major departure. No longer was there any reference to 1929 as the conceptual basis for allocation among the four core signatories and they now bore a much closer relation to current productive capacity. Had that been measured on the basis of actual performance in 1937 Malaya and the NEI would have gained considerably more but at least now they would benefit by nearly 10 per cent more than they had anticipated when the third

Table 12.2 Standard tonnages, January–June 1938

							Market share change cf	
	1937 ST	*Transfer*	*1938 1st*	*% 1937 actual*	*% 1937 ST*	*% 1938 1st*	*1937 actual*	*1937 ST*
NEI	36,330	4,994	41,324	22.1	18.3	20.0	–10.3	9.9
Bolivia	45,951	–8,406	37,545	13.9	23.1	18.2	23.6	–21.9
Nigeria	10,890	1,497	12,387	5.8	5.5	6.0	3.4	9.9
Malaya	71,940	9,891	81,831	43.1	36.2	39.7	–8.7	9.9
Siam	18,731		18,731	9.1	9.4	9.1	0.0	–3.3
BC	12,008	–200	11,808	5.2	6.0	5.7	9.2	–4.9
FIC	3,000	–330	2,670	0.9	1.5	1.3	30.5	–14.0
Total	198,850		206,296	100.0	100.0	100.0		

agreement was negotiated.[31] In the light of the contention that surrounded renewal, these results are quite remarkable.

On this occasion Bolivia was not in a position to press for a standard grossly in excess of capacity. The greater the level of underperformance, the more disruptive this was to the market and the greater the pressure for a permanent revision. It was far better to make a dramatic but temporary, concession which provided sufficient flexibility for a solution. The delegations had no obvious authority with which to change the de facto standards, yet this was the only basis on which a solution could be found. Within the wider arena, governments, Chambers of Mines and the press could introduce all kinds of symbolic considerations which exacerbated tensions; within the confines of a meeting of the ITC these could be all kept at a reasonable distance. Again, Antenor Patiño did well in deflecting Malayan pressure for an even more radical concession.[32]

The quota itself was also subject to some discussion as a result of varying estimates of anticipated consumption. It repeated earlier positions with Bolivia taking the more cautious approach but conceding to the more optimistic view which led to a final determination at 70 per cent. Although production would still be in excess of consumption, the price slide was at least temporarily halted.

Further progress was also made on the shape of an agreement for the buffer stock. It kept one principle from the 1934 agreement. The stock would be held in any grade of metal, so that contributions in Straits could be exchanged for standard warrants with the premia naturally repaid. That would again be contentious in Malaya since it implied an undermining of the role of Straits tin. However, it drew several lessons from the earlier experience. The day to day operations of the buffer stock would be under the control of a manager appointed from outside the industry altogether and the ITC itself would only receive aggregate information about the position of the stock. It aimed at a somewhat larger size than the previous one, 15,000 tons or 7.5 per cent of standard tonnages and a somewhat more modest price range of £200–£230.[33] In addition, it would run for the whole of the life of the current agreement. A draft was now ready for review by participating governments.

In the light of the history of the first buffer stock, getting Malaya's support would not be easy.[34] It meant going over the Malayan delegation, composed as it was of Glenister and Wilcoxson who were linked to the most vociferous opponents of this form of control. In addition, they had a new argument. A commitment to the buffer stock would simply pre-empt the possibility of a permanent and better redistribution of the standard tonnages.

In setting the quota for the second quarter of 1938 the ITC was most influenced by its estimate of the invisible supplies. The visible level remained sound especially in light of current consumption levels but it was widely assumed that much of the excess production towards the end of 1937 had gone into consumers' stocks. They were perhaps as much as 20,000 tons higher than at the beginning of the 1937 boom, sufficient to meet some two months' consumption.

It was also clear that a serious depression had taken root in the United States. Tinplate mills were operating at only 35 per cent of capacity in January, a figure far lower than even in 1931; steel was around 19 per cent. The position in the United Kingdom was just as serious where total tin consumption was running at levels well under those in 1932. In the first quarter some compensation was provided by heavy buying from the USSR but this was tailing off. A severe cut was clearly in order. Bolivia pressed for a cut to 45 per cent but naturally tied it to a revision of the concession made for the 70 per cent quota in the first quarter. Malaya and the NEI proposed 60 per cent and the ITC settled on 55 per cent.[35]

With a low quota, outside production took on an even larger significance. In the first quarter of 1938, outside production was just over the 12,500 ton threshold; the 55 per cent quota for the second quarter would take it over the 15 per cent threshold. The overall agreement now appeared to be purely voluntary since any member could terminate it on six months' notice. Just as the Malayans were asking why they should stay in when they and the Dutch appeared to be the only countries restricting, so the Dutch also asked why they should remain if the scheme lacked the flexibility of a buffer stock.[36]

The ITC was now paying a price for its negotiating strategy in 1936 which had been to maintain unity among the core members in order to present a common front in dealing with the Congo and Siam. In doing so, it overlooked the necessity to address internal issues such as the buffer stock and the problem posed by the long-delayed recovery of the Bolivian industry. They could no longer be postponed and since that made the very future of the agreement uncertain, the price of tin continued to languish around £180.

Buffer stock negotiations

The conception behind this buffer stock involved several modifications from its unfortunate predecessor.[37] It would be larger with at least 10,000 and perhaps as much as 15,000 tons. The range within which prices would be allowed to fluctuate was set at £200–£230, subject to modification in relation to a general price index. Although this range was nominally similar to that within which the 1934 stock operated, the rise in the overall price index meant that it was really 6 per cent lower. A more viable administrative framework was proposed: the ITC would set policy, a small subcommittee would appoint the Buffer Stock Executive which in turn would have full operational responsibility.[38] Opposition from the LME was limited by ensuring that all transactions would be conducted through its members, with the business being distributed equitably.[39] But the most important change was that it would be no mere experiment since it would only expire with the overall restriction scheme itself. In addition to moving towards a more co-operative posture vis-à-vis consumers and the LME, the ITC was making itself a permanent feature of the industry.

In the 1938 debate, the arguments for the buffer stock were much stronger than was the case in 1934. The overall problem of raw materials had become sufficiently pressing that the League of Nations established a committee to engage in a comprehensive investigation and its report was ready by the end of 1937. While the root source of the expressed problem had more to do with difficulties experienced by soft currency countries in securing access to commodities from hard currency areas and hence were not easily addressed, the committee also commented extensively on intergovernmental regulation schemes. By 1937 there were four such schemes – tin, rubber, tea and sugar – and their experience was reviewed very positively, not only as an effective counter-depression measure but also because of the role that their sponsors played:

> Thanks to Government control of these schemes, the power of determining the degree of restriction is placed in the hands of authorities who can look beyond the immediate interests of the producers to their ultimate interests, and also to those of the world at large. They must give consumers a confidence they could never feel if the whole control of the schemes was vested in persons actually engaged in production.[40]

The only point of criticism was the failure of these schemes to prevent price fluctuations and it therefore concluded:

> Whereas the professed object of these schemes is to maintain prices at a fair and equitable level, they considered that a particular obligation rests on the bodies administering such schemes to provide themselves with instruments which will enable them to prevent prices from rising to unreasonable levels, or, if they do, to create conditions which should lead to a fall.[41]

Buffer stocks were now being defined as essential features of such agreements. The roller-coaster of 1937 demonstrated the obvious advantages to both consumers and producers. Had such a stock been in existence, it would have moderated the price increase and cushioned the price decline. It would also have reduced the rate at which the production tap had to be turned off.

In spite of the strength of these arguments, the buffer stock proposal proved to be yet another site of controversy between familiar adversaries. Some brokers, such as Strauss, were opposed, though others, such as Fraser & Co, were in favour.[42] The TPA asked for details of the proposals from the Colonial Office but was told that, since the organization represented other interests in addition to Malaya and Nigeria, formal comments were inappropriate.[43] In spite of the fact that it was now defined officially as irrelevant, the TPA dutifully passed a supportive resolution. More relevant was support from the Nigerian Chamber of Mines which declared its support not only for the scheme but for an even higher target price.[44] Some local Nigerian producers were opposed but the government found their 'arguments unconvincing'.[45]

Malayan interests were more deeply divided. The London Chamber took a strong, although contested, position against any proposal and attempted to find confirmation in a set of criticisms of the 1934–1935 experience.[46] Given the way in which 1937 had exposed the inequities in the standard tonnages, this was an 'attempt to build a further structure on top of one, the foundations of which have definitely proved to be unsound'.[47] Division in London was reflected in the Executive of the FMS Chamber which rejected the proposal by the barest of margins, 7/6/2.[48]

However, the circles advising Thomas in Singapore were of one mind. Whereas in 1934 Clementi had been very supportive of the principle and simply wanted a referendum to cover himself politically, Thomas was very much opposed. In a long dispatch to the Colonial Office he laid out the basis of his thinking:

> I propose to try and explain to you the attitude which is held by the tin producers of Malaya in regard to a buffer pool. Those concerns which are British controlled are for the most part solidly opposed: those which are foreign controlled or controlled by financiers not interested in tin as anything but a speculative commodity are for the most part in favour: the Chinese will support any scheme which they are told will bring them more cash. They think only in the immediate present and their interest in tin as an asset to Malaya may hardly be said to exist. In what I am about to write they may be left out of account.

Such a tripartite division of the Malayan industry was common in 1931. The fact that it was operating in 1938, after seven years of working with the Dutch and three years since Howeson's elimination, indicates the profound insularity of the official Malayan perspective. Thomas continued:

> What is the reason for this antagonism on the part of British-controlled concerns? They are directed by reasonable and competent men who have successfully organised a great

> industry, men of sane judgement and balanced views. Their main reason is the fear that a buffer pool will tend to place the industry more and more in the hands of foreigners … On the International Tin Committee there are more foreigners than British.

In support of this position Thomas quoted from such 'a responsible producer':

> I am not surprised that a buffer pool is being so strongly canvassed. The failure of all other countries to produce their quotas has proved Malaya's contention that she is under-assessed, and has placed her in such a strong position that, now, it is only by the institution of some such thing as a buffer pool, that the Dutch and certain smelting interests can realise their ambition to control the world's markets. The fact that Malaya … should supply a larger proportion of the world's consumptive demand in the future is directly against what, I feel sure, was one of the objects of certain interests in establishing and control, viz: the forcing of consumers to use tin other than Straits tin.

In addition, this miner raised the spectre of the pool tail wagging the quota dog:

> Then there is the difficulty of finding some disinterested person to operate the pool efficiently in the general interest … Even if someone could be found who had the ability and experience necessary to run the buffer pool scheme *without having to rely on the advice of the Dutch and other professional operators* the position would still be entirely unsatisfactory. Such a man could not possibly discharge his duties properly unless he were give a free hand to act quickly … and he would be in reality, the dictator of the industry to whose instructions the I.T.C. (and even the Governments) would have to conform.[49]

Thomas simply reflected all the prejudices and anxieties that tin control produced in Malaya. At no point were the arguments in its favour addressed, nor would they be, since moving the buffer stock proposal to the centre of the future of the ITC meant displacing what was most precious to Malaya, namely the revision of the standard tonnages. However, Thomas was prepared to authorize a referendum among the producers which meant an intensive propaganda war between both sides in which Bagnall and the *Straits Times* played a prominent part.[50]

Thomas' definition of the situation was accepted by some in the Colonial Office as Cowell noted:

> The question whether the proposal is, in theory, to the best advantage of Malaya is for the experts; but if they cannot convince by their arguments those men of practical experience whose interests are bound up with the Malayan industry, it would seem highly undesirable to impose the scheme on Malaya by official pressure.

But the overall perspective in London was very different and Clauson responded:

> I do not share Mr Cowell's respect for the 'practical' man. In my experience people generally claim to be practical when they are being particularly mulish and bone-headed and can think of no answer to arguments which are obviously convincing.
>
> As I see the position those people in the industry who have the greatest intellectual powers and the widest experience of business and life are in favour of the scheme, while

> the opposition comes from people who are 'practical' (but not always efficient) miners and nothing else, or like Sir John Bagnall have some axe to grind.
>
> We may hope that the referendum will come out sufficiently in favour of the scheme to relieve us from further anxiety. If it does not we shall have, as so often happens in this office, to choose between two very disagreeable alternatives, either to do what we know to be right, but something that will cause us infinite trouble with a number of stupid and disagreeable people, or to choose the easy course of agreeing with them and inflicting lasting injury on a great Colonial industry.[51]

As the Colonial Office conducted its internal debate on the contingency of this vote being lost, it recognized that the future of tin control was at stake, especially since the Dutch had made it clear that they were prepared to invoke article 22. On this occasion, there was less support for overruling Malaya and preserving the agreement, as Campbell noted:

> Malaya's attitude to tin control has been unsatisfactory from the beginning; it has been an attitude of suspicion and resentment; of obstruction and dilatoriness. Malaya has not in fact worked what is essentially a partnership scheme upon a partnership basis; it has never fully trusted its delegation on the I.T.C.; and it has steadily refused to face the obvious facts of the position, or to get down to bed-rock in its examination of it. If the Secretary of State were now to impose a buffer stock scheme on Malaya there would be intense feeling regarding the matter; and all the practical difficulties in the operation of tin control would be accentuated. A refusal by Malaya to accept the buffer stock plan may quite easily lead to the abandonment of tin control; I should regret that, very much; but it seems to me that the continuance of that control is becoming increasingly difficult, in view of Malaya's general attitude. Put quite briefly, you cannot run a partnership if you dislike and distrust your main partner, if you constantly suspect his motives, and if you decide to indulge in the luxury of abusing him, violently, in public.[52]

While the thrust of the Cornish position was to block the buffer stock to bring the issue of the standard tonnages to the fore, it threatened to backfire, since division on the former could prevent unity on the latter. The President of the Selangor Miners' Association, Choo Kia Peng, now took a strong stand against the tactics of the Cornish.[53] Uncoupling these two issues was also supported by Anglo-Oriental which was now prepared to argue for a revision of Malaya's standard. While Choo denied the charge of a formal exchange of support with Anglo-Oriental, the fact that he was publicly critical of the main opposition to the buffer stock proposal certainly helped to reduce their influence among the Chinese.

The Department of Mines distributed a balanced brochure giving both sets of arguments to all miners together with the ballots. When these were counted, miners representing 1,358,522 pikuls of assessment had voted in favour and 715,217 against. The Europeans had voted 779,665 in favour and 531,924 against, while among the Chinese support was even more pronounced with 578,857 in favour and 183,293 against.[54] Given the unqualified way in which the Cornish interests had campaigned on this issue, it represented a signal defeat for their vision of the world tin industry and the role of Malaya within it. However, they had won the European sector which was not under Anglo-Oriental control handsomely, by a margin of nearly 2 to 1. It was not, therefore, a defeat that would change that vision. Nor would it be sufficient to undermine their influence with Thomas and in spite of the results of the referendum, the Malayan government refused to endorse the buffer stock, at least not without conditions.[55] Citing the overall importance of commodity schemes in general and the role of a buffer stock as an integral part of a well-regulated scheme, the Colonial Secretary, Malcolm MacDonald, overruled Campbell who continued to advise against overruling Malaya.[56] Unfortunately, the resentment now created would be even greater than on the previous such occasion four years earlier.

In the final draft of the buffer stock agreement one concession was made to Malaya and that was to uncouple participation in the buffer stock from the right to invoke article 22. The article therefore continued to stand for its original purpose as a symbolic threat against outsiders.

Otherwise, the key provisions in the draft that had been circulated to governments were retained, though one was the subject of some discussion and that was the question of the price range to be aimed at. In light of the course of prices over the past eight months Todd, the new US consumer representative, argued that the range aimed at was far too high and in this he was supported by Groothoof who informed the meeting that his government wanted one of £180–£200.[57] Patiño retorted that his government had only agreed on the basis of the proposed range of £200–£230[58] and Campbell noted that since the ITC could change the range at a later date, there was little point in altering it now.

While the ITC had an agreed draft, the process of formal ratification took many months. Siam did not do so until November and took even longer to decide whether to participate. Eventually all countries did in fact participate and on a strictly proportionate basis to their modified standard tonnages. The buffer stock therefore represented one of the few occasions on which complete symmetry among all members was realized.

As soon as the buffer stock began to be assembled, R. S. Mills was appointed as manager. He brought extensive experience from an even more volatile market, Brazilian coffee, and commanded an annual salary of £2,000.[59] This appointment was designed to mollify those who felt that people with experience in tin had a conflict of interest. However, at the same time Arthur Ellinger was appointed as Assistant Manager. Lyttelton knew of him as the 'best tin-market man there is' through his Metallgesellschaft contacts and managed to secure his release from Buchenwald in January 1939.[60] This buffer stock was therefore under far better technical management than its predecessor which augured well for its success.[61]

Assembling the stock took much longer than anticipated, largely as a result of delays from Siam and it was not until the beginning of July that it was completed. This was to be an active stock and, before market operations could begin, all the contributions had to be converted into warrants for standard tin.[62] In this respect it was therefore quite different from its predecessor, which had kept a high proportion in the United States and that would only increase American anxiety about its function.

United States Protest

The formation of the buffer stock led to an official intervention by the American government. It felt that prices over £200 could not be justified on the basis of production costs and it was also concerned about the level of stocks. Forcing the price up to £200 from its current level of £163 implied low quotas and immobilizing at least half of the stocks in the visible supplies.[63]

This was the first point at which the American government had taken an official position on a specific decision of the ITC. In the background was an increasing interest in acquiring a war stock[64] and the Navy had already taken the first steps in this direction in February. Unless a direct agreement were reached with the ITC, the buffer stock policy ensured that further purchases by the government would drive the market against itself.[65]

The position adopted was not particularly convincing, especially not to the American Ambassador, Joseph Kennedy, who would be responsible for communicating it. As it was being developed, Kennedy felt that the ITC would have the 'ammunition to back up' the proposed range of £200–£230 and asked for the evidence that warranted the claim that anything over £200 was patently unreasonable. Pressing such a claim would not be easy in the light of Hughes' decision to stop attending ITC meetings and in light of the fact that there had been no official protest against the much higher prices that reigned from 1934 to 1937.[66]

Kennedy's response prompted the State Department to take action to secure more effective consumer representation but he also passed on Campbell's concern about the effect of the American position on Bolivia and that was much more difficult to deal with. As the Economic Advisor, Herbert Feis, tried to think it through, the incoherence of American policy became evident. The £200 threshold had long been asserted by the *Economist* as representing the upper limit of any operational definition of the principle established at the World Monetary Conference which called for prices 'reasonably remunerative to efficient producers'. Applying this formula to the tin industry failed to recognize the need that had just been demonstrated in 1937 for reserve capacity by producers in all countries that failed to meet this criterion, especially those in Bolivia. Feis at least recognized the seriousness of the problem:

> This situation places the American government in a rather embarrassing position so far as discussion of the price objectives of the international scheme is concerned. If we support the idea of the operation of the tin scheme so as to bring a price not above that reasonably remunerative to efficient producers … we place ourselves in the position of saying that the scheme should not be operated so as to produce a reasonable return to producers in Bolivia.[67]

That did not prevent the State Department from moving ahead with its official protest. In doing so, it had turned the editorial positions of the *Economist* and the *Straits Times*[68] into official government policy without taking into account either the implications for its relationship with Bolivia or for the prospects of a domestic smelter.

A formal response soon came from the Statistical Office of the ITRDC. On the stock issue, it pointed to the high level of current stocks, especially those in the hands of American consumers. On the price issue, it started with a demonstration of the level required to ensure exploitation of poorer resources and ended with a comparison between tin and other commodities.[69] It failed to make any dent in the American position, which now became even stronger. As far as stocks were concerned, the primary obligation of the ITC was to avoid a repetition of earlier shortages, which meant that the buffer stock should be held above the level of normal stocks. That simply ignored the problem of how the buffer stock could then generate the cash required to support the floor. Furthermore, it took issue with the very idea of a stable price range, especially one that would apply in both periods of boom and slump. As far as price was concerned, this could not be justified on the basis of existing cost data since they 'may include several liberal estimates'. Nor could costs, however refined, at the margin be accepted. Finally, it seized on the fact that the price comparison had first used 1926 as a base and then included aluminium and nickel. For the State Department this confirmed its worst fears, since 1926 was, of course, a boom year for tin and both the other metals were 'unchecked private monopolies which do not serve the public interest'.[70] In short, the Department had no grasp of the actual problem facing the ITC that had emerged in response to its generous treatment of consumers in 1937 and no alternative solution to propose, except to put the whole apparatus of tin control on an entirely new basis. The rejoinder to the ITRDC therefore concluded: 'Virtual world monopoly achieved by action of governments cannot be expected to do so unless social responsibility of government is fully recognised and the legitimate interests of consumers are given equal weight with those of producers.'[71]

Contesting this claim to stake out the moral high ground would require a much fuller discussion of the total scope of the social responsibility of governments, especially those in poorer regions and of the assumptions that lay behind the purported equality of consumers and producers. Given the anti-monopoly ideology now prevalent in New Deal America, that was out of the question. Instead, the Department adopted an increasingly aggressive stance towards the ITC and even wanted the British government to intervene, stating that:

> My Government has been forced to the conclusion that the administration of the agreement has been unduly dominated by producing interests and from the standpoint of consumers, has been more arbitrarily administered than in the case of any of the other international control agreements. The particular form given the arrangements for a buffer stock of tin and the method in which this stock is to be controlled have not been reassuring and have not served to modify the general conclusions set forth above.[72]

It was not easy for the committee to find an effective way to respond. Since the US had provided no basis for its allegations, it is hardly surprising that Campbell felt that the ITC was 'being put in the dock', and refused to agree to any formal investigation into the question of costs and price.[73] While the ITRDC developed a second response which rested on a demonstration of the value to consumers of evening out the highs and lows of the commodity cycle, the Americans were not prepared to engage in a full discussion of all the dimensions to the price issue.[74] Moral rectitude had substituted for serious analysis[75] and the allegations that it spawned would prove impossible to shake off.

Revising the standard tonnages

Cutting production first to 70 per cent then to 55 per cent was not at all easy, especially in Malaya where it left 40,000 coolies unemployed.[76] This, of course, simply reinforced Malaya's determination to secure a permanent adjustment in the standard tonnages. Special but unorthodox measures were taken by Thomas to absorb surplus production in the form of metal in bond. Given the low level of stocks of concentrates, this was more technical than substantial but the Dutch used it in the overall struggle that was emerging between the two neighbours about the buffer stock. At the same time Bolivia also became an overproducer which reduced the extent of the exposure of her overassessment and she was therefore now less vulnerable to Malayan pressure.

Perhaps as a result of the stubbornness of Malaya on the question of the buffer stock, her demands for a radical revision did not meet with much sympathy in London. The Colonial Office felt that Malaya already had a good deal and that improving on it was hardly worth the trouble, as Clauson commented:

> Practical fact is that the large participants get more absolutely than the smaller ones and that in order to obtain this absolute gain it is worth while letting the smaller participants have greater relative gain. The position of Malaya in this matter reminds me of the Bulgarian who was told by the Almighty that He would give him anything he cared to

ask for provided that it was clearly understood that he would give twice as much to his neighbour and he replied 'Lord, put out one of my eyes'.[77]

The old tension between the regional and the international perspectives, which in the case of the buffer stock had focused on the special market for Straits tin, was now concentrated on the question of the standard tonnages. The aggressiveness with which Malaya pursued the issue on this occasion is in marked contrast with the positions she had taken but 18 months earlier.

When the ITC addressed the question of the standard tonnages in June it had three contradictory memoranda before it.[78] Malaya presented the case for an entirely new principle based on proved capacities, stressing the dislocation of the organization of her mines and labour force as a result of large changes on an underassessed standard; the NEI took a middle ground and claimed that if Malaya secured a revision, her own standard should reflect the ratio between the two countries that existed in the first agreement. She also recognized that the others had a case since they had signed the current agreement on the basis of the existing figures. Bolivia had the most difficult task but Patiño and Martinez Vargas pointed to the new conditions which would allow her to rebuild her 1929 capacity. With Campbell compelled to support the Malayan position, the impasse was broken by Gutt. He presented an immediate solution which would involve no concessions and no inquiries. He simply took the quotas in 1937 as the basis for new standards for both Malaya and the NEI and since they had been 107.5 per cent of the old standard, the new one was simply 7.5 per cent higher. It was an inspired initiative, since it probably went to the limit of what the ITC could approve without involving governments. The Malayan delegation was divided but was prepared to accept the proposal as long as there was unanimous support and following various caucus breaks the meeting came to an agreement.

Unfortunately, it was not one that Glenister could endorse since it went less than a third of the way to meet Malaya's demand and he therefore resigned lending further support to the notion that Malaya had been unfairly treated.[79] It was a charge that was quite inconsistent with the existing situation since it could only have been pursued with the threat of invoking article 22 and that in turn would have been considered very unfair by her partners who were on the verge of committing themselves to the buffer stock as the only way to restore stability. The results of the new allocation are presented in Table 12.3.

Since no explicit transfers were involved, the total standard tonnages rose by over 7,000 tons. Unlike the adjustments made for the first half of the year, these would remain for the

Table 12.3 Standard tonnages, July–December 1938

	1937	*1938*		*1937*	*1938*		*1938 2nd cf*	
		1st half	*2nd half*		*1st half*	*2nd half*	*1937*	*1st half*
	tons	*tons*	*tons*	*%*	*%*	*%*	*%*	*%*
NEI	36,330	41,324	39,055	18.2	20.0	18.9	3.7	-5.7
Bolivia	46,490	37,545	46,027	23.3	18.2	22.2	-4.5	22.3
Nigeria	10,890	12,387	10,890	5.5	6.0	5.3	-3.6	-12.3
Malaya	71,940	81,831	77,335	36.0	39.6	37.4	3.7	-5.7
Siam	18,731	18,731	18,628	9.4	9.1	9.0	-4.1	-0.8
BC	12,200	12,008	12,035	6.1	5.8	5.8	-4.9	0.0
FIC	3,000	2,670	3,000	1.5	1.3	1.4	-3.6	12.1
Total	199,581	206,496	206,970					

duration of the agreement. With the change in the total came a change in the ratios; the two winners each gained nearly 4 per cent and the losers each gave up over 4 per cent. Among the latter, of course, was Siam and this change was equivalent to a loss of 825 tons on her old standard. That was below the 18,000 tons for which she had been prepared to break the agreement in 1936! The fixation on absolute numbers disguised the fact that the real game was about ratios.

Recomposition of the Malayan delegation

Glenister's resignation simply revealed the extent of the tension that had long been simmering in the administration of tin restriction. Since the Colonial Office now had considerable experience with other restriction schemes, it attempted to bring Malaya into line with an extensive memorandum in which all the issues were thoroughly aired. It began by pointing out the

> extraordinary contrast between the tin scheme and the other regulation schemes with which we are concurrent in the Colonial Office. In the case of these other schemes what impresses me most is the extraordinarily harmonious spirit which prevails not only between the individual members of our Colonial delegations and also between the various delegations themselves. There is little of this in tin. There has been continuous discord both within the Malayan delegation and between that delegation and the other delegations – nine tenths of the trouble has come from the unofficial M rep and advisor. Matters have come to head with Glenister's resignation but this is only the end of a very long story in which the spirit of cooperation on his part has been conspicuous by its absence.[80]

For Campbell, this contrast was more pointed:

> Nothing is more striking than Malaya's sensible attitude towards rubber as contrasted with the very unsatisfactory attitude towards tin. That is due to the constitution of the Malayan rubber delgn on the IRRC; to the much higher quality of 'rubber' men generally as compared with 'tin' men; and to the support of the highly organised RGA which really brings all important decisions to London; and to the weakness of the Govr: in dealing with all tin matters. Tin started badly, under Sir Cecil Clementi; the industry was not adequately consulted; that has rankled ever since; and the essential basic condition – confidence in the scheme, and its treatment on a real partnership basis – has never been properly established. One sees, also, how much the tin scheme is weakened by the presence of Bolivia … with all its Governmental and other defects, its weak administrative system and its S American standards of morality.[81]

Independent confirmation of the Colonial Office view of the calibre of the Malayan tin men came from the head of an education committee sent to Malaya who reported that the:

> representatives of the tin industry were the only people who quite deliberately tried to deceive this committee and use them for their own ends. They claimed they could not create openings for educated natives unless they got more land and exports and then admitted even if they did, the employment created would be negligible. They were a thoroughly dislikable group of men.[82]

Lyttleton had received word that Thomas intended a more thoroughgoing reconstitution of the delegation, taking advantage of the termination of Lowinger's appointment as head of the Malayan Information Agency to replace him with Ward, the new Chairman of Pahang and a 'dyed in the wool opponent of tin restriction'.[83] It was therefore imperative to lay out the principles that should determine the composition of the delegation. Three elements should be represented: (1) Colonial Secretary, which meant someone in close touch with the general policy of the government and able to harmonize the individual policy of delegation with that policy; (2) colonial government, to watch over the interests of the country rather than the producers; (3) Producers. It was also clear that where the positions taken by these three elements conflicted they were to be resolved in the same order of priority and to ensure that outcome the vote held by the delegation would only be exercised by the representative from the Colonial Office.

As these principles were applied to the particular situation, much support was expressed for continuing with Lowinger as Malaya's representative: 'there may have been one or two occasions in the past when he has not been altogether wise, he has behaved with extraordinary wisdom and discretion in recent months'.[84] However, the Colonial Office would not press the point, though Thomas did eventually confirm his appointment. But as far as Ward was concerned, he would be 'as much out of place as a teetotaller on the board of a brewery company'. Glenister had been useless since:

> he has been unable to conceal his belief that tin control is wrong in principle and its operation should be impeded whenever possible. ... His successor should be prepared within the committee not to play to a particular gallery but to regard himself as one of the team, critical in a constructive sense where criticism is required but ready at all times to be helpful in carrying out what must be considered the deliberate policy of the British Government. In a word, he must be there to oil the works rather than throw sand into them and do his utmost to dissipate the existing atmosphere of distrust.[85]

Campbell's patience had long been severely tried by the obstructionism of Glenister and Wilcoxson and for him on the satisfactory resolution of this issue rested the fate of tin control. The Dutch had already lost their patience and had planned to make diplomatic representations, pointing out that, although Glenister and Wilcoxson were appointed to work a scheme endorsed by the Malayan government, they intended to do all they could to impede its practical working. The appointment of Ward would simply escalate this tension and Campbell added:

> The problem must now be squarely faced ... we cannot continue to work the tin control scheme, in any satisfactory way, unless we can secure reasonable cooperation on the part of Malaya. ... Common sense dictates that, committed as they are, most fully, to the control scheme, the only possible plan is to work that scheme on a partnership basis with the other signatory countries. Malaya must have a degn: which takes that point of view, or the whole thing will inevitably break down. I have struggled along, for seven years, on the present most unsatisfactory basis; if the position is now to deteriorate still more markedly, the continuance of tin control will in my view become impossible.[86]

That was sufficient to force Thomas to back away from Ward but since he had been told of his impending appointment, its withdrawal left him 'extremely sore'.[87] No other ideas arose, so Wilcoxson moved from his role as technical advisor to a full member of the delegation and his place was taken by Rich, from the Tronoh group. Campbell was guardedly optimistic, 'it looks as if Rich would be difficult; but we know how fast education can proceed in the committee, if a man is not wholly unreasonable'.[88]

As the air cleared about the composition of the delegation, the opportunity arose for the kind of clarification that should have taken place when MacDonald provoked the crisis by overruling Malaya in May. Thomas indicated that the original resentment about the lack of proper consultation in the formation of the ITC would have fallen away long ago but for 'the fear that foreigners are out to get control of the industry'.[89] While that issue had been flagged earlier in general terms, he could now spell out the basis of that fear.

The root source of the problem lay in the fact that restriction had resulted in excess capacity among the Straits smelters. This was compounded by the decision of Billiton to break away from STC in Singapore and tie itself more firmly to CTS in Arnhem. At the same time, the growth of Anglo-Oriental in Malaya resulted in more business going to ESC which then started to reduce its returning charges. The competitive pressure placed on STC was interpreted as a desire to either force it to close down or come to some arrangement with CTS. The increase in the power of CTS therefore threatened the relationship between Cornish-Malayan interests and STC and that threat was presented as one that affected the whole of the Malayan industry. CTS would then be in a position to dictate prices to the producer, close the Singapore market and sell directly into the United States. In Thomas' eyes, CTS was under de facto foreign control.[90] Now it was clear why the issue of the standard tonnages was so pressing. What was felt to be at stake was not just the general principle of equity but the fate of Straits tin and the Singapore market. STC and its allies were the only ones firmly committed to both. The more unequal the ratio between Bolivia and Malaya, the greater the handicap under which STC would operate. Regardless of how well such fears were founded, the fact that they existed would prevent Malaya from becoming an equal partner in any international tin restriction scheme, at least as long as its High Commissioner felt bound to accept this purely regional view.[91]

June–December 1938

With the question of standard tonnages resolved, the ITC had to address two problems at its June meeting. The quota issue did not generate much discussion and it was settled at 45 per cent of which 10 per cent would be for the buffer stock, should this be finally ratified.[92] At this lower level of production, the concessions made for the first half of the year no longer obtained. Bolivia, in particular, was interested in recovering some of the 1937 carryover since this would help wipe off her excess. A formula could not be developed internally to address the issue and it was agreed to refer it to arbitration, though later Campbell came up with a proposal the Bolivians found acceptable.

When the ITC had to consider settling the quota for the fourth quarter it again kept the same level of 45 per cent of which 10 per cent was for the buffer stock.[93] Getting production down proved difficult in both Bolivia and Malaya. By September, Bolivia had exported 40 per cent in excess of its quota for the quarter, Malaya 30 per cent. While Malaya was able to get her problem under control by the end of the year, Bolivia was not. Just as she had been a chronic underproducer during periods of high quotas so she was an overproducer during periods of low ones.

Coincidental with the formation of the buffer stock a recovery occurred on Wall Street. Just as the crash of 1937 had depressed all metal prices, so they all swung back up and whether the buffer stock had any independent positive effect is unclear.[94] Throughout this period there was little change in the fundamental features of the tin market. The basic policy of the ITC was to try to keep production below consumption to reduce invisible stocks, together with the proportion of visible stocks outside its own control. It was a policy that simply flowed from the way in which it had chosen to respond to the boom of early 1937 and its failure to turn off the production tap at an earlier point in time. It could not cut the quota below 35 per cent, especially since there was no assurance that it would be respected by Bolivia and the ITC simply had to wait for consumption to rise.

At the November ITC meeting, Todd provided an optimistic forecast of the consumption trend and this prompted Lyttleton to propose that the buffer stock be increased to 15,000 tons. That would mean a simple continuation of the current quota into the first quarter of 1939. The Dutch were not convinced and preferred to wait for further information about the state of consumption. In any case, they felt that Bolivian overproduction should be liquidated before the buffer stock could be put on a secure foundation. In spite of their objections, supported by Campbell, the meeting endorsed Lyttleton's position.[95] The ITC would therefore end 1938 on an uncertain note. It was now committed to taking control over the market through the buffer stock but whether and when that control would become effective was entirely dependent on a solution to the Bolivia problem.

The Bolivia problem

Campbell's frank comment about the limitations of Bolivia as a full partner in the restriction proved to be quite prescient. On this occasion the problem lay not only in administrative weakness in the face of the demands of the small miners[96] but also in an incoherent government policy in the face of a profound conflict among the large.[97]

Defeat in the Chaco had major consequences for the political economy of Bolivia. As political power shifted to a group of young army officers imbued with a strong sense of nationalism, the fiscal pressure on the mining industry increased considerably. Unwilling to give away its capital resources, the industry responded by suspending exports or moving production to lower grade sections. As production declined in 1936, the government was forced to negotiate a new set of incentives. They included not only reducing the proportion of the foreign exchange generated by tin exports which was sold at low rates to the Banco Central, but also a promise of an adequate share in future quotas in return for fresh investment. Both Hochschild and Aramayo then made substantial financial commitments, the former in developing a new mine, Colquiri, and the latter in refocusing his operations away from silver and back to tin.

As the international quota for the first quarter of 1938 was reduced to 70 per cent, Bolivia again became an over-exporter. No internal distribution was made for this quarter, though it was expected that retroactive adjustments would be made once a comprehensive basis for quota allocation was settled. By the end of the quarter, Bolivia was over by some 1,220 tonnes or 70 per cent of the monthly international quota then prevailing. Thanks to wildly conflicting conceptions as to the proper balance between Patiño and Hochschild held by various government departments, it was impossible to establish a satisfactory comprehensive basis for quota distribution. The issue was shelved for six months and in the interim quotas were to be based on 1937 output. The second quarter saw a similar level of overproduction, though a temporary solution was found thanks to the recognition on the part of the ITC that the surrenders of carry-forward of unused tonnages at the end of 1937 should be returned. Bolivia therefore ended the first half of the year with a more or less clean slate.

However, two serious problems were revealed. Neither of the two main beneficiaries of the 1936 incentive contracts, Hochschild and Aramayo, were prepared to produce at a lower level than they considered legitimate. The other was the familiar one of the small miners; in the first quarter they were producing at about the same rate as they had in 1929!

The breathing space provided by the interim quota arrangement allowed for the creation of a special commission to investigate the state of the industry and it argued for setting aside all previous contracts. Instead, domestic quotas were to be based on long term productive capacity. The commission largely endorsed Hochschild's position that Patiño was on the decline and that the future of mining in Bolivia rested on the development of his own mines. The result was a Cabinet decision that saw a major increase in the proportion to be allocated to Hochschild, from 18.9 per cent to 26 per cent, at the expense of both Patiño and the medium miners. The magnitude of this further concession to Hochschild is evident from the last column of Table 12.4.

If the ratios for the fourth quarter were to stand, then on the current 35 per cent international quota, Patiño would only be entitled to produce at 27 per cent of his 1929 output while Hochschild would produce at over 90 per cent of his. Patiño, however, was given a consolation prize. The group would receive the bulk of any increase following the raising of the international quota. A higher quota would maximize short-term revenue and this was a way of bringing pressure to bear on Patiño to use his role on the ITC to realize it.

Not only was Hochschild granted an extraordinarily generous quota but he was actually producing in excess of it. While Aramayo was also in excess, his case was considered to have been unfairly dealt with and he received a supplementary allocation to cover it. In addition, while the small miners actually reduced production, it was to a level far above their formal authorization. By the end of 1938, Bolivia was some 1,800 tons in excess, well over a month's allotted quota. There was clearly no interest on the part of the government in adhering to the international obligations it had inherited and only Patiño and the medium miners adhered to whatever internal obligations that were laid down. By the beginning of 1939 the consequences of this policy erupted into a major domestic and international crisis.

Patiño was not prepared to let Hochschild retain his advantage; nor was Hochschild prepared to let it go without a fight. It was a conflict that badly divided the government and each side openly used the press as the vehicle for their propaganda against the other. As a result of the political pressure, the government decided on yet another basis for quota allocations for the first half of 1939. This cut Hochschild back, to the benefit of Patiño, Aramayo and the small miners.

The new regime had supposed that it should be exercising the power of the state to facilitate the best long term course of development within the constraints imposed by restriction.

Table 12.4 Bolivia production, July–December 1938

	Quota	*Actual*	*Excess*	*Cum Excess*		*Quota cf 1929*
	tonnes	*tonnes*	*tonnes*	*tonnes*	*%*	*%*
Patiño	3,941	3,872	-69	57	9	27
Hochschild	1,837	2,091	254	671	190	93
Aramayo	417	650	233	391	574	17
Medium	1,158	1,072	–86			38
Small	831	1,462	631			36
Medium and small	1,989	2,534	545	685	217	
Total	8,184	9,147	963	1,804	132	

Source: Banco Minero, *Memoria Anual, 1938.*

Notes: Ordinary quota only, excluding buffer stock; cumulative excess as percentage of monthly quota at end of period

Unfortunately, without a comprehensive and expensive body of technical information at its disposal, the government was unable to justify, not least to its own departments, any particular pattern of distribution as superior to any other. That inevitably placed it in a position where the actual outcome simply reflected the balance of internal political forces. It was to distance itself from at least one source of those politics that the government created the Asociación Nacional de Mineros Medianos with the explicit responsibility of distributing quotas to the medium miners.[98]

Hochschild's domestic campaign had serious consequences for Bolivia's position on the ITC. Already the subject of unwelcome attention as a result of her overproduction, Hochschild's demonstration of the limitations of Patiño's productive capacity undermined all the work that Antenor Patiño had done in trying to reassure his colleagues that the underproduction of 1935–1937 was purely transitory.[99] While Hochschild could claim that he was more than capable of replacing PME, it would not serve to reduce the resentment caused in Malaya by yet another demonstration of the 'unfairness' of the overall standard tonnages.

At first, the Bolivian government thought it could regularize its position with the ITC by asking for a special dispensation which would tolerate its overexports. As Antenor Patiño canvassed the idea informally, he quickly learnt that there was no sympathy and advised his government to 'strictly enforce the quota with the object of avoiding the severe criticisms caused by Bolivia's overexports in the first semester'.[100] That was in July and similar injunctions were repeated in November, December, February and March. All that could be done was to 'urgently insist that measures be taken to reduce excesses. Other countries will not support heavy restriction to the benefit of those that do not observe the agreement'.[101] The fact that Bolivia's persistent disregard of her international obligations jeopardized the future of the ITC was not sufficient to constrain the internal politics. As is clear from Table 12.5, the excess actually increased during the first quarter of 1939.

A solution came when the ITC raised the international quota for the second quarter and then a sliding scale of distribution of the increases between fresh production and wiping off

Table 12.5 Bolivia production, January–June 1939

	Quota	*Actual*	*Excess*	*Total excess*	
	tonnes	*tonnes*	*tonnes*	*tonnes*	*%*
January–March					
Patiño	1,964	1,933	–31	26	4
Hochschild	859	1,049	190	861	300
Aramayo	286	365	79	469	491
Medium	532	564	32		
Small	450	711	261		
Medium and small	982	1,275	293	978	299
Total	4,092	4,622	530	2,334	171
April–June					
Patiño	2,245	1,906	–339	–313	–42
Hochschild	982	828	–154	707	216
Aramayo	327	280	–47	422	387
Medium	608	544	–64		
Small	514	657	143		
Medium and small	1,122	1,201	79	1,057	282
Total	4,677	4,215	–462	1,872	120

Source: Banco Minero, *Memoria Anual*, 1939.

arrears was negotiated.[102] At that point, Bolivia demonstrated her ability to comply but only through a disproportionate sacrifice made by Patiño. Even with more stringent controls, the small miners still produced in excess of their quota.[103]

Behind this apparent overall indifference to the ITC lay a deeper concern. The tension between state and industry was not simply grounded in control of resource rents and the desire on the part of governments to buy off a wide range of pressures for their allocation. Patiño's power domestically appeared to be reinforced by his power internationally, especially through his control over the largest smelter treating Bolivian concentrates. A more nationalist regime headed by Colonel Germán Busch, a pro-German President who had given himself dictatorial powers,[104] was anxious to explore a new role for Bolivia which would bypass Patiño and open up an entirely new developmental path.

Central to Busch's conception was a commercial agreement with Germany and his father was sent to Berlin to complete negotiations for one. Germany would provide capital equipment for the overall development of the mining, oil, agricultural and transport sectors, in exchange for which Bolivia would provide the whole of its mineral production. In order to deliver its side of the agreement, the Banco Minero would be granted an export monopoly and Bolivia would break away from the ITC.[105] Fortunately for the ITC and for Bolivia, the Germans had neither the smelting capacity, nor, more importantly, the market for more than a small share of her production. In addition, they worried about the impact of withdrawal from the ITC for Bolivia's overall solvency with respect to Germany.[106]

As the German commercial agreement fell through, Busch then explored two other options. One was to arrange a loan of £500,000 from the UK secured against 5,000 tons of tin to be shipped in excess of Bolivia's quota and held by the British government. While the Treasury was strongly opposed to the idea of a loan, it was quite prepared to accept the tin as security for an export guarantee for sales of British goods to Bolivia. Had it come to pass, such a move would have dealt a direct blow to the ITC, quite apart from the depressing effect that such stocks would have on price.[107] But there was nothing the Colonial Office could have done to prevent it nor to hold the ITC together in the face of yet another demonstration of Bolivia's disregard for her obligations.

The second option was to encourage the development of other outlets for tin concentrates. A military mission was dispatched to Japan to see whether she was prepared to expand her smelting facilities to treat Bolivian tin concentrates. More promising was a mission headed by the Minister of Mines, Dionisio Foianini, to the United States to explore her interest in developing a domestic smelter. Although that would not have represented an immediate threat to Bolivia's continued participation in the ITC, the terms could well have drawn Bolivia away from a co-operative arrangement with fellow producers to control the world market into one with consumers to destroy it.

There were several obstacles in the way of developing these ideas which were designed to bring about a fundamental change in the position of Patiño. However, they had emerged in response to the perception that he exercised far more power than was desireable for Bolivia. While that power was also seen as the basis on which the ITC functioned to the benefit of Bolivia, no regime which was concerned about its nationalist credentials could long tolerate it.

Busch's sense that there was a fundamental contradiction between the interests of Patiño and those of Bolivia led him to suppose that he used his control over the Bolivian delegation to the ITC to pursue his own interests, especially by opposing quota increases. Busch therefore attempted to curb Patiño's power by appointing a close friend, Gabriel Gosálvez,[108] to the delegation. On resigning from the cabinet in March 1939, Gosálvez was appointed Minister to the Vatican and was therefore available to attend ITC meetings on Busch's behalf. Since Gosálvez had no background that would enable him to contribute to the deliberations of the ITC, Antenor Patiño had no illusions about the implications of this appointment. He therefore ensured that Gosálvez was not notified in time to attend the one meeting held before Busch died.[109]

Busch's confrontation with the mining companies came to a head with a decree in June 1939, the most famous provision of which was that the companies would be required to turn over 100 per cent of their foreign exchange earnings to the Banco Central to which they would then have to turn for the remittances required to meet all their external commitments. Such a bold assertion of state power, without much consideration for the practical details of implementation, was not untypical. It was immediately followed by a decision to execute Hochschild for having dared to question another provision of the decree which would grant the Banco Minero a monopoly of ore-dealing and end that part of his business. Hochschild was only reprieved thanks to diplomatic intervention but these frustrations in dealing with the mining companies may have contributed to a tragic outcome. In August, Busch committed suicide.

While the new government that came to power retained many of the features of Busch's overall policy, it made one major modification in the principles of quota allocation. Quotas were suspended soon after the outbreak of war but should they be reinstated, they would be strictly based on past production performance.[110] In addition, the granting of a monopoly of ore-dealing to the Banco Minero offered the prospect of developing effective control over one source that had always undermined the viability of restriction, the small miners.[111] Had restriction continued, it is quite possible that her colleagues would have found that Bolivia had become a much more co-operative partner.

January–August 1939

The beginning of 1939 found the ITC still far from the point of stability that it was trying to reach with the formation of the buffer stock. Although price was well into the range envisaged, it was not high enough to allow the sales that would generate the cash required to support the floor. At the same time, members of the ITC were overproducing on their regular 35 per cent quota, visible stocks were rising and consumption levels were sagging. Even existing consumption levels were uncertain, given the possibility of another international crisis.[112]

Making the situation more complicated was the invisible stock position. The initial announcement of the formation of buffer stock attracted several private pools which helped drive up the price.[113] Quick profit-taking meant that the metal acquired soon passed into the hands of consumers so by the end of 1938 the market was again 'healthy',[114] at least in the sense of the absence of speculative accounts. As the buffer stock was nearing completion, consumers could be confident that the ceiling would be protected but there was no similar confidence in the floor. That simply added to the incentive to run down invisible stocks.

The ITC was facing perhaps the most difficult situation of its entire existence. If consumption did not rise, it could only reach the £230 target, by imposing yet higher levels of restriction.[115] That target, or something close to it, was the essential step towards a fully functioning buffer stock and the hitherto elusive point of stability. If it did not have 'sufficient guts' to do so in the face of inevitable American criticism, it would again lose complete control of the market.[116]

The preparatory moves for the meeting that would have to address this issue demonstrated the gulf in American approaches to the problem of tin control. One was represented by the State Department which tried to get Kennedy to force a renegotiation of the buffer stock agreement. That would mean dropping the range to £170–£200 and making the buffer additional to the level of normal stocks.[117] At the opposite extreme was Trench who argued against any slackening:

> American consumers, despite improvement in their business, are buying less tin now … for fear that tin control will not take drastic enough steps to balance supply and demand. American buyers do not complain about price … but they do not want to have large stocks at a time when there is a possibility of the situation getting out of hand. The fact which disturbs sentiment is that the visible has passed the 35,000 mark. Consumption eventually governs, but sentiment over quite long periods can make or break markets … The stablization area of £200–£230 ought to prove fair now that business is recovering, but the backing of the industry's customers awaits a clear indication that Tin Control will follow the course it set out upon last year, even if it causes a temporary embarrassment to producers in some regions.[118]

That translated into keeping the current quota at 35 per cent and it was reinforced by other market analysts with far less predisposition towards the ITC.[119]

Something of a middle ground began to emerge as Kennedy followed up on his instructions and initiated what would be the only serious direct discussion between the Americans and the British.[120] Setting aside the moral allegations allowed for a friendly review of the practical issues. Campbell wanted a reasoned case on the American stock position which he was quite prepared to put to the ITC and it was recognized that if Todd were given better data, the ITC could take a more liberal position which would bring the price closer to the floor. Campbell also agreed with the desirability of keeping half the buffer stock in the United States and pressed the case for building up an American stockpile.

On the question of consumer participation, Campbell pointed out the way in which consumers actually 'participated freely in the meetings of the committee on an equal footing with others', and saw serious practical difficulties in expanding the number of representatives. A third position would now have to go the USSR and any larger number would make the meetings quite unmanageable.[121]

The only sparring occurred on the question of production costs and their relation to price. Kennedy dutifully carried the State Department's moral position that, as a governmental agreement, the ITC had 'a special obligation to protect the consumer and prevent abuses, however difficult it might be to accomplish those objectives'. Any attempt to establish a fair price on the basis of comparative indices with other commodities was therefore flawed. Calder then raised the question of Bolivia and that became the vehicle whereby Kennedy could move away from a position that could only generate an exchange of moral brickbats. As a good diplomat, he felt he had found the basis of deflecting the price issue into the arena of practical politics and reported that unless the Department were prepared to strike an agreement with the Bolivian government, it would be unable to modify the policy of the ITC.

Kennedy accomplished more than he realized. As the meeting broke up, the Foreign Office representative, Leith-Ross, expressed the hope that the American government would give the tin scheme a fair trial but also indicated that he would continue to monitor the situation. Kennedy had laid the groundwork for further discussions which could have been even more productive, especially with the participation of another agency of the British government.

This, too, would come to naught but at least it is an indication of the fluid character of the context within which tin control actually operated.

Failure to establish a permanent basis for discussion of the practical issues involved in tin control had serious long term consequences. Two years later, Leith-Ross received a report of the

> deep-seated suspicion with regard to commodity control schemes ... found in almost every quarter in Washington. ... Complaints are emotional rather than logical and a great deal has been done in the past to meet complaints ... this emotional feeling is a serious fact in Anglo-American relations and some sort of concessions to the American point of view are necessary to soothe their feeling of grievance.[122]

As the experience with Bagnall and the Cornish testified, once emotion had become the basis of grievance, no concession, short of capitulation, would be acceptable.

When the ITC actually met in March, it also found the delegations divided. Malaya and the Belgian Congo argued for keeping the current quota at 45 per cent with the previous 10 per cent dedicated to the buffer stock now available for immediate sale. Bolivia argued for simply keeping the 35 per cent component but Patiño's ability to press this was severely compromised by the fact that her overproduction was partly responsible for the poor stock position. The Dutch argued for 40 per cent and were supported by Lyttleton which was sufficient to settle the issue.[123] One advantage of the new quota was that it would go some way to alleviating the accumulated arrears and that would reduce the extent of any real increase.

The ITC was most fortunate that there were now positive signs of a recovery in real demand and it proved to be somewhat larger than was anticipated, by some 23 per cent. Stocks could easily absorb this pressure and the visible supplies remained considerably above the target rate of 17 per cent of consumption. However, by this point the buffer stock held 55 per cent of the visible supplies and the residue was well under 10 per cent of consumption. With stocks tight, backwardation inevitably returned but on this occasion the problem was kept within bounds as a result of the willingness of the Buffer Stock Executive to lend spot metal to the LME brokers.[124] There was no repetition of the 1935 fiasco when backwardation rates became so great that they prevented effective hedging transactions.

Consumption levels in the United States continued to hold steady throughout the summer of 1939 and in Europe armament production began to have a multiplier effect on other industries. The market anticipated an increase in the quota for the third quarter to 50 per cent. In fact, no delegation was prepared to consider such a large increase. One of the factors dominating this discussion was the role of the buffer stock. The only cash at its disposal had come from exchanges with the brokers, through the fees charged for lending spot metal and through the realization of premia on higher grade brands. The third quarter quota was therefore set low in order to allow the realization of as much as a third of buffer stock. Eventually, of course, that would allow a higher quota since the risk of misjudging the market could be absorbed by the stock.

The lower the quota, the faster the rate at which the desired balance within the buffer stock would be reached and the delegations were divided on whether to keep it at the current level of 40 per cent or allow an increase to 45 per cent. The Dutch and the Belgians took the low position and Bolivia the high one and both the Malayan and Nigerian delegations were split. Campbell then sought consensus around 45 per cent for political reasons 'to avoid any action considered unfair to consumers', and managed to secure it, albeit over the objections of Lyttelton.[125] The market responded precisely as anticipated. The announcement of a quota below the anticipated 50 per cent boosted the price to the £230 threshold at which point the buffer stock became a steady seller. At the very moment when the programme of the ITC was beginning to generate its desired results, Europe started the drift into war and it would soon find the programme had turned into a serious liability.

Decision-making

The operation of the third agreement during peacetime generated far more controversy about quota levels than had the second. Now division was as common as unanimity but it was found as often within the Nigerian and Malayan delegations as between them and the others. However, as under the second agreement, there was no simple consistency. Each delegate responded to his own reading of the conjuncture of the market, domestic pressure and international politics. The results are summarized in Table 12.6.

The third agreement – an assessment

In reviewing the operation of tin control under the Third Agreement prior to its transformation by the war, several features stand out, of which the most obvious is the impossibility of regulating a commodity cycle without a large buffer stock.

Over the 32 months from January 1937 to August 1938, quarterly consumption had varied between a high of 44,500 and a low of 35,600 tons or an average variation of 7 per cent. Production had not only exceeded consumption but had fluctuated to a far greater extent, between a high of 57,800 and a low of 30,000 tons or an average quarterly variation of 19 per cent. Thanks to the termination of the buffer stock in 1935, the price boom of early 1937 could only be met by increasing the quota.

In serving the needs of consumers by preventing a runaway price, production was driven over consumption and then had to be cut below it. However, since this excess production went into invisible consumer stocks, it was difficult for the ITC to monitor the real state of the market and hence turn off the production tap soon enough, with the result that production remained too high for the whole period.

Table 12.6 ITC quota discussions, January 1937–June 1939

	Final Decision
Unanimous	6
Division	6
High Position	
Malaya	5
Nigeria	4
Bolivia	1
NEI	0
Belgian Congo	2
Low Position	
Malaya	1
Nigeria	1
Bolivia	5
NEI	1
Belgian Congo	2

Table 12.7 Excess production, July 1938–June 1939

	July 1938–March 1939		*July 1938–June 1939*	
	tons	*%*	*tons*	*%*
NEI	185	1.8	1,346	9.5
Bolivia	2,584	23.5	2,195	14.2
Nigeria	7	0.2	32	0.8
Malaya	602	3.0	539	2.0
Siam	637	7.1	184	1.6
BC	1,323	36.0	555	11.2
Total	5,338	9.4	4,851	6.2

Note: Percentages are of entitlement.

Given the magnitude of the difficulties in forcing producers to reverse the momentum that had been built up during 1937, overproduction was to be expected during the first half of 1938. But it re-emerged during the nine months from July when the regular quota was only 35 per cent and was only slightly eased with the 40 per cent quota of the second quarter of 1939. Table 12.7 demonstrates the size and distribution of the problem.

With the exception of Nigeria, all members failed to comply with the strict provisions of the agreement. Bolivia, of course, was the biggest culprit and the internal political reasons for her violations have already been considered. However, expressed as a proportion of entitlement, the most serious offender was the Belgian Congo and Siam too was a problem. However, by the end of the full 12 month period, Siam had effectively liquidated her excess and both Bolivia and the Belgian Congo had made considerable progress towards dealing with theirs.

Moral pressure was the only resource available to the ITC to ensure compliance, at least as long as the threat of non-renewal was still some time away. But the problem had to build up to serious dimensions before it had any effect. The case of the Belgian Congo reveals its limits.

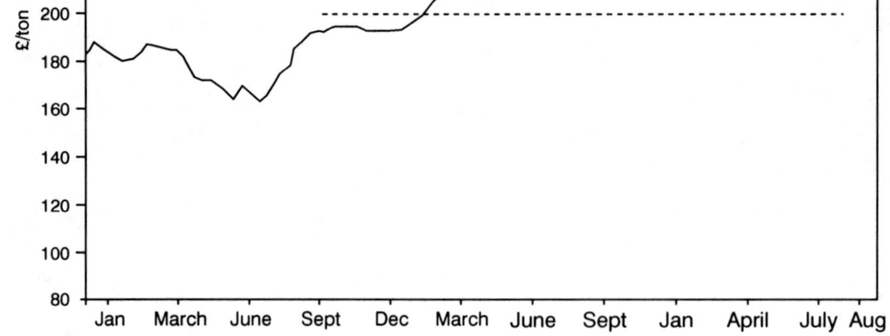

Figure 12.2 Tin prices, January 1938–August 1939

Given the plans the companies had been making on an expanding entitlement, the reduction imposed during 1938 had a more serious effect in the Belgian Congo than elsewhere. Of the four large companies only two actually reduced production and then only by 5 per cent from their 1937 performance; the nine small companies increased theirs by 36 per cent. The Belgian Congo had no internal machinery for quota allocation, so the moral pressure could not be directed towards a specific administrative failure. As the problem grew to alarming proportions, the four large companies then worked out a system of allocation among themselves. That would enable the small companies to continue to grow and the large would then reduce themselves to the extent necessary to bring the Congo into compliance. By June, the problem was well on the way to being solved. This self-policing also indicates that not only did the ITC get something from the Congo in the form of restraining its rhythm of growth but real restriction. Production of the four large companies in 1939 was cut by 2,200 tons from their level in 1937.

By mid-1939 the buffer stock was functioning well and the problem of compliance was in the process of being resolved. Had war not shattered the assumptions on which the ITC operated, it is quite possible that this would have been defined as the point at which the top of the learning curve had finally been reached. Figure 12.2 demonstrates the effect of the formation and operation of the buffer stock.

The outbreak of war posed yet an entirely new set of issues to be addressed. However, the industry that also had to confront them had changed markedly under the aegis of the ITC. Those developments, first on the production and then on the consumption side, will be reviewed before examining the way in which the ITC responded to the challenges of war.

13 Development under restriction: the producers

With the formation of the ITC, producers found themselves with a new set of constraints and opportunities which would shape the overall course of development of the industry. One condition governed all producers and that was the prospect of a return to the free market on the expiry of the agreement. Those who wished to survive an inevitable, though perhaps temporary, decline in price could not simply respond to the existing constraints and opportunities without considering their potential implications for the future. The market, albeit in partial suspension, would impose its own dynamic. Miners outside the scope of the ITC, of course, were free to develop without constraint and whether that freedom was primarily granted by the price recovery is one issue that must be considered in more detail. Those subject to its production quotas had to explore such opportunities as remained within an administrative framework that became more politicized as governments discovered they had new powers with which to shape the industry. How well they managed to secure their objectives in the face of these new constraints will demonstrate the extent to which restriction simply preserved the industry or determined a new course.

The outsiders

Tin producing countries were affected by the ITC in two different ways. The effective members were the five (Bolivia, Nigeria, Malaya, NEI and Siam) whose current production was restricted, plus the Belgian Congo, where restriction measures largely retarded the rate of growth of production. While some others (Portugal, Cornwall, French Indo-China) were formally members of the ITC, the terms of their adherence did not in fact make any difference to their actual production. For the purposes of this analysis they can be included with all other producers as outsiders.

The specific effect of the ITC on the position of the outsiders cannot be analysed in the absence of a counterfactual model of the world tin economy and the cost conditions facing the producers. In considering the effect of the price recovery secured by the ITC, the best that can be done is to compare the base year of 1929 with the period from 1933 when the ITC had finally stabilized the tin market. Table 13.1 indicates that outsiders posed a serious problem:

In order to look at the effect on the main five members, the Belgian Congo has been subtracted from the figures for world production. While the remaining world supply shrank by 25 per cent, outside production grew by 44 per cent, so that the outsider share of the market virtually doubled from just under 10 per cent to just under 20 per cent. Had the ITC secured complete coverage and imposed the same level of restriction on all its members, then the main five would have produced over 11 per cent more. While granting the fringe

Table 13.1 World distribution of tin production, 1929–1939

	World	*ITC core and Siam*		*Outsiders*	
	tons	*tons*	*%*	*tons*	*%*
1929	194,994	175,819	90.2	19,175	9.8
Average 1933–9	145,665	117,958	81.0	27,708	19.0
Change	–49,328	–57,861		8,533	
	–25%	–33%		44%	

Source: International Tin Council, *Statistical Yearbook,* 1959.

Note: World without Belgian Congo.

producers a free ride is an inevitable feature of commodity agreements, it can be reduced by ensuring that the price policy does not actually encourage outside production.[1] Had the outsiders been kept to their absolute levels established in 1929, then the main members would have produced 7 per cent more. It is therefore not surprising that the rise in outside production was a cause of some concern and resentment among those inside the ITC.

The key question is whether this rise in outside production can be attributed directly or indirectly to the ITC itself. There are two lines of argument which suggest that it can. The first rests on a simple marginal cost analysis. If these producers were high cost, then they could only expand as a result of the price umbrella being held high. The second is more indirect and rests on the desire of tin capital and displaced tin miners to exploit their particular comparative advantage by seeking new outlets when other opportunities are blocked. However, these factors must be examined against other equally plausible explanations. One has already been encountered in analysing the experience of the Belgian Congo where the boom of the 1920s led to the foundations of a new industry whose results were only evident a decade later. Others may return to the question of price but focus on two other kinds of prices. Where tin is produced as part of a process involving other minerals, then the price of the other mineral may determine the production of tin. Finally there are the competitive advantages secured through currency depreciation. These arguments will be reviewed to see the extent to which each explains the rhythm of growth of each of the outsiders.

The position of each outsider is presented in Table 13.2.

Table 13.2 Outside production

	1929	*1933–9 average*	*Change*		*ITC induced*
	tons	*tons*	*tons*	*%*	*tons*
China	6,800	10,443	3,643	54	
Burma	2,586	4,715	2,129	82	
Japan	869	1,793	924	106	
French Indo–China	829	1,358	529	64	
Mexico	0	300	300		200
Argentina	0	626	626		200
Other America	40	413	373		400
Cornwall	3,271	1,903	–1368	–42	1,000
Portugal	600	834	234	39	200
Other Europe	111	321	210	189	
South Africa	1,218	563	–655	–54	
Other Africa	612	1,236	624	102	
Australia	2,239	3,205	966	43	2,000
Total	19,175	27,708	8,533	44	5,000

Source: International Tin Council, *Statistical Yearbook*, 1959. *Note:* Induced is estimate

Burma

The easiest of these cases to settle is that of Burma where a high proportion of the tin was produced in conjunction with wolfram. From 1933 wolfram prices rose at a far faster rate than did those of tin. In 1933 the price was 27/6 a unit and over the whole period from 1934 to 1939 it rose to 48/6 a unit. Wolfram production rose at a similar rate and where the two ores were mixed together it was technically impossible to concentrate mining operations on the wolfram at the expense of the tin. Under such conditions, the tin could well have entered the market for simply the cost of shipping and smelting.

That does not explain the whole of the Burmese story. Anglo-Oriental imposed a voluntary level of restriction on its dredges[2] but Austral-Malay did not. At first the experience of its Thabawleik dredge appears to confirm the impression that it was only preserved because of the ITC. Although the dredge had started in 1927 on ground so rich that it could cover the costs of working in difficult clay conditions, they continued to rise to the extremely high level of £130/ton and it closed down in 1930, only to reopen in 1933 once the price recovery had been secured. However, when it did so it had overcome its cost problem which dropped by two thirds, making it just as efficient as most of the dredges in Malaya.[3] While Thabawleik clearly took a free ride since it produced at twice the level that would have been authorized had it been located in Malaya, it was clearly not a marginal producer only preserved by the ITC.

One new producer entered Burma specifically because of the absence of restriction and that was Burmese Hydraulic Tin.[4] The Malayan and General group formed this company in 1935 and it is an example of tin capital looking for fresh outlets. However, it can only have produced very little before being liquidated in 1937. At best, the policies of the ITC played an insignificant role in explaining Burma.

China

Two minor producing districts, Kwangsi and Hunan, benefited indirectly from tin restriction. They were the destination for many of the coolies repatriated from Malaya and the price encouraged the local government authorities to initiate a general programme of modernization.[5] However, any long term effect on the overall tin market was pre-empted by the Japanese occupation of the coastal region in 1937. This not only cut off the supply of around 3,500 tons[6] but it also stopped the plans of the central government to encourage American investment in the industry.[7]

The overall condition of the Yunnan industry in 1933 was not strong. A manager of one of the smaller mines considered that most were operating at a loss and that despite high prices there was 'little hope' for its future.[8] Even with the assistance provided by more or less continuous devaluation from 1935[9] and the availability of extraordinarily cheap, though not very efficient, child labour,[10] they remained very marginal.[11] This marginality was a function of the organization of the industry, rather than the quality of the ores. It was the prospect of a discovery of a particularly high grade deposit that kept most of the mines going and there were enough 'tin kings' with fabulous wealth to reinforce that dream.[12]

War with Japan encouraged the national government to stimulate production in Yunnan from 1937. The number of mines more than doubled, many with financial support from Chinese capitalists in Malaya.[13] More disturbing from the perspective of the ITC was the prospect of modernizing the smelting side of the industry. From 1935 a former general manager of STC was having some success in producing around 2,000 tons to international standards and the national government then approached BMC and CTS for assistance in developing a much larger smelter to raise the quality of the whole of local output.[14] Apart from the attractive profit rate, Lyttleton saw more strategic advantages. It would pre-empt an American initiative and would secure Yunnan's membership on the ITC.[15] Neither of these arguments persuaded Campbell who saw it much more as a threat to any prospects of renewal of the agreement.[16] Nonetheless, Lyttleton persisted and by early 1939 the Colonial Office was again being canvassed to lend its support.[17] In spite of the intimate association on the part of both BMC and CTS with the ITC, neither was particularly concerned about the way in which their Yunnan venture could undermine its future. The initiative was not theirs and they could take comfort from the fact that it would eventually occur regardless of their position, perhaps with far greater danger to the ITC.

This episode reveals something of the fragility of the ITC, though it is not one that it brought on itself as a result of its price policy. China was the most backward of the important producers and a programme of consolidation and modernization was bound to come at some point. The high level of profits secured by the Yunnan Tin Trading Corporation which operated the one modern mine at Malaga suggests that China would eventually find a way out of the marginal position she had long occupied.[18]

French Indo-China

There is little detailed information about the production conditions of the companies operating in French Indo-China and they may well have been weakened since France remained on gold until 1937 but from then on the producing companies reported substantial profits. They were generally aided by the simultaneous exploitation of wolfram and were therefore quite viable without the support of the ITC.

Japan

Devaluation of the yen by 40 per cent against sterling in 1931 and other measures taken by the government to stimulate tin production are sufficient to explain the increase in Japan[19] without any reference to the ITC. Anglo-Oriental's Toyo Tin which operated the largest mine was the primary beneficiary of these measures and recorded large profits in the mid-1930s.[20]

Mexico

Although there are extensive tin deposits throughout Mexico, they are generally of very low grade and found in association with other minerals that make them difficult to treat. Production was generally undertaken on an extremely small scale basis and must be regarded as quite marginal. One important exception was the San Antonio mine where a deposit of tin was found in a lead-zinc-silver mine and this accounts for over half the Mexican production of the 1930s.[21] Since its peak production was in 1931 and 1932, it was clearly not stimulated by the ITC, though the remainder almost certainly was.

Australia

Following the devaluation of the Australian pound by 28 per cent against sterling in March 1931, Australian producers enjoyed a considerable cost advantage. Stabilization of price provided the additional incentive necessary for the formation of several new companies from

1933,[22] though most appear to have been both small and economically marginal.[23] Existing small producers found that neither devaluation nor the price recovery provided any relief,[24] though the few large producers such as Aberfoyle and Briseis were sufficiently strong that they needed neither crutch, especially where they also produced tungsten.

One ominous development was the formation of Tableland Tin Dredging in 1937. This was floated with the support of Malcolm Newman from Alluvial Tin and was designed to operate on the same large scale as the most advanced dredges in Malaya but at even lower cost.[25] Fortunately it did not become fully operational until 1940 but it represents a good example of the way in which existing tin interests could take advantage of the price security offered by the ITC to develop new productive capacity quite capable of competing with most existing low cost producers.

As the industry confronted the depression of 1938, pressure was brought on the government to cushion the producers by imposing a tariff 'so that the price may be fixed at a payable level in keeping with Australian costs of production'.[26] Overall, therefore, Australia is best considered as a marginal producer with a high proportion of her output only made possible thanks to the existence of the ITC.

Portugal

Perhaps around a third of Portuguese tin output came from mines jointly producing wolfram. The Portuguese-American Tin dredge continued as an efficient producer of tin alone but the rest must have been very marginal indeed. Lagares Tin Dredging resumed at the end of 1933 but ran losses as frequently as it turned profits until it was abandoned in 1939.[27] Over the period from 1930 to 1933 there were, on average, 34 mines working only for tin with an output of less than one ton per month. With price stabilization this number jumped to 128 in 1934, rising and falling thereafter with the change in price,[28] which suggests they could hardly have survived at all without the ITC.

Argentina

In 1933 cassiterite was discovered in the dumps of an old gold dredging operation in a region just south of the Bolivian border. More thorough prospecting revealed the existence of high quality alluvial and lode tin and the following year Patiño, in collaboration with National Lead and ASARCO, formed a company to exploit it.[29] By 1938 it was producing concentrates with around 1,000 tons of metal content but this proved to be its peak and from then on it rapidly declined. Such production would almost certainly have occurred without the support offered by the ITC.[30] However, the remainder of Argentine output, perhaps 200 tons, was produced under the same small scale marginal conditions that existed in Bolivia.

Cornwall

At first sight this appears to be the most clear cut case of a marginal producer only kept alive because of the ITC. The point was often noted in the technical press[31] and the Chairmen of two of the largest of the Cornish producers, Geevor and East Pool, admitted that tin restriction had 'saved the industry in Cornwall'.[32] Its marginal character is most evident in East Pool which showed continuous losses from 1935 until its closure in 1940.[33] South Crofty, however, was in a strong enough position to survive without the ITC,[34] as probably was Geevor and these two accounted for around half of Cornish production. One disturbing development was the formation of Mount Wellington in 1938 by a subsidiary of BMC, British Non-Ferrous Mining, which planned to overcome the traditional weakness in Cornish mining by consolidating operations for really large scale mining.[35] Although this was a complete failure, it confirms the willingness of existing tin capital, closely linked to the forces behind the ITC, to operate in a way that could reduce its effectiveness.

Other producers

The less important the producer, the less information is available about the conditions under which it operated. It may be surmised that the increase in both Germany and Italy, which account for virtually the whole of the other Europe category, was related to the desire to develop a modicum of domestic supply. The decline in South Africa is related to the recovery of the gold industry. Of the six other African territories recording tin production, two, Southern Rhodesia and French Cameroons,[36] were entirely new, presumably as a result of the same kind of prospecting momentum that was evident in the Belgian Congo and they account for half of the increase in that group. There is no reason to suppose that the rest of the producers were sufficiently marginal that they only survived because of the ITC.

Once the necessary adjustments are made to the total production figures for the group of outsiders, it would appear that the ITC either protected or induced an average of some 5,000 tons annually over the period from 1933 to 1939. That element of pure free riding reduced the production of the five main producers by only 4 per cent.

The experience of the members of the ITC will be divided into three categories: (1) the fringe producers of the Belgian Congo, the UFMS and Malacca; (2) the established producers of the FMS, Siam, NEI and Nigeria which best demonstrate the economic effects of restriction; (3) Bolivia, which remained a special case on account of the economic and political implications of the Chaco war.

Belgian Congo[37]

The Belgian Congo was particularly badly hit by the general collapse of commodity prices. Copper accounted for two-thirds of exports and its value shrank by 86 per cent between 1930 and 1932. The ensuing economic crisis immediately became a fiscal one. Declining profits triggered various guarantees on private capital and they had to be met from a shrinking tax base. The tin industry therefore found itself in an extremely favourable environment. Expansion was now encouraged to help offset the collapse of copper. Constraints on labour supply were also relaxed thanks to an extension of compulsory crop cultivation by indigenous communities. Intensification of rural poverty created a powerful incentive to seek employment in the mines. Given the pattern of prospecting that had been undertaken in the 1920s, the tin industry was well placed to take advantage of this new environment.

One pattern of development was set by Géomines and that was to implement a comprehensive programme of mechanization of all phases of production, from extraction to smelting. Expansion of production permitted an economic local smelter, especially given the high transport costs and the availability of cheap hydroelectric power. That in turn provided its own incentive for further expansion.

While labour was plentiful, it was not cheap, especially in comparison with Nigeria. It was inexperienced, inefficient and required high levels of supervision by expensive European engineers. Mechanization would cut the total size of the labour force but make it more stable and experienced. When restriction began to bite for the first time in 1938, Géomines found itself in a much more flexible position, since it was now far easier to vary the amount of equipment employed than that of labour. While the technology was no more advanced than that in place elsewhere, it was being used on a wide scale and on a high-grade deposit. This translated into a high level of profits, of around one third of the value of the metal when sold in Europe, or 15 per cent on its inflated nominal capital. Its expansion could therefore be easily financed internally.

The experience of Géomines was closely followed by Symétain, to even better effect. Its production schedule was considerably delayed until a road was completed in 1934 and its growth fortuitously coincided with the high level of prices then prevailing. Its remote location meant that it had to establish a stable labour force with high levels of investment in social capital and the ensuing costs provided a powerful incentive to mechanization. Symétain soon found itself in a position where the scale of production justified the development of a local smelter. Overall, the productivity of mining labour, however, remained very low, especially by comparison with Nigeria. But this was more than compensated by a grade of ore which was nearly four times as high and this provided the basis for another large, very profitable company.

As far as the rest of the Congolese industry is concerned, the experience was very mixed. UMHK reopened its tin mine at Busanga in 1933 but resumption of production was prompted as much by the opportunity it provided to retain European staff that its copper operation had now made redundant, as by any price incentive. It was economically quite marginal and serious thought was given to suspending operations during the 1938 depression. However, as the company began to review the role of the mine in its overall strategy, it decided to invest considerable sums in developing its underground deposits. One other large producer, Sermikat, was very successful but the few medium and small producers were very marginal.

As was the case elsewhere, the tin industry in the Belgian Congo was highly differentiated. Here there was a distinctive basis to that differentiation, in the variety of systems of property rights in force. Both of the strong companies, Géomines and Symétain, developed within the framework of a Chartered Company which held a monopoly of the mineral rights in its respective domain. Most of the weak companies operated in Ruanda where the rights were allocated to prospectors indiscriminately.

The major producers were sufficiently strong, technically and financially, that their pattern of development does not bear any obvious relationship to the fact of the existence of the ITC at all. With the exception of 1938, they shared none of the burdens of restriction and enjoyed all its benefits but whether that is sufficient to simply treat them as free riders depends on yet another judgement. Among the objectives of the ITC was the establishment of the conditions under which producers could secure an adequate return on their investment. The Chartered Company model had induced considerable investment in prospecting in the 1920s on which a return could only be realized by production in the 1930s. One criterion of fairness is therefore met, while another is violated, since these producers did not bear the costs of reduction of the utilization of current equipment that were incurred by all the others. Finding a universal criterion of fairness is no easier here than elsewhere.

Unfederated Malay states and Malacca

The standard tonnage allotted to Malaya and the consequent quotas, had to cover not only the FMS but all of the other states. While the Senior Warden of Mines for the FMS had primary responsibility for ensuring that the total Malayan production was consistent with her international quota, he had to do so within a complicated administrative framework. At first, each state was allotted a standard tonnage based on its production in 1929. In the case of Perlis and Kelantan, these tonnages were considered as a flat rate; the standard for Kedah, Johore and Trengganu was subject to the international quota. When Malacca was fully included within the restriction agreement, she was subject to the lower FMS domestic quota. Since these states were generally stagnant, their standards bore a much closer relation to actual productive capacity than was the case in the FMS, where the production momentum had continued until 1931, and as a result the incidence of restriction was far less severe.

The generally stagnant condition of the industry had a further consequence. As some mines became exhausted, the remaining miners were not in a position to take up their quotas, so states would often experience surpluses. Where other states found new opportunities for growth, then interstate transfers could be arranged; otherwise the surpluses were transferred back to the FMS to ensure that Malaya always met its quota. As each of these transfers, often involving quite small amounts, required consultation with the Senior Warden of the FMS, such a system entailed a considerable amount of administrative work.

Following the signing of the third agreement when it was clear that restriction would remain over a much longer period, the administrative structure underwent substantial modification. The other states were now defined as a single unit with its own supervisory committee, independent of the FMS.[38] The total tonnage, subject to one important modification in relation to Johore, was maintained at its 1929 level but this was distributed among the various states on the basis of the ratios of actual production in 1934 and 1935. That meant ending the flat rate principle from which some states had benefited during the worst of restriction.

The logical development of this new system would have been to grant these other states their own international tonnage.[39] Instead, the new supervisory committee continued to treat the UFMS as a single unit and routinely monitored new developments on the basis of which it recommended transfers of standard tonnages between the states. Restriction there would be much more closely based on an assessment of current capacity but administration would be far more complicated as a result.

Perlis

Perlis included some quite geologically distinctive tin deposits, which were found in large pockets within limestone caves. Extracting them required blasting extensive tunnels at great expense over several years. It was an extraordinarily speculative business which generated some colossal fortunes but many failures[40] and meant for an erratic overall level of production in the state. Finding some principle of equity in restriction that would also preserve the long term viability of this form of mining was extremely difficult.

Perlis was included in the initial round of consultations conducted in early 1931 on the desirability of tin restriction for Malaya. The miners were generally opposed; at least one to the very principle and the others to its practical implications for Perlis.[41] The basis of this opposition was conceded by the Senior Warden of Mines for the FMS who simply wanted Perlis to establish controls sufficient to prevent smuggling from neighbouring Kedah and to ensure that at least 1929 production was not exceeded.[42] No system of formal restriction was therefore put in place and the only contributions that Perlis made to tin restriction was to impose a cess on exports to support the ITC and to avoid granting fresh leases.

Throughout 1931 the lack of restriction posed no particular problem and exports continued to fall away in the face of low prices, exposing the overall marginality of mining in Perlis. One of the European miners, also a Sir John Campbell, commented: ‘The slump here is terrible and the distress here is appalling. Everyone is bankrupt and Chinese mining coolies are working for food only costing 20c a day.’[43] However, he was not discouraged from continuing to explore for one of these pockets and by 1932 he was producing 200 tons. Exports for July alone were 109 tons, just under the annual level that would have been allotted to Perlis had she adopted the same principles of restriction that operated in the FMS. Such a disparity was bound to create resentment among miners in the FMS working on a 25 per cent quota and Perlis was therefore expected to adopt a restriction scheme similar to that already implemented in Kedah.[44]

Perhaps because by 1932 the ITC had demonstrated the effectiveness of tin control, the miners changed their overall attitude. They ‘proved very reasonable in the matter and agreed that some form of control was necessary to avoid complications’,[45] but needed a system that was not based on past production performance, otherwise there could be no incentive to continue exploring these distinctive pockets. Negotiations with Kuala Lumpur produced a formula. Production between 1923 and 1930 had averaged 350 tons per year, so 300 tons was taken as a flat rate of which 50 was defined as a reserve available only for sudden increases.[46] Internally control was exercised not on the basis of production but on exports, with a complicated formula that would ‘cut the large producer more than the smaller men’, and it met with the unanimous approval of the miners.[47] Excluded from all these discussions were the dulang washers and, in order to provide more room for the tin pockets, their passes were not renewed.[48]

Although this formula meant that for 1933 Perlis would restrict to 75 per cent of her 1929 output, while the comparable figure in the FMS was 33 per cent, it was proved to be quite insufficient to solve the distinctive problem posed by the pockets. Whereas the financial resources necessary to support the suspension of dredges and gravel pumps were quite modest, those that stood behind these highly speculative ventures were not. Unless adequate incentives were provided there was a danger that the whole industry would be permanently dislocated. Perlis therefore continued to press for an increase in the flat rate and even tried to resort to a little blackmail by claiming that she was not in fact a party to the overall International Tin Agreement.[49] But any revision that would deal with the distinctive problem of Perlis had to be suspended until the overall severity of restriction was lifted.

Relief came in two forms. Following the renewal of the overall agreement in 1934, the flat rate was revised from 300 to 315 tons and in 1935 200 tons of the standard tonnage allocated to, but not needed by, Trengganu was transferred to Perlis. Since this was a standard tonnage, it would be subject to the domestic FMS quota. But just as the right to produce increased, so the pockets began to dry up and Perlis found herself unable to utilize her full quota and 103 tons of her entitlement of 430 tons for 1935 was then reallocated to the FMS.[50] While production increased in 1936 and 1937, it still remained insufficient to cover the increases in quotas.

Trengganu needed its tonnage returned, with the result that in 1938 Perlis was only granted a permissible quota of 258 tons. Unfortunately, this was precisely the year when the results of fresh attempts to locate these rich tin pockets initiated in the boom years of 1936–1937 were coming to fruition. At first it was thought that they would just as suddenly cease and temporary overexports were tolerated by the advisory committee but the main 'offender' was:

> on to an extremely rich thing which shows no sign of giving out. They could actually, if permitted, produce the whole exportable monthly allowance of the state in a few days. Their find is a result of great labour in the past and their cave is liable to flooding.[51]

By the end of the year, Perlis was 400 tons over her quota and these overexports would make it difficult for the Malayan delegation on the ITC.

Not only was Perlis in violation of her commitment to control exports but she had also allowed stocks to accumulate. By June 1939 they had reached 543 tons, or 113 per cent of her standard. While some of this was offset by transfers of stock entitlement from other states, it appeared as though Perlis would have to shut down its whole mining operation for the next 30 months in order to conform to its obligations.[52] Given the overall level of restriction being imposed elsewhere no one had any sympathy with the predicament facing Perlis[53] and it was only thanks to the outbreak of war that the straitjacket in which she had been placed was loosened.

Except for a brief period in 1932–1933, no effective restriction operated in Perlis. By one set of standards applied elsewhere, she had therefore been very generously treated. However, there was another set, namely the need to preserve the conditions under which the long term viability of producers was sustained. In the case of the FMS a system of transfers allowed producers to buy those conditions where allotted quotas were insufficient but no such system could operate among the other states, at least not unless one state were prepared to compensate another for its loss of revenue. The distinctive structure of mining in Perlis, therefore, forced a choice between development and co-operation. Without specialized administrative familiarity with the nature of mining, it is not surprising that the British Advisor was content to let development win and leave the consequences to another round of politics.

Trengganu

Of all the other states, Trengganu was the only one to have shared in the tin boom of the 1920s. Annual production levels from 1900 to 1923 were around 300 tons but by 1928 they had quadrupled, reaching 1,118 tons. As in the FMS, the increase was a result of the development of dredging but here it proved to be a transitory phase. In 1928 the state was thoroughly 'prospected by agents of powerful mining groups looking for alluvium for dredges', but they came away empty handed.[54] However, the existing production levels of the late 1920s served to grant Trengganu a standard tonnage of 976 tons.

Restriction regulations were established in April 1931, following the pattern in the FMS, but with two important differences. Producers were subject to much tighter constraints. No consolidation of production was permitted through formal grouping arrangements and at first quotas were expected to be produced on a monthly basis with no carryforward. On the other hand prospecting continued to be permitted. While new entrants were not allowed, existing producers were free to expand.

Bundi was Trengganu's largest producer, with two dredges, which together accounted for 62 per cent of the state's production in 1929. The company was not in a strong financial position[55] and one of the dredges soon exhausted its ground, closing down in December 1931. In spite of its success in lobbying the state council to get the bulk of the assessment released transferred to its other dredge,[56] the company remained unsatisfied and began to explore other opportunities. In March 1933 the dredge still working was dismantled and eventually shipped to New Zealand to work a gold deposit. At that point Trengganu found itself unable to produce its quota even on a 33 per cent basis.[57] For the following four years the state had a surplus to transfer to the others.

With the high prices prevailing in 1937 fresh prospecting was undertaken and Trengganu was encouraged to rebuild its earlier capacity.[58] However, without an adequate transport infrastructure to attract large companies, mining remained largely in the hands of small, inefficient miners.[59] The Bundi lease was acquired by Lui Ah Hon who rehabilitated the remaining dredge but his murder in 1938 prevented any fresh development, especially since his estate remained unsettled for another two years. The only lasting initiative of any significance came from a Japanese iron ore company, Ishihara Sangyo Koshi (ISK), for which tin was an important byproduct.[60]

With the low quota levels prevailing in 1938 and the emergence of fresh productive capacity, Trengganu again had to impose restriction. Since the extent could be reduced, if not eliminated, with the restoration of the 1929 entitlement, Trengganu pressed for an increase in its standard tonnage. However, since 1929 had ceased to have any relevance to the distribution within the UFMS, Trengganu had to argue its case based on comparative capacities. The emergence of an important new producer in the form of ISK considerably strengthened its hand and the standard was raised in four separate stages from 476 at the end of 1937 to 950 tons by the end of 1939.

The problem with Trengganu was revealed as soon as restriction was lifted in 1940. Her miners only managed 352 tons, a mere 37 per cent of her standard! That should have given considerable pause to any plans to link entitlements with estimates of capacity.

Johore

A preliminary review of the production figures for Johore would suggest that she, too, was stagnating. During the war, Johore had become quite an important producer with an average of 2,135 tons of tin-in-concentrate but during the 1920s output generally declined to the point where it was only 672 tons by 1929. However, this overlooked the particular way in which the development of dredging technology, which had transformed the FMS, would also affect Johore.

One of the factors accounting for the decline was the closure of the oldest mining district, the Pelepah valley. Lack of government control over mining had resulted in an 'accumulation of tailings in the river bed and acres and acres of land ruined by the river silting and flooding'.[61] The situation was continuing to deteriorate but the prospect of large scale dredging offered a solution to the silting problem and the government became interested in an initiative by Simms on behalf of the Tronoh affiliate, Sungei Besi. In May 1929 prospecting started[62] and while the initial results showed a rather low grade proposition, nine months later Simms felt there was sufficient to justify the investment of £130,000 in 'one of the largest and most up to date' dredges in the peninsula.[63] On the basis of a promise to issue mining leases covering 800 acres,[64] Simms left for England with plans to order a dredge. Twelve months later, in March 1931, the dredge was ordered and Sungei Besi floated a subsidiary, Pelepah Valley Tin Dredging.

While the interest of Johore in becoming part of the new mining industry was clear, that of the Cornish in promoting yet another large producer at the very time when restriction had become imperative is not. Fortunately, these moves did not become public until 1934.

Otherwise they would have provided irrefutable confirmation of the Dutch thesis that the cause of the problem in the industry could be firmly laid at the door of the rapacious Malayans.

Tin restriction began on the assumption that it was designed to bring about a short term correction. When the FMS started to formulate a detailed administrative scheme in June 1931, Johore was allotted a standard tonnage equivalent to her 1929 production of 667 tons and there was no thought of the necessity of making some provision for Pelepah since the dredge would not become operational for another two years. Actual restriction was not implemented until late 1931, when it was feared that the cut in the international quota might leave Johore overproducing. Restriction provisions followed the model of the FMS and became effective on 1 December.[65]

This fear was not well founded. Even when the international quota was cut to 33 per cent in July 1932, the miners were still not prepared to produce their full allotment. This was set at 27.5 per cent of their assessments which brought them into line with the 25 per cent domestic quota operating in the FMS and the remainder of Johore's entitlement, a full 25 per cent, was allocated to dulang washers.[66]

Johore, like many others, had assumed that should restriction be renewed, there would be an opportunity to revise the initial arrangements and that her standard tonnage would, therefore, be increased to accommodate the needs of the Pelepah dredge. However, the context within which the agreement was renewed pre-empted any such opportunity, so Johore's standard remained unchanged. Meanwhile, Pelepah had been concerned to secure some commitment which would allow the dredge to operate. In November 1932 the dredge was formally granted an assessment which amounted to 700 tons or more than the entire standard tonnage of the state. Given the overall level of underproduction among the existing miners, a viable quota could have been allotted to Pelepah had the international quota reached 75 per cent but that was a long way off.

No immediate solution could be found without an increase in Johore's standard. Pelepah had a legal right to a CoP but since all holders had the right to the same level of quota, redistribution would leave the dredge with a trivial amount. The FMS was approached for a share of its standard but was flatly rejected. The dredge had been ordered a year too late to be eligible for an assessment under FMS rules and there was no sympathy for an exception for Johore.[67] Thomas tried to get the Colonial Office to intervene but could not overcome its antipathy to the Cornish. Campbell noted that there were allegations of bribery in securing the lease, confirming his impression that 'Mr Thomas is the sort of person one uses a long spoon when supping with'.[68]

Pelepah had one further card to play and that was to threaten a law suit. While the restriction regulations barred any suit in relation to a decision arising from their implementation, it was not clear that a court would refuse to hear a case based more on 'the defiance of the Rules than from their operation'.[69] That could call into question the good faith of the Johore government and perhaps the whole legal basis on which Johore was obligated to implement tin restriction.[70] It was to avoid this embarrassment that the High Commissioner decided to force the FMS to make the necessary concession.

The concession was far short of what Pelepah had asked for. The dredge was initially assessed at 700 tons but since the amount transferred was half of this, at 350 tons, the actual assessment was correspondingly reduced to 400 tons. However, as other Johore producers found themselves unable to produce their quotas, their rights were transferred to Pelepah. Whether even Pelepah was effectively restricted is unclear since the dredge produced only 75 per cent of its authorized quota in 1939–1940.[71]

Malacca

In 1929 Malacca produced concentrates with around 5 tons of metal, a quantity so insignificant that she was ignored in the initial restriction arrangements. However, in 1932 the Straits government insisted that all lessees employ their full labour force and by 1933 production had grown over tenfold to 55 tons. Given the importance of ensuring that Malaya's total production did not exceed her quota, it was imperative that Malacca be brought within the framework of control. This would not only provide a more predictable basis on which to set quotas within the FMS but it would reduce the prospect of Malacca being used to smuggle ore from neighbouring Johore. Following the renewal of the second agreement, the FMS government therefore approached the Straits government to secure its own international quota and pass its own restriction legislation.

The lessees naturally referred to the terms received by the Belgian Congo and Cornwall and claimed that the machinery installed and the labour employed to avoid forfeiture of their leases in 1932 entitled them to an annual rate of 150 tons![72] In this they were supported by Bagnall who used them in the ensuing debate on restriction legislation to give vent to his opposition to tin control, suggesting that Malacca negotiate directly with the ITC for its own flat rate on terms similar to those recently granted the Belgian Congo. The Acting Governor contemptuously dismissed this as 'little less than absurd'.[73] There could be no question of negotiating a separate quota with the ITC, nor of going back to a local standard based on 1929 and the four producers were given a flat rate of 58 tons. As in the case of Johore, the FMS was forced to absorb the implications of continued development after restriction.

At first this flat rate proved to be too generous but by 1936 Malacca produced her quota and was allowed to exceed it during the boom of 1937. In the readjustment of tonnages among the other Malay states that occurred in 1938, Malacca was granted an assessment of 200 tons and her actual quota of 104 tons meant little, if any, restriction.

Kelantan

Kelantan followed a similar path to that of Malacca. Total production for the whole of the 1920s had barely reached 15 tons, yet this was the annual flat rate allotted to the state on the commencement of restriction. So marginal was the tin industry in the state that it simply shut down entirely in March 1931[74] and did not resume until the price recovery of 1933. Considerable tin prospecting activity had been stimulated in 1933 and as the price continued to rise in 1936 production increased to the point where it could only be accommodated by a transfer of quota from Trengganu.[75]

This production momentum was rewarded with an assessment of 100 tons in the allocations for 1938 but the confident hope 'that Kelantan producers can now look forward to years of steady development'[76] was quickly dashed. By 1939 she could barely produce 20 tons. Kelantan therefore is yet another case of a complete free ride.

Kedah

As in Johore, tin mining in Kedah was constrained by an overall concern to protect agricultural land[77] and it was not in a particularly strong condition during the 1920s. In 1929 the situation changed with the decision of Kampong Kamunting to move one of its dredges over the border from Perak, so that production in 1930 reached 342 tons, which entitled the state to an initial assessment of 318 tons.

Table 13.3 Other Malay States, tin production, 1929–1938 (tons)

	1929	*1931*	*1932*	*1933*	*1934*	*1935*	*1936*	*1937*	*1938*
Johore	672	426	305	219	393	604	755	814	784
Trengganu	952	493	400	252	366	399	480	429	332
Kedah	268	148	119	133	158	200	308	341	259
Perlis	421	321	483	312	432	323	458	456	651
Kelantan	1	2	0	1	3	6	20	34	28
Total	2,314	1,390	1,307	917	1,352	1,532	2,021	2,074	2,054
Authorized		1,579	990	775	1,155	1,539	2,423	2,815	1,637
Assessment		2,269	2,269	2,269	2,366	2,619	2,619	2,619	3,000
UFMS total % assessment			58%	40%	57%	59%	77%	79%	69%
FMS authorized % assessment			33%	24%	34%	42%	67%	78%	41%
UFMS : FMS ratio			1.8	1.7	1.7	1.4	1.2	1.0	1.7
Malacca	3	15	37	57	49	51	58	72	113

Source: Sir Lewis Fermor, *Report upon the Mining Industry of Malaya*, Kuala Lumpur, 1940, p. 60.

Although Kedah established the same kind of system of administrative regulation as in the FMS, at first it proved to be quite unnecessary. Prices were simply too low for Kampong Kamunting which kept its dredge idle until late 1934 and sold its quota to the remaining mines. When it resumed, the dredge was granted a rather low assessment in light of its previous performance, presumably to give greater protection to the other 11 mines. But with the dredge in operation, Kedah was in a position to pick up some of the surplus quota from other states, for which she was eventually rewarded by an increase in her assessment. By the end of 1938 this had reached 480 tons which considerably reduced the extent to which the dredge was required to restrict.[78] Kedah is therefore a case where low prices during the first agreement were sufficient to eliminate the need for artificial restriction; for the second and third agreements restriction may have been administered effectively but at a very modest rate.

Table 13.3 demonstrates the extent to which producers in the other Malay states enjoyed a far more favourable position than their counterparts in the FMS.

When quotas were low, 1931–1933 and 1938, they exceeded their authorized entitlements by a considerable margin; when they were high, 1936–1937, they were unable to produce them. Although producers in the FMS often felt that they were carrying other countries on their back, at no point did they raise any concern about the inequities in the internal distribution of the overall Malayan assessment. There was not even a protest against the granting of a share of the 1938 adjustment, even though it had been earned entirely by the FMS miners. Had they paid more attention to their neighbours, they would have developed a greater appreciation of the inherent power that is enjoyed by fringe producers.

Federated Malay States

Administrative regulations governing tin restriction established three categories of producers: (1) those where tin was incidental to other mining operations, (2) dulang washers, (3) all other miners. Tin produced incidentally to other mining operations was not subject to any effective restriction. Since the amounts involved were very small, this group created no administrative difficulties.[79] Both of the other categories posed their own problems which will be reviewed separately.

Table 13.4 FMS, dulang washers, 1929–1939

	Licences		*Production*	*cf 1929*	*FMS quota*	*DW : FMS ratio*
	no	*kattis*	*tons*	*%*	*%*	
1929	8,947	22	1,060			
1930	7,784	23	960			
1931	8,738	30	1,410	133	68	2.0
1932	9,000	20	970	92	33	2.8
1933			740	70	24	2.9
1934		15	760	72	34	2.1
1935			870	82	42	1.9
1936			1,030	97	67	1.5
1937			1,010	95	78	1.2
1938		20	1,030	97	41	2.4
1939		20	850	80	53	1.5

Sources: Yip, p. 211, Khalid, p. 96

Note: katti is 1.33 lbs and refers to quota of monthly concentrate; production metal content.

Dulang washers

Given the nature of dulang washing, there was no need to determine productive capacity; instead each licensed washer was granted a specific quota. Dulang washers occupied a dual position both as workers and as property owners. As owners, they enjoyed the same basic right as all other owners, namely the right to a quota level that would maintain their existence. Unlike other workers, therefore, they were not subject to unemployment. However, their property rights were severely circumscribed. They could not transfer their licences, except by administrative permission, not even to their own daughters on their retirement.[80] Nor could they transfer their quotas. A strict prohibition against such transfers was designed to prevent leakage from the regular mining sector.

In shaping a policy for dulang washers, the Mines Department had to address several issues. In addition to determining an appropriate quota level for each washer, the number and allocation of washing licences was also subject to administrative discretion. Table 13.4 indicates the extent to which they enjoyed a far lower level of restriction than did regular miners. However, the cut to 20 kattis of ore per month in 1932 would have meant an income of only around 30 cents per day, or about the same level to which unskilled workers were reduced.

From 1934 the number of licences was increased in response to the demand for even minimal income opportunities and they were also awarded to Malay and Tamil women.[81] While illegal, the practice of selling quotas appears to have been ‘common in Perak and Selangor’.[82] However, the Mines Department was not in a position to take greater control over this informal sector and the most that appears to have been done was to add a photograph and other identifying details to the pass, so that at least the passes themselves were not sold.

Regular mining operations

The administrative system that was designed in 1931 in order to preserve existing mines did not take full account of the particular pattern of property rights prevailing among the Chinese. Key to this was the separation between the lessee, who held the legal right to mine and the actual miner.

The sublease then transferred that right in return for a percentage of the ore extracted. Actual mines could work land held under different leases and the arrangements often became quite complicated, since the owner of the mine could be subleasing one part and sub-subleasing another.

Although all these modifications to the original lease were duly recorded by the Mines Department and entered on each CoP, it was unclear as to what claims lessees held over mines that wished to group and work other land. In Perak they were discounted entirely and miners were free to form groups without their consent. That, of course, meant that some lessees were deprived of any opportunity to derive an income from their capital. While others, on whose land production was now concentrated, would find their capital depleted at a faster rate than they may have wished. But in Selangor it had become the practice by mid-1932 to require lessee consent and lessees began to insist on a 'definite statement of opinion from government that it is sound and equitable that the lessee should receive a share of the tribute'.[83] Several officials in the Mines Department were opposed to a policy which would expose producers to extortion by lessees. The District Controller for Perak noted that:

> lessees will at once realise that they have the producers completely at their mercy and many I fear will not hesitate to make use of their opportunities. The producer must receive prior consideration since he bears the risk, whereas the lessee in the majority of cases is a parasite whose financial risk is negligible by comparison.[84]

In spite of such opposition, supported by the FMS Chamber of Mines, the Senior Warden of Mines insisted on granting lessees rights since 'even a parasite is allowed to nibble till it is dead'.[85] In making this concession, he overlooked the fact that without actual extraction there would be no death and the lessee could continue to transfer the same production right over and over again. In spite of the best efforts of the Department, a gap began to emerge between the original assessments and current productive capacity, one that could only be closed by a laborious and controversial complete reassessment. The size of that gap and the extent of parasitism, became ever more evident as the amount of quota sold increased, generating yet another parasite, the quota broker.

Depletion without development meant that a parallel gap also opened up between current capacity and Malaya's ability to produce her standard tonnage, one that was exposed by the increase in the international quotas in 1937. The government's response was three-fold. In June the restrictions on the alienation of mining land that had been imposed as a conservation measure in 1929 were lifted.[86] Fresh assessments were granted to a few dredging companies. The most important was to create a new class of Chinese miner who was granted a nominal assessment of one pikul and who was then in a position to buy sufficient quota to add to his CoP and develop a new mine.

The overall development of the industry in the face of restriction and these new incentives is best appreciated against the statistical picture presented in Table 13.5.

Three interrelated shifts in the structure of the industry can be identified, all of which represent continuations of the trends that had been evident at the end of the 1920s: (1) decline in the proportion of production under Chinese control; (2) rise in the proportion met by dredging; (3) increase in the influence of Anglo-Oriental.

Table 13.5 FMS, distribution of production, 1929–1939 (%)

	Chinese	*European*		*Dredge*	*AO*
		Other	*Dredge*	*European*	*Dredge*
1929	37.9	23.3	38.9	62.5	15.9
1930	37.1	24.9	38.0	60.4	15.9
1931	34.1	24.3	41.6	63.1	21.2
1932	34.5	22.7	42.8	65.4	20.7
1933	33.5	21.3	45.2	67.9	18.3
1934	34.0	19.6	46.4	70.3	
1935	34.6	20.6	44.9	68.5	
1936	32.6	20.2	47.1	70.0	21.0
1937	32.5	19.3	48.2	71.4	21.4
1938	32.8	22.3	44.9	66.8	23.0
1939	30.5	21.5	48.0	69.0	
1940	28.5	20.4	51.1	71.4	
1940*				77.4	39.4

Source: FMS, Mines Department, *Monthly Bulletin of Statistics*.

Note: 1940 * assessment from Fox papers.

Chinese sector

The largest proportion of production by the Chinese came from gravel pump operations, and they varied considerably. This variation occurred along four dimensions: (1) technical, since gravel pumps had nozzles which ranged from 5 inches to 8 inches; (2) scale of operations, which was a function of the size and number of pumps; (3) ownership; (4) cost. Some disaggregation is therefore necessary to provide a picture of the position of the Chinese.

While no comprehensive body of data is available, there is a set covering most of the districts in Selangor for the years 1935–1937 and one set for the Batu Gajah district in Perak for 1936. The former encompasses 96 separate operations and the latter 30. Together they account for some 40 per cent of the Chinese sector for the whole of the FMS. Table 13.6 presents the distribution by size of original assessment.

The extent to which production diverged from original allocations is presented in Table 13.7.

By 1936 four new mines had been authorized with one-pikul assessments but they were only responsible for 3 per cent of the total gravel pump production of 137,381 pikuls of concentrates. Those with larger assessments tended to transfer half their quotas; those with smaller assessments tended to acquire as much quota from external sources as from their own entitlements. As a result, those with smaller assessments were producing on a larger

Table 13.6 Distribution of Chinese mines by size

	Batu Gajah			*Selangor*		
	No.	*Total pikuls*	*Mean*	*No.*	*Total pikuls*	*Mean*
<1,000 pikuls	7	4,918	703	39	9,760	250
1,000–1,999	11	15,300	1,391	24	37,490	1,562
2,000–2,999	5	12,530	2,506	18	45,425	2,524
>3,000 pikuls	7	41,790	5,970	15	53,620	3,575
Total	30	74,538	2,485	96	146,295	1,524

Sources: BG484/36; SelgG57/35, Form A1.

Note: the medians for the totals are 1,580 and 1,440 pikuls respectively.

Table 13.7 Selangor Chinese quota transfers,1936

		Total		*Mean*	
	No.	*Quota pikuls*	*Transfer pikuls*	*Quota pikuls*	*Transfer pikuls*
Sellers of >10% of quota	23	36,542	–18,126	1,589	–788
Other sellers	9	7,869	–40		
Buyers	64	53,573	52,982	837	828
One–pikul miners	5	5	4,580		
Total	101	97,990	39,394		

Source: SelgG57/35

scale than those with higher assessments. This is one important indication of the extent to which the assessments based on 1929 and 1930 had already become obsolete by 1936.

The primary form of transfer was through the grouping system and was adopted much more extensively in Perak than in Selangor. The pattern of grouping in Batu Gajah is presented in Table 13.8.

The Kam Beng pool was organized by a European miner, Cummings, and included not only 63 mines from Batu Gajah but 26 from other districts in Perak, so that its scale of operations was even larger than is presented above. Similarly, many of the other groups not only included mines from other districts but also mines from other states where the links were through ownership. One of the largest of these personal groups was organized by Foong Seong with a total of 24 mines and 13,500 pikuls of assessment, or about the size of a medium dredge.

However, in spite of efficiencies secured by consolidating mining operations with particular members of the group, by 1936 the Chinese sector had become increasingly marginal. Table 13.9 identifies the extent of that marginality.

Even in the boom year of 1937, there was still a significant proportion of production that was uneconomic, 10 per cent if the costs exclude tribute but closer to 20 per cent if they include it. Under boom conditions costs rose considerably and the 1936 costs can be regarded as representative of those obtaining over the long term. In that case, these figures rise to 14 per cent and 35 per cent respectively. However, the price of over £200 was a function of restriction and had this dropped to a 'normal' £150[87] then at most 22 per cent could have survived. The inevitable conclusion is that a large part of the Chinese sector was entirely dependent on restriction to secure its viability.

The sector, however, came under particular pressure when production was cut in 1938. The new one-pikul miners who had been able to buy quota cheaply at $6/pikul[88] in 1937 faced ruin when the cut raised the price to $40/pikul a year later. In spite of the fact that they had become quite an important part of the Chinese sector[89] and had enabled Malaya to meet her overall production targets in 1937, the government refused to use the readjustment in

Table 13.8 Batu Gajah Chinese assessments

	Groups	*CoPs*	*Pikuls*	*%*
Kam Beng	1	63	67,426	35
Other	23	84	93,859	49
Independent		27	30,147	16
Total			191,432	100

Source: BG311/25

Table 13.9 Selangor Chinese Financial Position

	1936	*1937*	
	£205/ton	*£242/ton*	*£150/ton*
S$/pikul gross	61.7	73.7	37.1
to sublessor	–6.2	–7.8	–3.7
S$/pikul net	55.5	66.3	33.3
S$/pikul mean cost	49.5	54.1	
% viable at gross	86%	89%	22%
% viable at net	65%	82%	15%

Source: SelgG57/35, Form A1

Note: $/pikul of concentrate after royalty and smelting charges.

the standard tonnages to grant them some degree of security. In light of the increase in the standards of the UFMS, none of which had earned them, this was a flagrant injustice.

The response of the one-pikul miners was to form a new organization, the Perak Miners' Association, to press the case for control over quota transfers. It was taken up by a retired Senior Warden of Mines. The best solution would be an end to quota sales altogether and a complete reassessment but he was prepared to settle for a system regulated by the Mines Department which would set a fixed price.[90] At one level neither the Colonial Office nor the FMS government had any sympathy with their case since they had all been warned about the risks when they took up these assessments in 1937.[91] Nonetheless, the sale of quota at high prices was one indication that the administrative system established in 1931 had become obsolete and the government now began to give more systematic attention to the pattern of transfers.

As the problem became acute in 1938, a subcommittee was struck which drafted a set of rules for transfer of quota to outside groups but none of the other associations of miners accepted them and the government was certainly not prepared to enforce a scheme against the will of the industry.[92] The most that could be gained was a commitment to undertake a complete reassessment should the agreement be renewed in 1941.

Investigation of the pattern of transfers revealed two important features. The net balance of transfers showed that overall it was Europeans who transferred quotas to Chinese.[93] Since European miners were not, at least not yet, in a parasitic position of holding assessments without being able to produce them, they always faced a decision as to whether to produce the whole of their entitlements or sell part of them. Any system of regulation of quota sales which fixed a price could simply reduce the amount available to that unpredictable quantity released as a result of unanticipated breakdowns.

The other important feature revealed was that the rate of quota transfer was a function of the rate of the overall domestic quota. In 1938, with a domestic rate of 37 per cent, 6 per cent of total quotas were transferred outside the established groups; in the first half of 1939 with a domestic rate of 25 per cent, the figure dropped to under 4 per cent.[94] The implication was clear. Raising production rapidly in times of high demand required an influx of fresh miners, who were then expelled when it dropped away. Where the ITC distributed the burdens of the commodity cycle among all producers at the international level, the system of administration in Malaya concentrated them on one particular group. While Malaya protested against the perceived inequities that resulted from the principles adopted at the international level, she was oblivious to the same problem that arose domestically. Indeed, it would be repeated when production again had to be raised quickly in 1940.

Such insensitivity rested on a marked preference for stability. As Thomas explained to the Colonial Office:

> The existing tribe of miners are on the whole an efficient body who know their work and get good results. It is always open to an outsider to buy out a miner and become eligible to carry on mining operations, but such an event is rare and seldom leads to anything but loss of money by the purchaser. Government cannot cause an old established miner to close down merely because the land they may have been working in 1931 is exhausted. If such mines were not given more land, new miners could take their place, but this system would cause a serious dislocation of technical organisation, and prospecting staff, and a large quantity of valuable plant would have to be sold at scrap prices.[95]

That principle pre-empted the possibility of a considerable expansion in the European sector and concentrating the costs of bearing the excess capacity there. In addition, it is worth noting that most of Thomas' arguments could be replicated at the international level and it is unfortunate that analysis of the difficulties of administering restriction domestically did not lead to a greater appreciation of parallel problems internationally.

European sector – dredging

Given the way in which dredging had reshaped the structure of the industry in the late 1920s it is not surprising to see a continued momentum. Following the shaking out of a few non-dredging European companies in 1930–1931, the size and structure of that sector remained largely unchanged throughout the whole period of restriction. Dredging, by contrast, saw continued growth from around 40 per cent of total output in the early 1930s to around 50 per cent by the end of the decade. In part this was a continuation of the momentum that had existed at the point when restriction was introduced, but the sector received an additional stimulus following the renewal of the third agreement.

Dredging technology underwent several modifications during the decade. When quotas were low during the first two agreements, older dredges were laid up as production was concentrated on the newer and more efficient ones. As production recovered, these older dredges were often reconstructed before being put back into service. At the same time they were converted from steam using coal or wood to electricity which also tended to drive down costs. It was in the area of the construction of new dredges that the most dramatic technical developments were to be seen. In the late 1920s new dredges were designed to extract around 200–250 cydm; a decade later, new dredges were designed around 300–400 cydm. Not only were they larger but they could dig deeper and thereby make marginal deposits profitable.[96]

The impact of these technical innovations will be considered on the basis of the experience of some representative companies.

Tronoh

As one of the largest and oldest dredging companies, Tronoh provides a good opportunity for the examination of the impact of restriction on costs and of the overall pattern of technical innovation.

Table 13.10 Tronoh dredges, comparative costs

	No 2	*No 3*	*No 4*	*No 5*	*No 8*	*No 1*	*Total*
Commenced	1915	1921	1928	1928	1936	1940	
Annual average:							
Capacity cyd/hour							
1928–1930	130	159	212	226			727
1937–1940			230	229	250	456	1,165
Direct costs cents/cyd							
1928–1930	20.8	16.6	12.9	11.7			15.5
1937–1940			13.8	13.6	12.7	13.6	13.4
Grade katties/cyd							
1928–1930	0.44	0.55	0.34	0.34			0.42
1937–1940			0.37	0.52	0.42	0.46	0.44
Direct cost £ per ton metal content							
1928–30	130.2	81.9	102.2	94.8			102.3
1937–40			102.9	71.7	81.9	80.5	84.3
Capacity tons concentrate at 7,000 hours per year							
1928–1930							1,262
1937–1940							2,143

Source: Tronoh Mines, Annual Reports

By 1929 Tronoh had five dredges: No. 2 was prewar; No. 3 had started in 1921, while Nos. 4 and 5 had started in 1928 and No. 8 was under construction. When restriction commenced, Nos. 2 and 3 were stopped and production concentrated entirely on the more modern Nos. 4 and 5. That immediately brought about a considerable reduction in cost. In 1929 the two older dredges treated 2.2 million cubic yards at a cost of 18 cents/cyd, while the two more recent ones treated 2.7 million cubic yards at a cost of 12.4 cents. Eliminating that obsolescent capacity meant a reduction of 20 per cent in average costs.

An overall picture of the way dredging at Tronoh changed is presented in Table 13.10.

Four shifts are worth noting. One is the existence of improvements to the technical capacity of the new dredges of 1928 and a second is the fortuitous improvement in the standard of ground being treated. Most important is the introduction of additional dredges. Following the renewal of the second agreement, Tronoh decided to complete the erection of No. 8 which, while of the same style as Nos. 4 and 5, had a somewhat larger capacity. Following the renewal of the third agreement, Tronoh bought a much larger dredge, new No. 1 which only became operational at the very end of 1939. These developments had a profound impact on costs. In spite of technical improvements to the existing Nos. 4 and 5, their costs rose over the decade and that reflected a general increase in the cost of inputs especially of fuel and labour. But replacing Nos. 2 and 3 with 8 and 1 would more than compensate for such increases.

The improvement demonstrated at Tronoh was only possible thanks to its financial strength and the operational flexibility provided by several dredges. This allowed No. 2 to be sold to David for a new company and No. 3 to be transferred to Tronoh's own Siamese subsidiary, Tin Lay, while at same time developing the new No. 1 dredge.

The overall financial health of the company at the end of the 1930s is demonstrated in Table 13.11.

Tronoh had three major sources of income and hence profits: (1) from its own dredging operations, (2) from lands subleased to tributers,[97] (3) from its investments, mainly in other

Table 13.11 Tronoh finances

	1929			*1937–1939 annual average*		
	tons concentrate	*£ per ton*	*£*	*tons concentrate*	*£ per ton*	*£*
Production						
Dredges	1,324			949		
Tributers	539			883		
Total	1,863			1,832		
Total profit			138,861			140,768
Distributed:						
1. Dividends			93,367			92,500
2. Retained			45,494			48,268
Source						
1. Depletion ore						
a) Tributers		19.9	10,719		18.9	16,706
b) Own		19.9	26,330		18.8	17,797
2. Extraction		27.7	36,672		55.1	52,327
Increase in price					11.5	10,946
Increase in efficiency					15.9	15,087
3. Other			65,140			53,938

Source: Tronoh Mines, Annual Reports

mining companies. The tribute paid was for the value of the ore in situ. Applying the same rate to the ore extracted by the dredges permits a disaggregation of the first kind of income, between depletion and extraction. The effect of restriction on profits from dredging can be seen in three ways: (1) increase in price by some £11 per ton; (2) efficiencies in production with the consequent reduction in costs by some £16 per ton; (3) increase in the number of tributers thanks to their need to secure fresh land to keep their assessments. In spite of a reduction in permissible production, the overall level of profit remained about the same at the end of the decade as it had in 1929. In fact the overall financial position of the company was strengthened. Whereas it had to turn to a share issue in 1928 to finance the construction of dredge No. 8, it could pay for the more expensive new dredge No. 1 ten years later entirely out of its own funds.[98]

An important turning point in the history of the company was marked in 1937. It took advantage of the end to the limitations on prospecting but since it found nothing worth developing in Malaya, it took advantage of the new opportunities in Siam. The bonanza of that year provided the financial support necessary to pay for the Siamese property and to start construction of the new dredge. Renewal of the agreement, especially given the difficult circumstances that attended that process, made such expansion both possible and essential. Five years ensured a reasonable period of time over which much of that investment could be recovered; more important was the fact that there was no assurance that it would be renewed for a further term. In that case only a strong company with modern equipment could expect to survive any period of disruption that termination would inevitably entail. Restriction therefore had a paradoxical long term effect. The prospect of termination encouraged the shift to low cost but capital intensive production which maintained excess productive capacity which simply reproduced the conditions that led to the formation of the ITC in the first place.

Yukon Gold

The Guggenheims had planned to expand their operations in the late 1920s but were discouraged by the high price demanded for dredging land. With the price fall of 1930 land now became much more reasonable and properties were acquired for which a new company, Malay States Tin, was formed as a subsidiary of Yukon Gold. Restriction forced a deferral of development but following the relaxation of the alienation policy in 1937, Malay States was allowed to proceed to build its dredge and was granted an appropriate fresh assessment with which to work it.[99]

Anglo-Oriental

At the beginning of the decade, Anglo-Oriental had two kinds of properties: operating dredges acquired from the Alluvial Tin group and two dredges that it had developed under its own auspices, Kampar Malaya and Lower Perak. Its subsequent evolution would repeat this process, both in the development of new properties and in the acquisition of existing companies. But acquisition would now be tied to a much more to a comprehensive process of consolidation. In addition, Anglo-Oriental extended its managerial services to several other companies. Restriction provided Howeson with the opportunity to realize his vision of rationalization of the Malayan tin industry. The overall evolution of the Malayan section of the Anglo-Oriental empire is presented in Table 13.12.

These processes had a dramatic impact on the structure of the Malayan industry. Four of the groups based in the FMS, Miles, David,[100] Sime and Henggeler, were now part of Anglo-Oriental. By the end of the decade there were only five dredges under local control. Adding Kamunting and part of the Guthrie group reduced the size of the British sector not under Cornish control. Of the other major groups that existed at the end of the 1920s, only Yukon Gold and Austral-Malay were left more or less intact. Not only did Anglo-Oriental now surpass the two Cornish groups combined but, in eliminating many of the smaller groups, the industry became that much more polarized.[101]

Anglo-Oriental also found that its overall focus was now redefined. Malaya was now the largest component of its tin interests by a considerable margin and it could no longer be accused of using Malaya to support its ventures in Nigeria and elsewhere. In 1934 the Prattens decided to sell off their remaining holdings in the Alluvial Tin group. Anglo-Oriental arranged for them to be picked up by Dean Finance (£100,000), BM (£125,000), CGF (£125,000) as well as by London Tin (£106,000).[102]

In addition to simple acquisition, Anglo-Oriental embarked on an ambitious programme of consolidation of existing properties, of which the most important was undertaken by Southern Kinta Consolidated. It brought in one of the best dredges from the original Alluvial Tin acquisition, Changkat, together with the two Yuba dredges from Kampar Malaya and Lower Perak. Of the seven other dredges brought into the company only two were built before 1928. On its formation, Southern Kinta became the largest operating company in the British Empire and with direct costs of around £45 per ton of concentrate in 1939 it was capable of meeting any competition. The evolution of the company followed the pattern set by Tronoh: acquisition of fresh land and the building of an even more modern dredge. However, here the fresh land was not only in Siam but also in Malaya and the dredge would be added to the fleet. Together with the cost of the land, the new investment amounted to £360,000. The whole sum would have been financed internally had restriction not been so severe in 1938, which meant that shareholders were asked to cover the cost of the land at

Table 13.12 Anglo-Oriental in Malaya, 1940

		Corporate Identity Retained		*Consolidation Ampat–S. Kinta*		*Total*	
		No of companies	*Assessment*	*No of companies*	*Assessment*	*No of companies*	*Assessment*
Control by 1930							
Acquired from Alluvial Tin		8	6,936	1	681	9	7,618
Developed by Anglo-Oriental				2	1,445	2	1,445
Expansion from 1931							
Control over							
Kamunting	1931	1	1,324			1	1,324
Henggeler	1933	5	1,809	1	333	6	2,143
Other	1933			3	704	3	704
Miles	1934–1937	2	1,074			2	1,074
Sime	1934	3	720	1	703	4	1,423
David	1934	1	114	2	1,387	3	1,500
Tongkah Harbour	1935	1	514			1	514
Development							
Berjuntai	1937	1	617			1	617
Lower Perak	1937	1	771			1	771
Total		23	13,880	10	5,254	33	19,133

Notes: Assessment in tons metal; Sime includes one company in Johore; dates are of commencement of management control or of initiation of development.

£87,000. The Cornish now had good reason to be concerned about their fate. Not only had Anglo-Oriental overtaken them in size but its flagship company was just as strong as any of theirs.

Anglo-Oriental's response to the incentives provided both by the renewal of the agreement and the availability of fresh mining land was even more extensive. The residue of the Lower Perak property was equipped with two modern dredges. They were also introduced into four other companies in the group: Kamunting, Rawang Tinfields, Taiping Consolidated and Kuala Kampar. At Kuala Kampar this released an older dredge which was reassigned to work a new property for which a new company was floated in association with the Prattens, Berjuntai. This dredge had not operated since restriction began but it was thoroughly reconditioned and, when it was ready in late 1938, its capacity had almost doubled from 180,000 to 300,000 cydm; as a result it was granted a generous assessment of 857 tons of concentrates.[103] Additional assessments were also granted to Kuala Kampar and Taiping, so that the overall programme of modernization was well supported by the Department of Mines.

Of the 13 new dredges ordered after 1937, Anglo-Oriental was responsible for seven, followed by Malayan Tin with two and just one each for the Redruth, Tronoh and Guthrie groups and one for an independent company based in the FMS. That would establish an overall technical superiority for Anglo-Oriental in the most important sector of the Malayan industry. However, one of the fears concerning the influence of the group did not materialize. Yuba was granted no more contracts, so that five went to the British engineering firm, Payne, and two to United Engineers based in Singapore.

Pahang Consolidated

In the late 1920s Pahang had discovered an extraordinary rich lode in its Willinks mine and at the same time its overall costs of production declined from £188/ton metal content in 1926–1927 to £134 in 1929–1930. It therefore entered the 1930s in an extremely optimistic mood, with more efficient working methods about to be applied to a new lode. That optimism was reinforced as further development work showed that Willinks had proved reserves of 515,000 tons at 3 per cent or 15,700 tons of concentrate.[104] Although production in 1930 had only amounted to 2,650 tons, the new lode must have been the basis on which the assessment awarded was as high as 3,760 tons. Even so, restriction meant a slow rate of depletion, especially given the capacity of the seven other mines operated by the company. Unlike dredges with modest overhead costs during periods of suspension, as a lode mine, Pahang had heavy pumping costs and its total overheads amounted to some 30 per cent of its total costs under normal conditions.

In addition, Pahang was a very large but remote mining complex, supporting a population of around 8,000. For the paternalistic management committed to providing good facilities to a permanent labour force, laying off workers would be very difficult.[105] It is, therefore, not surprising that Frisby was anxious to secure exemption from restriction. In 1932, as he reported on his failure to secure satisfaction from the Privy Council, he stated that restriction had raised costs to £95 per ton concentrate whereas under normal working conditions they would be as low as £60.[106] With unrestricted production and a price of £100/ton metal, the company could turn a profit of £100,000 but even at a price of £230 it could not pay a dividend on its ordinary shares.[107]

The company's performance after the price recovery of 1933 is demonstrated in Table 13.13.

Table 13.13 Pahang Consolidated, finances

Average	Crushed tons	Grade %	Concentrate tons	Costs £/ ton	Profits £/ ton	Price £/ ton	Profits £000
1923/4–1928/9	177,567	1.34	2,388	113.8	53.5	160.5	127.8
1933/4–1938/9	134,883	1.56	2,101	83.3	61.9	142.9	130.2
Change	–24.0%	15.8%	–12.0%	–26.8%	15.8%	–11.2%	2%

Source: Pahang Consolidated, *Sixty Years of Tin Mining: A History of the Pahang Consolidated Company, 1906–1966*, London, 1967, p. 67.

By comparison with the mid- and late 1920s, the 1930s saw a considerable reduction in costs, only partially explained by the higher grade of ore. That reduction more than compensated for the drop in price so the profits were slightly higher even in absolute terms. Most important was the fact that the company's capital asset was being depleted at a much slower rate.

The £100 claim was seized on to demonstrate the artificiality of tin restriction[108] and it continued to circulate very widely in spite of the fact that it is prima facie absurd, not only as the basis for stabilizing the world market but even for Pahang. The £100 for metal in the market translates into approximately £62 for concentrate at the minehead and the company would have to get its costs down to £35 to reach the profit target of £100,000. Even if the costs were reduced to zero, the total profit over the lifetime of the mine under the free market desired by Frisby would be less than that secured by the ITC.[109] The actual experience of Pahang suggests that Frisby was simply carried away by an irrational desire to smash the ITC. Taking his claims seriously reflected an even wider irrationality.

Fermor Report

The experience of unlimited production of 1937 forced the FMS government to recognize that it could no long administer tin restriction on the basis of the framework established in 1931 and that it also had to be prepared for the termination of the ITC. It also had to face the fact that the Mines Department was not in very capable hands. The current head was considered to be 'neither very strong, nor very intelligent', and lacked 'vision as regards the future and coherence as regards the present'.[110] Fortunately Harris was about to retire and Thomas saw this as an opportunity to bring in someone of real ability. Without anyone suitable from inside the Department, Thomas turned to the Colonial Office for advice and it passed on a recommendation from the Board of Trade that Sir Lewis Fermor pay a visit to Malaya with a view to becoming a permanent part time consultant.

Fermor was a geologist with the Indian Geological Survey but his assumed mandate went far beyond the sphere of his professional competence. As he arrived in Malaya, he found himself embroiled in the disputes that still continued about the buffer stock, standard tonnages and the principles of tin restriction. Almost as intense were complaints about the restrictions imposed by the Drainage and Irrigation Department on mining operations.[111] All these issues, together with those that had justified his fee of £1,500, had to be considered within the initial six months of his contract. Even that had to be cut short by an illness in early 1939. It is, therefore, not surprising that the final report proved controversial.

The overall approach adopted by Fermor was designed to strengthen the role of mining as the mainstay of the Malayan economy.[112] The first step was to absolve the industry of the 'undeserved obloquy' it had received for its apparently modest contribution to environmental damage.[113] Setting environmental criteria aside allowed Fermor to argue that the overriding consideration in the utilization of land should be its potential in generating revenue from the unexploited tin deposits.[114] With reserves of around a million tons in land already alienated, the industry only had a life of 16–20 years at current production rates. Expanding those reserves was therefore imperative. Several specific recommendations flowed from this position. Some tried to curb the power of the Drainage and Irrigation Department. Its ability to insert conditions into mining leases and establish other expensive constraints on mining operations was to be severely limited. Conflicts between the Departments of Mines and Drainage which had been resolved at the political level by the Resident of the state were now to be settled by a technical expert, preferably drawn from the FMS Geological Survey. Above all, there should be a much more extensive pattern of prospecting and that would involve making available lands hitherto reserved for various purposes and limiting the power of the officers in control of them, power which had rendered much of the prospecting effort undertaken in response to the new regulations of 1937 'largely infructuous'.[115] Development would therefore involve changing a whole pattern of administration and it is not surprising that the report was welcomed by the industry[116] but condemned by those who had other administrative priorities.[117] Particularly offended was the Sultan of Perak who took issue with the assumption that the resource belonged to the state, rather than the people, which allowed Malays to be dispossessed from their land without any provision for their incorporation within the mining industry.[118]

Fermor also felt obliged to tackle many of the issues that had arisen out of tin restriction. Some were purely domestic but others concerned the operation of the ITC and Malaya's role within it. Unfortunately, he was not restrained by the lack of any relevant expertise. There were two key domestic issues and both concerned assessments. The fact that existing holders of assessments were becoming a 'hereditary tribe', and blocking new entrants had been causing some concern since 1935.[119] Here, Fermor recommended that the assessments belonging to leases that were exhausted should not be automatically transferred to another lease held by the same lessee but instead put into a pool, from which a portion could be allocated to new entrants.[120] No consideration was given either to the principles of allocation between existing and new miners, or to the desirability of expanding the number of different mining operations. Nor was any attention given to the real problem of the one-pikul miners who were finding it very difficult to sustain their operations.

The second key domestic issue was that of eliminating obsolete assessments and while the recommendation was obvious, that there should be a reassessment *de novo* prior to the renewal of the fourth agreement, the basis for the new assessments was poorly thought out. They were to be based on a prospective estimation of the total production over the life of a new agreement, assuming a 100 per cent quota. Not only did such a method involve several arbitrary assumptions about a mine's capacity but it could hardly be applied to the many Chinese miners who had yet to secure access to leases covering such a period.

At the international level there were two policy questions: should Malaya enter into a fourth agreement and if so what should be the basis of the standard tonnages? On the first, Fermor formally recommended that not only should she continue as a member of the ITC but that, if necessary, she should do so on the basis of the existing standards.[121] But at the same time, he repeated the standard Malayan line that 'for a restriction scheme to work equitably, the countries therein must be assessed on an equitable footing', and considered that Malaya could use the results of domestic reassessment as a basis for negotiation, as though her partners would regard the data as uncontaminated by self-interest! In case that argument failed, Fermor turned to historic production trends and concluded that since Malaya had produced over 50 per cent of the world's supply of tin in 1898–1903 she was not responsible for the excess capacity at the end of the 1920s when her share was down to 34 per cent![122] If taken seriously, these musings would make negotiations very difficult.

It is not, therefore, surprising that Fermor's report caused considerable consternation in the Colonial Office. Campbell considered that its economic arguments were 'jejune, defective often, and dangerous in fact'. The extent of the danger, however, was limited as he concluded: 'If Malaya accepts the report, we shall clearly have difficulties; but I have little doubt that control will in fact continue, all the same. I should like Malaya to do its own negotiating next time; if that were practicable.'[123]

War pre-empted the possibility of implementing any of Fermor's recommendations and, although the report was published and widely circulated, the context within which its diagnosis of the position of Malaya and tin restriction could be debated had disappeared. In some respects this was unfortunate. There are many points where Fermor provided ample confirmation of Campbell's scathing dismissal. The most striking is to be found in a brief discussion of the price at which unrestricted production would generate better results for producers. Basic cost data suggested that was around £150 but there was no consideration of the implications for the Chinese, for government revenues, nor indeed of the actual level of profitability by comparison with a price around £215.[124] Otherwise, he tended to undercut the basis of the criticism levelled against the ITC that had been so forcefully pressed by the Cornish. On the question of the buffer stock, he recognized that this had some plausibility and found arguments on which to justify the price range of £200–£230. On the question of the discrepancy between domestic assessments and the standard tonnage, he certainly felt that 1937 had exposed an important gap but that it was not nearly as large as represented by the assessment figures, especially after the modifications of 1938 were taken into account.

There is a considerable amount of incoherence in the report, perhaps as a result of the hasty way in which it was prepared and Fermor certainly lacked the analytic skills necessary to address the issues in a systematic fashion.[125] However, it is a pity that the report was not reworked to find those principles of coherence, since they would have exposed the weaknesses that lay at the heart of the Cornish position and prevented the distortions in the way in which the experience of tin control was perceived in Malaya.

Department of Mines

The superficiality of Fermor's report soon became evident when a full-blown corruption scandal erupted. Officials in the Department of Mines were granted considerable discretion in the way in which they administered the wide range of regulations and few were able to resist the temptation it provided for personal enrichment. While the fact of petty but widespread corruption was generally known, it was not until 1940 that the full extent emerged. Successful prosecutions were brought against two senior officials, Macdonald, Controller under the Tin Restriction Enactment and Kershaw, Deputy Controller for Perak, and a Commission of Inquiry was struck.[126]

The culture of corruption was rooted in the desire of many miners to remain on friendly terms with officials and presents were routinely offered without being tied to any particular favour. An investigation of the books of the Perak miners revealed that over 70 per cent were making regular payments to officials in the Department. Senior officials reinforced this culture by offering their cars and household furniture at auction as they were about to depart on furlough and the winning bidder expected to stand in their good graces on their return. Even a figure as senior as Choo Kia Peng, who was a member of the central committee regulating tin restriction and an unofficial member of the Federal Council and therefore presumably familiar with the norms governing public servants, complacently admitted that he gave one official S$2,000 (£230) since every miner had 'to oil the wheel before it would move smoothly'.[127]

Restriction simply added to the opportunities for abuse. Much of the evidence in this area was heard in camera but one particularly compromising episode was reported by David. Although he had made a 'free gift' of shares in Teja Malaya to Macdonald, he had been unable to get an adequate assessment for his Kuyoh dredge and the company went into liquidation.[128] Macdonald then advised Coates, who subsequently became the local head of Anglo-Oriental, to take a closer look at the property. Tambah Tin Syndicate was then formed in 1935 between Coates and two Miles brothers, with a capital of S$46,000, of which S$5,000 represented vendor shares to Coates who promptly presented Macdonald with 3,000 shares as a reward for his advice. Four years later, Macdonald sold back the shares at S$7,500. On the formation of Tambah the old Kuyoh assessment of 1,200 was raised to 2,500 pikuls.[129]

It is impossible to determine the extent to which widespread corruption actually shaped the implementation of restriction. The Commission confined itself to making a set of recommendations which would dilute the power of the mines officials on the Central Committee. The Committee would be chaired by a member of the Malayan Civil Service and include some independent lay members. State committees would be revitalized and recover their responsibilities in the area of assessments, with clear appeal procedures from their decisions.[130] However, the generally close relationship between officials and Chinese miners would have reinforced the reluctance to undertake a comprehensive revision of assessments. In the light of the advances in dredging technology such a revision would have seen an even greater shift towards the European sector and a corresponding reduction in the rentier income that many Chinese enjoyed.

Siam

The political changes taking place in Siam went far beyond the formation of a state prepared to take a much more aggressive stance vis-à-vis her partners on the ITC. It also created one which was prepared to intervene much more actively in shaping the future development of the industry. No longer would Siamese mining policy follow the pattern set in Malaya.

In March 1933 the first break occurred with the reopening of prospecting and a relaxation of the restrictions on the alienation of mining land.[131] Soon new producers would be allowed to enter the industry and that gave the government considerable power over the pattern of expansion. Next came an increase in the fiscal pressure on the industry. Import duties were raised and additional taxes imposed on salaries and dividends and this was followed by an increase in the scale of royalties by 30 per cent in February 1934.[132]

Overall, the policy of the new regime was to encourage the growth of the local mining sector.[133] Dulang washers were granted a fixed quota of 36 kg of concentrate per month, or about twice the level of their Malayan counterparts,[134] with the result that around 6 per cent of Siamese production came from this source. Otherwise, the growth of the local mining sector saw a number of new miners operating gravel pumps and the first dredging operation under Sino-Siamese control.

One feature of the pre-revolutionary regime was exacerbated and that was the taint of corruption surrounding governmental decisions.[135] Many new leases were granted, less with a view to exploitation by their owners and more with a view to selling their quota rights and sharing the proceeds with the officials.[136] Following the renewal of the third agreement,

Table 13.14 Siam, tin production, 1929–1939

	Total *tons*	*Dredge* *tons*	*Dredge* *%*	*Other* *tons*	*Assessment* *tons*
1929	10,516	5,199	49	5,318	
1930	11,526	6,425	56	5,101	
1931	12,493	8,076	65	4,417	
1932	9,266	5,731	62	3,535	15,071
1933	10,274	6,311	61	3,964	
1934	10,650	5,927	56	4,723	
1935	9,744	5,454	56	4,291	18,760
1936	12,731	7,716	61	5,015	18,608
1937	15,447	9,751	63	5,696	19,078
1938	14,817	8,918	60	5,900	
1939	15,444	9,420	61	6,025	

Sources: *The Record, Bangkok Times.*

restrictions on the alienation of mining land were re-imposed,[137] though not on government itself. The Ministry of Defence secured its own quota, much of which was used to support a new mine operated with convict labour.

The result of this reorientation of the industry was to reduce the dominance of the European dredging companies as is evident from Table 13.14.

Most of the growth within the dredging sector was simply a realization of the capacity already in place by 1931. Austral-Malay's Pungah was one of the few companies to undertake extensive improvements and there productive capacity was effectively doubled. Of the new companies floated, only the Tronoh subsidiary, Phuket Tin Dredging, started with a large dredge in 1934 but otherwise the few fresh dredges that appeared were obsolescent units shipped from the FMS. The dredge worked by Tronoh's other subsidiary, Tin Lay, and the one sold by Ipoh Tin Dredging to a Sino-Siamese miner in 1936, were of this kind. Siam therefore saw none of the investment in really large modern dredges that occurred in the FMS and the NEI.

There was one important technical innovation which offered considerable promise and that was the development of a lode mine in Yala.[138] This was thought to have an extensive but complicated, ore body containing some 60,000 tons of metal.[139] Exploiting this deposit required the application of a new technology, one which would produce metal locally through an electrolytic process.[140]

The operating company was Siamese Tin Mines, wholly owned by British American Tin Mines, in which the promoter, P. K. Horner, brought together a variety of existing interests: British, such as Anglo-French, Simms and Broadbridge, and American, such as Pardners Mines and Easley.[141] However, it had some difficulty in raising all the capital required and in 1934 it issued £100,000 in debentures. While this was sufficient to enable production to start in 1935, it was not enough to bring the project to completion. In 1938 most of the shares in British American Tin Mines were sold to a consortium made up of BTIC, Anglo-Oriental and Billiton which then arranged for a further £200,000 in debentures to be subscribed by CTS.[142] As a condition, CTS required that Siamese Tin cancel its contract with STC for the refining of the metal into Straits tin, which in future would be treated by ESC in Penang.[143] If the mine had been able to realize its anticipated annual production of 1,000 tons at the cost of £100 per ton of metal, it could have matched the kind of technical innovation taking place elsewhere in dredging. However, its development received a major setback when the invading Japanese herded the engineers into a hut and murdered them with grenades.

While there was little other technical innovation, there were significant changes to the structure of the dredging industry. In 1932 the King-Munro group sold its dredges at Huey Yot and Haad Yai to Anglo-Siamese Tin Syndicate, which was yet another joint venture by Anglo-Oriental and Billiton.[144] When Anglo-Oriental assumed management responsibility for Kamunting in the FMS it also acquired its Siamese subsidiary, Pangnga River Tin Concessions. But the most important change came with the acquisition of the flagship Australian company, Tongkah Harbour, in 1935 which eventually absorbed Anglo-Oriental's original flotation, Talerng. While this turned Anglo-Oriental into one of the major groups operating in Siam, it was far from the dominant position it held in Malaya. However, it did displace the Australians from the position they had held since the inception of dredging in Siam.

The shifts in the balance between the various groups is illustrated in Table 13.15.

The obsolescence of the existing schedule of assessments and the allotment of fresh assessments 'to new producers whose output capacity is practically nil',[145] meant that there was no sound administrative basis on which to ensure that Siam could meet the higher quotas for 1937. While a complete reassessment was clearly required, the government decided on a rather curious method of accomplishing it. Instead of allowing unlimited production for the first six months, the government retained 28 per cent of the international quota and the remaining 72 per cent only permitted producers to work at 66.5 per cent of their current assessments. Once productive capacity had been determined on this basis, most of the retained quota would be distributed and producers would then catch up to their new overall entitlements.[146]

Turning on the production tap was no easier in Siam than elsewhere and deferring the point at which it would be opened wide meant that Siam would not be able to produce its

Table 13.15 Siam, dredging groups, 1931–1941

	1931	*1941*
	No. of dredges	*No. of dredges*
United Kingdom		
Anglo-Oriental	1	6
Kamunting	2	To AO
Siamese Tin Syndicate	6	6
Malayan & General	2	2
Tronoh	0	2
BMTS	1	0
Total	12	16
Australia		
King–Munro	2	To AO
Tongkah Harbour	4	To AO
Austral–Malay	2	2
Burma–Malay	2	2
Kerry	5	5
Palfreyman	9	8
Total	24	17
Other	1	7
Total	37	40

entitlement. For the first quarter she was producing at 84 per cent of her entitlement and that proportion dropped to 72 per cent for the third quarter. While production recovered in the fourth quarter it was still short of what she should have produced even on a 100 per cent basis. As with her partners, the extent of Siam's underproduction was masked by running down minehead stocks but the delay in raising production cost the Treasury some £177,000. Worse, it could cost Siam the standard tonnage she had risked so much to gain.

Crosby commented on the lack of wisdom in the Siamese approach to mining issues:

> Not to be supposed that the arrangements were made with the deliberate intention of inflicting injustice … they afford yet another example of that ineptitude which is so characteristic of the present day Siamese in administrative questions, and which makes one despair of their ever being able to run their national household efficiently once they have carried out their ultimate intention of dispensing with Western assistance and advice.[147]

Had he but come to that judgement six months earlier, he might have overcome his own ineptitude that so badly distorted negotiations for the renewal of the agreement.

Nigeria

The initial economic response of the Nigerian industry to restriction was to mothball much of the freshly installed machinery and revert to hand-working.[148] This was a function of the economics of scale, the collapse in African living standards[149] and perhaps the desire to keep the essential elements of the labour force intact. Initial changes in the structure to the industry saw a substantial reduction in the number of small independent miners and companies and some consolidation of existing ones. Over the period from 1930 to 1934 eight companies in the Latilla group went through various consolidations to become Gold and Base Metals. In 1934 three companies in the Naraguta group were consolidated as Naraguta Tin with a much more appropriate capital structure at 40 per cent of their combined original nominal value.

Anglo-Oriental had reached the limits of its growth in Nigeria by 1930 and the only further development was an extensive programme of investment in 1937 and 1938 in a new dredge and six large draglines for a total cost of over £500,000.[150] This programme was capped by bringing both ATMN and LNTM together as Amalgamated Tin Mines of Nigeria which was completed in 1939. The extent to which Anglo-Oriental and other sectors of the Nigerian industry improved their competitive position is presented in Table 13.16.

Table 13.16 Nigeria, production costs, 1929, 1938

	1929			*1938*		
		£/ ton metal			*£/ ton metal*	
	tons	*Direct*	*All–in*	*tons*	*Direct*	*All–in*
Anglo-Oriental	4,761	110	150	5,566	79	120
Large/medium	3,960	101	141	3,479	90	131
Other				3,337	110	151
Total				12,382	90	131

Source: Nigeria, Mines Department, Annual Reports.

Notes: All–in costs do not appear to include depreciation; large/medium refers to six major companies.

Table 13.17 Nigeria, labour productivity, 1929, 1938

	African labour per ton			*European labour per 1000 tons*		
	1929	*1938*	*Change %*	*1929*	*1938*	*Change %*
Anglo-Oriental	2.54	2.29	−10	13.14	15.85	21
Large/medium	2.75	3.95	44	11.04	12.69	15
Other	6.02	6.14	2	38.36	24.31	−37
Total	3.35	3.67	10	17.9	16.78	−6

Anglo-Oriental was able to bring about a substantial reduction in its working costs by nearly 30 per cent and the six other large and medium companies for which data is available show a much more modest reduction of 11 per cent. The remaining producers may have been marginal in any return to an unrestricted market but the bulk of Nigerian production should have been able to survive.

However, such strength as the industry showed was entirely dependent on the availability of cheap African labour. With the exception of Anglo-Oriental, the overall improvement in cost conditions was not a function of a simple process of mechanization. Table 13.17 indicates the way in which the labour inputs changed:

Mechanization at Anglo-Oriental reduced the need for African labour but expanded the role of European technical and supervisory staff. The reverse occurred in the small sector but the six other large and medium companies saw substantial increases in both. Since the sector reduced its costs it must have been primarily at the expense of the reduction in living standards on the part of its African labour force and that condition would not be easily sustained over the long term.

One other development provided a cushion for at least some producers in Nigeria. Interest on the part of the American military in improving the construction of aircraft fastened on the use of columbium in the manufacture of critical components. Tin mining in Nigeria had produced columbite as a byproduct which was simply discarded. From 1936 shipments took place to the USA and the fact that columbite was both valuable and cheap to recover gave some Nigerian producers a unique advantage.[151] Fortunately for the ITC, the demand for columbite was not so large as to make tin the byproduct and undermine the commitment of the Nigerian tin producers.

The administrative system that had created so many complications during the first agreement proved to be more manageable as quotas rose and many small miners began to exhaust their deposits. Indeed several were unable to take full advantage of the minimum guaranteed by compensating allowances and the amounts released were then redistributed among all producers. When the international quota reached 80 per cent the quota committee expected that a 'considerable amount might be returned for redistribution'.[152] Nigeria now had to worry that it could not produce its full standard tonnage.

With the renewal of the third agreement came a revision of the entire basis of domestic allocation. compensating allowances were converted into admitted claims with a minimum of 20 tons. The position of the small miners was therefore improved, at least as long as the domestic quota did not drop below 60 per cent. Holders of admitted claims under 20 tons had them raised to that level. The much more vexatious question of the status of the special claims was less easily settled. Naraguta still wanted them abolished altogether but the debate within the Quota Committee turned more on the rate at which specials would be converted into ordinary admitted claims and that was eventually settled at 34 per cent. This was the current rate of conversion when the domestic quota was set at 70 per cent. Moving from a sliding scale of conversion meant that the position of holders of special claims would be improved when the domestic quota was above that level but reduced when set below. Freezing the special claims at such a low rate was justified on the grounds that five years of restriction had been sufficient to recoup the fresh investment made in 1929 and 1930 which served as the basis of the special claims.[153]

On this occasion the politics were largely confined to the Quota Committee. Its restructuring in light of the advice of the Maxwell Committee[154] had served to give each group an opportunity to debate its case fully within the confines of the Committee without involving either the Governor in Lagos or the Chamber in London. Although this process had produced a bias towards the small miners and the older companies, it did make provision for reduction in admitted claims should producers be unable to win their quotas.[155] That was at least a gesture towards placing domestic allocation on a basis more in line with actual productive capacity.

Formally, the decisions of the Quota Committee on these issues remained advisory but the Acting Chief Inspector of Mines, Wright, was not prepared to press any position of his own. He was too conscious of the fact that 'We are still heirs to the dissension and bitterness caused by assessment of potential production from development in 1930 …' to be prepared to undertake the task of comprehensive reassessment of all producers. While this 'would be very excellent if it were practicable', he felt that: 'The assessing authority would require the wisdom of Solomon and the guile of a serpent to arrive at a fair decision on many of the claims which would be put forward.'[156]

This confession of administrative weakness in the face of tin politics is understandable in light of the difficulties encountered by his predecessor. But at some point such a reassessment would become imperative and Wright failed to lay the groundwork for any new system which could accommodate the long term development of the industry.

Such a reassessment was in fact necessary to ensure there would be no repetition of Nigeria's embarrassing failure to produce her quota in 1936. Although Wright recognized that ATMN would have to bear the major responsibility in filling Nigeria's quota, he was more concerned to ensure that 'justice may be done to the claims of the lesser companies'.[157] ATMN tried to get the Quota Committee to change its mind but that only produced a temporizing formula and the Secretary to the Northern Provinces proved unwilling to substitute his own judgement for that of a 'well-constituted' committee. The sacrifice that ATMN and other holders of the special claims had made to buy peace in 1931–1932 now became permanent.

Capitulating to the politics of resentment soon proved to be unwise. In 1937 Nigeria failed to meet her quota by 1,260 tons or nearly 11 per cent of her entitlement. As was the case in Malaya, the level of exports was only reached by running down minehead stocks. When the 1,563 tons of stocks are taken into account, Nigeria was only able to produce 82 per cent of her international standard tonnage, or 63 per cent of the total claims admitted in 1931. Such a figure has to be further discounted by 725 tons which came not from the facilities that supported those claims but from fresh investment stimulated by the high quotas of 1937. That programme experienced considerable delays since Nigerian miners found themselves at a considerable disadvantage in competing with the armament programme for equipment but when it was complete a further 3,900 tons of fresh capacity had emerged.[158] Nigeria's overall standard tonnage was therefore not jeopardized by the poor performance of some of her established producers and the case for a complete domestic reassessment had become even more compelling.

While there was no willingness to revisit the principles of domestic allocation, the Quota Committee reviewed the experience of nine companies and 14 independent miners who accounted for 1,064 tons of the shortfall with a view to reducing their admitted claims.[159] Each had to give reasons and although they often included factors outside the control of the miner, such as lack of labour and rain. They also revealed some serious weaknesses, especially among independent miners who found it difficult to make satisfactory arrangements for their European leaves.[160]

Of the delinquent companies only one, Ex-Lands, had previously held substantial special claims and they were, therefore, drawn disproportionately from those to whom 'justice' had been granted at the end of 1936. As the shortfall became evident, companies were authorized to export in excess of their initial quotas and ATMN contributed its share with an additional 112 tons. Since the company's production was 75 per cent of its previous total of special and ordinary claims, it confirmed the position of those who supposed that the special claims were largely inflated. However, it also indicated that the Quota Committee had been quite ungenerous in recommending the conversion rate of 34 per cent when a strict principle of equity in relation to performance would have produced 50 per cent, a level even higher than that pressed by the company itself.

In early 1939 the Mines Department recognized that the negotiations for the fourth agreement would open up the question of standard tonnages and was anxious to establish claims based on potential production at as high a level as possible. With the current quota at 35 per cent, pressure was being felt by two categories of producers and it was feared they might be forced out altogether.

Over the decade many producers had moved to lower grade ground, so that whereas in 1929 each acre had generated 154 lbs of concentrate, by 1937 this had dropped to 131 lbs.[161] Since the government had to share royalties with the United Africa Company, Nigerian producers had paid lower royalties but higher ground rents. As the grade of ground declined these rents were considered to be a serious obstacle to maximizing production since they reduced the profitability of those who worked low grade deposits. It was therefore proposed to reduce rents to the lowest level that would prevent the tying up of land and increase royalties.

The second category overlapped with the first and that was the independent miners. With the current levels of restriction those with the minimum admitted claim were only entitled to produce half a ton of concentrate per month 'which does not secure them a living wage'.[162] Since their productive capacity was still needed to support Nigeria's international claim, it was also proposed to revert to the 1931 system of a basic quota of 12 tons per annum. The calculation of living wage was based on the assumption that since these miners were mainly skilled engineers they could draw an annual salary of between £600 and £900. In 1938 they had received a net income of between £300 and £400 and 'therefore can fairly be considered to have worked at a loss!'[163] That bias towards the independent miners would again have to be at the expense of Anglo-Oriental. However, on this occasion there was a much stronger case.

Taken together, ATMN and LNTM held land that was far more productive than the rest of Nigeria. It was estimated at 234 lbs per acre, while the remainder worked ground that only produced 94 lbs. Changing the ratio between rents and royalties would therefore increase their costs, from £20 to £24 per ton metal. Since both the working and rent/royalty costs were lower than the Nigerian average, the companies could easily absorb this transition to a more equitable set of fiscal principles. Nonetheless it was expected that there would be considerable opposition.

Although the outbreak of war prevented the implementation of this new system, it indicated that the Nigerian government was now prepared to take the lead 'in the interest of the development of Nigeria's mineral wealth'.[164] There was now no trace of the sentiments that shaped the attitudes of previous administrations. There was no contempt for the industry by virtue of the corruption that surrounded its birth; no suspicion of the ITC; no lingering desire to return to a free market; above all, no assumption that the task of government was simply to facilitate collective action on the part of the producers and bottle up their conflicts as smoothly as possible. One factor prompting this change of attitude was a recognition that the government had been too generous in reducing rents at the beginning of the depression which had cost it over £250,000. But the most important was the recognition that the ITC was likely to become a permanent institution. If Nigeria were to derive maximum benefit from its tin industry under such conditions, the government had to play a much more active role.

Anglo-Oriental/London Tin Corporation

Anglo-Oriental's programme of further investment in Malaya and Nigeria was undertaken following a complete reorganization of the parent finance houses. Howeson had become a liability after the pepper debacle and in March 1935 he resigned from AOMC and LTC. Given the unstable nature of the overall commodity market, especially after the House of Commons debate, the Governor of the Bank of England, Montagu Norman, offered his full support in shoring up the central element of the Anglo-Oriental empire. The first step was to reconstitute the Boards of both companies, eliminating those who could be identified with Howeson and that placed Lyttleton and three of his friends in control.[165]

Lyttleton had already realized that the finances of both companies were in disorder but he was not prepared for the extent of the mess that awaited him. LTC had failed to keep proper track of transactions in the tin pools, engaged in some quite creative accounting and had a portfolio of 59 very miscellaneous companies under its control. However, a preliminary review came to the judgement that the major problem was with AOMC, rather than LTC and that not only was there a 'solid foundation on which future business can be built', but that it was 'desirable (if not essential) from the point of view of the tin industry that both companies should continue'.[166]

A more comprehensive evaluation of all the subsidiary companies found that many had outlived their usefulness. Most of the finance and prospecting companies were simply wound up, as were the inactive operating companies. A few, including the original flagship Tavoy Tin Dredging, were to be reconstructed but the bulk were sufficiently successful to warrant rebuilding LTC.[167] The whole process took place under Norman's scrutiny, who served notice that no mistakes would be tolerated.[168] The new LTC would therefore shed any lingering suspicion inherited from the Howeson years. Once the overall fate of the two companies was clear, Norman put an accountant, Bunbury, in charge to complete the reorganization.

Amalgamation of AOMC and LTC proved to be a complicated process, since it involved assessing the comparative claims of both preference and ordinary shareholders in the two companies.[169] It was further complicated by the inclusion of a third company, Metal Securities. This had been created by Howeson in 1934 and was capitalized at £800,000, drawn mainly from LTC, BMC and Dean Finance. Metal Securities was a vehicle through which a further £250,000 was invested in AOMC and which also held some of the shares recently acquired from the Prattens. Once the cross-shareholdings had been eliminated, the valuation of the assets left the position of the LTC shareholders more or less intact but severely reduced the claims of those in AOMC and Metal Securities. When the new LTC emerged, the original nominal capitalization (net of cross-shareholdings) of the three companies had been reduced from £4.7 to £3.6 million. With this water squeezed out of the capital, LTC could then return to the dividend lists. Since the major asset of AOMC was AOGIT, the management and technical support side of the group continued unchanged.

Bunbury managed to deflect one initiative from Billiton which wanted to become part of the new LTC and secure representation on its Board. It was blocked on the grounds that it would compromise the effective functioning of the ITC. Linking the two major companies in two of the most important countries could generate criticism from both shareholders and political quarters and Bunbury was able to persuade van den Broek that a 'sturdy independence was a necessary prerequisite for the satisfactory functioning of the International Tin Restriction Scheme'.[170]

Once Bunbury had 'brought reasonable order out of complete chaos'[171] he was ready to return to his old firm. Even though Howeson was now in prison, there was a strong movement afoot to restore him. Given the uncertainties around the renewal of tin control in June 1936, many felt that his leadership was still required both to bring negotiations to a successful conclusion and to preserve British dominance.[172] Neither Campbell nor Norman could countenance such a proposal[173] and Norman began to look for a strong name with which to block Howeson. Norman's encounter with tin politics was brief but sufficient to recognize the most important qualification for the position as he commented: 'Of course, this is no job for a fellow who cannot play poker and argufy all day and through the early hours of the next morning!'[174] The appointment then went to Lyttleton.

With Lyttleton as Chairman of both LTC and BTIC, the question of the relationship between the two companies arose, especially since the reconstruction of LTC had expanded BTIC's interest.[175] A proposal to amalgamate the two would cement the dominant position of LTC and incidentally dilute Patiño's influence in BTIC. While Patiño supported the principle of amalgamation, he objected to the conditions. Lyttleton was fearful that too visible a Patiño presence in LTC would upset Malaya, so he refused to allow Patiño a place on the Board and wanted him to limit his participation to a maximum of 25 per cent of the equity. When Patiño desisted, Lyttleton approached Norman to intervene with Patiño. Campbell was very nervous that Patiño could not be trusted to keep any such overture confidential and, while that would not seriously affect the future of tin control, it would certainly reinforce the antagonism in Malaya that was causing so much trouble for the Colonial Office.[176] That reinforced Norman's own apprehensions and the matter was dropped. Without the need to navigate in the contentious waters of tin politics, LTC would have become considerably larger and the core financial alliances that much firmer.

Netherlands East Indies

Whereas tin restriction facilitated technical and structural changes that were already underway in Malaya, in the NEI it served as a powerful stimulus. Dutch participation in all the negotiations leading to the agreement had been undertaken under the leadership of Billiton. Although Banka was considerably larger and stronger, it was based in Batavia and hence was excluded from the formation of the strategic alliances that lay behind the ITC. The Dutch, therefore, found themselves in a more complicated position than did any of the other delegations. In order to negotiate with the others, they first had to complete a set of negotiations between The Hague and Batavia. While London found itself in a parallel position in dealing with Kuala Lumpur, they could always overrule the colonial authorities on the grounds that they were too much under the influence of local politics. But The Hague could not simply disregard Batavia when it directly controlled the largest sector of its industry.[177]

This division between the metropolis and the colony overlay two others. Billiton, especially following the construction of the Arnhem smelter, was in close touch with the tin market. Banka sold in Batavia which made it more remote. More important was the fact that, as a government department, Banka was considered to be much less efficient since its economic position was based solely on the richness of the ground that it worked. In July 1932, Billiton proposed a dramatic reorganization of the entire industry, the centrepiece of which was a single organization following the form of the joint venture already adopted when the company was reorganized in 1924.

The proposal for fusion of all the Dutch producers identified three major advantages. The emergence of a political dimension to the industry had shifted its centre of gravity to Europe and a single organization based in The Hague would find it much easier to negotiate with Howeson in London and Patiño in Paris. A single selling organization would not only realize some savings but also generate a more informed assessment of the state of the market. Finally, and most important, it claimed that, freed from direct government control, considerable efficiency gains could be realized on Banka.[178]

One element of the fusion plan could be put into place without difficulty and in 1933 Billiton took over Singkep.[179] However, the proposal ran up against some deep prejudices held by the head of Banka. De Iongh had little but contempt for both parts of the Billiton group and was not prepared to admit that a change of management would generate an increase in profits, a position largely endorsed by the Governor-General. However, the Billiton case received full support from Colijn and it was soon strengthened.

Billiton management brought about immediate savings on Singkep, by nearly 30 per cent, thanks in part to selling in Europe and smelting the concentrates at Arnhem rather than Singapore. In addition, it was clear that the NEI would have to meet stronger external competition. Not only was there concern about the implications of the rise of the Belgian Congo but Anglo-Oriental now had to be treated with much greater respect. In 1934 a Billiton delegation visited Batavia to discuss the fusion proposal and it found considerable interest on the part of the new engineers. Some members went on to Penang to see ESC's smelter and returned via Kuala Lumpur where they met with Anglo-Oriental. The visit made a profound impression. The strength being shown by Anglo-Oriental as a result of the consolidation process underway shook the complacent assumption that Banka could easily be victorious in any resumption of a battle for the survival of the fittest. With the prospect of an end to restriction, Colijn increased the pressure on Batavia for an immediate decision.

The context was now set in which the technical, financial and administrative details of the proposed fusion could be addressed. Central to any assessment of the financial position of the two companies was an estimation of their comparative ore reserves. This was complicated by a number of factors, including the differential implication of various rates of quota and prices. Once the work of a financial and technical commission from both parties was complete, a detailed proposal was circulated. It was not well received. The Council of the Indies advised against and not only was the technical detail subject to criticism but also the underpinning assumptions about the desirability of maintaining restriction and moving from public to private enterprise.[180]

Following the renewal of the third agreement, the issue was pressed yet again. On this occasion there was greater government support but the debate in the Volksraad was unable to get much beyond the symbolic implications of losing colonial control to a metropolitan

Table 13.18 Banka Tinwinning, 1930–1939

	Production	*Dredge*	*Costs/metal*		*Grade*
	quintals	*%*	*f/qtal*	*£/ton*	*lbs/cuyd*
1930	219,428	15.3	109.2	88.8	1.46
1933	80,640		117.0	110.2	1.29
1934	116,060		74.2	95.0	1.53
1937	221,424		49.1	52.5	1.46
1938	182,007	28.7	63.2	67.2	1.22
1939	180,846	33.3	62.8	75.1	1.21

Source: A. L. ter Braake, "Rationalisatie der Banka Tinwinning," *Koloniale Studiën*, vol. 25, no. 2, 1941, p. 281.

Note: costs include depreciation and 6% return on capital.

company. It was prepared to concede privatization and to encourage co-operation with Billiton in the area of selling but not the central principle of fusion.[181] As the expiry of the agreement loomed larger, further negotiations were undertaken in 1939 but were suspended with the outbreak of war.[182] Restriction had simply suspended competition at one level, only to replace it at another. Always present was the need to prepare for the possible demise of the ITC and hence to strengthen the position in an open market. Countering this inexorable economic logic was the complacent colonial sentiment that was always nervous about surrendering power to those who claimed new responsibilities. The Hague had proved even less capable than London in its ability to resolve this tension.

Although the marriage between Banka and Billiton was not solemnized, a platonic relationship developed.[183] In 1933 two Banka engineers were seconded to Billiton to study working practices and Billiton lent two of its dredges to Banka. The results were quite dramatic and are presented in Table 13.18.

While much of the reduction in costs in 1934 is a function of the increased scale of production and the treatment of higher grade ground, the comparison between 1930 and 1937, which controls for both variables, indicates the extent of the improvement which can be attributed to new production techniques.

Much poorer ground on Billiton meant that it did not get its costs down to a level where they matched those of Banka. However, significant reductions were achieved and are attributed to the increase in dredging. They are presented in Table 13.19.

Table 13.19 G.M.Billiton, 1929–1939

	Production	*Dredge*	*Costs/metal*		*Grade*
	quintals	*%*	*f/qtal*	*£/ton*	*lbs/cuyd*
1929–1930	132,693	17.0	152.85	124.3	
1932–1933	45,960	32.4	137.81	129.8	
1933–1934	52,071	27.5	138.69	167.8	
1936–1937	121,765	34.9	75.04	80.2	
1937–1938	119,948	49.4	81.32	84.5	
1938–1939	96,911	44.0	86.29	103.2	.63

Source: GMB, Jaarsverslag, 1939.

Note: metal costs include depreciation but not 6% return.

By 1938 the number of dredges had increased from seven to 12 and in 1939 two larger dredges were introduced.[184] With another dredge built at Singkep in 1937,[185] Billiton was able to keep up with the technical pace set by Anglo-Oriental in Malaya.

Bolivia

The rhythm of development in Bolivia followed a very different course from that elsewhere. Tin restriction had caught the industry at the point where it was undergoing a major shift as a result of the decline of her premier mine at Llallagua. Hochschild paced out one developmental path which was to reinvigorate Unificada and expand Colquiri.[186] Patiño took another, which was to make Llallagua even more efficient.[187]

These developments had to take place within a new financial and fiscal framework. In 1936 the boliviano became inconvertible as the government saw control over foreign exchange as a mechanism of resource transfer, to itself to meet its own fiscal needs and to the urban population in the form of cheap imports of food and raw materials.[188] It is only possible to examine the impact of the imposition of exchange control for PME. In order to examine the development since 1929 Table 13.20 presents two sets of accounts, at the actual exchange rate prevailing in 1938 and at a realistic hypothetical rate.

Actual costs to the company rose considerably but that is a function of two separate processes. Lower production levels meant higher depreciation charges [189] Compulsory delivery of foreign exchange raised the immediate cost of placing metal on the market by 14 per cent. Under a different fiscal regime costs under the company's control would have declined by 8 per cent which demonstrates its underlying strength, one that would preserve its overall competitive position.

Neither of Hochschild's two major projects turned out to be very successful. Colquiri was a new mine but relied on existing technology to process a high grade deposit. Unificada required a new technology to process very low grade deposits in order to become viable. Hochschild certainly responded to the technical imperative and invested over $2 million in a new process which was developed in a laboratory in Baltimore but failed when applied at an altitude of over 4,000 metres.[190] The earliest comparative data available is for 1945 and that shows the direct costs at each of these mines were around 30 per cent higher than at Llallagua.[191] Without an entire reorganization of the fiscal and legislative framework within which the tin industry operated in Bolivia, it is most unlikely that Hochschild and many other mines which were under less capable technical management could have survived the collapse of the ITC.

Table 13.20 Patiño Mines & Enterprises costs, 1929, 1938

	1929	*1938*	
	£/ton	*£/ton @76 bos/£*	*£/ton @120 bos/£*
Mining costs	88	97	81
Realization	31	34	34
Total	119	131	115
Depreciation	19	41	40
Taxes	11	17	13
Total	149	189	168

Source: PME, Annual Reports, 1929, 1938.

Notes: 1929 production 21,600 tons, 1938, 7,700; Taxes on imports included in mining costs.

Table 13.21 Major world groups, 1931–1940

	1931	*1940*	*Change 1931–40*	
	tons	*tons*	*tons*	*%*
Anglo-Oriental	18,190	28,860	10,670	59
Banka	21,595	23,230	1,635	8
Patiño	26,920	21,050	–5,870	–22
Billiton	13.430	19,320	5,890	44
Redruth/Tronoh	15,645	17,065	1,420	9

Note: Patiño is 1941.

Major international groups

The overall pattern of development under conditions of restriction had changed the relationship between the major groups that had been active in the initial drive towards restriction in quite a dramatic fashion. Their comparative position is presented in Table 13.21.

Whereas Patiño was the largest group, by a considerable margin, in the early 1930s, a decade later it had declined to third place. Its position was now taken by Anglo-Oriental. Only Billiton showed a comparable growth rate; the two groups that had taken the primary responsibility for the formation of the ITC had clearly flourished under its auspices.

Smelting

Against the background of a major shift in the organization of mining, a similar shift occurred in that of smelting. As Anglo-Oriental expanded in Malaya and Siam, concentrates were diverted towards ESC and away from STC. In 1933 Billiton expanded its smelter at Arnhem, drying up most of the supply of NEI concentrates that had hitherto gone to STC. The following year the whole market for Straits tin came under pressure as Williams, Harvey raised its grade of English refined to 99.9 per cent which considerably increased its competitiveness in the United States.[192]

STC had ample financial resources and fought back in two ways. One was to form British Tin Smelting (BTS) to build a new smelter at Litherland which became operational in 1937. Although it was hoped to offer some competition for Nigerian concentrates with CTS, it relied mainly on supplies from East Africa. One minor source was secured by investing in the Kyerwa Syndicate which operated a mine in Tanganyika.[193] By 1938 BTS produced around 5,000 tons[194] but any further growth was pre-empted by the war.[195] The second was to move beyond a passive investment in the equity of mining companies into more active promotion along the lines of a conventional mining finance house. In 1938 Straits Mining and Finance Corporation was formed as the overall financial holding company, followed by General Mining and Agency to undertake its managerial and secretarial side. Under its auspices several properties were acquired in both Malaya and Siam from which four small operating companies emerged.[196]

None of these developments served to re-establish the position of STC as 'the leading tin smelter in the world'.[197] In 1929 it was over twice the size of ESC; by 1938 ESC was nearly one and a half times the size of STC.[198] Its international competitive position deteriorated to an even greater extent. In 1929 it had smelted nearly a third of the world's tin; by 1938 this proportion was reduced to 18 per cent. Where STC had been 20 per cent larger than the CTS-Arnhem smelting complex, by 1938, even including BTS, it was only just over half the size of its rival.

The experience of STC largely paralleled that of the Cornish-Malayan interests with which it was so closely aligned. They were both financially very strong and enjoyed considerable prestige and influence. Both saw themselves being marginalized as a result of the emergence of Anglo-Oriental and Billiton as the most dynamic and dominant forces in the industry. In the ITC they saw an expansion of that power and their resentment and frustration at their marginalization lay at the source of their antagonism toward the ITC itself.

While tin restriction introduced several rigidities, they did not extend to the most important dimensions of the industry, technical innovation and corporate reorganization. The dynamism that had characterized the industry in the 1920s continued through the 1930s.

14 Tin consumption and research

The ITC saw itself as obligated to administer restriction in such a way as to meet the needs of both consumers and producers and its guiding principle was stability, both over the short term in the elimination of price fluctuations and over the long term in moderating the commodity cycle. However, the competitive position of tin in the ultimate commodity market was far less stable. One potential source of that instability was, of course, the ITC's own price policy and its critics often alleged that it was encouraging the search for substitutes. Shoring up the competitive position of the metal, on the other hand, was the responsibility of the ITRDC. These processes of market erosion, expansion and consolidation will be reviewed primarily on the basis of the evidence available from the USA. Not only was this the largest market but it was also the most dynamic and any trends visible here can be extrapolated to provide a picture of the problem facing the ITC.

Competition works at many levels and tin suffered from several disadvantages. Since it came from hard currency sources, in 1934 Germany included tin among the base metals whose importation was to be severely restricted.[1] In developing a metallurgical technology based on domestic aluminium, Germany not only limited its consumption of tin but it also served as a demonstration effect for substitutes.[2]

Much of the prejudice against tin in the United States derived from strategic considerations. In late 1932, when the price was still languishing at under half its 1929 level, the American representative of the ITRDC found the Bureau of Mines staff very concerned about the possibility of war in which the interests of the United States and those of Great Britain were opposed, cutting off the supply of metal.[3] Regardless of the existence of the ITC, tin would inevitably operate under a cloud. However, it was not one that cast such a shadow that the demand for tin suffered by comparison with other traditional base metals, all of which were drawn from domestic sources. If the consumption level for 1937–1938 is compared with that reached in 1928, then tin did as well as zinc and considerably better than either copper or lead.[4]

Consumption trends

The consumption figures for tin can be disaggregated into two categories: products which are directly purchased by consumers and other products most of which are simply inputs into various industrial processes. At a strictly economic level, competition operates on two quite different levels: for the final product and for the intermediate product. These processes can be seen most clearly with respect to the position of tinplate in the food container industry.

Tinplate

Tin was fortunate in that the most important end use was tied to the most stable of all sources of final consumer demand, food. However, food in tin cans is generally an inferior product and demand ultimately rested on the limitations on the availability of fresh food.[5] The 1930s saw the development of refrigerated railway wagons which extended northern access to southern fruit and vegetables. Inferiority was also being exposed in some lines, with the development of frozen foods. In others, inferiority was a function of pure prejudice. Glass containers for baby foods appeared to be cleaner and safer and here tin cans were entirely replaced.[6]

However, none of these processes were sufficient to erode the dominance of the tin can. While glass had some advantages in presenting the food contents, its technical disadvantages as a result of weight and breakage confined it to high quality items where presentation was essential to securing a premium price. The only area where there was direct competition between tin and glass was over beer but that was a function of the response of the glass industry to the inroads that tin was making on its traditional market, following the introduction of the beer can in 1935.

As a food container, the tin can suffered from one technical imperfection and that derived from the nature of the tinplate. Bonding the tin to steel to form tinplate had traditionally been undertaken by dipping steel plates into baths of molten tin.[7] It was an imperfect technology since it occasionally left pinprick holes in the finished can. Acidic foods could then dissolve traces of iron through these holes with an adverse effect on their taste. A shift from a hot to cold reduction technology in the steel plates resulted in a much better bonding process and a saving of around 10 per cent in tin.[8] Not only did this improve the can but it transformed the tinplate industry. The capital requirements of the new technology were such that they led to an extreme concentration of production which matched a similarly extreme concentration at the level of manufacture of the cans. As far as the steel manufacturers were concerned, tinplate was their most reliable and profitable line and there was, therefore, a very substantial commitment to research that would sustain the industry's competitiveness. At the same time, the foundations were laid for an entirely different bonding technique based on the electrolytic deposition on a continuous steel strip.[9] This shift from hot-dipped to electrolytic tinning reduced the percentage of tin in the tinplate by two-thirds. While the change was very slow in the 1930s, this eventually became the dominant form of tinplate production. With the commitment of the steel industry to tinplate, tin had a secure future but the extent of it would depend far less on the price of the metal than expansion of demand for cans.

Tin foil

Price competition alone was responsible for the most serious case of market erosion. In 1927–1928 tin foil had consumed an average of 4,600 tons; a decade later this had dropped to only 1,870 tons. Research undertaken by the aluminium industry had come to fruition in a much cheaper foil and although tin could have met the competition, it was only at a price of around £45/ton.[10]

Pewter

The 1930s saw a modest revival of interest in the use of pewter and the most interesting case occurred as a result of an initiative by the manager of Anglo-Oriental in Malaya. Chinese pewtersmiths had manufactured traditional ceremonial artefacts. But as tastes changed the industry was in decline. Hutton advised one family business to take advantage of the low price of tin in the early 1930s and start a new line of manufacture of more practical items which would be attractive to Europeans.[11] That marked the beginning of what became a very successful international enterprise

Intermediate products

A high proportion of intermediate products went into the manufacture of automobiles. Although the tin contents added at most \$2–\$3 to the cost, there was considerable research into ways of reducing usage, with the result that the tin contents of the average car dropped from 6lbs in 1929 to 4lbs by 1940. One area of elimination was well underway by 1932 and that was in the development of a cadmium-based alloy for bearings which was designed to cope with the greater pressure from higher piston speeds. However, it was retarded when the 1937 boom exposed a weakness in the cadmium market where there was an actual shortage and extremely high prices demanded.[12] The other major shift was in the composition of solder and that was certainly affected by comparative prices but here, too, no reasonable tin price could have halted it.

International Tin Research and Development Council

The research side of ITRDC had three major foci: (1) advice; (2) innovation; (3) propaganda and this was the one area where, with one minor exception, where complete harmony prevailed, even as far as the Cornish were concerned. Rich visited the headquarters and reported that, in his opinion, 'extraordinarily good work was being done'.[13]

As soon as the research programme was organized, a technical data base on the applications of tin was constructed. Inquiries from manufacturers were invited and advice was freely offered. This was perhaps the most important function since the technical education of manufacturers in the best uses of the metal increased their level of confidence in existing technologies.[14] This method of strengthening the market was reinforced by the American Tin Trade Association which formed a Tin Research and Development Committee in 1934 to bring together brokers, consumers and foreign producers.[15] That assured the Director of the ITRDC considerable co-operation in making his visits to the United States a success.[16]

Both product and process innovation required specialized laboratories. At first, work was contracted out and much use was made of the Battelle Memorial Institute in Ohio but in 1938 the ITRDC decided to build its own laboratory in Greenford, Middlesex. However, it only became operational in early 1939 and the war would soon deflect the research agenda. There is inevitably a long interval between the formation of a research project, its development in the laboratory and its application to general industrial practice, so there were few areas where the ITRDC could show that it had contributed to the expansion of tin usage, at least not during the life of the ITC.[17] Of more immediate significance was its work in shoring up existing markets by overcoming some of the technical deficiencies in tin-based products such as bronze.[18] However, the long term value of this research programme was clearly evident many years later when it was considered 'one of the most sophisticated on behalf of one metal [which] can claim credit for a number of new applications for tin as well as for process innovation'.[19]

The work of the ITRDC in the above two areas lent itself to the publication of technical reports which were freely distributed.[20] In addition, it published non-technical reports designed to promote a positive image of the metal. Some were of a general promotional character, such as 'Tin and its Uses', and 'Tin and Civilisation'. Others had a more specific focus, in promoting new products such as canned beer, or in countering some of the prejudice about existing ones such as canned foods.

Perhaps as a result of concern about the lack of immediate progress on the technical side, the ITRDC decided to make a more substantial commitment to this general propaganda. A concerted scheme was worked out by Pearce and Baddeley which they put through at a meeting at which Campbell was absent but in the full knowledge that he would be opposed. Campbell considered it a 'complete waste of £5,000' and resigned.[21]

From 1938 Lowinger became Chairman and continued to guide the ITRDC until after the ITC had itself folded. Reducing Campbell's responsibilities may have been wise, since Lowinger was in a position to put more effort into ensuring that as the research programme grew it would generate more tangible results.[22]

Even if the work undertaken by the ITRDC had produced more immediate results, it would have been quite insufficient to outweigh the active research programme into reduction and substitution undertaken by American industry. However, that was well underway by the late 1920s, fuelled by the price boom and fears of an impending tin famine. The fact that tin was the only base metal imported from overseas and considered subject to 'monopolistic manipulation of its production', may well have been a further incentive for American consumers to turn their attention 'toward research for substitutes'.[23] However, even Strauss, no friend to the ITC, reviewed the growth of substitutes and concluded: 'The total effect of these developments and substitutions is not serious; they are mentioned as evidence of a tendency.'[24] While that tendency would gather considerable pace under the pressure of World War II and largely transform the demand side of the postwar industry, it was grounded in factors outside the control of the ITC. Only if the tin industry were prepared to live with a price well under £150 could the ITC have exercised any restraint on this tendency.[25]

15 The International Tin Committee and World War II, 1939–1942

The outbreak of war in Europe reshaped the structure of the international tin market and the ITC had to respond immediately with both instruments at its disposal, the buffer stock and production quotas. Two other issues were soon added to its agenda. The United States was now anxious to end its dependence on that market and that meant direct negotiations both with Bolivia and the ITC. Since any reshaping of the structure of the industry under wartime conditions spelt a potential repetition of the dislocation that followed in the aftermath of World War I, it was imperative to see a renewal of the overall agreement. At least there would be some machinery in place to cope with the anticipated postwar depression. War would not suspend the politics of tin but simply give them a fresh inflexion.

While tin is considered a strategic metal, in the sense that it is indispensable in the production of various goods which are either necessary or useful in the waging of conventional warfare,[1] the experience of the previous war suggested that overall demand is not significantly affected. The general shift from civilian to military production simply reallocated, rather than increased, final use. War affects tin in three more important ways. The first is the reorientation of commercial relationships in light of new political ones. A second is the creation of a new set of administrative controls which determine both price and allocation, domestic and international. The third is stock policy. Tin has to compete with other commodities for shipping space; supply routes become more circuitous and are vulnerable to interception. As a result, consumers must carry much higher levels of stocks.

The tin market went through three quite separate phases as it responded to the impact of war in Europe. The period from September to December 1939 was marked by considerable uncertainty about the implications of the war. From January to the fall of France in June 1940, the ITC was in a position to adopt a coherent policy. By mid-1940 the United States had formulated its own policy in association with the ITC which lasted until the extension of the war to the Pacific at the end of 1941 and that made the ITC itself redundant. These three phases posed quite different problems and will be treated separately.

September–December 1939

Declaration of war had several immediate outcomes. Shipping difficulties on the West Coast of South America meant a suspension of shipments from Bolivia. Controls on the allocation and export of tin were implemented under the authority of the Non-Ferrous Metals Control (NFMC) division of the Ministry of Supply.[2] Finally, tin was included in the embargo on all shipments to German-occupied territories.

To compensate for the shortfall from Bolivia, the Ministry of Supply ordered Nigerian mines to ship all their stocks. Once they were converted to metal and Bolivian shipments resumed, the situation in Britain would rectify itself but in the interim her stock position could only be preserved by severe limitations on exports.[3]

Lyttleton was appointed Controller of NFMC while still remaining a member of the ITC which ensured that the policy of the ITC would be increasingly shaped by the overall objectives of the metropolitan government. The process whereby the ITC would lose much of its autonomy was underway.

The embargo reshaped the structure of the industry. Although the Netherlands remained neutral, Billiton decided to close its Arnhem smelter as a precaution against the requisitioning of its stocks following a possible German invasion.[4] Henceforth, Germany could only rely on small quantities bought in Portugal and Spain and on any shipments from the East via the USSR. Curbing supplies to Germany and Central Europe not only meant a reduction of some 15 per cent in world demand but it also increased the domination of the remaining market by the United States.

During this first phase of the war, the American market operated without any constraints and it immediately panicked. Potential demand had increased as a result of the opportunity to capture fresh export markets; supplies were reduced not only as a result of the British export ban but also by the coincidental curtailment of shipments from China. Above all, American consumers were now very worried about their stock position. That was not simply a function of the desire to carry higher levels of stocks to compensate for any delays or interruptions in supply. It was also a function of the stock policy induced by the programme pursued by the ITC since the formation of the buffer stock in 1938.

The formation of the buffer stock had encouraged consumers to run down their private stocks, offloading their own carrying costs. When the price began to approach the ceiling, all but immediate purchases were suspended since the price could not rise above £230 and was bound to fall below it.[5] By August, visible supplies to the United States had fallen to a total of 8,200 tons, the lowest level since November 1935. In itself, this was far from being a critical figure but if invisible consumer stocks had declined at a sufficiently rapid rate, the situation was potentially quite alarming.[6] What made it even worse was the fact that Mills and Ellinger had full discretion about the way in which the buffer stock would be managed and preferred not to incur higher charges by storing any part of it in the United States.

The commencement of sales from the buffer stock in June had alerted the ITC to the necessity of reviewing the quota. By August, it was clear that although production was formally in excess of authorized quotas, it was now approximating current demand. Campbell then endorsed a Dutch proposal to wipe off the accumulated overexports with a retroactive increase in the quota from 45 per cent to 60 per cent, a level that would be continued for the final quarter.[7] When the communiqué was finally issued on 4 September it appeared as though it was a direct response to the outbreak of war.

This announcement was far from meeting the situation that had arisen in the United States and in New York prices for spot metal rose from 49 c/lb on 1 September to 65 cents a week later. Campbell then proposed a retroactive increase to 80 per cent while still holding the fourth quarter at 60 per cent. This was designed to address the American stock problem with the release of most of the accumulated minehead stocks, while avoiding a production excess over the longer term.[8] A communiqué to that effect was issued on 12 September. The terse language of official communiqués did not serve to communicate the way in which the ITC was reading the state of the market and the decision to continue with a quota of 60 per cent for the fourth quarter struck many Americans as quite unreasonable.[9] At the same time, the

British and Dutch governments informally authorized both Malaya and the NEI to produce freely, at least to the point where the current labour force was utilized fully.[10] That would at least build up minehead stocks.

While metal brokers such as Trench issued statements trying to reassure consumers that the situation was well under control,[11] they were not supported by the AISI's Tin Advisory Committee which urged the ITC to extend the 80 per cent to the fourth quarter. The new American consumer representative, Todd, went even further and argued for unrestricted production. This would not only satisfy the need to 'build up war reserves' but meet the specific demand for Straits tin, even at the expense of the interests of other countries on the ITC.[12] Industry pressure was reinforced by the State Department and Hull added that 'the real consumption of tin should not be the principal factor influencing decisions at this time'.[13]

On 18 September the Ministry of Supply issued a series of control orders setting maximum prices for all non-ferrous metals. In the case of tin, this was the buffer stock ceiling price of £230, with no recognition of the substantial increase in costs borne by the producers under wartime conditions.[14] As long as the control order was in effect, it was illegal to sell tin above £230.[15] At the same time, the Ministry authorized the LME to reopen for dealings in tin[16] though while the control order lasted, the institution was simply a device for the allocation of the metal. The physical metal remaining in the buffer stock was largely exhausted while the LME was open in September and, until December, the metal supplied came mainly from the repurchase by the buffer stock of metal it had previously sold forward.[17]

With the reopening of the LME, the American problem took on a new dimension. There were now two prices, a benchmark world price set by the British government and the ITC and a much higher price that had to be paid for spot metal in New York.[18] That gap between two metal prices was being widened by the gap between two prices for sterling.

On the outbreak of war the value of sterling was officially fixed at \$4.03, a devaluation of 14 per cent. However, that rate only governed approved transactions in London and producers who needed to sell their sterling for dollars could only do so in the free market in New York and there sterling began to sink. Unless the gap between the dollar price in New York and its equivalent to the sterling price in London could be closed, there would be a strong incentive to build a smelter in the United States and divert all the Bolivian supplies. Not only did that pose a serious problem for British wartime economic policy but it threatened to destroy the whole international structure in which Britain occupied such a privileged position and which she hoped to preserve for the postwar period.

Britain moved quickly to try and solve one part of the problem and that was to concede the right of producers to convert a portion of their sterling earnings into gold which could then be sold for dollars at the official rate. However, there was a quid pro quo, a long-term contract with British smelters and most producers preferred to take a wait-and-see attitude. The prospect of an American smelter had become considerably more likely.

The existence of a dual price system created three sets of resentments. The Bolivians resented their exclusion from the higher prices paid in New York. Those shipping to the Straits smelters saw brokers able to pocket the price differential. Americans saw themselves gouged by those able to take advantage of the opportunity to buy in one market and sell in the other.[19] The extent of the divergence in price that remained even after the September panic had eased is presented in Figure 15.1.

The perpetuation of the dual price system even after the initial panic had subsided and after the ITC had reaffirmed that its policy was always to provide the market with the tin it actually needed made the quota decision for the fourth quarter the most difficult of all.

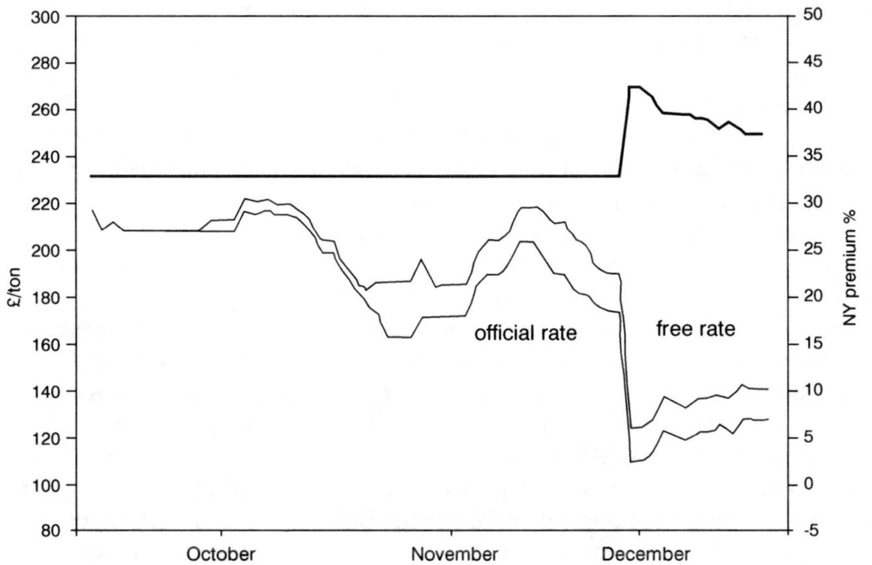

Figure 15.1 Tin prices, October–December 1939

Conceptually, this decision was no different from any other. The problem was that now estimates of consumption and of the normal level of stocks to support it were far less reliable than before. Increasing that uncertainty simply increased the inherent conflict as to who should bear the risks of misjudgement. If such production proved to be in excess of requirements, the immediate result would be excess stocks which consumers would draw down at the expense of demand, forcing a cut in subsequent quota levels and lower prices.[20] Shifting the risks to producers in this way simply meant the prospect of an even sharper roller-coaster than that of 1937–1938. Given the inherent lag in raising production in Bolivia, the market would be further depressed as a result of the overhang of Bolivian entitlements, exacerbated by the degree to which Bolivia's standard tonnage remained unrealistically high.[21] While the cash resources of the buffer stock remained available to modify the full impact of such shifts, the prospective costs to producers were far greater than the costs borne by the consumers.

Once a quota decision had been announced, it could not be easily revised downwards and it set the base for the following quarter. Caution therefore dictated that the quota for the fourth quarter be set low and any revisions required take the form of a retroactive adjustment for the third quarter. At a certain point such adjustments to the third quarter became ineffective, so the same logic dictated that the decision focus on the first quarter of 1940 which would be set low with adjustments to the fourth quarter. Caution also dictated that the decisions be deferred as long as possible to allow for greater information. The vacuum created was inevitably filled by an intense pattern of politics.

By early October some of the uncertainties concerning consumption trends were being reduced. Lyttleton reported that the increase in absorption reflected not just restocking but a 'genuine demand for consumption greatly above the rates we were calculating', and sought the support of the Dutch for an increase in the quotas for the third and fourth quarters to 120 per cent and 70 per cent.[22] The Dutch responded with a statistical memorandum supporting this assessment but neither was sufficient to convince Campbell and Calder. Trench had reported that while the tin was going into consumption, in the sense that it was being transformed into tinplate and other intermediate products, they, in turn, were being stocked rather than passing into final use.[23] In effect, the stock problem remained but in a more intractable form. Additionally, the New York forward market was showing signs of severe weakness, with prices dropping as much as 5 cents below their prewar levels.[24] A rise for the fourth quarter to 65 per cent was the most that could be contemplated but that decision was to be postponed until the situation became much clearer.

The cautious approach adopted in London was not supported by Malaya. While the Singapore price remained fixed, there was no mechanism in place for the allocation of the available metal, parallel to that operated by the LME. The gap between buying orders and offerings therefore became very evident[25] and by early October all minehead stocks had been cleaned out. It was therefore clear that Malayan capacity would have to be expanded to meet the particular demand for Straits and on 5 October the High Commissioner acted unilaterally in announcing that the domestic quota for the fourth quarter would be 60 per cent, equivalent to an international quota of 85 per cent.[26]

Before London had an opportunity to react, it had to confront the fact that the announcement was read as settling the still undetermined fourth quarter. Trench reacted in the strongest of terms. A quota of 85 per cent was far too low in light of consumption trends, the desire of consumers for additional reserves and the entry of the American government into the market, buying for its own account. Since it could not provide much detailed statistical support, it had to rely on rhetoric:

> We are confronted with a condition not based upon any theory based upon statistical calculations of past performances. Perhaps you cannot understand this and neither can we, but action should be taken according to the realities of the situation. … An error involving excessive supplies would be a detail; an error involving shortage of supplies would be a calamity.[27]

The realities defined here were those of feeding the American market on its own terms and postponing any day of final reckoning. Campbell conceded, though in the same way in which previous modifications had been made. The third quarter quota was raised retroactively to 120 per cent and the fourth set at a modest 70 per cent and the communiqué was issued on 11 October. Again, *The Economist* predicted increased levels of stocks, even as high as 35,000 tons.[28] These quota decisions meant, at least temporarily, unlimited production and several Malayan interests wanted to make it permanent,[29] invoking article 24 if necessary.

Article 24 gave governments the right to seek exemption from production limits in the event of hostilities and some have expressed surprise that the British government did not pursue that option.[30] To have done so would probably have meant the end of the ITC and with it any machinery to cope with an uncertain postwar situation. A return to unrestricted production would adversely affect Britain's dollar earnings at the point when she needed them most. In any case, the pressure on tin came not from Britain as a belligerent but from a neutral, the United States, and invocation would therefore be inconsistent with the spirit of the article.[31] In spite of these arguments, pressure to end the ITC continued from certain Malayan producers who now found new voices to express their position at question time in the House of Commons.[32]

Final resolution of the long deferred issue of the quota for the fourth quarter came at the end of October. Lyttleton provided a very thorough assessment of both the demand and supply situation from which he concluded that the Americans had not in fact been 'so wild as first appeared', and recommended a quota of 100 per cent for the fourth quarter and a minimum of 60 per cent for the first quarter of 1940. As before, Campbell remained sceptical and but he could only agonize over the horns of the dilemma on which he was caught, between 'intense resentment if action on our part resulted in insufficient supplies … against letting out a flood of excess tin'.[33] Six weeks into the war he could do no more than restate the issue that had been evident at the beginning. In response to Todd's pressure, he wrote:

> I cannot agree that consumers' interests in the USA must be regarded as paramount. The interests of consumers as a whole must be taken into consideration, as also the interests of the producing countries. It is not in the interests of the consumers as a whole that any action should be taken now … which may lead to a very large drop in price later, nor is it in the interests of the producing countries that production should be very rapidly forced up to a high figure if that will compel the Committee to fix very low quotas in the future.[34]

Not only was Campbell as far as ever from determining the figures that would serve to balance these two interests but he seemed oblivious to the fact that the American consumers were now in the majority.

American tinplate manufacturers were being singularly unhelpful. Although the AISI had established a tin advisory committee to provide Todd 'with detailed information on consumption and stocks',[35] manufacturers were loath to provide any of the data required. That placed Todd in a difficult position. He simply cabled Campbell with demands but was unable to provide the factual basis to support them and 'the net result is that US consumers are in a weakened position vis-à-vis the ITC. They have made rather strong demands on the ITC, always without offering a statistical basis, and have destroyed confidence in their claims.'[36]

However, one source in which Campbell had some confidence was Trench and he continued to send sharply worded messages which supported Lyttleton's position: 'Unless British producers are given absolute free hand in production, exports and sales at least this quarter, industry is headed for greater dissatisfaction and greater trouble. You may reckon foregoing as real facts.'[37]

Campbell backed away from any attempt to find the right point of balance and sent out a recommendation to the ITC delegates endorsing his 100 per cent/60 per cent formula. There was still hope that sometime soon a point of stability could be identified and the American government was asked to refrain from entering the market until it could be reached.[38] The market was officially informed of the quota decision with a communiqué on 1 November.

One point of stability, the fixed sterling price, was now called into question. When it was set, the Dutch were prepared to co-operate but they were now getting restive in the face of the rise in real costs and the depreciation of sterling. By November they had stopped supporting the Singapore price and were selling in New York.[39] The Bolivians experienced a further problem, a rise in smelting costs in the UK, and naturally began to look again at the prospects of selling into the unregulated American market. If the ITC could not generate the supplies that would bring about parity between London and New York, the London price had to be revised. Pressure for that revision also came from natural market forces. Many Chinese in Malaya were withholding supplies in anticipation of an increase in price and a bull position of about 1,500 tons was even built up in London itself.[40] However, as long as the buffer stock held out, transactions could be completed at the fixed price and there remained two overriding commitments to it. The Ministry of Supply felt that it continued to send an important signal to the United States about Britain's desire to curb overall wartime inflation, especially in view of a likely spillover effect of a rise in the price of tin on that of other metals. The American government continued to regard the £200–£230 limits of the buffer stock as the range which it considered the ITC remained obligated to defend. Anything in excess of £230 represented overcharging![41]

The November communiqué sent two sets of signals. Since the minehead stocks that had been depleted in September and October could now be rebuilt, there were no effective restrictions on the utilization of current productive capacity. However, producers were being warned not to undertake fresh investment. At the same time, consumers were being told that while stocks under the control of the ITC were being replenished, the committee was still uncertain about the long term pattern of consumption that it was required to meet.

The message was not well received in New York. Todd stressed that the current level of consumption warranted an immediate announcement of a quota for the first quarter of 1940 of at least 100 per cent and impatiently added:

> Estimates of American quarterly consumption reported to your committee have been proven correct in all instances but have not been accepted and acted on promptly by your committee and due to this we are now faced with serious situation and responsibility for this rests with your committee regardless of any information received from other sources.[42]

Trench reinforced this advice pointing to the return of high forward prices as indicative of a real concern and added:

> No wonder consumers are excited and no wonder they fail to understand why steps are not taken to encourage production and effect prompt release of supplies. … Personally cannot understand why industry should be jeopardized by adherence to theoretical notions of what are adequate supplies. Again plead for immediate and effective action.[43]

The response came with a summoning of the first actual meeting of the ITC since June, though in the absence of both the Bolivians and the Dutch it could only make a tentative decision to continue with the 100 per cent with the prospect of raising it to 120 per cent should the absent members agree.[44] In preparation for this meeting, the British delegation had secured the agreement of the Ministry of Supply to rescind the price control order when the new quota level was made public.[45] On 8 December the ITC announced that quota for the first quarter for 1940 would be 120 per cent and when the LME reopened the following Monday the spot price jumped to £271/10/–, settling back by the end of the year to £247/15/–.[46] Convergence with New York was now in sight.

The failure of the ITC to keep control over the market for the first four months of the war was a simple result of the unwillingness of the Americans to address the root source of their stock problem. Instead of negotiating a direct agreement with the ITC to supply a reserve stock under governmental control, they insisted that their views of the level of current demand be respected without being able to provide adequate statistical backing. Such posturing was similar to the interventions the State Department had tried to make during the formation of the buffer stock. On this occasion, it provoked a more serious rebuke from Kennedy:

> I believe it is absolutely a psychological mistake to proceed the way we are proceeding. This method of practically once a week making suggestion to the government as to what change should take place in the production quotas … is to me very bad trading policy. I went to see MacDonald yesterday and he pointed out to me our methods were the most childish methods he had ever seen in an attempt to influence a change in government policy, and I wish to add that I think this statement is a typical British understatement. I think them worse than childish.[47]

This was but a prelude to a 'diatribe' delivered by Campbell to Kennedy and passed on to Washington with little comment:

> There is no rational justification, even with your increased demand, for the panicky buying which has been going on. They are competing against themselves to push quotations up and they will merely find themselves in the end with extensive inventories obtained at absurd prices … It is incomprehensible to me how the Department of State, who must read the Hague Statistical Bulletin … can be induced by Todd to follow the line it does … Lord knows why you want all this tin, but have it you will and you must take the consequences … The contrast between the case of tin and that of rubber I consider most interesting. It may be explained by Todd's refusal to comply with my request to issue a statement similar to that issued by Viles to calm down speculation. The result has been that prices in rubber have been broadly reasonable, but those in tin completely crazy.
>
> Todd is rather new to the field, has only attended a couple of meetings and then went back to the United States. I don't consider that he has shown himself a very useful representative of American consuming interests. He is completely under the thumb of the steel people. They tell him he must get more tin from those 'bloody Britishers' and he just runs to the nearest telegraph office and repeats like a parrot what he is told. He fails to act in any advisory capacity and has exercised no moderating influence based on a broad understanding of the underlying factors of tin production and consumption … If I had Todd here I'd tell him exactly what I'm telling you and I only wish he'd give me the chance.[48]

The system of consumer representation had broken down precisely when it was most needed. In spite of his initial efforts to participate actively in the affairs of the ITC, Todd had come to tin too late to be of any use and his little knowledge had proved to be very dangerous.

At no point in this extensive exchange between the ITC and the USA was there any recognition that it might be in the ultimate interest of the latter to ensure that the former play a constructive role in the formulation of the kind of overall economic policy that would enable Britain to wage war effectively. Indeed, in spite of all the fuss about supplies, there was no coherent economic policy. The absence of any controls on American exports meant that at least 2,000 tons almost certainly went 'into undesirable hands'.[49] Britain, as a belligerent, was expected to sacrifice its military interests for the sake of the private interests of certain Americans.

Both the price and physical indicators offered retrospective support for Campbell's cautious approach. Apart from a brief period in mid-1940, following the fall of France, spot prices in New York never again reached the levels of the autumn of 1939. The physical indicators are presented in Table 15.1.

By the end of December stocks of metal, including the publicly unreported smelters' stocks, had been built up to 51,000 tons, at the low end of the range that was considered

Table 15.1 Tin production, consumption and stocks, September–December 1939

	tons metal
ITC members authorized	90,551
ITC members actual	85,374
Others	10,926
Total Production	96,300
Consumption	63,500
Stock increase	
Visible and smelters	16,000
Consumers	1,000
Concentrates in transit	15,800
Total	32,800

Source: ITRDC, *Statistical Bulletin*, February 1940

appropriate under wartime conditions. But the lag in smelting the shipments of concentrates meant that another 16,000 tons was about to enter the market, of which perhaps as much as two thirds would be considered excessive. The ITC would then face a more familiar situation.

January–June 1940

The first quarter of 1940 saw one of those rare moments when production and consumption were more or less in balance, though not at the levels anticipated when the quota had been set at 120 per cent. As the American tinplate mills dropped from working at 95 per cent of capacity to 65 per cent consumption naturally fell away. Only Nigeria and the NEI, both of which carried forward large excesses from their 1939 shipments, were within their respective production targets.

Bolivia's failure was not unexpected, though the extent of her shortfall was more a function of political than technical factors. Although the new government that emerged after the death of Busch ended the public ideological challenge to the large mining companies, it was far from formulating a policy that would give them the confidence they needed to rebuild productive capacity. The exports at the end of 1939 had drawn on minehead stocks and the companies began to slow down production to bring pressure on the government to revise its fiscal policy.[50] However, Bolivian shipments had been sufficiently large that the British tin position was now sound and export licences were granted freely. The fall in price, therefore, appeared even sooner than Campbell had predicted and discussion of the quota for the second quarter had to return to a familiar issue – how far actual production should be cut.

Lyttelton laid out a comprehensive analysis and defined the major objective as that of stopping an upward rise in price and the repetition of the problem that arose in the previous autumn. The quota should therefore 'err on the side of liberality', especially since the cash in the buffer stock together with a retroactive cut remained as weapons that could be used to arrest a fall. In effect this meant a change in ITC thinking and priorities and he added that 'market psychology must enter into our calculations and not merely statistics'.[51] A quota of 75 per cent was recommended with an upper limit of 80 per cent. This proved to be the basis of a compromise position.

Todd maintained his extraordinarily extreme position and advocated keeping the current quota of 120 per cent. At the other extreme were the Dutch who became so alarmed at the way in which the current quota was building up unmanageable stocks that they wanted a retroactive cut to 90 per cent. The Colonial Office was divided. Campbell supported the Dutch in wanting a reduction in stocks but others were mindful of the problem that the severe cut had posed for Malayan producers in 1938 and were more sympathetic to the views of Lyttleton and NFMC. They then encouraged the Americans to bring pressure to bear on the Minister of Colonies at The Hague to modify the Dutch position.[52]

Unfortunately, American diplomacy was conducted no better at The Hague than in London. A meeting was hastily arranged where the Americans argued an extreme position for 100 per cent to which the Minister was prepared to listen sympathetically. He had not been properly briefed, however and his aides soon retracted any implied commitment on the grounds that he 'really did not know what he was talking about'.[53]

None of these background diplomatic efforts served to change the Dutch position. At the next meeting of the ITC, Campbell, Houwert and Patiño all thought a cut in the quota to 70 per cent would be appropriate. However, the 'political and psychological factors' stressed by Lyttleton overrode the more straightforward statistical arguments and the meeting settled on 80 per cent.[54] Given the production deficit carried forward from the first quarter, the effective quota level was closer to 90 per cent but the fact that the industry was returning to real restriction was sufficient to prevent a further drop in price.[55]

While the tin market had generally anticipated a cut to this level, it was not viewed with favour among some of the Cornish interests in Malaya. Rich broke confidentiality and alerted Simms to Lyttleton's thinking about the appropriate level of the quota prior to the ITC meeting. Simms sought a meeting with the Colonial Secretary to protest against it, taking over an hour to lay out his overall position:

> It is difficult to place a rambling and disconnected set of malicious statements in any sort of logical order, but broadly speaking he started from the point that the Americans were now buying all the tin which offered and that public interest demand that Malaya should be allowed to continue to produce as much tin as she could. … with the possible exception of the Dutch, no other party could produce much above 80 per cent, and therefore any cut to any point between 120 per cent and 80 per cent represented a cut of Malayan production for the benefit of foreigners.
>
> He then proceeded to give a brief resumé of events in the Malayan tin industry … since 1932 in which the sinister figure of Mr Howeson was, as might have been expected, painted in the darkest colours. … Mr Simms then proceeded to give his views on the personalities of the British Delegation. Sir John Campbell and Mr Lowinger … had never been in the tin industry and knew absolutely nothing about it. Captain Lyttleton was a crypto-Bolivian and the only righteous men in the place were Mr Wilcoxson and Mr Rich, whose opinions were steadily overruled. In wartime it was obvious that the interests of the British Empire must prevail over everything else and Sir John Campbell was wholly unfit to lead the delegation because he had stated that an international scheme must be run in accordance with international considerations.[56]

The tone that Simms adopted well expressed the long felt frustration about the way in which the ITC operated: it was too international in its perspective and was under the control of those with neither experience nor commitment to the industry. At this point, of course, the war only engaged Nigeria and Malaya and therefore appeared to represent the perfect opportunity at which to link the interests of their producers to those of the empire.

As Simms developed his argument for unlimited production, he was quite prepared to see the price drop as low as £166. Leaving aside the dubious assumption that Malaya could actually produce nearly 100,000 tons at such prices, this policy would have resulted in a further 50,000 tons added to the extraordinarily high level of stocks of 70,000 tons. Had Britain been successful in its defence of Southeast Asia, the consequences for the postwar tin market would have been devastating. In addition, the Malayan producers would have sacrificed 30,000 tons of their mineral resources without any benefit, apart from the purely emotional satisfaction of seeing some of their supposed rivals thwarted.[57]

Simms was far from satisfied with his interview and continued his campaign against the ITC well into June. At the Sungei Besi AGM he declared:

> Those responsible for the administration of Tin Control ... have made blunder after blunder and have rejected and resented the sound advice from the Malayan Chamber of Mines ... which would have placed British tin resources in the position of making the maximum contribution to the war effort. In other spheres of national effort men who have failed less conspicuously have been replaced by others more competent, and until this procedure is followed in connection with the mobilisation of our tin resources they cannot be used to advantage.[58]

Simms was also successful in presenting himself and his colleagues as 'responsible businessmen above reproach', in several political quarters, not least the House of Commons where Stokes continued to ask pointed questions of the Colonial Secretary only to receive bland assurances.[59] Since 'tin restriction is such a large and amorphous subject that a full discussion of it would cover a great deal of paper', the Colonial Office simply engaged in trading insults, trying to fob off the criticisms as emanating 'from a small minority of European mine-owners in Malaya who are remarkable alike for their thickheadedness and their malevolence'.[60]

The price history of this second phase of the war is presented in Figure 15.2.

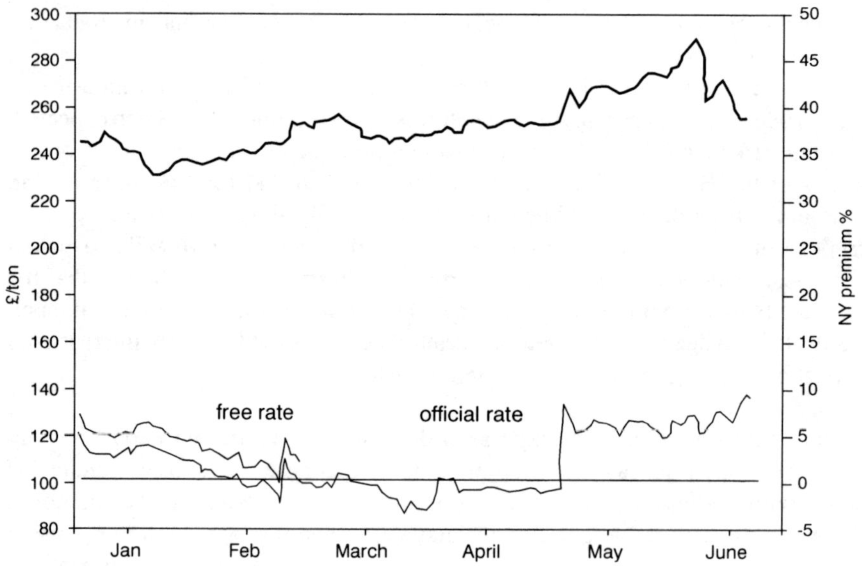

Figure 15.2 Tin prices, January–June 1940

While it was clearly impossible to bring about a perfect alignment between New York and London, the gap had narrowed to the point where it became insignificant.[61] At the same time, the price was kept reasonably stable. Until May, the price ranged from a low of £231 to a high of £258 around a mean of £247, figures that were quite comparable to those of the prewar buffer stock, adjusting for the inevitable increase in costs. Over this period, the ITC had demonstrated its ability to regulate the market effectively even without the support of the buffer stock.

With the German invasion of Western Europe in May 1940 the stability that the ITC had brought to the market was shattered. German control of the Netherlands naturally raised concern about the future of supplies from the NEI and the ITC met the ensuing demand for increased levels of stocks by raising the quota for the third quarter to 100 per cent and encouraging producers to start exporting at that level.[62] Reestablishment of tighter export controls in Britain meant a reopening of the gap between the London and New York markets and with it the threat of diversion of shipments from Bolivia. It was within this new context that the United States could actively pursue its plans for a stockpile and a domestic smelter.

Building the American stockpile: phase I

Although the issue of tin supplies to the United States in the event of war had been a matter of official concern since at least the McReynolds hearings of 1935, little preparation had in fact been done.[63] The Navy Appropriations of Bills of 1938 and 1939 authorized the acquisition of tin as part of a programme of stockpiling of strategic materials but only small amounts were bought. In June 1939, Congress passed a Strategic Minerals Act which endorsed a more comprehensive stockpiling programme but by the end of the year only 3,680 tons had even been ordered.[64] Actual stocks in government hands at the end of 1939 were a mere 1,955 tons, no greater than 12 months before.[65] The obstacle, of course, was that in trying to buy stocks in the market the government would be competing with private consumers and moving the price against itself. A solution to the tin problem would only be found through direct negotiations with producers and that would involve both Bolivia and the ITC.

Building a substantial stockpile needed a special agreement with the ITC to guarantee the additional supplies. For the United States, the main issue was one of price; for the ITC, the terms under which the metal would be released. As preparatory discussions got under way, it was expected that the US would want a cut price[66] within the old buffer stock range of £200–£230 but there was no willingness to settle for anything less than 50c/lb, equivalent to £267/ton in London. In fact, the Americans were not interested in protracted negotiations and since the 50 cents was about 10 per cent lower than the average New York price after the invasion of Western Europe, it was considered cheap.[67] This price set a floor to the New York market so that normal commercial transactions were conducted at a premium, reinforcing the gap between London and New York.

To make available the minimum of 75,000 tons that the Metals Reserve Corporation (MRC and a subsidiary of the Reconstruction Finance Corporation, RFC) had committed itself to buy between 1 July 1940 and 30 June 1941, the ITC declared that the quotas for this period would be 130 per cent.[68] In return, the MRC ensured that should the stocks not be needed in the event of an emergency, presumably defined as the inability of the ITC to meet normal commercial demand, they would be not be liquidated before January 1944 and then at a rate of not more than 5 per cent to a maximum of 5,000 tons per quarter. Since that was equivalent to 10 percentage points on the international quota, liquidation would not be painless unless the postwar market were particularly buoyant.

The MRC/ITC contract marked an important departure for the ITC. The MRC had only committed itself to buy at the floor. With no prospect of an early end to the war, consumers had every incentive to continue to build up their own stocks. As long as they were prepared to pay over the 50 cents floor, they were free to do so and the excess supplies released went into private rather into public stocks. By the end of 1940 consumers' stocks had risen to 45,500 tons, more than twice the level 12 months earlier and more than twice the level of stocks controlled by various government agencies.[69] The ITC was therefore losing control over the most important feature of the market and when this restocking was complete it would again have to face the necessity of imposing severe restriction.

As long as the American market remained open, the MRC was confined to the role of a residual buyer. It was therefore interested in extending the contract. Since American consumers were beginning to explore the possibilities of substitution more vigorously, Campbell encouraged the MRC to buy a further 75,000 tons.[70] In May 1941 the contract was renewed for another year on the same terms.[71] Producers were therefore given a reprieve with which to continue to enjoy their wartime bonanza. One important condition was added: the renewal of the ITC agreement itself.

The United States and Bolivia

War also created an opportunity for a different kind of shift in the structure of power relations in the industry. London's role as the commercial and financial hub was dependent on the existence of a large domestic smelting industry and that could only be fed by concentrates from Bolivia. This structure created two sets of resentments. One found its best expression in the McReynolds Report which considered that the absence of a domestic smelter prevented the United States from exercising the influence it deserved. The other was in Bolivia itself where Patiño had long been seen as exercising far too much power, in part because of his control over Williams, Harvey which treated the bulk of Bolivian concentrates. There had, therefore, always been a mutual interest in developing an American smelter but it could never come to fruition under peacetime conditions thanks to the competitive superiority of Williams, Harvey. It was a superiority based not only upon lower costs of labour and fuel but also on access to the higher grade concentrates supplied by Patiño, mixed with even higher quality concentrates from Nigeria.

Intensification of hostilities in Europe provided the arguments necessary to ensure that government funds were made available to finance the construction of an American smelter and, if necessary, subsidize its operation. Although strategic considerations suggested that it would be wise to send Bolivian concentrates direct to the United States to prevent possible submarine interception, they were not paramount. A large American smelter would tie up more stocks of concentrates and actually reduce the supply of metal. If it could not be made competitive when the war was over, then the amortization costs meant that the United States would be paying very dearly for its tin. Strategic arguments simply provided the cover necessary to realize a long-standing dream.[72]

Britain quickly saw the threat posed by the American smelter and moved to take pre-emptive action. Increasing stringency of exchange regulations provided the incentive. Under the arrangements introduced in the autumn of 1939, Bolivian producers received 50 per cent of the gross value of the metal in their concentrates in sterling and the remainder convertible into dollars at the official rate. The inconvertible proportion would be reduced to 33 per cent on the signing of a ten-year supply contract with a British smelter.[73] While this was of no particular interest to most companies, it was to the Patiños, much of whose fortune was held in sterling. They were then able to bargain for a special concession whereby they could convert a high proportion of their personal sterling assets into dollars at the official rate, in return for which they signed the required contract on behalf of their companies.[74] With the Patiño concentrates committed to Britain, it was hoped that the American project would be halted.

American companies had been undertaking experimental work with low grade Bolivian concentrates for some time. Hochschild was an assiduous promoter of the prospect of expanding Bolivian production on the basis of more effective treatment of low grade ores and provided much of the encouragement required for these experiments. The dream that lay behind the American smelter was much more ambitious than the British recognized. The Americans were therefore not at all deterred by the announcement of the Patiño-Williams, Harvey contract.

When the American government announced that it was prepared to receive proposals for the construction of a domestic smelter, six American firms indicated their interest, as did two foreign ones, Williams, Harvey and Billiton. Patiño was quite prepared to break his recently signed contract with the British if he could secure better terms from the Americans. In spite of the obvious advantages of using the technology of Williams, Harvey and securing the supply of Patiño concentrates, the Americans were not interested in Patiño's proposal. The State Department was concerned about the effect on the overall relationship with Bolivia, which had seen the American smelter as the opportunity to break Patiño's control over smelting, not to reinforce it.[75]

The Patiño initiative seriously backfired. Not only did it upset the British but it also upset van den Broek, since he had proposed a co-operative arrangement with CTS.[76] Since none of the American firms had the necessary expertise in treating low grade Bolivian concentrates, the contract was eventually awarded to Billiton and the smelter was built at Texas City. However, it was not designed as a commercial venture. While Billiton supervised the construction of the smelter, it remained the property of the RFC and the metal it produced was for the account of the MRC. Billiton simply took a management fee for its service and any subsidy for the smelter could be effectively disguised.

While the Americans delayed finalizing arrangements for the smelter, they moved more promptly to secure its main source of feed. In November 1940, the MRC announced that it had signed a contract with organizations representing most of the non-Patiño production. For each of the next five years, the MRC was prepared to buy concentrates with 18,000 tons of metal, at a price tied to the ITC agreement. Since these concentrates could go as low as 18 per cent, the MRC was opening up new possibilities for supplies which were otherwise not commercially viable.[77] Although the contract stipulated that, should the ITC re-impose restriction, then it would apply equally to these sources, the fact that such a large volume was now directly integrated into smelting and marketing posed a serious threat to the remaining unity among the producers.

By the end of 1940 the structure of the tin market had undergone a fundamental transformation. The non-Axis world was awash in tin. During the course of the year, stocks of metal in the United Kingdom, excluding those in the hands of consumers, had risen from 8,228 to 10,457 tons, or sufficient for 18 weeks supply. In the United States, they rose from 5,300 to 30,600 tons, or sufficient to meet demand for 31 weeks. When consumers' stocks and shipments afloat are added, the United States had a full year's needs in hand.[78] Ample stocks set upper limits to price fluctuations and the MRC/ITC agreement had set lower ones in New York.

This pattern was simply repeated in 1941 with one modification. As contracts with British smelters expired, Bolivian concentrates were shipped to the United States. Since this threatened to raise the LME price over the New York floor, the Ministry of Supply used the buffer stock organization to sell its own stocks to keep the London price just below parity with New York.[79] However, the supply position was sufficiently good that by the end of the year, stocks in the United Kingdom rose by a further 6,200 tons or 25 weeks' supply. In the United States they rose by nearly 78,000 tons, sufficient for almost two years.[80] In no other commodity was 'the position so comfortable'.[81] Since 1941 was the last year of the third agreement, the ITC also had to turn its attention to its own renewal.

Negotiating the fourth agreement

Discussions about the shape of the fourth agreement began in late 1940 and started on a particularly angry note. News of the MRC/Bolivia contract caused a great deal of alarm about Bolivia's overall commitment to the ITC, especially since there were indications that the State Department saw it as a means of breaking tin control.[82] Old wounds, especially those suffered by the Dutch, were reopened. Pearce served as the messenger of the 'bellicose attitude' now taken by van den Broek:

> He was all out to 'down' Patiño and Bolivia; he meant to wreck both, if he possibly could; and in particular (he was tin king in the NEI and the Govt: there was, for this purpose, van den Broek) he meant to sell Dutch ores to the USA, for their smelter; and he was determined to break up tin control, because tin control helped Bolivia more than it helped any other country. ... He is making a profit of two million guilders, or so, a month; and he says openly that, that being so, commercial considerations – loss on the sale of the ores; additional freight; and so on – just do not matter in the least. He means to dominate the situation; and his primary motive is to pay back Patiño for a host of old grievances, and some new ones; and to punish Bolivia. van den Broek often talks wildly – Pearce admitted that; but he thinks he really means what he says this time.[83]

Just as in 1936, van den Broek could only be restrained by calmer Dutch voices and by a recognition of the bigger issues at stake than particular rivalries. One such voice was the Dutch Minister in London, Welter, who was 'personally strongly in favour of continuance of control', and it was hoped that when van den Broek could return from New York to London face to face discussions might at least 'weaken his opposition'.[84]

The first formal step towards a renewal was taken at the December ITC meeting which simply provided the opportunity for a tentative and preliminary discussion as a prelude to a comprehensive scheme to be endorsed in March with formal ratification to take place soon thereafter. It did not go well. No one from the Bolivian delegation was present and its technical advisor, Pearce, could only repeat the formal instructions he had received, which authorized him to sign a simple renewal of the existing agreement keeping the existing standard tonnages unchanged.[85]

Van den Broek managed to attend the meeting and treated Bolivia's proposal as preposterous, especially in the light of the fact that the NEI had now overtaken her by a considerable margin. He even went so far as to challenge Pearce's right to speak on behalf of Bolivia and asked him to withdraw from the meeting but did not press the matter on Campbell's pleading.[86] The attack on Bolivia continued, by denying her the right to base her tonnage on current production, since such a high proportion sold to the MRC included concentrates that were normally unsaleable. Fortunately, Pearce found an argument to resist this and van den Broek had to back off. In any case, determining this proportion would be quite impossible, especially since the US smelter was a long way from even being constructed.[87] However, van den Broek met with no resistance when he raised yet another of his concerns about Bolivia, namely the need to have some sanction against excess production when quotas were low, probably in the form of a threat to terminate the scheme.[88]

The existing standard tonnages and minimum guarantees reflected a pattern of political negotiations loosely linked to arguments about the differential evolution of various tin producing regions. Those arguments had inevitably weakened with the passage of time and the fact that production was currently unlimited placed everyone on a more or less equal footing. Previous negotiations had been framed by the need to preserve an agreement among the four core members in order to deal with the fringe members. War had changed the position of both French Indo-China and the Belgian Congo. Since falling under control of the Vichy government, French Indo-China had ceased to participate in the ITC. The Belgians recognized that the Congo had now matured as an established producer and was no longer in a position to make any special claim. Wilcoxson proposed a formula, actual exports for the 12 months of the 130 per cent quota, July 1940 to June 1941, which met with general assent on the part of delegations.[89]

When the ITC met again in March it was able to complete a draft of a new agreement.[90] It would run for five years and was expected to cover at least the start of the postwar period. Several important changes were introduced. Since the experience of the war had demonstrated the need for a direct relationship with the American government, article 13 expanded the number of consumers' representatives to include a government nominee. Article 12 introduced the concept of 'important violation', and specifically identified an excess of 10 per cent over quota for one quarter. Article16c gave members the right to withdraw from the agreement in the case of such violations.[91] At least the Bolivian problem had been formally defined. The need for flexibility under uncertain conditions was met with Article 3i which kept open the option of considerable revision during the life of the agreement.

The only point of contention was, of course, the standard tonnages. The Wilcoxson formula was expressed in more technical terms, as the actual production from 1 July 1940 to 30 June 1941, capped at the rate of 130 per cent on the existing tonnages and subject to an earlier termination date should hostilities in the East be intensified. It was adopted as part of the draft to be recommended to signatory governments with only Bolivia dissenting.[92]

Bolivian resistance to any modification received support from the State Department which wanted to mitigate the impact on supplies for the American smelter should restriction be reimposed.[93] This was yet another poorly thought out intervention since the MRC was simultaneously expressing its interest in extending its contract with the ITC which would continue unlimited production. However, the potential impasse provoked some panic in the Colonial Office. Clauson was quite prepared to concede the American position and see Bolivia in 'on an unreal basis' as the price of an agreement between Malaya and the NEI. He then explored the prospect of a joint diplomatic initiative between Britain and the United States to bring pressure on the Dutch to overcome van den Broek's position.[94] Campbell managed to rein him in and let 'bargaining between the Dutch and the Bolivians follow its natural and probably protracted course'. Clauson then tried to reassure the Americans that the position of their smelter would not be jeopardized since he would advise his own government to reduce its proportion of Bolivian supplies in the event of restriction.[95] It was an extraordinary concession, not least because of the difficulties it would create for the Patiño-Williams, Harvey contract and the attitude then revealed would become a serious liability when Clauson assumed greater responsibility for British tin policy.

Bolivia was far from the only problem facing the Colonial Office. While the NEI and Nigeria were producing at the 130 per cent rate, Malaya was only at 115 per cent and that was below her domestic assessment. The excuse offered, that the excess profits tax (EPT) regulations discouraged maximum production, was rather thin in light of the performance of the Nigerian companies subject to the same regulations. Malaya therefore tried to find another basis for the standard tonnages. Still festering was the experience of 1937 and at first that was proposed as the basis for the new standard tonnages. It was so absurd, especially in light of the recovery of Bolivia and the expansion in the Belgian Congo, that her delegation would not even present it to the ITC.[96] She then attempted to move the basis to production for 1940, since that would also place in her in a privileged position but at the expense of Nigeria and the NEI.

Just as difficult was the Malayan position on the proposed length of the agreement. She was not willing to accept anything for more than a one year renewable agreement, arguing:

> Apart from necessity of preserving freedom of action on account of threats to market Straits tin after the war from new smelters in USA, it appears most undesirable during the period of post war trading negotiations tin production within the empire should be subject to foreign control.[97]

The official position simply followed the overall line taken by the Cornish[98] and it seemed quite innocent of the potential problems posed by the huge accumulation of stocks, the likelihood of severe cuts once this accumulation had been completed and the fact that the distinction between empire and foreign had become quite obsolete. Small wonder that Campbell commented: 'It is like arguing with a fog to get clean-cut and sensible proposals out of Malaya.'[99]

Dealing with Malaya followed a familiar pattern. MacDonald indicated he was quite prepared to overrule Malaya if she could not 'be induced to see reason by gentler methods'.[100] Several arguments were advanced. The longer period was required to avert dislocation after the war and to secure Dutch support. There was no logical basis to 1940 since it was not completely unrestricted and quotas had been settled on a quarter by quarter basis in the first half of the year. That had prevented producers from committing themselves to fresh investments. Abandoning tin would jeopardize plans to extend regulation to cocoa and cotton. Clauson went on to comment:

> It is very clear to HMG that international regulation of trade must play a very important part in post war economic reconstruction. Hitherto one difficulty has been traditional American antipathy to any form of international regulation, but this is breaking down now, partly no doubt because Americans are now themselves possessors of very large quantities of tin and rubber bought at high prices and it would be very regrettable if American conversion to the principle of regulation was accompanied by British recession from it.[101]

Whether the Malayan government recognized the force of these arguments is unclear but it could easily see the threat behind them and conceded, though the industry continued to be 'busy devising reasons why the period chosen is unfair to it'.[102]

With the support of the High Commissioner, the FMS Chamber of Mines returned to its campaign on the composition of the Malayan delegation. It wanted Campbell out of the Malayan delegation so that it could represent Malaya more effectively without being constrained by the 'obligation to strive for unanimity'.[103] None of the points made during the thorough airing of the issues in 1938 seemed to have registered since Thomas again proposed Ward. It found the Colonial Office even less sympathetic:

> The teams work together reasonably well, and Messrs Wilcoxson and Rich have behaved well lately, although they are potential wreckers, since both in their heart of hearts would sooner there was no scheme. The purpose of the present proposals is to strengthen the unofficial element in the Malayan delegation and with it the wrecking element, since Mr Ward is an out and out anti-restrictionist. They seem to me directly contrary to the spirit of the times. It is quite clear that in the immediate post-war period it will be more than ever necessary to keep firm governmental control of these schemes and ensure that they continue. Neither of these objects will be served by adopting the Governor's proposals.[104]

Thomas dutifully recognized the argument for continuity and retention of strong governmental control and accepted the recommendation that the current delegation be reappointed.[105]

The composition of the Nigerian delegation also posed some difficulty. It, too, reopened old issues. When Lyttleton joined the Cabinet in December 1940 he resigned from the ITC and recommended that his position be offered to Burgess, the managing director of a firm of metal brokers.[106] This would be a problem under peacetime conditions but not for the duration of the war and it was supported by Campbell who wanted a 'really good man on ITC at this stage, for we must now take up the extremely difficult question of the renewal of control'.[107] Unfortunately, the Nigerian Chamber of Mines proposed that the appointment go to Mooney, from the Kaduna group, and the Governor was in practice committed to accept the recommendation.

A few months later, the Chamber proposed that the delegation be strengthened and recommended Burgess. Campbell consulted other members of the ITC but they recalled some of the difficulties created by involving Lazarus in the early years. In spite of the 'quite strong view' of the ITC, he was quite prepared to go ahead.[108] The rest of the Colonial Office was divided. Clauson considered the ITC position 'farcical', in light of the limited role of the LME during the war but Calder became more nervous, especially since another and more fundamental issue surfaced.

The Nigerian Chamber pressed its nomination on the following grounds:

> The intention was not that Governments should control the industry, but secure the fair and impartial administration of the scheme. If producers make any suggestion which is not incompatible with this, the Government ought to give effect to it.[109]

Since the claim that the 'producers should run the scheme is not acceptable', Calder rejected Burgess' nomination. Nor was he prepared to reconsider in the light of pressure from the Governor who had formal responsibility for the decision.[110] While Calder was simply stating the standard Colonial Office position on the governmental character of the ITC, the power he was asserting could prove to be particularly important when the fourth agreement in turn came up for renewal.

By August Bolivia felt more relaxed about her position. The results of the base year were available and the MRC agreement had been renewed. She was still 2,000 tons short of her current standard but remained committed to the actual figure for reasons of domestic prestige. However, she made an important concession. The proposed basis of 1940–1941 would be used to determine the ratios but the actual numbers would be inflated so that Bolivia could retain her existing nominal standard.

That left Siam, now renamed Thailand,[111] whose position had not substantially changed since the last round of negotiations in 1936. In spite of the fact that she had only produced 17,043 tons during the base period, her demands included: keeping the current nominal standard of 18,500, with a minimum of 11,000 for the first three years and then increasing it to 20,000, with a minimum of 12,000 tons, for the following two years.[112]

There could be no question of any increase in the standard, at least not one that was not granted to all other members. At first, it was thought the Thailand problem could be solved through the same arithmetical solution as was contemplated for Bolivia. Since Thailand was underperforming at twice the rate of Bolivia, that would entail an even greater inflation.[113] When this formula was communicated to Bangkok, a favourable response was sent back to London through Crosby, suggesting that 'Thailand has taken the whole thing – hook, bait and sinker'.[114] The results of these computations are presented in Table 15.2.

One minor detail of form proved to be a major obstacle. On each of the three previous occasions, the Colonial Office had breached the protocols expected by the Foreign Office concerning agreements with sovereign powers. In its defence, the Colonial Office had claimed that pressure of time to complete these agreements had simply made it impossible for them to be vetted by the Foreign Office but promised that it would follow proper procedure in the future. Now there was no such excuse. The Foreign Office was insisting that the agreement be signed on behalf of the United Kingdom, rather than its colonies, and in normal times this would have presented no problem, especially since it simply followed the model of the second rubber agreement to which Thailand was a signatory.[115] But in late 1941 the Thais were nervous about the possible response of the Japanese[116] and refused to sign an agreement with the British as opposed to the Nigerian and Malayan governments since it would 'otherwise give a political complexion which it is not meant to possess and which might lead to misunderstanding in the present state of world politics'.[117]

Even without the distraction of the Japanese, negotiations would not have been as simple as Crosby had supposed. He had received a premature response in Bangkok and when the official one came through the Thai Legation in London, it was clear that the only concession that Thailand was prepared to make was on the issue of the increase of 1,500 tons in the standard for the last two years. She now proposed returning to the existing standard tonnages on the grounds that Malaya and the NEI had enjoyed privileged access to mining equipment since 1937 and that this had placed them in a superior position during the base twelve-month period.[118]

While there was no evidence to support that allegation,[119] the war itself had an adverse affect on production. Until June 1941 it was always possible for Germany to acquire tin concentrates in Thailand and she was prepared to pay a premium price to do so.[120] Japan, too, was anxious to increase her imports from Thailand and the Thai government was only too pleased to oblige.[121] Overcoming the recalcitrance of the producers to see any diversion to politically objectionable destinations required the introduction of extraordinary measures and many would have been prepared to forego the incentives they offered.[122] On the material side, production had been adversely affected by a chronic

Table 15.2 Fourth Agreement, standard tonnages

	3rd Agreement		*4th Agreement*		*July 1940 – June 1941*			*4th compared to 3rd at 100% quota*	
					130%	*Actual*			
	tons	*%*	*tons*	*%*	*tons*	*tons*	*%*	*tons*	*%*
Malaya	77,335	37.4	95,474	38.0	100,536	87,955	37.5	1,265	1.6
Bolivia	46,027	22.2	46,768	18.6	59,835	43,985	18.8	–7,524	–16.4
NEI	39,055	18.9	55,113	21.9	50,772	52,441	22.4	6,318	16.2
Thailand	18,628	9.0	18,500	7.4	24,216	17,043	7.3	–3,397	–18.3
BC	15,035	7.3	20,178	8.0	19,546	18,589	7.9	1,577	10.5
Nigeria	10,890	5.3	15,367	6.1	14,157	14,253	6.1	1,761	16.2
Total	206,970	100.0	251,400	100.0	269,062	234,266	100.0		

Note: Third Agreement Final Operating Tonnages

shortage of oil following the decision taken in the summer of 1939 to rely on Japan as the sole supplier.[123] Negotiations would have determined the extent to which these factors warranted modification of the results of the base year and hence whether there was a real basis on which Thailand's position could have been accommodated within the overall principles governing the new agreement. But, as was the case in 1936, that would have required a meeting between principals in Bangkok rather than intermediaries in London and in late 1941 it was quite out of the question.

There was some thought of simply capitulating to the Thais, though Calder wondered whether they could be bullied into the necessary concession by ordering the smelters to refuse to treat their concentrates. Fortunately, the Foreign Office was not anxious to see any complication to an already very delicate political situation.[124] The choices were therefore whether to renew without Thailand or drop the proposal for a fourth agreement altogether. As these were reviewed by the ITC, it was decided to recommend acceptance of the agreement as negotiated thus far, allowing Thailand the option of joining later and the members the option of abandoning later.[125] To keep open the option of Thailand's subsequent adherence, the inflated standard tonnages were retained and the agreement was now ready for formal ratification. Within the next few days, Japan took effective control over Thailand as she moved forward in her conquest of Southeast Asia, thereby changing the whole position of the tin industry.

Although the new standard tonnages never had an effective operational significance, they mark an important development in the history of the ITC. For the first time there was a completely clean agreement with no compromises of any kind. Current productive capacity, albeit subject to rather different operational definitions, was universally accepted as the basis on which the standards were to be determined. The war had made redundant all the earlier debates about who bore responsibility for the development of excess capacity and about the special claims from Bolivia and the Belgian Congo. However, the speed with which the war was extended to Southeast Asia prevented the ITC from testing the extent to which it was committed to its new principles in dealing with its most recalcitrant member, Thailand.

These new tonnages also cast some of the old points of contention in a fresh light. Table 15.2 provides a comparison between the entitlements under the old and the new agreements. The Dutch, the Nigerians and the Belgians could all feel that their various cases were now vindicated. By contrast, Malaya, where complaints about the inequity of the standard tonnages had been the most strident, gained very little, just a 1.6 per cent increase. Indeed, had the new tonnages been strictly based on actual performance, she would have only gained a trivial 0.05 per cent. While she had always been correct in pointing to the overassessment of both Bolivia and Thailand, she had failed to recognize that endorsing the principle to which she was always committed would simply result in a redistribution, far less to her advantage, than to that of her neighbour. Furthermore, that redistribution might have other consequences, since it not only meant a slight decline in the supply of Straits tin but also the prospect of Dutch brands being produced in sufficient quantity to challenge the Straits premium in the USA.

At the eleventh hour the State Department again made one of its rather clumsy interventions. It asked for details of plans to protect consumers and to be given the opportunity to review them before any agreement was signed.[126] All it elicited was a polite comment pointing out the increase in government representation and that the ITC considered the existing consumer representatives as quite content since they had not made any criticism.[127] The Department then asked the ITC to consider the following text:

> The United States Government expresses its expectation and confident anticipation that the operations of the Committee will at all times be so conducted as to assure adequate supplies of tin at a price reasonable to consumers and producers alike. Today, when the question of the terms of access to raw materials in the post war period is so important ... this government feels that a useful purpose may be served by a reiteration by it and by the International Tin Committee of the intention always to strive for these objectives.[128]

Hull was trying to define the standards to which he could hold the ITC to account but since these were always those of the ITC itself, Campbell could reclaim the moral high ground:

> The committee welcomes this opportunity of expressing their agreement with the message. They have at all times been anxious to conduct their operations in such a manner as to ensure that adequate supplies of tin, at a price reasonable to consumers and producers alike, will be available to all countries on a non-discriminatory basis.[129]

These moves came at the very point when the ITC was working in even greater harmony with the MRC in expanding its stockpile. It is unfortunate that neither the State Department nor the ITC saw the need to move beyond these empty gestures, governed by the conventions of diplomacy, to a more serious discussion of the way in which their relationship should evolve.

Building the American stockpile: phase II

Once the MRC had started the process of building the largest ever stock of metal, its appetite continued to grow and with it further disruption to the structure of the tin market. The contract with Bolivia was revised to extend the annual maximum from 18,000 to 24,000 tons and the commitment to the British which had enabled them to continue to treat the residue of the Bolivian concentrates was cavalierly set aside.[130] For Campbell, it violated many norms, as he noted: 'their action is neither friendly, nor considerate, nor morally justified. It is in most marked contrast with our action as regards emergency supplies of tin and rubber to the USA.'[131]

Although the original conception of the American smelter was to be based entirely on Bolivian concentrates, van den Broek persuaded the Americans that smelting would be easier with high grade concentrates from the NEI. There were both commercial and political issues at stake. The Billiton concentrates enhanced the prospects of making the smelter commercially viable and hence surviving as a competitive force in the postwar period. In addition, the Dutch were 'obsessed with the idea that once the US started getting Bolivian ores', they would lose political interest in the fate of the NEI.[132] The concentrates would therefore cement a broader relationship. It also appeared that van den Broek had done a direct deal with the MRC, probably at a cut rate, to supply Dutch metal and 'establish a firm market for Billiton tin in the States for many years to come'.[133] With both the MRC and van den Broek going back on their repeated assurances, Campbell saw it as nothing less than 'the spread of Hitlerism'.[134] The trust that had emerged from the long working relationship between him and Campbell and which had served as an important basis of the ITC was now at an end.[135]

With an increase in the supply of concentrates from Bolivia and the NEI, the MRC revised its stockpile plans. It now wanted a total of 100,000 tons to be bought as metal and for this it needed a third agreement with the ITC. Lowinger had been sent to Washington to explore the position that would arise, as seemed increasingly likely, should the ITC not be renewed and he found MRC prepared to make an attractive contract.[136] The MRC would immediately become the sole buyer and distributor of tin in the USA. It would then buy all the tin offered until its own stock reached 100,000 tons. As the stock reached its upper limit, the MRC agreed to consult with the ITC regarding future quota levels which were required to meet current consumption. Temporary imbalances between production and consumption would be absorbed by both the buffer stock and the MRC stock. Liquidation provisions remained unchanged, as did the price, 50c/lb. The agreement would run until 30 June 1943 or any earlier cessation of hostilities.[137]

The MRC found itself dependent on the ITC for two reasons. One was to ensure that governments would enforce the sole buyer provision by only issuing licences for the USA to exporters consigning metal to the MRC. The second was to guarantee that, should the stockpile not be needed as a contingency supply, it would not be liquidated at substantially below the cost of assembly. With a price now fixed for what appeared to be the duration of the war there was another powerful incentive to renew the overall tin restriction agreement and it was endorsed without qualification by the ITC.[138] The day of reckoning was being postponed to another year.

Although this agreement was designed for the war emergency, it came close to providing the stability that the ITC had been looking for since its inception, a stable price and one guaranteed by the capacity of stocks which were sufficiently large to buffer any fluctuations in production and consumption. The Japanese occupation of Southeast Asia meant that this form of tin control was never tested.[139] However, it is unlikely to have served producers' interests at all well and the fact that the ITC was prepared to endorse it is indicative of the extent to which the war had shifted the overall balance of power in favour of the USA.[140] The best that can be said is that it was far superior to the policy advocated by the 'thickheaded' Malayans, which would have seen an even greater shift in power, with the tin supplied at fire-sale prices.

One familiar discordant voice argued a position not heard for over a decade. Mair criticized the revised ITC/MRC agreement in the following terms:

> Purpose of control is to prevent the prodigal exploitation of vanishing asset …now asked to produce as much as possible of our capital assets not for the sinews of war, but to create a strategic pool in another country, probably for post-war use … to do this we have to deplete our reserves and our shareholders derive no financial benefit, thanks to the royalty and the Excess Profits Tax … prodigal exploitation of a vanishing asset as far as our shareholders are concerned and such a procedure is completely divorced from any sane control scheme.[141]

This was the same language that Mair had used to promote the case for restriction in 1929. In spite of all his criticisms of the activities of the ITC, he now revealed his fundamental sympathy with its original aims. Like his previous criticisms, this one failed to recognize the political context within which an intergovernmental body had to operate.

Ratifying the fourth agreement

Ratification proved to be far more complicated than was anticipated. Malaya did so under protest, still wanting a short term agreement.[142] Bolivia was even more reluctant since she saw withholding support as creating an opportunity for strategic negotiation with the MRC.

The MRC/Bolivia contract had tied its base price to the MRC/ITC contract at a discount of 1.5 cents. While this may have been reasonable when it was signed in late 1940, 12 months later many producers claimed that rising costs made the 48.5 cents unprofitable. There was, therefore, considerable pressure on the Bolivian government to renegotiate the price terms of the contract.

The MRC/Bolivia contract would also directly affect the Patiño/UK contract. Immediately following the Japanese occupation of Malaya, the Ministry of Supply took direct control over UK tin supplies and the LME was then closed. In the absence of a LME price, the Patiño contract had to be redefined as one of strict parity with the terms of the MRC/Bolivia contract. Bolivia now considered that signing the new ITC agreement would bind her to the 50 cents clause in the MRC/ITC agreement. At the same time she felt that the MRC needed the agreement in order to hold the Dutch and the Belgians to that clause.

Bolivia had a good excuse to hand. With the pre-emptive destruction of much of the tin industry in Southeast Asia in the face of the Japanese advance, the standard tonnages accorded to Malaya and the NEI ceased to have even potential operational significance. Since Patiño would or could not attend meetings in London where the issue might have been thrashed out, inconclusive cables went back and forth between him and Campbell. From January to March 1942 Campbell's patience was sorely tested by the way the Bolivians had continued to raise every sort of difficulty which pushed him to the point where he considered terminating any further plans for renewal.[143]

In the absence of immediate ratification, the MRC concluded that the agreement would not in fact be renewed and it simply set about negotiating directly with the Dutch and the Belgians, expecting to hold the 50 cents as the basic price for all its contracts.[144] From then on the ITC was largely irrelevant to US tin policy. The MRC was in a strong position in dealing with Bolivia. Construction of the smelter would not be completed until early 1942 and it would be many months before the MRC began to feel the pressure of any shortfall in supplies. Bolivia's leverage against the US was primarily political[145] but the MRC also had to deal with the considerable pressure from the UK to prevent the price increase that it would have to match. However, by June the Bolivians had won their struggle in Washington, the MRC/Bolivia contract was amended and Patiño was now authorized to complete the process of ratification.

It was not until September when Patiño could reach London and physically put his signature to the document which would make the fourth agreement at least a legal reality. It contained a supplementary protocol which simply stated that standard tonnages of both the NEI and Malaya would have to be revised in light of a new assessment of their productive capacity. Such a statement had already been conceded many months earlier, so Bolivia had achieved nothing as a result of this delay, except to confirm her reputation as an untrustworthy 'blackmailer'.[146]

With formal ratification complete, the ITC reminded the world of its existence and not only announced the terms of the new agreement but added that the quota would continue as before, at a nominal 105 per cent on the higher tonnages:

> The policy of the Committee is to encourage during the war period the maximum possible production of tin in all territories not under enemy control. The quota has been continued (until further notice) at 105 per cent of the standard tonnages. The Committee have, however, decided that, if any territory should be able to exceed this amount, the quota will be increased, retrospectively and generally, in order to cover the actual production of that territory.[147]

With that the ITC went into 'hibernation' for the rest of war.[148]

The announcement of the ratification of the agreement caused both surprise and some temporary consternation both in London and Washington. While the Foreign Office had kept the US Embassy informed of the laborious process of negotiations with Bolivia, it was too routine to pass on. Since the relevant British ministries were also out of the loop, so too were their representatives in Washington.[149] With the signature of the British government and with no change in the overall objectives of the scheme, the Americans were naturally suspicious that this represented an attempt on the part of the British to 'maintain in the postwar world, the status quo', regardless of the new forms of Anglo-American co-operation in tin policy that had now been implemented for many months.[150] Such an impression was reinforced by an article in the *New York Times* which claimed that the ITC was incompatible with Article IV of the Atlantic Charter,[151] where Roosevelt and Churchill declared: 'They will endeavour, with due respect to their existing obligations, to further the enjoyment by all States ... of equal access to the trade and to the raw materials of the world.'[152]

While the British were eventually able to dispel the impression the announcement had created, at least in some official circles, it did not augur well for the continuation of the ITC which had failed to think through the implications of changes in the structure of the industry for its effective regulation. Had the agreement been ratified before December 1941, it would have been easily seen as a measure required to ensure continued supplies during the war and to liquidate any remaining tin in the American stockpile when it was over. But in September 1942 a termination date of 1946 was regarded as a serious bid to continue control without any serious modifications in the structure of the agreement.[153] Bolivia had not only irreparably harmed her own reputation but also that of the ITC itself.

Buffer stock operation and termination

By the end of 1939 the buffer stock was entirely in cash and, in accordance with article 23 which required that amounts over £2.5 million be distributed to the producers, just over £1 million was paid out between November 1939 and February 1940. With prices controlled and sterling further depreciating, the Dutch sought some compensation and argued for distributing the remainder. In this they were supported by other producers such as Hochschild. This would not terminate the scheme itself since they offered to pay the money back should it be needed!

Administrative protocol required that the Dutch proposal be presented to Nigeria and Malaya. The Governor of Nigeria dutifully consulted the local leaders of the industry and found them unanimously opposed to a further distribution. He reported that 'They contend, rightly in my opinion, that the existence of this fund is essential to the restriction scheme and that the restriction scheme is absolutely essential to the healthy existence of the tin mining industry in Nigeria.'[154]

Not only did this reveal a fundamental change in attitude to both the principle of restriction and the role of the buffer stock within it but it also elicited the following comment from the

Colonial Office: 'Fortunately Nigeria and Malaya have produced the right answer which we should have had to impose on them anyway for exchange reasons.'[155] Tin politics were clearly subordinated to metropolitan interests.

In early 1940 the Dutch dropped their request. Holding sterling was now more attractive and as the ITC moved towards re-imposing restriction, the remaining cash would be a useful resource in case it had to re-enter the market should the quota turn out to have been excessively generous.[156] A year and a half later, the case for retention was even stronger, since this would work in tandem with the MRC. Co-operation with the MRC meant a joint responsibility for stocks in which the MRC would use its stockpile to remedy the mistakes it made on the consumption side and the ITC would use its buffer stock to remedy mistakes made on the production side. If the MRC stockpile were completed by June 1942 but the MRC were still committed to buy at 50 cents until June 1943, the ITC buffer stock would prevent the reopening of a dual price system at which the 'Americans would be very reasonably annoyed'.[157] The buffer stock was therefore retained in the proposals for the fourth agreement.

Within a few weeks, the Japanese invasion of Southeast Asia made the case for the buffer stock redundant. The prospects of needing the cash were now remote and British companies operating in Malaya could use the proceeds. Since, even on a care and maintenance basis, the buffer stock cost £3,000 per annum, the ITC recommended that the whole scheme be simply wound up.[158] Where possible, distributions were made in the summer of 1942. The final payment brought the total distribution to £230/16/– per ton which would have shown a substantial profit but only to those who kept their accounts in sterling.[159]

Terminating the scheme raised some difficult legal issues. The ITC took the position that since the buffer stock agreement was between member governments, it was governments that had the responsibility of distributing the proceeds to the producers. Since no transfer could be made to either Thailand or French Indo-China, the proceeds to their producers had to be held in abeyance.

British companies operating in Thailand decided to challenge this decision in court. It took another two years before the case was heard but it clarified an important constitutional issue. The government was not an agent of the companies. The judgement went on:

> This was a Government controlled scheme providing for the marketing of the tin. The Government intended to deal fairly with the producers, but the producers had to rely on the Government for that. They could not go behind the Government and claim on their own behalf the proceeds of the sale.[160]

Legally, therefore, the tin was the property of the government. This was not a dispute to which the Thai government was a party and it would have had no recourse had the ITC authorized the buffer stock executive to make the cheques payable to the companies concerned. It is therefore the best test of the principle on which Campbell had always insisted, governments were in control. For the sake of pure form, he was quite prepared to see real hardship.

While the proceeds were the legal property of the two governments, they could not be transferred to them until the war was over. When the Thailand case was finally adjudicated, they could be transferred to the Custodian of Enemy Property. But he was not prepared to receive those on behalf of French Indo-China, which he considered outside his jurisdiction, as this was not an enemy territory but rather enemy-occupied. Until that issue was resolved, the buffer stock executive had to remain in existence,[161] which meant that Mills and Ellinger continued to draw their salaries for some time even though they had absolutely no work to do. Not only were the funds in legal limbo but so was the executive, since it had been created under the authority of the third agreement, which was now defunct and the fourth agreement was a legally separate instrument. Resolving the anomaly of employees without an employer took up much of 1942 and it was not until 1949 that the Thai producers received their cheques.[162]

Since the ITC would only continue to meet to discuss its own future, the resolution of the buffer stock problem marked the end of its active life. With its completion, it is now possible to undertake a review of both that life and its assessment.

16 The International Tin Committee and its critics

Throughout the whole of its life, the ITC was the subject of considerable controversy and the issues in contention can be viewed from three different vantage points, depending on the status accorded to the principle of cartelization itself. The first suspends judgement on that principle and simply reviews the experience of the ITC against its professed objectives. The second applies the analytic categories of neo-classical economics to challenge that principle. The third steps outside those categories which results in casting the principle in a more positive light. All of these three positions have a predominantly normative objective and this chapter will consider the strength that lies behind the main arguments they have supported.

The ITC and its objectives

The ITC had a conceptually simple task: match production with consumption around a normal level of stocks. A price level which was both stable and reasonable would then emerge in the interests of both producers and consumers. Since the critics always challenged this principle of legitimation by referring to the limited role played by consumers in the deliberations of the ITC, that issue will be considered first.

Consumer representation

Although consumers are just as heterogeneous as producers, the ITC operated with a rather narrow definition. Any interest of the ultimate consumer was discounted on the grounds that tin was an insignificant proportion of the final products. Consumer meant industrial user, in particular the fabricator of tinplate, and that interest was defined primarily as one of price stability.[1] Even the Anti-Trust Division of the US Department of Justice was forced to conclude: 'So far we have not found much consumer dissatisfaction with the cartel's price, so long as the range remained narrow and stable.'[2]

However, in spite of the absence of explicit consumer criticism and the many measures taken by the ITC to stabilize prices, including those taken in 1937, it was always subject to the criticism that it gave inadequate weight to the views of consumers. The focus of the criticism was constitutional. Consumers had no voting power and were simply invited to tender advice that others would take into account when they made their decisions.[3] As a result the ITC could be simply defined as a producers' cartel. The issues proved to be far more complex.

A fuller discussion of the role of consumers occurred in the Colonial Office in 1941 when a regulation scheme for sisal was being developed by the NEI and the UK. This took place against the background of the prejudices constantly revealed by the Americans about the operation of both the ITC and the IRRC and the limited role that consumer representatives played within them. Unless consumer representatives were granted the same voting rights as producer representatives, the schemes were inevitably defined as unable to meet their paramount objective of serving consumer interests. Such symmetry has a superficial attractiveness since it corrects some of the apparent deficiencies of regulation schemes exposed by the neo-classical position.

Caine, however, considered that granting any vote, let alone an equal one, could create some serious difficulties. The problem is that the symmetry is quite specious since there is a more fundamental asymmetry of risk.[4] Producers only mine tin; tin is a tiny fraction of consumption. Producers suffer heavy financial penalties if the quota is set too high, both in the lowering of prices and the costs of the subsequent adjustment. Industrial consumers rarely, and then in very special circumstances, faced an actual shortage of metal and any undesirable rise in price as a result of setting quotas too low could be far more easily absorbed.

Voting allows for coalition formation. The distribution of votes on the ITC ensured that no two countries could either force or veto a position. That effectively meant that a vote as such was normally redundant, since the effort required to broaden the winning coalition generally included everyone. Granting consumer representatives votes would allow for a new kind of coalition formation, in which a minority of producers could outvote a majority and 'create an atmosphere of discord and suspicion within the committees which would gravely imperil their smooth working'.[5]

One response to Caine's position endorsed his judgement about representation as it applied to manufacturers but wanted to see the ultimate consumer play an active role in shaping policy. Since technical innovation tended to lower costs, cleavages emerged on both the producer and the consumer sides. Manufacturers had no interest in lowering prices since that would reduce the value of their inventories. Instead, Rowe wanted the dominant coalition to be formed by representatives of the final consumers and low cost producers which would translate lower costs into lower prices.[6] These proposals illustrate the limitations of the economist's perspective in dealing with the practical problems posed by an unstable commodity market. Assembling such a coalition would itself be difficult given the different roles played by governments and producers but translating it into the basis of a coherent policy would require far more information than was available to the ITC.

For Campbell, the practical issues were naturally paramount and considered that 'it may be desirable to have Gvt reps on the control cttee to safeguard consumers' interest, but Gvt reps will have to be supported by trade reps – manufacturers'.[7] Unfortunately, manufacturers tended to overestimate consumption patterns and the experience of both tin and rubber suggested that estimates by American representatives were generally less accurate than those of the producers. However, Campbell recognized that there was an overriding case for better consumer representation:

> Ideal would be to recognise the consumer interest fully and devise such measures as will associate consumers closely as possible with control and to place on them a clear responsibility for the decisions taken … quite willing for US representative to have vote. … in practice – and one can hardly insist too strongly on this – the pivot of the whole thing is the personnel of the cttees. The utmost care shd be taken to get good cttees and to keep them good. Rubber had consistently good press, tin consistently bad press; special reasons for this; but the one is a good cttee and the other is a bad cttee. Need enough really broad-minded and able men; quick in debate; honest in their outlook;

> versed in affairs; and willing to look steadily at the broad issues where they conflict with smaller sectional interests. There has never, at any time, been effective Amn consumer reps on the tin cttee despite the scheme's provisions.[8]

That assessment was fully shared by the State Department.

As long as Feis remained in the Department, the American position on tin was always ambivalent. It only had one consistent principle and that was to support the construction of an emergency stockpile. Otherwise, concern about the power held by the ITC generated a debate which shifted between devising ways of reducing that power and ways of influencing its exercise. Reduction was contingent on a side deal with Bolivia to feed an American smelter but that ran into an insurmountable problem. If prices dropped substantially, then so too would supplies from Bolivia.[9] Increasing influence was a more promising strategy.

Although the State Department always invoked the abstraction of 'consumer', its ultimate reference was the American purchaser of final products.[10] Given the extensive scale on which regulated capitalism operated in the United States, other Departments offered little support. Agriculture was directly involved in domestic price maintenance schemes; Commerce recognized the value of the export cartels that operated in steel and copper.[11] It was only in tin and rubber that the consumer issue could become politicized. In the case of rubber, the existence of a comprehensive trade association, the Rubber Manufacturers' Association, ensured an active interest in the work of the IRRC. The absence of a similar organization for tin meant that the State Department felt it had to take a particularly prominent role. Feis summarized the difference:

> The iron and steel industry which is the largest consumer of tin has not in the past been particularly interested in this question and the burden of consumer protection has fallen almost wholly on this Government. The representative of the tin consumers has not been particularly active in carrying out his functions. In this connection the tin consumers in this country stand in decided contrast to the rubber consumers who in recent years have maintained a real and active liaison with the IRRC. Despite the foregoing, we have no suggestion at the moment for revision of consumer representation on the committee.[12]

Consumer representation started on a poor footing since Hughes felt that the issues before the first ITC he attended in 1934 had been largely settled informally before it commenced and then simply sent his regrets to the next 20 meetings. After the meeting in June 1937, he indicated he had 'no intention of attending the ITC or becoming more active in his position'.[13] This indifference on the part of both the organizations which had nominated him, US Steel and AISI, extended to an antipathy to any dealings with Washington. Hughes was eventually persuaded to talk to the State Department but 'gave the impression that tin consumers had only one desire and that is to be let alone by agencies of the government'.[14] Even the Department's efforts to try and define a reasonable price were treated with suspicion and it was 'only with the greatest difficulty that Veatch got him to understand that the Department was in no sense acting hand in glove with the Governments represented on the ITC',[15] to get consumers to endorse the buffer stock price range.

This lack of interest and support from the immediate consumers did not, of course, deter the State Department from attempting to articulate the interests of the ultimate consumers. However, diplomatic representations about lack of consultation and about price levels remained at the level of vacuous generalities and could not be pressed effectively without hard evidence.[16] The Department therefore moved to get the AISI to replace Hughes and managed to solve one problem. The European representative of the Steel Export Association of America was based in London and Todd then started attending meetings of the ITC from June 1938. Todd worked closely with the American Embassy and at least now the Department received copies of all official ITC correspondence including minutes of the meetings.

Todd found Campbell's conduct of the ITC meetings rather inhibiting as they 'were stiff and formal', but Lowinger was a frank and helpful guide to the various personalities and he began to 'develop friendships with members outside of the meetings'.[17] Had Todd stayed in London and had the ITC remained focussed on regulating the industry under peacetime conditions, it is possible that a functioning relationship between American consumers and the ITC could have been built up. War put an end to both.

Although the war created an even greater interest in the ITC, no attempt was made to provide for better representation. Todd continued to serve on the ITC and, even after he had severed his relation with the AISI, he managed to attend one meeting when he found himself back in London. However, once the Department of Justice began to take a more assertive position on the whole question of international cartels, the AISI began to fear that any association with the ITC might bring about an anti-trust suit[18] and Todd eventually resigned. Indeed, when Justice and State actually met to discuss the issue, the former took the extreme position that 'any participation by this government might eventually prove embarrassing and that even the government representative … might well be withdrawn'.[19]

By that point Feis had left the State Department and was no longer in a position to shape tin policy but his observations suggest one direction it might have taken, had the prejudices of the Justice Department been set aside. In an interview covering a wide range of issues concerning tin, Feis made it clear that he had no respect for Campbell and that the operations of the ITC, at least up until 1940, were 'quite bad'. However, he thought the ITC could be reformed:

> Mr. Feis feels that this country should at least begin the post-war period by attempting to cooperate with the Tin Committee. He said our active participation on the Committee would be welcomed by the British and the Dutch and he thinks that if we made an early agreement with them we could just about write our own ticket. He thinks we should ask for representation on the Committee, which representation should be evenly divided between the Federal Government and consumer industries. He sees no point in having a delegation of more than two or four men and he sees no point in asking for formal voting power inasmuch as the producer groups are comparatively numerous. He feels that to ask for a veto power would be rather out of order, and suggests that a similar effect could be had from an arrangement under which it would be contemplated that if the U.S. delegation dissented at a specified number of meetings, it would be understood that American participation would thereby be ended.[20]

Critics of the ITC have correctly identified the absence of effective consumer representation as a serious weakness but have mischaracterized its nature. Hughes' perfunctory participation meant that there was no one who enjoyed the confidence of either the American government or industry who had any grasp of the debates that went on within the ITC. A dangerous situation then emerged, to which neither Campbell nor Feis responded creatively. Campbell simply ignored it and Feis first encouraged the McReynolds Committee to engage in a condemnation of the ITC and then sent diplomatic protests about its operations. Since neither drew on any hard evidence, Campbell could effectively ignore them too. That evidence could only be obtained from an industry which was prepared to supply information to a comprehensive representative trade association and one willing to co-operate with the government. Those conditions were far from being realized in New Deal America. The convergence between the views of Campbell and Feis during the war suggests that had they engaged in a frank exchange when the failure of the Hughes' appointment was first apparent in 1935, at least the problem would have been clarified, without being diverted into the irrelevant constitutional question of what voting rights consumers' representatives should enjoy. Had it been solved, the ITC would have found itself much stronger.

Stock control

Defining and then stabilizing normal stock levels was the greatest challenge for the ITC. There were four critical ratios, of which the most important was that between the stocks held by consumers and their anticipated rates of depletion and augmentation. Since these rates were a function both of actual consumption and speculation by consumers on likely price trends, the ITC was always in a position of second-guessing consumer behaviour. That ratio then determined the second, between invisible consumer stocks and stocks in each of the consuming centres, though this was also affected by speculative decisions on the part of their holders.

From the perspective of the ITC, rational stockholders should build up their own stocks when prices are low and visible stocks high and then run them down when prices are high and the visible low. The costs of holding stocks would then be more than compensated by these price changes. However, American behaviour was often just the reverse and that made it very difficult for the ITC to estimate real requirements.[21]

The other two ratios involved an element of time: one was between the level of stocks in the pipeline and that currently in consuming centres and the other was between the level of stocks in LME warehouses and the volume of forward trading on the LME itself. There is no reason to suppose that there is any point of stability at which all of these ratios can be perfectly aligned with each other. Price fluctuations simply reflect the fact that as any one of these ratios reaches a point of normalcy it often upsets the others.

While the existence of a buffer stock tied to the ITC could absorb some of the imbalances between these ratios, it created problems of its own. Unless the ITC were to sponsor a central selling organization, a certain proportion of the stocks in the consuming centres had to be held outside the buffer stock. The question then becomes one of determining the right ratio between these two sets of stocks. In addition, the existence of a buffer stock itself affects the first of these ratios as consumers let the buffer stock carry the costs of stockholding and run down their own invisible stocks.[22]

The most important indication of the presence of a real problem with stocks is the inability of consumers to obtain supplies. That appears to have arisen on very few occasions and then it was confined to a few 'hand-to-mouth' users.[23] Otherwise, the problem with short stocks was experienced by the LME dealers as they closed out their speculative positions.

Once the initial excess stocks had been worked off, there were two periods when stock levels were out of line with consumption trends, from mid-1935 to 1936 when they were too low and 1938 to mid-1939 when they were too high. Only for the 12 months of 1937 did the visible supply approximate its normal level and even then the invisible level was somewhat below its norm. Both discrepant periods were a function of the same problem, an inability to change production levels sufficiently quickly in response to changing patterns of consumption. While much of that rigidity was a function of political factors, the ITC was always handicapped by the absence of reliable information on consumption trends.[24] While more responsible consumer representation might have improved the quality of information, there is an irreducible uncertainty that only a properly functioning buffer stock could absorb.

Price stability

There is little point in trying to assess the performance of the ITC in this area against an artificial target. Here, the best that can be done is to compare fluctuations in tin prices with those of other non-ferrous metals.[25]

Since tin has a unique market structure, comparisons can only generate overall impressions but they serve to put some of the issues in a wider context. The experience of lead and copper[26] serve as the basis for suitable comparisons since they were both widely traded on the LME. However, unlike tin, the LME price for these metals was not a world price since American producers enjoyed substantial tariff protection behind which they could develop a system of administered prices. There are, therefore, three kinds of comparisons that can be made. The first is from the perspective of American consumers which focuses on prices set in the various American markets. The second is from the perspective of the ITC and other consumers which focusses on prices set on the LME. The third is more relevant to this study which is the comparison between the price for standard tin on the LME and that for Straits in New York, together with one which shows the effect of price stabilization mechanisms.

Since a functioning commodities market, such as the LME, requires some range of fluctuation, it is important to set a threshold beyond which fluctuations are considered problematic or 'violent'. Such thresholds are inevitably arbitrary and the one selected here suggests that fluctuations of up to 5 per cent from one month to another should be considered within a normal range.[27] Table 16.1 presents a set of comparisons, over four different periods. The first covers the whole period from July 1933 to August 1939. The second and third control for the effect of the pools linked to the ITC. The last looks at the period of particular turbulence from July 1936 to December 1937.

The results can be summarized as follows. On the LME, tin shows a consistently better performance than either lead or copper. The one exception is the measure of fluctuations in

Table 16.1 Monthly price fluctuations, 1933–1939

	USA			*LME*		
	Tin	*Lead*	*Copper*	*Tin*	*Lead*	*Copper*
A: 7/1933–8/1939 (74 months)						
Average monthly	3.4%	3.1%	3.0%	3.0%	5.4%	4.0%
maximum	21.2%	18.3%	17.5%	20.9%	20.7%	18.8%
No. Months over 5%	17	14	19	13	31	18
Average over 5%	9.0%	9.7%	8.8%	9.9%	9.4%	8.8%
B: Pool support (35 months)						
Average monthly	2.4%	2.2%	2.1%	1.5%	4.1%	3.7%
No. Months over 5%	4	2	6	0	11	7
C: No Pool support (39 months)						
Average monthly	4.2%	3.8%	3.7%	4.3%	6.4%	4.6%
No. Months over 5%	13	12	13	13	20	11
D: Most unstable (17 months)						
Average monthly	5.6%	5.7%	5.8%	5.8%	9.0%	7.1%
Maximum	21.2%	17.5%	17.5%	20.9%	20.7%	18.8%
No. Months over 5%	8	8	9	8	13	8

Source: Minerals Yearbook.

excess of 5 per cent for the whole period. Although the number of months on which such fluctuations occurred is less than either lead or copper, the size is much greater than lead and slightly less than copper. Tin, therefore, demonstrated its volatility on fewer occasions but when it did, it proved to be much stronger.

On American markets the picture is entirely different. Tin was generally worse than either lead or copper. This is to be expected, since the dollar price of Straits reflected both the competition between Straits and standard and the fluctuations in the dollar/sterling exchange rate. However, there was one important exception and that is during the particularly turbulent period when tin did slightly better than either lead or copper. In light of the fact that this period covers the uncertainties associated with the renewal of the third agreement, that is not an inconsiderable accomplishment.

These results suggest yet another basis to whatever cleavage emerged between the ITC and American consumers. The ITC was inevitably focussed on the London market and could take considerable comfort from the fact that tin outperformed the other non-ferrous metals. The Americans, however, were more likely to measure the price stability of tin against that of other metals in their domestic markets in which it did poorly. The American problem was a combination of two factors, greater stability on the lead/copper side and greater instability on the tin side. The comparison between the LME and the American markets demonstrates the former and the latter was a function of the absence of any mechanism to stabilize the price of Straits as opposed to standard tin.

The effect of the existence of specific price stabilization measures is clearly demonstrated by these data. During the periods in which they operated there were no months with fluctuations over the 5 per cent threshold and the average was as low as 1.5 per cent. However, these were periods in which fluctuations were much lower in the other metals as well. A situation which was generally more stable was, therefore, stabilized to a greater extent.[28]

The neo-classical critique of tin control

The beauty of neo-classical economic analysis lies in its ability to identify an optimal pattern of allocation of resources, against which any particular distribution can be evaluated. The Archimedean point of optimality is to be found in the position of consumers whose welfare is to be maximized by acquiring goods at the lowest price. Producers are only granted entrance to the economy by virtue of the 'hidden hand' assumption,[29] whereby their search for profit in a competitive market leads not only to the provision of these goods but to innovation and an eventual lowering of prices.

An inelastic demand curve allows producers to increase their incomes by cutting production and raising prices. If there is no difference in cost conditions, the result is a simple cleavage between producers and consumers in which the resource transfer raises the income of the former at the expense of the latter. Since reducing production entails a rise in costs, total welfare is actually cut. Since the elimination of competition reduces the incentive for innovation, the total expense to consumers increases over time. These problems are compounded where there are differences in cost conditions among the producers and when restriction levels are set at the point where the price preserves those with the highest costs. The net value of the resource transfer is then reduced.

The greater the gap between high and low cost producers, the higher the level of restriction. For low cost producers, this increases the gap between their optimal position and the actual result. That is both a result of the higher costs imposed by lower production and the way in which it inhibits further innovation. As long as they judge that the actual suboptimal result is less than the one produced by a competitive market, there is basis for co-operation between all producers. However, since the low cost producer will always seek to try and change the terms of that co-operation in order to get closer to their optimal position, an inevitable cleavage is set up between those that are burdened by high costs and those that enjoy low ones. Should they judge that an unfettered market would produce the least suboptimal result, they will seek to break away and point to their greater mutuality of interest with consumers. Given all these considerations, it is not surprising that neo-classical economists are strong critics of restriction agreements.

No one who has relied on this form of analysis to criticize the tin cartel was under any illusion that the market solution to the problem of tin would be painless but advocates were divided on how it should be borne. For Eastham, this raised no moral issue. His diagnosis of the problem of tin rested on the assumption that high cost production had been induced by fresh speculative investment prompted by the artificially high prices of the mid-1920s. The market would therefore drive out those who should not have entered it in the first place.[30] Others recognized that the problem had far wider causes which inevitably raised a moral issue. At one extreme was Copeland who advised:

> We must … look facts squarely in the face. History is strewn with examples of the suffering of those who could not … adapt themselves to revolutionary change … The hardships borne by those submerged by these changes are one of the prices paid by society for progress and the attainment of higher standards of living for those who accommodate themselves to the changes.[31]

At the other was Rowe who assumed that a free market price would have been £50 in 1931 which prevented the necessary operation from being undertaken, since 'as a surgeon, laissez-faire cuts unnecessarily deep, while its treatment in general is so rough that the whole constitution of the patient will be undermined, and in any case is so brutal that the patient will run away'.[32]

Instead, the necessary cuts should be deferred until demand has recovered, at which point low cost producers will find it in their interests to return to a free market. Prices will be high enough to cushion the blow for those expelled from it and the problem is finally solved under the most optimal circumstances. All that needed to be done to ensure that it was actually brought about, was to build into the ITC agreement a further clause that would have triggered its demise.[33] Here we have a more formal statement articulated within the neo-classical framework that justifies the ITC as a temporary phenomenon.

Much later Rowe returned to this issue and assumed that the free market price would have been considerably higher at £100 in 1931. In that scenario, the ITC was redundant since excess capacity would have been eliminated had the market been allowed to operate for just a 'year or two longer'. His sympathy with the victim was now confined to Bolivia which would have to been given some breathing space, on purely social grounds, with which to move her economy away from dependence on tin.[34]

Before considering how both versions reveal an inability to recognize some important practical and conceptual problems, it is worth comparing Rowe's assumptions with the results of a counter-factual model of the tin market in the early 1930s. A Dutch economist constructed one in order to determine the longer term price level.[35] Price was treated in the first instance as a function of the interrelationship between three variables: demand, stocks and production. Since price affects each of them, Schut computed five scenarios. While they produced very different results as far as price is concerned, none envisaged a solution to the problem of excess stocks. The best case scenario predicted that the conditions of 1931–1933 would have resulted in a 1934 price at around £155 and stocks at 42,000 tons; the worst case one, £40 and 80,000 tons. A more intuitive reading of the results led Schut to conclude that the final price range was likely between £130 and £140, a level at which few producers could meet their costs.[36] This simply confirms what the key participants in the decisions to establish and maintain the ITC sensed, that they simply could not afford to pay the price exacted by the market since there was no prospect in sight that it would ever deliver any compensating benefits.

Different practical problems attach to each of Rowe's scenarios. The first fails to recognize the difficulty in determining the soundness and duration of the new level of demand that would trigger a termination clause. The second offers no solution to the obvious question. How is Bolivia to be granted a breathing space, especially one that would have to last around two decades to enable her to overcome the constraints imposed by her economic geography?

Both suffer from the same conceptual problems. Even if were possible to predict the capacity needed over the full life of the commodity cycle, the fact that minerals are depleting resources makes it impossible to determine what part of current capacity is permanently obsolete. If the knife has cut too deep, repairing the damage will come at higher costs. While alluvial deposits remain intact, lode ones do not. Lode mines normally work different grades of ore as part of the same extraction process and price determines the actual cut-off grade. As price declines, the economic cut-off grade rises and the lower the proportion of the total deposit that is extracted. If it is required in the future, working it will entail much higher costs.[37]

Judgements about the future are notoriously unreliable in the case of mining and producers make them in different ways depending on their own level of optimism and risk tolerance, coupled with the degree of financial and political support they can draw on. Although Unificada was a high cost mine and an obvious candidate for elimination, Hochschild had sufficient faith in its long-term potential that he would never shut it down until all its financial resources were exhausted. While that is an extreme case, it highlights the fact that strong financial support may enable high cost mines to survive while some low cost ones are driven out. The same principle applies to political support. Some countries may decide to protect their industries by reducing costs through cuts in taxes, or even subsidize them by acquiring stocks. Unless all follow suit, higher cost operations may survive longer than lower cost ones. The most obvious candidate for elimination, Bolivia, had the most powerful cost-cutting weapon and the greatest incentive to use it. As an independent country, she could reduce the external value of her currency. While the market may produce a neat solution, it rests on a set of contingencies, the outcomes of which cannot be assumed. Survival may be just as much a function of greater faith, patience and support, as of cost.

The most important contingencies lie beyond any assessment. Natural disasters, such as prolonged drought and excessive rain, can severely interrupt supplies from alluvial operations. Political ones, such as the wars over the Gran Chaco and Southeast Asia, result in even longer interruptions. Since the market is innocent as to who bears its pain, in these cases it would be borne entirely by consumers in the form of 'famine' prices. The costs of absorbing capacity that is in excess of current needs may be far less than those incurred when it proves to be deficient.[38]

Constructing tests to determine the extent to which neo-classical analysis reveals some real problems in the key areas of costs and prices is difficult on the basis of existing data. But those which are available suggest that they were not particularly serious. The issues will be reviewed for each of these areas.

Costs

When the problem of excess capacity is solved by idling all productive resources, rather than permanently eliminating a portion of them, producers are prevented from realizing economies of scale. Not only do fixed costs have to be covered by fewer outputs but those costs rise when idle equipment deteriorates. During the first agreement, when restriction was particularly severe, dredging companies, in particular, experienced a major increase in their working costs to order to maintain their equipment.[39] But for remainder of the 1930s, it is impossible to determine the specific impact of the reduction of economies of scale by comparing costs during periods of low production with those of high. Since the cost of all local inputs varied with the level of demand, the costs associated with low production levels were lower than those associated with high levels.[40] Nor it is possible to extract any information about the role of indirect costs from corporate accounts. Directors of British companies were anxious to provide their shareholders with as stable a stream of dividends as possible and that required them to present an annual report at their general meetings which showed a stable net income. Managers could offset a rise in direct costs by deferring expenses that were not immediately required and catching up when total income increased. These equilibrating factors vitiate the utility of any of the available data.

Focus on economies of scale overlooks the way in which restriction generates its own incentives to cut costs. One immediate result was to process a reduced amount of material extracted more efficiently. An increase in the proportion of cassiterite captured in the concentrate also facilitated an increase in the proportion of the metal captured by the smelter. Consolidation of production on fewer units also serves to reduce costs. In this case, the increase in total income is shared between those who concede production rights and those who exercise them. This is most evident when quotas rights are sold but is also present in the case of corporate consolidation, since some of the property rights of shareholders in the high cost company which is absorbed are preserved.

But the most distinctive contribution derives from the fact that a restriction agreement is for a specific period of time and the producer, therefore, must be prepared for a particularly severe pattern of competition should it not be renewed. That prospect, coupled with a firm date, served to concentrate the entrepreneurial mind on cost-cutting measures.

All these considerations mean that the cost approach to a critical evaluation of the ITC requires considerably more work before it can generate any useful results. The price approach appears to offer much more promise.

Prices

While it was always easy to claim that prices were too high, establishing the criteria by which to make that judgement is far more complicated. Some arguments avoid that challenge by pointing to certain unfortunate results and then attributing them to excessive prices. Both the issues of substitutes and free riders play that role.[41]

The issue of substitutes is easily disposed of on empirical grounds alone. There was certainly considerable effort devoted to finding substitutes but this was driven by many other factors apart from the price of tin and in any case did not produce many results. The analytic arguments are equally powerful. If price affects long term demand, then this should only be of concern to producers as they determine an optimal depletion path for their mineral resources. It does not concern final consumers since their actual needs continued to be satisfied. Where substitution of product means displacement of manufacturers, their complaints may well stress the role of price but they carry no more weight than those from any other loser in competitive markets.

There is considerably more empirical support for the assertion that the ITC increased the level of free riding. The free rider problem takes a particular form when dealing with depleting mineral resources, since the level of free riding simply determines two separate depletion paths. It only becomes an economic problem for those who are subject to restriction when their revenue stream is cut to the same level that would have been produced without restriction. Indeed, under some assumptions about the comparative depletion paths, free riding may benefit those who restrict. Free riding certainly strained the moral commitments that lay behind the ITC but the actual quantities involved never came close to the point of eliminating its economic benefits.

Comparison with benchmarks established by other commodities was also used to support the claim that tin prices were too high. Simple statistical observations carry no particular weight unless it can be shown how the benchmark selected is appropriate for the task. The issue is most clearly illustrated around the selection of a benchmark date.

When Campbell responded to the American assertion that prices were too high he referred to pre-depression prices, using the government's own objective of restoring the overall level to1926. Tin was still only at 80 per cent of 1926, while other commodities were around 70 per cent.[42] If the overall thrust of macroeconomic policy was to restore prices to pre-depression levels, then tin was setting a lead that other commodities should follow. If, however, there were more structural shifts in the economy which allowed recovery but at a lower price level, then tin was out of line.

Such an argument was most easily made by pointing to the condition of other non-ferrous metals, which remained low even during the period of 1934–1938. At that stage, tin had reached 90 per cent of the 1923–1939 level, while the other major metals were still only around 66 per cent. Whether this solves the benchmark problem depends on a deeper analysis of market structures.

For tin, the interesting comparative price is that for tinplate. Indices for both Pittsburgh and Welsh tinplate prices show that these were maintained at a level about 10 per cent higher than those for tin over the period from 1934 to 1938.[43] Lower tin prices could therefore have simply resulted in higher profits for tinplate manufacturers.

In engaging in a defence of price along comparative lines, the ITC defined itself as solving a problem of market failure and measuring itself against the results that markets should have produced had they been in a position to work under normal conditions. It could still be criticized on the grounds that it misjudged those results and that it should have pursued a different policy. Before the war, the criticisms from the American government always took that form.

For the ITC, the overriding consideration in relation to price was that of stability and it was always stressed that manufacturers are less concerned about the absolute level of price than in ensuring that their competitors pay at the same rate. The question then becomes one of the conditions under which it is possible to establish stability at a lower price level. This is not a matter of finding a lower point on the demand curve, since fluctuations in absolute levels of demand at any given point can be far greater than changes between points. If there is a solution, it is to be found through stock policy. The issue arises at two quite different moments, when prices are already high and to be lowered; when prices are too low and to be raised.

Prices were too low on two occasions, 1931 to mid-1933 and in early 1938. Prices began to rise out of the first of these troughs as a result of the recovery in consumer demand. The

ITC was not in a position to open the production tap any wider given the need to eliminate excess stocks and could not therefore establish a price target. The situation in 1938 was entirely different since it led to the formation of the buffer stock On that occasion the ITC could have settled on a lower range and it almost certainly would have done so had it realized that stability could only be found at the top or bottom of the range.[44] But the advent of war pre-empted the possibility of rectifying any mistake.

Lowering current prices poses a much more difficult problem. It can only be done by increasing stocks to a point above the top of the normal range of stocks. Unless these are held within a buffer stock, the strategy carries enormous risks, since a slight reduction in demand will result in excess stocks and price will again be too low and fluctuations too high. The experience of 1937 demonstrated the extraordinary difficulty in lowering prices for consumers without at the same time imposing undue burdens on producers. The implications for any assessment of the ITC are clear. For most of its life, the ITC simply did not have the machinery that would it enable it to set and hold price targets. When it finally did, it did not have enough experience to be able to decide precisely how to use that machinery. It must always be remembered that the ITC was an innovation and one that faced a steep learning curve.

One set of relevant benchmarks is given by the operation of the stock exchange which attempts to equalize anticipated returns to capital at a particular point of time. Any variation between different classes of shares over time should reflect the different histories of their respective industries. If prices were too high, then that should be reflected in a better performance on the part of tin companies than others. An overall impression[45] of the comparative position of holders of equities in tin companies is provided in Table 16.2

Different kinds of assessments can be made in relation to the various base years. The average of 1927–1929 covers the dramatic expansion of investment and here it is clear that the ITC not only signally failed to sustain the share values reached then but the companies did worse than any other category, with the exception of rubber. The ITC only managed to sustain the share values of 1929 during the life of the second and third agreements. However, here the performance of tin is slightly better than both the overall index and that of industrial shares for the latter part of the decade. Tin's performance improves if 1923 is taken as a base year, as the one prior to any artificial distortion of share values as a result of the boom of the mid-1920s.

Table 16.2 London Stock Exchange, equities, 1931–1938

					Tin cf		
	All	*Business*	*Rubber*	*Tin*	*All*	*Business*	*Rubber*
1927–1929=100							
Average 1931–1933	69	72	30	43	1.62	1.68	0.71
Average 1934–1938	95	94	61	80	1.19	1.18	0.76
Average 1931–1938	86	86	49	66	1.29	1.30	0.74
1923=100							
Average 1931–1933	87	93	43	72	1.22	1.30	0.59
Average 1934–1938	120	123	86	135	0.89	0.91	0.64
Average 1931–1938	108	112	69	111	0.97	1.01	0.63
1929=100							
Average 1931–1933	73	74	36	54	1.35	1.36	0.67
Average 1934–1938	100	97	72	101	0.99	0.96	0.71
Average 1931–1938	90	88	59	84	1.08	1.05	0.70

Source: Investors' Chronicle.

Several conclusions are consistent with these data. The ITC saved the tin industry from a far worse fate than that which awaited rubber. If the ultimate objective of those who formed the ITC was to preserve the value of their investments during the tin boom, then it was a failure. If that period is discounted, then tin shares did more or less as well as the overall index, which is precisely the result that a competitive market is expected to produce. In that case, the ITC was hardly distorting that market but rather realizing its objective of equalizing the rate of return to capital.

These indices disguise a considerable variation in the performance of particular countries and particular groups. They are best revealed by comparing the actual rates of return to shareholders' equity. This method solves the fundamental problem of disaggregating the portion of the revenue stream that represents a return on the investment, as opposed to that portion which represents the depletion of the capital asset. It rests on the assumption that such depletion is taken into account in the valuation of mining equities on the stock exchange. There are many different ways of calculating these rates of return but the method adopted here rests on a simple pattern of investment behaviour in which a share is bought in the market at a particular point in time, all the dividends received are consumed rather than reinvested and the share is then sold at a later point. Gains and losses on the sale are then added to the dividends and the whole pattern is reduced to an annual rate.[46]

All companies within the British financial sphere of influence with listings on the London, Melbourne and Sydney stock exchanges provide the basis for the following analysis. Only shares that were listed in 1929 and which remained in some form or other throughout the decade are included. The sample is of 103 companies with a capital value of £18.6 million in 1930 and the experience of the various fractions of that capital is presented in Table 16.3.

Overall, the companies experienced a positive rate of return of 7.4 per cent on 1930 share values but much less at 3 per cent on their 1929 values. Companies operating in countries outside the ITC experienced both very high and very low returns but overall their return on both the 1929 and 1930 measures was much lower than average, which suggests that, while free riding benefited some companies, it was generally not a very profitable position. Of those subject to restriction, Malayan companies benefited most on the 1929 measure but least on the 1930 measure. That is largely attributable to the fact that their 1929 share values held up much better in 1930 and the absolute returns were on a much higher low.[47] The imputed weakness is most evident among the Siamese companies, whose share values in 1929 were double those in 1930, so that the apparent advantage on the 1930 measure may be less a function of the extent of any free ride they enjoyed and more a misjudgement about their earning capacity. These data also shed some light on the perennial controversy about the role of Anglo-Oriental. Restriction was not sufficient to do more than preserve the capital values of 1929. For the group as a whole, all it did was generate the same rate of return on 1930 values as for this section of the whole industry.

Although a number of companies did very well on both measures, it was very misleading to point to their experience as representative of the industry as a whole. Unfortunately, comparative data for other industries are not available for either of the two periods used here. But data covering 1929–1939 for all equities on the London Stock Exchange give an average rate of return of 2.9 per cent;[48] for this group of tin companies the return over the same period was 1.1 per cent, or –1.8 per cent if those companies that failed after 1929 are included. While those who bought tin shares in 1930 may have been reasonably content with the return on their investment, those who did so the year before must have regretted that the ITC did not have more power to generate a return at least equal to far less risky investments.

Table 16.3 Internal rates of return to British companies, 1929–1939

	Capital Values		*Distribution of Capital Values by Internal Rate of Return*						*Internal Rate of Return*	
	1930 cf 1929	*1930*	*<0%*	*0–7.4%*	*7.4%+*	*<0%*	*0–7.4%*	*7.4%+*	*1930–39*	*1929–38*
	%	£000	£000	£000	£000	%	%	%	%	%
Other	–45	1,131	691		441	61		39	4.9	–3.0
Siam	–50	1,425	174	222	1,030	12	16	72	12.1	2.7
Nigeria	–45	2,623	350	709	1,565	13	27	60	8.3	1.1
Malaya	–29	13,404	2,514	4,981	5,910	19	37	44	6.9	4.0
Total	–35	18,583	3,728	5,911	8,945	20	32	48	7.4	3.0
Malaya										
Redruth	–19	4,574	1,897	1,369	1,308	41	30	29	4.3	3.7
Tronoh	–15	2,157	345	1,270	542	16	59	25	7.5	7.5
Malyan Tin	–20	1,288		368	919		29	71	7.9	6.4
Anglo-Oriental	–41	2,829		795	2,034		28	72	9.7	2.9
Other	–39	2,556	272	1,179	1,106	11	46	43	6.4	2.1
AO total	–43	5,111	947	1,143	3,022	19	22	59	7.5	0.1

Sources: Stock Exchange, Official List of Making–Up Prices, *Mining Yearbook*.

Note: Capital values, average of March and October.

Beyond the economics of tin control

The first step outside the categories of neo-classical economics is the Keynesian one. That relocates the problem of market failure to the level of the economy as a whole. The collapse of primary commodity prices is seen as one of the causes of the depression and their restoration as one of the important means of priming the pump of general economic recovery. This position was commonly argued by politicians[49] though it was subject to some challenge in the case of tin.[50] Testing the multiplier effect of one pattern of distribution of resources against another, however, is quite beyond the scope of this study.

A much more important move is to fill the abstract categories of producer and consumer with historical actors and insert them into concrete social structures. The following modifications then occur:

First, the only consumer-buyer in the tin market is another profit-making enterprise. Those enterprises produce a product which itself is acquired by another profit-making enterprise and so on down a long chain before a product enters into final domestic consumption. The extent to which prices in any of these markets are translated into the final price paid is a function of their structures. Complaints about high prices may simply be a move in a zero-sum game played by corporations in which none can occupy a privileged position.[51]

Second, as soon as the concepts of final consumer and producer are unpacked, a wide variety of quite different occupants are revealed. On the consumer side they range from the purchaser of a luxury automobile to the purchaser of a can of fruit to supplement a plain diet. On the producer side the range is even larger, from wealthy shareholders living off ample dividends, to shareholders supplementing meagre incomes with small dividends, to well-paid engineers, managers, directors, to poorly paid labourers and so on. Once this variety is recognized, then any discussion of distribution of resources has to be qualified by its social implications.[52]

Third, one of the distinctive features of the tin industry is the fact that it is produced in countries that became the major focus of economic theory in the postwar period. That theory does establish a point of privilege, namely poverty. In the hands of theorists such as Raul Prebisch,[53] the analysis of these market structures provides the basis both for an explanation for poverty and for policy formation. The assumption that lower costs engendered by technical progress in the production of primary commodities should at some point be translated into lower prices enjoyed by consumers is simply the basis of one form of unequal exchange. The welfare gains are reaped by consumers in rich countries without any reciprocal transfer of the benefits of technical progress in manufacturing industries. The oligopolistic structure of those industries ensures that those benefits are not passed on in lower prices but captured in the form of higher incomes. From this perspective, all the ITC was doing was engaging in overt collusion and thereby going some way to countering the covert forms of collusion which lie at the basis of international inequality.

Since this whole body of analysis[54] is couched at a level which tries to grasp the generic properties of market structures, it misses one feature that makes a commodity such as tin quite distinctive and which therefore calls for an entirely different approach. The concept of depletion identifies just a part of that distinctiveness. The most important part is only captured by a fuller discussion of the concept of property.

Property rights and the role of the state

Mineral rights are a distinctive form of property. In most of the world of tin, they were private property, owned by a specific legal entity, the state, with the formally unfettered legal right to decide whether to grant access to the resources and if so under what conditions. Although private property, this is not capitalist property since it was not acquired in the market and it is therefore subject to no economic norms. Just as there are no normative criteria by which to establish whether a distinctive work of art or the license to use a patent is fairly priced, so there are no such criteria by which to determine whether the owner has placed too high a price on access to mineral resources. Owners therefore have a right to decide on whatever price they choose as a condition of making their irreplaceable resources available to the economy and it is a right that permits collusion with other such owners.

One of the first articulations of the distinction between these two forms of property emerged in a dispute between Great Britain and Sicily over the price set for brimstone sulphur in 1838. Sicily defended her policy first by reference to her material needs but concluded by asking: 'If the Island exclusively possesses sulphur, why is she not to benefit by a good which *not man but nature* has granted to her?'[55] Whereas in the nineteenth century, the question was settled by force,[56] in the twentieth, international jurisprudence has enshrined that right as an absolute condition of political sovereignty.

These considerations make the state the starting point for the analysis of a mineral economy such as tin. It grants the state an analytically privileged position without at the same time granting it a normatively privileged one. Separating the two dimensions moves the normative into the political arena.[57] There are many different positions as to how the state should exercise its legal right and each may appeal to various forms of analysis, including that which derives from neo-classical theory, to demonstrate the wisdom or otherwise of particular policies. But, in the last resort there are no universal criteria that can resolve these differences; they have to be left to the standard cut and thrust of politics. Ultimately, the evaluative question reduces itself to a simple one. Did the ITC serve the needs of the governments that sponsored it? Nothing suggests that they acted against their own interests in signing the agreements.

Governments secured two kinds of advantages, the terms under which they participated in the international economy and in their fiscal position. Bolivia provides the best basis on which to examine the effect of the ITC on the terms of trade; Malaya the best basis on which to examine the fiscal impact.

The change in the terms of trade for Bolivia over the period from 1927 to 1938 are presented in Table 16.4.

The net barter terms of trade refers to the purchasing power of tin as measured in terms of imports; actual income is a function of the quantity of mineral exported.[58] The combination of declining barter terms and quantities produced a severe depression in the early 1930s but, once its worst had been weathered, Bolivia was considerably better off, at least in macroeconomic terms, than she had been at the end of the 1920s.

The data also demonstrate the extent to which the ITC was able to reduce international inequality. The British economy showed the same overall pattern of decline and recovery but since the decline was much lower, the gap between Bolivia and the UK widened considerably. Bolivia's faster rate of recovery then narrowed the gap and that was sufficiently strong to make this sector of the international economy slightly less unequal than it had been at the end of the 1920s.

Analysis of the fiscal position of the state has to take into account both the current and the anticipated revenue stream over the whole life of the deposit. Both have to compare the results under restriction with a return to a free market on the basis of some reasonable assumptions about both scenarios. The actual level of production (63 per cent) and price over the period from 1934 to 1938 (£218) is treated as a set of conditions to be expected to prevail

Table 16.4 Bolivia, tin terms of trade, 1927–1938

	Barter ToT	*Income ToT*	*UK GDP*	*UK GDP cf Bolivia*
1927–1929	100	100	100	1.00
1931	48	36	93	2.57
1932	89	43	91	2.13
1933	203	74	93	1.25
1934	190	98	98	1.00
1935	184	137	102	0.74
1936	104	78	108	1.38
1937	141	141	115	0.82
1938	121	83	119	1.44
1931–1933	113	51	92	1.81
1934–1938	159	115	108	0.95

Sources: Import indices from R. M. Swagler, "An Analysis of Variations in Alternative Terms–of–Trade Measures. A Case Study of Bolivia, 1925–1965," Ph.D. dissertation, Ohio State University, 1971, pp. 111, 115, 134. UK GDP, <www.eh.net/hmit/ukgdp>.

Note: 1936 excluded from 1934–8 since Chile's blockade of Arica artificially inflated import prices.

under restriction until the deposit is exhausted after a period of 30 years. The free market scenario assumes an average price of £150 and full utilization of existing capacity, which reduces the lifespan of the deposit.

Such assumptions fit the Malaya case. The effect of restriction on current revenue is quite modest, around 2 per cent, but spreading that extra income over a longer time period increases it considerably. When future income is discounted at an annual rate of 4 per cent, the net present value of the resource is around 30 per cent greater. A rational assessment of the ITC from the fiscal perspective of the FMS government would therefore see it not simply as a bulwark against immediate disruption but as the best way of securing its own interest even after that disruption was over.

Summary

Neither theory nor evidence provides much support for the manifold criticisms to which the ITC has been subjected. Once those assessments of the overall operation of the ITC are set aside, it is possible to recognize some serious mistakes in the actual decisions taken. Three stand out: the failure to cut production sufficiently quickly in 1931–1932, the termination of the first buffer stock scheme in 1935 and the maintenance of unlimited production in late 1937. All were a result of the political environment within which the ITC had to operate and none left any lasting damage. However, as the more general economic criticisms became incorporated into ideology and politics, the fate of the ITC itself would be sealed.

17 The demise of the International Tin Committee, 1945–1946

By 1943 the broad contours of the postwar tin position were clear. In the immediate aftermath of the war, there would be neither a surplus of capacity nor of stocks. However, once the dredges and gravel pumps were reinstalled throughout Southeast Asia, the industry would again face the need to deal with excess production. While the ITC would have no role to play before the expiry of the fourth agreement, the issues that it had been designed to address would inevitably recur.

In March 1942 Campbell retired from the Colonial Office and the tin portfolio was then handled by Clauson. Campbell continued as Chairman of the ITC but in December 1944 he died. With his departure, the ITC was left without its most able advocate within British official circles. Thinking about tin now had to be shaped without his detailed command of a wide range of issues and it was governed by several factors, none of which pointed in the direction of preserving or even modifying the ITC.

Postwar commodity policy

The overall tone of thinking about postwar reconstruction was extraordinarily optimistic. It looked forward to a world free of the restrictive practices that had emerged in the Depression and which had fuelled nationalist rivalries. As Keynes began to consider the whole structure of institutions to regulate the postwar international economic order, he returned to the problem of primary commodities which had played such an important role in trying to shape counter-depression strategies and sketched out the elements of a new, comprehensive model.[1] It was not designed to address the problem of chronic surplus capacity but rather the need to establish price stability. The burden of control would be born by a buffer stock, with production restriction available only as a last resort. Administration of each separate commodity would be in the hands of a body representing producers, with 40 per cent of the voting power, consumers with 30 per cent and the balance in the hands of independents. Each controlling body would in turn be represented on a General Council of Commodities to resolve disputes and help set prices, all subject to the overall supervision of the United Nations. Of all the prewar forms of intergovernmental commodity control, the one that had come closest, though not very close, was the ITC's third agreement.[2]

The general issues were first joined around the question of the future of the rubber agreement which had formally expired at the end of 1943. Franks, from the Ministry of Supply, argued for 'the advisability of breaking away from the past rather than trying to graft the post-war schemes on to the pre-war set up'.[3] While there was a 'confused battle' on this issue, Franks won the day since 'discussions on rubber with the Americans through a body which retained vestiges of the old IRRC might prejudice not only talks on rubber but all future discussions with the Americans on issues of commodity policy'.[4]

Rubber had become an extraordinarily sore point for the Americans, especially in 1940–1941 when the IRRC had moved far less quickly than the ITC in raising quotas and prices soared far above their prewar levels.[5] The IRRC was therefore unceremoniously wound up and future discussions about rubber would be held within a broad group bereft of any power, the International Rubber Study Group. The ITC would suffer the same fate.

These early discussions marked some important shifts not only about the way in which the problem of primary commodities would be addressed but also in the institutions that would take responsibility for policy in this area. While the interdepartmental discussions in London always included the Colonial Office, it had ceased to have primary administrative responsibility.[6] Its perspective was now shaped less by a concern for the overall welfare of Nigeria and Malaya, than by one determined by several other Ministries: Supply, Production, Board of Trade and the Treasury, all of which took a metropolitan view.[7] Without any real experience with the distinctive problems of tin, it was too easy for those other agencies to fit the ITC experience into the simplistic economic model and conclude: 'In effect, the object of the ITC was to protect the inefficient by curtailing the activities of the efficient producer. This object was accomplished at the expense of higher prices to the consumer.'[8]

Within the Colonial Office an extensive debate occurred about the new direction of government policy, prompted by a memo which argued:

> The present state of British official opinion has been noted, particularly the heavy bias against regulation schemes. It is not based on a correct appreciation of the facts of several of the industries affected, and that the policy is liable to have very serious consequences for the Colonies ... official policy requires drastic revision in the direction of reinstating regulation as a legitimate means of dealing with commodity problems. From this it follows that existing regulation schemes should not be allowed to lapse in favour of hazy, untried, and questionable programmes, but should be kept in being ... Rather, they should be reviewed in detail, in the fullest confidential consultation with those who framed and have operated them, in order not only that the voice of the industries concerned may be heard before their fate is decided for them, but also that policy may even now be framed in the light of some acquaintance with the practical issues involved.[9]

While sound, this advice failed to address three obvious problems. One was practical. The schemes had no mechanism with which to police the 'external parasites and the internal parasites', and Clauson felt that if consumers had substantial responsibility for the success of the scheme, they could enforce the kind of discipline that had always eluded the ITC. The second was ideological. They were simply inconsistent with the spirit of the times which was one of 'making a new start in the hope of creating a better world'.[10] More specifically, they helped shareholders, rather than workers and were hardly part of a general commitment to an expansionist trade policy.[11] Finally, no 'appreciation of the facts' could affect the position of the United States which would exercise a de facto veto over all matters of international economic policy.

The results of the debate within the Colonial Office were eventually leaked to the press. Under the heading 'Tin Control Killed', the *Daily Express* reported a rumour that both Bolivia and Nigeria were afraid of competition with a re-equipped Malaya and were pressing for renewal and that the Colonial Office had made it quite clear there could be no renewal of the old form of the ITC.[12] Since few[13] were considering a simple renewal without a reconceptualization of the role of consumers, that was a rather ambiguous message. The TPA put out a brochure which tried to put the experience of the ITC in a positive light[14] but that simply provided a target for much adverse criticism in the press.[15]

By early 1945 it was open season on the ITC. An editorial in the semi-official publication *Crown Colonist* reproduced the old assertion which neatly caught two complaints that had always resonated among certain sections of the Malayan industry: prices were too high and Malaya's share too low. Frisby's charge that the Orient could have supplied the entire world at £100 and still make a profit proved to have considerable propaganda value. In reply to a letter criticizing the criticism, the paper published a more scathing attack which included a quote from Simms:

> whether the effect of the scheme be considered from the standpoint of the producing miner, the mineworker, the consumer, the Government of Malaya, on trade and industry there and in this country, the result is the same. It was bad for all.[16]

Any trace of the earlier vilification of Simms and the others who shared his views had long disappeared.

The general principles designed to govern postwar trade policy were published in early 1946.[17] These were presented as an American initiative and were endorsed as official British policy.[18] They made it clear that the commodity schemes of the 1930s were obsolete and prejudice against them was reinforced by a series of scholarly studies sponsored by the Food Research Institute at Stanford. With financial support from the Rockefeller Foundation, Knorr completed the one on tin in November 1944[19] and its timing could not have been better. It was short, clear and authoritative, just the kind of study that politicians and bureaucrats could rely on as tin issues were again coming to the fore. Knorr provided ample support for Simms' terse assessment and it therefore 'became the "bible" for Washington's tin experts'.[20] The *American Metal Market* recommended the book as 'well worth reading', but considered its conclusions as 'unconvincing to the many who are familiar with the industry'.[21]

Such familiarity was not relevant to the overall debate that emerged about the nature of international cartels. The war left two indelible impressions. Cartels had played an important role in the emergence of aggressive nationalism in both Germany and Japan. Worse, their restrictive practices had inhibited the development of American technology and limited supplies. In short, cartels were both a cause of war and an obstacle to winning it. This provided the opportunity the Anti-Trust Division of the Department of Justice needed to develop a comprehensive programme to dismantle cartels[22] and it soon received the full support of the President and the rest of the administration.[23]

The campaign against the ITC reveals the extraordinarily superficial level at which the Anti-Trust Division operated. Circumstances compelled it to come to quick judgements on complex matters made by people with no background familiarity. The file started with the fourth ITC agreement. It was read as though it was actually negotiated in September 1942 and was considered to be:

> One of the most remarkable cartel agreements which the Anti-Trust division had examined. Japan had overrun the principal world sources of tin in Malaya and the Dutch East Indies, yet the ITC solemnly met and not only recognized the quotas of Malaya and the East Indies, but quite substantially increased them over the remainder of the

> world. Bolivia, on which the Allies must place the main reliance for tin, was granted a negligible increase.[24]

The view that the ITC was indifferent to the war was then widely promulgated.[25] Not only was the significance of the standard tonnages misrepresented but there was no grasp of how the ITC actually operated. It was supposed that in the postwar period the British and Dutch interests could use their power to distribute production entitlements in such a way as to benefit smelters under their control, at the expense of Bolivian supplies to the Texas smelter.[26] Such power could clearly not be tolerated. A review of the delicate way in which negotiations had been conducted between the British and the Americans over the allocation of Bolivian concentrates dismissed members of the State Department as 'perfectly willing to elevate fiction over reality', and led to the following conclusion:

> Obviously … the effect of such thinking amounts to acquiescence in the continued control of the international tin cartel. The perpetuation of this scheme of control makes ludicrous the avowed purpose of the war. Though the United States is … the world's largest consumer, its own public officials here accept a situation which means continued subservience by this country to a cartel in which consumer representation is nothing more than a farce.[27]

One of those public officials smeared was Feis.[28] Within a few months he would resign in frustration at the 'intrigues, cabals, and falsehoods', that had emerged within the State Department and which prevented the formulation of anything 'resembling a coherent economic policy'.[29] Justice welcomed his departure and considered that State's 'official stand on the subject of cartels is now pretty much as we would desire it'.[30]

In case there were any lingering doubts as to precisely where that point should be, the new head of the Commodities Division at State, Donald Kennedy, was told quite forcefully by the Mead Committee. The discussion started with a review of the ITC which led to the following exchange:

Senator Ferguson What about the price? What did they do on the price with the cartels?

Mr. Kennedy I have no doubt, sir, that there were price stabilization activities of the international committee.

Senator Ferguson You call it stabilization. What did it mean? How much more did we have to pay?

Mr. Kennedy I have no information as to how much we had to pay.

Senator Ferguson Don't you think that is material if we are going to do anything in the future? How are we going to solve the problem in the future if you don't know what happened in the past?

Mr. Kennedy I believe, sir, one of the very efficient ways of attacking that problem, is the provision in our proposed convention that there will not merely be observers from consuming countries but they will have equal voice.

Chairman Mead Does that indicate the State Department program, then, is for us to become part of the cartel, and have some influence on it, rather than bust the cartel?

Kennedy then lamely tried to define the ITC as an international agreement between governments rather than as a cartel but Ferguson cut him off:

MR. KENNEDY I wouldn't call it a governmental cartel and we certainly would not want–
SENATOR FERGUSON (interposing). What do you call it?
MR. KENNEDY An international commodity arrangement.
SENATOR FERGUSON That is a nice-sounding phrase. I don't suppose many people would object to that because they wouldn't know what it was.[31]

Congress was in no mood to receive any advice on how international commodity policy should operate and certainly not from someone who could only prevaricate with platitudes. Simon Strauss also appeared at these hearings. His experience with MRC had taught him much about tin and he spoke to one of the great accomplishments of the ITC in preserving Bolivian capacity. Since the future of the Texas smelter was dependent on the continuation of that capacity, the ITC or some functional equivalent was required. But the earlier exchange had made it clear that the issue was 'theological', rather than one of practical politics.

Debating the future of the ITC

The first postwar meeting of the ITC was not convened until September 1945. Van den Broek had been appointed Chairman on Campbell's death and the first order of business was to pay him a suitable tribute:

> He had a wonderful knack of smoothing out difficulties by talking to the various members until a good result was obtained … He was a man, who through the long years of service had taken a completely objective point of view. Although he was also a representative of Malaya and Nigeria and the voting member of the British delegations, nevertheless as Chairman he was always ready to listen to and consider the views of others and, after expressing his own personal view, as Chairman maintained an impartial attitude, never furthering the interests of Malaya and Nigeria to the prejudice of others, and always giving full weight to the various opinions uttered.[32]

Such an accolade would have provided much of the confirmation that the Malayans had always been looking for in complaining that Campbell had subordinated their interests to his responsibilities as Chairman. It is unfortunate that so few had the experience of seeing Campbell at work. Otherwise, they may have appreciated the wisdom that lay behind Houwert's more widely circulated tribute:

> Sir John Campbell was one of those rare personalities whose horizon far transcended national boundaries and who are thus pre-eminently suited not only to appreciate international problems but also to give guidance to their solutions. He knew how to keep united the delegations of the various countries, often with conflicting interests, on the International Tin Committee. He succeeded in the first place, because he himself knew how to subordinate the interest, more particularly represented by him, to pursuing the great lines of general co-operation. But it was no less his personal gifts and his personality that enabled him always to be successful in his attempts at reconciling parties whenever difficulties threatened.[33]

Given the strength of the various centrifugal forces, the fact that the ITC functioned as well as it did must be largely attributed to Campbell's capacity to dissipate them.

There was only one substantive issue to be addressed at this September meeting, the future of the ITC. No one would have been surprised at Clauson's announcement that while he had not received final instructions, 'there was no prospect that HMG in the UK would be prepared to sign another agreement'. But only Wilcoxson was prepared to give him any support. Otherwise, van den Broek strongly argued in favour of renewal and his position was endorsed by Bolivia, Belgian Congo and Lowinger. It was generally recognized that the scheme had to be modified to give consumers more influence and, although the US Embassy sent a representative, there was no willingness to discuss how this might be accomplished. Rubber had clearly set the course: replace what was considered as a producers' cartel with a study group, bring the Americans into the discussion and then, as a last resort, establish a new organization with control shared equally between producers and consumers.[34]

While the Anglo-Oriental and Stephens groups continued to support the desirability of restriction, they were without strong leadership.[35] Nor was the TPA in a better position to get its advice taken seriously.[36] While the Colonial Office felt that, since there was to be 'a certain clash with the industry', a representative of the British industry should be included on the ITC subcommittee charged with making specific recommendations for the future, even this gesture was blocked by the Ministry of Supply.[37] It was therefore left to the three non-British members to press the producers' position and they would not give up without exploring all the available arguments. Peñaranda, the Bolivian Chargé d'Affaires in London, found three: (1) the need for a body capable of taking action quickly; (2) the inconsistency between removing the structure of production control but retaining wartime price and other controls; (3) the contribution made by the producers to the war effort, especially by accepting administered prices.[38] A strong statement from all the Belgian producers reinforced the Bolivian position. The pressure for maximum production during the war had meant that 'they did not hesitate to work their deposits on a high grading basis ... so that several of the producers are now facing impoverished deposits, several of these having ceased to be rentable'.[39]

In setting aside these claims, the British government found itself in a contradictory position. In the name of producer/consumer co-operation, the implications of one form of such co-operation, the wartime tin contracts, were to be ignored. In its name, too, the voice of the anonymous consumer was to be preferred, while that of the actual producer ignored, or worse, banished from the discussion. Also excluded was the voice of the colonial governments. Since it was the British government whose signature was on the fourth agreement, the fate of the ITC was to be settled without regard to the implications for its colonial dependencies. In the past the Colonial Office had felt itself forced to overrule the expressed positions of Nigeria and Malaya as being inconsistent with the long term interests of those territories. Now it was being compelled to subordinate those interests to the overall politics of Anglo-American economic co-operation.

The Colonial Office had a complicated game to orchestrate. It wanted the ITC to accept its fate and see the Study Group as the natural transition,[40] while at the same time demonstrating to the Americans that in endorsing the Study Group the producers had effectively repudiated the ITC. At the March meeting the ITC received the report of its subcommittee on renewal and passed the following resolution:

> In the light of these [Anglo-American] proposals, the committee feels that it would be inopportune to continue the International Tin Agreement in its present form and they recommend that the Government of the United Kingdom be authorised by the other Governments parties to the agreement to invite the USA, the USSR, France, China and

> Siam … to confer with them with the least possible delay with the object of framing an international agreement in regard to tin acceptable both to consumers and producers.[41]

The world was then officially informed that, while a recommendation had been made in regard to the future of the agreement, it was confidential. Confidentiality solved the problem of the Americans. They were 'frightened to death of what Congress might say and it was only by presenting the summons to the Conference as a spontaneous act of HMG not the ITC that we have got their consent to come'.[42] The actual text allowed the producers to suppose that the Conference would provide the forum at which the agreement would simply be modified to make it 'generally acceptable to all parties concerned'.[43]

It was on this assumption that Lowinger and Houwert then set about developing a negotiating document addressing the issues raised in the Anglo-American proposals in light of the impending recreation of surplus productive capacity. They were 'personally convinced US consumers were not antagonistic to ITC regulation', and foresaw 'production excess and repeat of the condition that led to the ITC', for which co-operation would be required and warned that 'if we destroy the partnership atmosphere it will be difficult to regain it'.[44] The fundamental flaw in the Anglo-American position was that in defining production control as the last weapon in the struggle to stabilize commodity markets, a period of chaos had to be tolerated before regulation was again authorized. That period could turn out to be very extended since authorization only came after several governments had ratified an agreement negotiated at a special conference called on the recommendation of a study group.[45]

Clauson was not prepared to rethink the issues that Lowinger raised and tried to shut him down by at first refusing to accept the demand and supply projections and then by rehearsing many of the old arguments about the ITC. It preserved high cost producers, created inefficiencies, prices were too high and he added:

> I can assure you that the UK Government are just as passionately interested as the US Government in seeing that their public is not exploited by a combination of producers. This does not of course mean that either the British or the American Governments wish to rub the producers' faces in the dust, but both of them have an uneasy feeling that the public were made to pay more for their tin than they need have in the past and they draw considerable support in this view from the large profits which many of the tin companies made.[46]

The producers were being asked to trust the consumers, even though the consumer position was based on an extraordinarily superficial analysis of the operation of the old ITC and one which was largely irrelevant to the current situation in which the producers were simply asking for a reformed ITC operating in a quite different context.

Such trust was not willingly granted. Van den Broek even proposed a meeting of the ITC to review the Lowinger/Houwert document and develop a producers' position to present to the Conference. Nothing could have been more damaging to Clauson's plan to secure American co-operation and he only managed to block it by threatening a boycott by the British delegations.[47]

Just as the producers were finding it difficult to co-operate with the Colonial Office so were the Americans. At first, they insisted on several conditions before they could participate in the conference, including an unconditional agreement to jointly recommend that the ITC be dissolved. That would avoid any further embarrassment with Congress which continued to insist that there could be no truck with the cartel.[48] While discussion of the future of the

ITC could not be avoided at such a conference, such an agreement would upset the British plans to let the decision to dissolve the ITC emerge from the conference itself and focus discussion on the Study Group as the best way of weaning the producers to that position. It was not until June that the Americans were persuaded of the wisdom of Clauson's strategy[49] and invitations to a conference to be held in October in London could then be sent to the USA, USSR, China, Bolivia, France, Belgium, Netherlands and Thailand.

With the exception of the USSR, all attended and the conference went according to plan.[50] The International Tin Study Group (ITSG) was established with a permanent secretariat in The Hague[51] and the Americans 'even seemed anxious that it should make plans to hold in readiness for intergovernmental discussion as soon as serious disequilibrium seemed probable'.[52] It proved to be van den Broek's last play. He continued to argue the case for a 'remodelled commodity agreement based on the ITC, but it was a last ditch stand without any prospect of success'.[53]

The final meeting of the ITC was the occasion for several tributes. One went to van den Broek who had recently died. A second went to the ITC itself, since the total cost over the 15 years of its existence was just under £20,000.[54] It was also an occasion for some regret. Although the Dutch had lost the battle for the continuation of the ITC at the Tin Conference, Houwert could not refrain from stating that:

> he felt the demise of the ITC deeply, and with considerable regret … He had attended nearly all the meetings and he had always found the British producers and hitherto the British Government on his side. They had always worked together in the greatest friendliness. He must confess however (and he hoped it would not give offence) that he felt very strongly about the attitude now adopted by the British Government in refusing to consider the reconstruction of the scheme on a wide basis so as to include consuming interests as well as those of the producers. The Committee had worked together most successfully for fifteen years and they, the Dutch, on their side had always shown great spirit of co-operation.

Clauson's response simply vindicated Houwert's charge that the British government had taken its position for political reasons:

> It is no use blinking the fact that the ITC was no more popular in the United States than the IRRC had been. The Executive Branch of the American Government had made it quite clear that even if they themselves had been willing to join the committee, they could not have persuaded Congress to approve an international agreement by which the US would be represented on the Committee. … same situation had arisen in rubber in 1943. In those days there was continuous warfare against the RRC and all its work. Here again an attempt was made to bring in the American Government to the Committee. They refused, but did come to a conference in August 1944 at which the term 'Study Group' was invented …Nobody who had not lived through the experience of Rubber could have conceived of the difference in atmosphere which had taken place … where formerly there had been continual sniping criticism of rubber in the United States, the friendliest possible relations now existed … appears likely would be another rubber control agreement. They could never have got this personal atmosphere if they had forced a reluctant set of American officials into the Rubber Regulation Committee.[55]

On that optimistic tone the ITC came to an end.

Clauson was in fact being quite disingenuous. Friendly relations were simply a matter of courtesy, easily granted when all the political issues were no longer on the table. But three months earlier, Clauson had a very different view of the inherent co-operativeness of the Americans. Commenting on a strategy to get them to close down their Texas smelter on the grounds that it was not economically viable, he added:

> But if they prove bloody-minded and insist on keeping it open, we shall have to make an agreement with them governing the rate of its operation and laying down what proportion of ore is to be taken from each of the producing countries.[56]

Only some version of an ITC could have the muscle required to enforce any such agreement. In trumpeting 'The International Tin Committee is dead – Long Live the International Tin Study Group',[57] Clauson was deceiving no one that the path from the old style cartel to the new style of producer-consumer commodity agreement via a Study Group would be at all smooth.[58] Just how rough it proved to be is the subject of the next chapter.

18 From the International Tin Committee to the International Tin Council, 1946–1985

Tin continued to be subject to control for another 50 years after the end of the ITC. In the immediate postwar period two political imperatives ruled: economic reconstruction of Western Europe and contingency planning against World War III. Tin played an important role in both but the terms on which producers participated were largely dictated by the United States government. Once these objectives were realized in the early 1950s, it retreated from the industry, leaving the producers to confront the problems it bequeathed. To deal with them, producers turned to a new kind of commodity agreement. From 1956 to 1985, the International Tin Council brought producers and consumers together to develop a common policy. The experience of the Council, the source of the problem it inherited and of that it, in turn, left to the industry is the focus of this chapter.

Tin control under the auspices of the United States: phase I, 1945–1949

After 1942 tin was subject to the following controls. At the international level, the UK-US Combined Raw Materials Board (CRMB) allocated the bulk of the tin metal and concentrates available to the United Nations between Britain and the United States. Each had separate contracts with the producers, with the result that a wide range of prices prevailed. Each exercised complete control over domestic end use and established a similar price on manufacturers authorized to buy the metal. Since the producers, Bolivia, Nigeria, Belgian Congo and Cornwall, were all committed to maximize production for the war effort, the result was equivalent to a command economy.

The transition to peace modified these features in several ways. Each of the three main wartime suppliers saw production decline and while that was compensated with the recovery of production in Malaya and the NEI and the seizure of metal in Japan and Thailand, production was not sufficient to meet demand. As long as Western Europe needed tin for its own reconstruction, a return to a free market was out of the question and international allocation controls were, therefore, retained. The CRMB was wound up and in its place a new body, the Combined Tin Committee (CTC) was created.

The CTC allocated not concentrates but metal. All countries were entitled to make an application but were required to include details of the end use. The CTC was a hybrid organization which not only contained representatives of the main countries that controlled metal, which now included Belgium and the Netherlands, but also representatives of some consuming countries, France, Canada and India. It met twice a year to assess the availability of supplies, review the applications and allocate the amounts that the applicants were entitled to acquire. The same set of rules were supposed to apply to all applicants, regardless of whether they were members of the CTC or not. However, these rules were not acceptable to the USSR which preferred to keep details of the extent to which it had developed its own tin mining and smelting industry during the war a secret. The postwar period began with the exclusion of the world's third largest consumer from the participation in the core institutions that provided some degree of stability.

The supply side was governed by a series of bilateral contracts. The British Ministry of Supply was a monopsonist buyer of all metal produced in Penang and Singapore and of all concentrates produced in the other British colonial territories. Negotiations proceeded with representatives of each on price but they were governed by a de facto most-favoured-nation clause which ensured uniformity. In addition, Britain continued to draw on the Patiño concentrates under the terms of the contract signed in 1940.

In the United States, the RFC remained the sole buyer of both metal and concentrates. Malaya continued to be the primary source of imported metal which was simply acquired in Singapore at the price fixed by the Ministry of Supply. Both the Belgians[1] and the Dutch signed contracts for the supply of metal and concentrates which specified quantities and prices. From the American side, it was hoped that their high grade concentrates would enable the Texas smelter to be placed on a more efficient foundation. Bolivia, however, remained the primary source of supply and here negotiations were particularly difficult.

During the war, Bolivia had been the primary beneficiary of the system of unrelated bilateral contracts which granted her preferential terms on both price and smelting charges. With its end, the RFC was anxious to raise the latter and hold the former especially since increases in price would set the context for the renegotiation of other contracts. The wartime contract expired in June 1945 and the Americans would only renew with a staggered set of price decreases. Although Bolivia still provided around 50 per cent of American current consumption, stocks of concentrates at Texas City were sufficiently high that the RFC was in a strong position to adopt a take it or leave it approach.[2] It was designed to force Bolivia to recognize the urgency of placing her industry on a basis that would enable it to survive when the whole system of controls was dismantled.

For Bolivia this was a serious blow and one that she was determined to reverse when the 1945–1946 contract expired. She had several moral arguments to present: compensation for the sacrifice made during the war and the importance of tin prices in maintaining political stability and facilitating economic development. None had any cogency in the political atmosphere of Washington. As she lost patience with the RFC, she extracted a commitment from Argentina to take 8,000 tons from the amounts that would otherwise have gone to Texas at an acceptable price of 76c/lb. For the Americans this was the economic equivalent of Pearl Harbor and on this occasion she was forced to match the Argentine price.[3] Whereas negotiations between the Ministry of Supply and the producers over which it had control went reasonably smoothly, those undertaken by the RFC were always fraught with tension.

Tension at the international level was mirrored at the domestic level. Just how international allocation operated is unclear but its domestic implementation rested on a series of regulations determining which end uses were considered acceptable. American manufacturers, in particular, chafed at the perpetuation of these restrictions, especially since they were relaxed much earlier in the case of other non-ferrous metals. Pressure to dismantle this system and return to a free market continued to build with support from both consumers and producers.

Continuation of this system was a result of the increasing international political polarization. In July 1946, the prewar legislation authorizing the creation of stockpiles was amended as PL 520.[4] When specific objectives were formulated they assumed a worst case scenario, a five-year war during which the United States would be cut off from Asia and subject to severely reduced shipping across the Atlantic. As these were translated into the case of tin, they amounted to 210,000 tons and would later be substantially increased. With perhaps some 12,000 tons remaining from the stockpile built under prewar authorizations, the gap could only be filled by absorbing the whole of the surplus available once consumer demand had been met. Perpetuation of domestic controls increased the level of this surplus, so that in 1948 30 per cent of the metal supplied to the United States, including that from the Texas City smelter, was diverted to the stockpile. Complementing this was the role of the CTC which effectively prevented the USSR from building its own stockpile by acquiring metal from either European or Eastern smelters.

The system was brought to an end as soon as production again exceeded consumption. The pressure was first felt by the Ministry of Supply when the demand met by British smelters was exceeded by the supplies it had contracted to buy. Although Britain had its own stockpile requirements, they were very modest. Since the government was certainly not prepared to modify the purely administrative role of the Ministry of Supply, it was compelled to reopen the London Metal Exchange which occurred in November 1949. Much to the chagrin of the Americans, the CTC was forced to abandon international allocation forthwith.[5]

The orderliness of the system of control that operated through the regulations governing international and domestic allocation was not matched on the production side. The war had divided producers into two groups. One had been able to continue production in an uninterrupted fashion; the other had seen production capacity either destroyed or abandoned. Over the first four postwar years, the former group was producing at around 90 per cent of the level reached in 1940. The latter group encompassed all the Asian producers and they were only able to reach 40 per cent.

Rehabilitation of Asian production proceeded in a very uneven fashion. The country that made the fastest progress was the NEI. War had provided the Dutch with both the resources and the incentive to play a much more important role in the industry. Furthermore, since the independence movement that had flourished under the Japanese did not include the tin islands, they were the obvious basis from which to re-establish the Dutch empire. In 1946 GMB committed $19 million to a fleet of eight new dredges that would operate in all the islands and by 1947 they were in place. There was no interest in a prompt rehabilitation of the smelters and all concentrates were shipped to either Arnhem or Texas. In 1948 the company was awarded a five-year contract to manage Banka in a modified form of the fusion plans that had been frustrated in the 1930s. The transition to independence in 1949 made no immediate difference to these arrangements.

When the European companies returned to their Burmese properties they found a far less supportive environment and production languished at around 30 per cent of its prewar level. China also fell far short and would not rccover until after the formation of the People's Republic in 1949 and at that point most of its production was removed from the world market and sent to the USSR. Thailand's rehabilitation was delayed until arrangements were completed which would compensate companies for the mineral extracted from their properties during the war.

Malaya experienced the most difficult problem of all. Since the producers were financially weaker than their Dutch counterparts, they were dependent on government compensation grants and loans for the resources required to make the necessary investment. Administration of this programme inevitably took time and the results were often 'considered grossly inadequate and inequitable'.[6] Furthermore, these funds could only be spent within the sterling area and tin companies had no priority in the competition for steel and other essential components. Malaya's dependence on imported rice meant that she suffered from the dislocation to that market. Food supplies were therefore both expensive and inadequate which resulted in high labour costs and low productivity. Compounding these difficulties was the transformation of the war against the Japanese into a civil war in which European companies became primary targets of the insurgents. Although Malaya returned to her position as the largest producer, she could no longer stake any claim to be the cheapest.

War had left a distinctive legacy in Bolivia in the form of a militant miners' union supported by the Movimiento Nacionalista Revolucionario (MNR), determined to use any means to recapture the power it lost in 1946. PME became the centre of class conflict with major strikes in 1947 and 1949, the last of which escalated into civil war. While the state was prepared to repress the most overt forms of labour conflict, it was subject to other political pressures which could only be met by raising taxes on the industry. Production inevitably declined.

Of the main producers, only the NEI, Nigeria and the Belgian Congo operated within a secure political environment which permitted reasonably confident predictions about the level of production. That left 60 per cent of the world's supply in the hands of countries which were politically unstable.

Although projections of both supply and demand in the early postwar period were unreliable, the ITSG, from its inception, was fully aware that the problem of excess production would inevitably recur and that it was best addressed with another intergovernmental agreement. However, it had to reflect current thinking about such agreements as defined in Chapter 6 of the Havana Charter.[7]

Chapter 6 broke from the assumptions that lay behind the prewar agreements. One was clearly obsolete, since there was to be no world-wide depression for which such counter-cyclical measures were required. Nor were there any echoes of the commodity cycle thesis which had made such agreements a normal way of maintaining a suitable level of capital investment and productive capacity. Instead, commodity agreements were considered to be a temporary solution 'to prevent or alleviate the serious economic difficulties which may arise when adjustments between production and consumption cannot be effected by normal market forces alone as rapidly as circumstances require'.[8] These difficulties were then conceptualized as a 'burdensome surplus' and the only relevant burden was one that took the form of 'serious hardship to small producers who account for a substantial proportion of total output', or 'widespread unemployment or under-employment'.[9] A five year, renewable term was allowed to give the industry the breathing space required to overcome the constraints of an inelastic supply curve. The only echo of the prewar agreements was the desirability of preventing 'pronounced fluctuations in the price of a primary commodity'.[10]

While the Chapter was silent on the acceptable forms of commodity control, it was explicit about procedure. An agreement could only be ratified after a draft had been endorsed at a general conference to which all countries were to be invited. Its administration was to be in the hands of a body on which consumers were to have equal votes with producers.[11]

Although the International Trade Organization envisaged by the Havana Charter never came into being,[12] the ITSG proceeded on the assumption that the principles of Chapter 6 continued to apply. They were first used as the basis of criticism of the current system of international allocation and domestic controls. Thereafter, they were considered to provide the minimal conditions under which the Americans could be persuaded of joining an agreement.

The negotiating process was quite complicated both in form and substance. It began at the second full ITSG meeting in April 1948 which struck a Working Group whose results were reviewed at the third full meeting in October. A Drafting Committee was then struck which developed a detailed set of proposals in December. These were largely modelled after the draft International Wheat Agreement.[13] However, once the United States started to negotiate long term bilateral contracts, the problem of an impending surplus lost much of its urgency. Those commitments would both allow more rapid construction of the stockpile and eliminate any immediate problem.

As the urgency of the global problem receded in 1948, that of Malaya intensified. Initial thinking about the political structure of postwar Malaya attempted to end two sets of anomalies. One was long standing, the fragmentation into three types of dependent units; the other emerged from the role the Chinese had played, under the leadership of the Malayan Communist Party (MCP), as the basis of resistance to the Japanese occupation. The Malayan Union plan addressed both. All territories, apart from Singapore, were to be included in a single state with universal citizenship rights. The plan was imposed in 1946 but Malay opposition led to it being replaced by the Federation of Malaya in 1948 which restored Malay privileges. As the Chinese reacted to their loss of rights, they turned to the MCP. In June 1948, inspired by the success of Mao Tse-Tung, the MCP launched a campaign of armed struggle, termed by the British and their Malay allies, the 'Emergency'.

Since the British could not agree with the Americans about the terms of a bilateral contract, Malaya was particularly vulnerable to any early completion of the stockpile and the return to a condition of production surplus. Without any visible progress in the civil war, uncertainty in tin could result in the 'loss' of Malaya and eliminate a major source of dollars for the sterling area. The ITSG therefore returned to the question of a commodity agreement in April 1949. A statistical assessment in June showed that production would be around 21 per cent higher than commercial consumption over the next four years. Another working party started to draft the terms of an agreement in October which were modified to try and take account of American objections in March 1950.

Apart from the obligatory equality in voting rights between consumers and producers, this draft bore no resemblance to the Wheat Agreement. Although it could not be acknowledged, its essential features were taken from the ITC Third Agreement. At its core was a range of prices established on the LME, with two instruments, a buffer stock and production control. The ceiling would be protected by the buffer stock, the floor would be protected by controls on exports together with the buffer stock. It had four major modifications. Implementation of production control required a two thirds vote by both constituencies and entitlement would be based, not on an agreed standard tonnage but on actual performance. The buffer stock manager (BSM) was expected to try and stabilize price around the midpoint of the price range. Finally, orderly liquidation of the stockpile was to be governed by general principles rather than specific monthly percentages.[14] In spite of the objection of the American delegation, the ITSG felt it was ready to ask for the United Nations to convene a full tin conference.

Tin control under the auspices of the United States: phase II, 1950–1952

The second phase began as the outbreak of war in Korea in June 1950 disrupted the transition to a fully functioning free market following the opening of the LME in November 1949. Then the Ministry of Supply was the only source of free tin with a stock of over 25,000 tons. Once it realized that a stable price could be reached at around £600, these stocks were fed into the market at a rate that would preserve this price and created a situation not unlike that of 1939. Consumers were content to let the Ministry bear the cost of carrying stocks and did not build up their own.

The Korean War came at the wrong moment. Fearful of any extension into Southeast Asia, consumers began to restock. In August, the Ministry announced its stocks were being depleted to the point where it could no longer guarantee supplies of spot tin. In addition to these stock movements was an increase in demand, especially from the USSR. A combination of these factors was sufficient to prompt a bull market but what turned it into a runaway one was the decision of the United States to expand its stockpile acquisitions. Prices rose at the monthly rate of 13 per cent from £602 in June to a peak of £1,470/ton in February 1951. Although all commodity prices rose after the outbreak of the Korean War, most were still subject to regulation. On the LME, tin was still the only metal traded and it attracted a disproportionate amount of speculation.

In this context, no progress could be made on a commodity agreement. Although the planned tin conference was held in Geneva from October to November, rising prices reduced the sense of urgency and rising costs made it difficult to fix a range within which the buffer stock would operate. Unfortunately, the Americans seemed determined to vindicate their opposition to the conference and presented a series of demands about the buffer stock. Its target was 30,000 tons and this had to be filled before any production cuts could be implemented to protect the floor price. They would then be convinced of the existence of a burdensome surplus. Half of the stock would be reserved to protect the ceiling price. Particularly galling was the demand that a stockpile authority would have the right to buy the buffer stock at the current market price. The conference simply adjourned without settling any of the outstanding issues.

A serious problem of market instability escalated into a major crisis thanks to the way in which it was used by two individuals, Lyndon Johnson and Stuart Symington. Johnson was a freshman Senator in 1950 and saw the war as an opportunity to advance his career. Truman had established his reputation in the Senate and the country as an active Chair of a Senate Investigating Committee into Defense Mobilization in 1941. Johnson managed to get the Armed Services Committee to strike a similar committee, Preparedness Investigating Subcommittee, and secured the Chairmanship and a large team.[15] It was put to good political use with a series of reports that were 'mundane in substance', but 'long on phraseology that was grist for a reporter's typewriter'.[16] The one on tin was released in March 1951 and was particularly provocative.

The tin report had two targets, the National Production Authority (NPA) and the producers. The strictures on the supposedly dilatory way in which the NPA had exercised its power to restrict domestic consumption in the second half of 1950 indicated a lack of coherence to American tin policy. While the report recognized that the ensuing competition between consumption and the stockpile was the immediate source of the price problem, it went further and laid ultimate blame on the producers.

A simple comparison between the average output of 240,000 tons in 1940–1941 and that of 164,000 tons in 1949–1950 was sufficient to establish the case. Although part of the failure to reach wartime levels was attributed to depletion and postwar reconstruction, it was not sufficient. For that, the report reviewed the high level corporate concentration in both mining and smelting and then the various ways in which the producers had controlled the industry beginning with the Bandoeng Pool. Particularly relevant was the experience of 1939–1941 when the ITC pursued a narrow policy designed to prevent a stock overhang at the expense of the strategic interests of the United States. Since the war, there had been continued pressure for a return to production control, most recently at Geneva which taught the following lesson:

> The most salient fact which we must note is that even as recently as November 1950 the representatives of the producing countries which were on the side of the free nations,

> appeared still to be in greater fear of surplus tin capacity than of Communist imperialism, and showed relatively little concern for the need for the rapid accumulation of a United States stockpile to be used in building up an arsenal for them as well as ourselves ... The conclusion seems inescapable that we cannot rely upon private producers alone to make adequate supplies of tin available to us to fill our strategic stockpile and at reasonable prices.[17]

The report then laid out a comprehensive programme explicitly modelled after the system erected during the war. It involved the following steps: (1) closure of the LME and return to international allocation; (2) domestic allocation coupled with stringent conservation measures; (3) persuasion of allies to reverse their priorities and commit to higher production; (4) provision of more explicit assurances about both stockpile targets and eventual liquidation; (5) negotiation of bilateral agreements on price which would give a reasonable profit to producers, recognizing the differential costs both between Bolivia and the East and between the normal level of production and the extra required for the stockpile. Until this system could be put in place, it added a further recommendation: suspension of all further stockpile purchases.

Immediately following the release of the report, the system of domestic allocation was re-established. The RFC again became the sole importer of tin and the NPA established a fresh set of rules about permitted uses. Not only were stockpile purchases suspended but imports for industrial consumption were reduced to the barest minimum, with the explicit aim of driving prices back down. The RFC exercised its option under the Belgian and Indonesian contracts not to buy metal if the price were over 103 cents. Negotiations with Bolivia failed to reach an agreement on price and purchases were terminated in May.

At that point, the monthly intake of tin in concentrate for the Texas smelter was around 1,200 tons. The NPA had managed to cut monthly consumption by 25 per cent from its 1950 level to 4,740 tons but it was reaching the limits of what was politically feasible short of an all-out war. With around 40,000 tons in free stocks[18] to cover the gap, the United States had less than 12 months in which to force the producers to capitulate.

From the Geneva conference to the Johnson report, American tin policy had undergone a fundamental shift. The State Department had been marginalized and with it the prospect of a commodity agreement that would meet all interests. Now the RFC and Johnson were in charge and for them such an agreement was dismissed as 'what you might call an old-style cartel arrangement'.[19] Instead, they had embarked on a course that was not only very aggressive but fraught with risk. It was one that received the full support of the new head of the RFC, Stuart Symington.[20]

Central to the new thinking about tin was a judgement about the operation of the Singapore market and Symington testified to the Johnson committee:

> I think it is important for us to present here that all tin operations all over the world are based on the Singapore price, which is actually no price at all. They do not publish the transactions in any detail and it is simply an announcement of a price controlled by the cartel.

One member of the committee then interrupted, saying: 'I have head the word 'cartel' used several times. Now, what does the cartel consist of?' to which Symington simply replied: 'I think that your report, Senator, is the best illustration of exactly what it consists of. Mr Cook will show you the graph of the interlocking directorate.'[21]

From this perspective, the world of tin was very simple. One member of the committee summed it up as an 'international tin racket', designed to take advantage of the security needs of the United States and 'gouge' the American taxpayer, already overburdened with the cost of defending the free world. With the chief racketeers considered 'certain very powerful individuals', it was impossible to grasp the way in which the tin industry actually operated. Corporate concentration was a sufficient condition of the existence of a cartel and all the problems Americans experienced from prices to the constraints on the ability of the Texas smelter to obtain high grade Bolivian concentrates were laid at its door. The tendentious way in which the history of tin had long been represented in American political circles continued to be fed into the shaping of policy.

As was often the case in tin politics, the stakes were higher than for those of the industry alone. For Symington, the tin policy was 'an experiment – a test case – to decide whether this country can acquire needed amounts of [other] critical materials at a fair price'. If it failed, then the 'heavy and unnecessary taxation' could lay 'the foundation for the ultimate destruction of free enterprise', in the name of which the war was being fought.[22]

The producer response was swift and unequivocal. Particularly outraged were the Malayans who were producing under the threat of assassination by the MCP. Considering themselves on the frontline of the struggle against communism, the notion that they were insensitive to their responsibilities to maximize production to meet the defined needs of the United States was one they had to reject. In doing so, they pointed to the way in which the MCP 'bandits' raised their production costs and constrained their operations.[23] Nor were they restrained in the terms used to describe the Tin Report which 'contains many mis-statements of facts, many allegations against producers which are groundless, many half-truths and is probably one of the most unauthoritative reports that has ever been published by members of a responsible legislative body'.

Its motives were also impugned since it was 'carefully timed to influence international negotiations on tin', and may have been designed to 'cover up errors in bulk buying policy prior to the Korean War and then to try and redeem these errors by means of a deliberate programme of exploitation of the Tin Producing Countries'.[24]

The Indonesian and Bolivian responses invoked another moral principle, one to which Truman was publicly committed: provision of the assistance required for 'the achievement of economic and social progress for the underdeveloped countries'. High commodity prices were the best way in which that objective was to be accomplished. Once the moral high ground had been staked out, it then followed that 'the attack on high prices of tin has been based on moral grounds which are wholly non-existent … [and] worthy of moral disapproval, inasmuch as it contributes to the perpetuation of an economic form of colonial exploitation'.[25]

Both parties were now locked into a moralistic view of the world of tin. Although no cracks were possible in the image that each held of the other, only the RFC was in a position to change course and it had every incentive not to do so. Even if it were prepared to conduct a serious analysis of the tin market and drop the assumption of a nefarious cartel, the pressure to which it was subject should be deflected to another agency of the United States government. If the American taxpayer was expected to subsidize the development of these countries, then providing the necessary resources through high tin prices was the least efficient way of doing so.[26]

The State Department made a valiant effort to alert the RFC to the risks it was running to overall American interests and missions were dispatched to both Bolivia and Malaya. The one to Malaya was simply a goodwill gesture; that to Bolivia was expected to review the real costs of production and assess her case for 150 cents as the basis for a new contract. On the basis of its report, Symington revised his fixation on 103 and offered 112 cents.

Bolivia had become even more unstable thanks to a military coup in May 1951. Without a tin contract, the fate of the new government was uncertain and it responded politically. Defining itself as the innocent victim of the United States, it secured the support of many other Latin American countries. The State Department now had a major diplomatic problem on its hands and attempted to break the stranglehold held by the RFC by seeking Truman's intervention.

By late 1951 the RFC gamble had failed. Domestic opposition to the restrictions on consumption was on the rise.[27] Bolivia could continue to produce with finance provided by the New York banks secured against the eventual value of the concentrates. The Texas smelter was now running short of supplies and unless the RFC returned to the Singapore market, the United States would face an immediate shortage of metal. However, there was one further card that could be played and that was to turn to the strategic stockpile. It is not clear whether permission would be granted but the mere thought that it might made the British very alarmed. It would violate the explicit, albeit verbal, commitments made to both the Indonesians and the Belgians. Turning to the stockpile to cover a civilian deficiency in these circumstances would make it difficult for the producers to trust the United States in any future agreement, even when tin had ceased to be a source of political grandstanding.

The rearmament needs of the United Kingdom presented the opportunity to break the deadlock and in January 1952 a quasi-barter deal was struck. Over a million tons of American steel would be exchanged against 15,000 tons of aluminium and 20,000 tons of tin. The tin would be supplied at the specified price of 118 cents, fob Malaya or United Kingdom. Part of this would come from stocks still held by the British government but the bulk would be bought in Singapore with an inevitable market risk.[28] This could only be negotiated between Truman and Churchill, bypassing the RFC altogether and Symington's resignation followed.[29] The agreement also established transitional arrangements. As the British withdrew from Singapore, the RFC monopsony was terminated and American private trade allowed to return.

The British contract price was equivalent to 121.5 cents cif New York, so the RFC immediately raised its selling price from 103 cents and then set about negotiating new contracts with both Indonesia and the Belgians on the basis of the new price. In September, the RFC agreed to take the accumulated Bolivian production at the same price, 117.5 cents, fob Pacific ports. The supply position in the United States had been restored and stockpiling could be resumed.

Although Johnson welcomed this overall result, he was angry about the way in which the administration had undercut his ally in the RFC and its case for 112 cents. In deliberate defiance of the State Department which warned about the counterproductive use of provocative language, Johnson escalated the ideological battle:

> One of the more disturbing aspects of the cold war is the seeming inability of a few international combines to recognize that their long range interests in the survival of western civilization transcend their short-range interests as businessmen seeking the highest possible profit … They greeted the Korean war as the lighting of a fuse which would skyrocket prices into the stratosphere … A considerable share of the history of the past few years has consisted of the struggle to bring those prices back down to earth. It has been a depressing story – a story of men seeking to reap exorbitant profits from programs designed solely to protect free western civilization against the

> onslaught of its militant Communist enemies. The fact that a triumphant communism would crush not only liberty but their own economic position did not deter these combines from their folly … The United States has recently won an important victory against one of the largest and most ruthless of these combines – the international tin producers.[30]

In the eyes of the committee, it was not only a victory over the cartel but one that saved taxpayers the enormous sum of $500 million. This figure reflected the depth of the commitment to the notion of a powerful cartel, since it was based on a counterfactual model in which the price would be held at the peak of the equivalent of 202 cents reached for one day on 14 February 1951 which would be held for the five years during which 280,000 tons would be acquired.[31] Not only were these assumptions without foundation but the model failed to take account of the costs borne by American manufacturers and individual consumers. In fighting a chimerical cartel, Johnson and Symington had not only done enormous political and diplomatic harm to American interests in several countries but had inflicted serious damage to the domestic economy.

In the United States, only the trade press pointed out the immediate lesson of this experiment but that has left no dent on the congratulatory tone with which this episode has been treated.[32] Such a major mistake may be attributed to a fatal combination of three sets of factors. One was the ideological atmosphere of the Korean War which allowed any claim, however ill-founded, that national security was at stake to trump mundane economic and political considerations. A second was the fact that, for the whole of the previous decade, the RFC had played the central role in American tin policy. Domestically it worked within a command economy; internationally it was always engaged in conflictful, politicized negotiations with suppliers. It had neither the experience of the way in which the tin industry actually worked, nor any ability to tap that of those who did. For Johnson and Symington, such expertise was either contaminated or irrelevant.[33] That they could take such a position so naively was a result of a third factor, the tendentious way in which the history of tin from 1921 to 1941 had been constructed. The ITC continued to cast a long shadow, less because of what it actually did but more because of the commentary it had generated. That proved all that Johnson and Symington needed to know: corporate concentration meant a cartel and a cartel meant producer selfishness at the expense of the common good. In the 1930s that was consumer welfare; in the 1950s it was Western civilization. It is not surprising that, armed with such knowledge, Johnson and Symington acted as 'Bulls in a Tin Shop'.[34]

The formation of the International Tin Council, 1953–1956

With the resolution of the immediate crisis, both the RFC and private traders were supplying domestic consumption. A uniform price of 121.5 cents was therefore imposed on all sales to domestic consumers. Consumption remained subject to NPA regulations until January 1953, when they were all relaxed with the exception of reporting requirements. By April private trade was in a position to meet domestic needs. The RFC then withdrew from the market and the New York price was again determined by the world price.

Henceforth production from the Texas smelter, together with metal that continued to be bought under the Indonesian and Belgian contracts, was earmarked for the stockpile. In March 1953, the American government announced that it had sufficient supplies on hand or under contract to meet its stockpile requirements. The problem of surplus production was now at hand and was reflected in an immediate drop in price. By April the New York price had dropped to 97 cents, by mid-summer it was down to 85 cents, at which level it remained until there was the prospect of a commodity agreement.

Unfortunately, it was not easy to wind down the Texas smelter as the stockpile was completed. Pressure to keep it going came from Johnson who pointed to its strategic importance. It also came from Bolivia and Indonesia and the State Department was even more concerned about the political implications of tin policy in both countries.

Failure to reach a tin contract with the RFC in 1951 had severely undermined the legitimacy of the military government in Bolivia and in April 1952 it was overthrown by the MNR. Since one of the primary bases of support of the revolution was the mineworkers who had long fought the large mining companies, the new regime nationalized the properties of Patiño, Aramayo and Hochschild. The Corporación Minera de Bolivia (COMIBOL) was formed in October 1952 and immediately encountered four major problems. Two were technical/administrative, as most of the foreign mining engineers left and the miners' union was granted extensive power under a system of workers' control. Two were financial and resulted in the decapitalization of COMIBOL. Until an agreement was reached with the expropriated companies, Williams, Harvey demanded an additional payment for a compensation fund. State ownership eliminated any restraint on the diversion of foreign exchange earnings. Highgrading enabled production to be raised in 1953 but in 1954 it fell off dramatically. The repercussions were sufficiently seriously that the State Department persuaded the Eisenhower administration to provide substantial amounts of foreign aid, one element of which was continued purchases for the Texas smelter.

Indonesia also saw an important administrative change. In 1953, the management contract with GMB was not renewed and Banka returned to government control. It, too, was anxious to keep a secure market and since the country was also considered politically volatile, the State Department offered the same kind of support. The RFC therefore continued to accept shipments even though they were surplus to the requirements of both the stockpile and consumers.

In November 1953 the tin conference was resumed in Geneva. To the obvious problem of low prices and surplus capacity, the RFC added a further note of urgency since it indicated it may have to sell some of its surplus tin. Although the State Department was now much more sympathetic to a participation in a commodity agreement, it was aware of the considerable domestic opposition it would face from other agencies, especially Commerce and Interior, which were opposed on both ideological and pragmatic grounds to one of any kind.

Negotiations continued for over three weeks and settled four contentious questions. Voting rights and regulations were designed to prevent de facto vetoes on the part of either the United States in the consumer bloc or by the United Kingdom, as both consumer and sympathetic to Malaya and Nigeria, in the Council as a whole.[35] Should production restriction become necessary, quotas were to be determined in the first instance on the basis of performance over the preceding three years.[36] The buffer stock target was reduced to the equivalent of 25,000 tons, of which 25 per cent would be in cash. Production quotas could be instituted on the basis of separate majority votes once 10,000 tons were in the buffer stock.

Without any agreement on a methodology to determine the initial floor and ceiling for the buffer stock, consumers and producers simply presented ranges that reflected their respective positions. Bolivia attempted a range that would break the impasse and, much to the consternation of the State Department, it was endorsed by the head of the American delegation. That was a businessman, Figgis, a Director of American Can, who had become very sympathetic to the problem facing both Bolivia and the Belgian Congo. So 'without any apparent logic' the floor was set at 80 cents (£640), just below the current market price and the ceiling at 110 cents (£880).[37]

The conference ended by specifying the proportion of both consumption and production represented by participating governments that had to ratify the agreement before it could come into effect. Although there was now an agreement that could be sent for ratification, it could prove to be the prelude to yet another conference. As the Americans made it clear that ratification by the United States was doubtful and, as some other delegations indicated, they would follow the American lead, that seemed the most likely outcome.

On its return to Washington, the American delegation pressed the case for participation. It would entail no real cost but would serve important foreign policy interests in all the producing countries. However, such pragmatic considerations could not overcome the ideological opposition to any apparent violation of the principles of unfettered markets that had become even stronger in both Commerce and Interior following the publication of the Randall Report in early 1954.[38] However, the State Department then reported that the government had adopted a position of 'benevolent neutrality', and that allowed the British to recruit enough countries representing consumer interests prepared to ratify the agreement.

On the producer side, some countries had to weigh their various options. In the case of Malaya, this was done through a referendum which received sufficient support to allow ratification.[39] Bolivia soon discovered that it could no longer expect any further contracts with the Texas smelter and the window for ratification was extended to allow the Indonesians to come to the same conclusion. By 1955, Bolivia, Belgian Congo, Indonesia, Malaya, Nigeria and Thailand had ratified and the new International Tin Council commenced work in July 1956.

The results of the process of ratification reinforced the asymmetry between producers and consumers. On the consumer side only 36 per cent of that market was represented and 60 per cent of that figure was made up of countries such as Australia and Spain whose consumption was largely met from domestic mines and from countries such as Belgium, the Netherlands and the United Kingdom, each of which retained close political and economic ties to producing members. The pure consumer interest was therefore very weak. By contrast, the producer interest was very strong. As a result of the diversion of Chinese production and the reduction in output from the minor Asian producers, the producers on the new Council represented 96 per cent of the capitalist section of the world market.

The unseen but omnipresent element was the American stockpile. Although its actual size was a secret, it was not difficult to determine its approximate extent from the statistics of consumption and imports. What was not known was that the announcement in March 1953 of its impending completion was based on a reduction in its target from 350,000 to 245,000 tons. However, the extension of the Bolivian and Indonesian contracts that were based on other political grounds meant that the excess over this reduced target continued to grow. When details of the stockpile were publicly announced in 1962, it had reached the figure of 361,000 tons, or over two years' world production. Over the intervening period, the United States had absorbed more than 15,000 tons a year, most of which would otherwise have spilt over into the market. In addition to the American stockpile, there were other, albeit much smaller, stockpiles held by Canada, the United Kingdom and Italy that had been built up or maintained during the Korean War. The industry that the Council had to regulate therefore contained an enormous quantity of excess stocks, the proportion of which could enter the market grew in inverse relation to the decline in the prospects of a communist take-over of Southeast Asia and the outbreak of war in Europe.

Reshaping the structure of the tin industry, 1956–1985

Analysis of the period during which the Council attempted to regulate the tin market needs to take account of several shifts that marked a definitive break with the structure of the industry that had been emerging in the first half of the twentieth century. They occurred along three different axes: technology of consumption, geographical location of both consumption and production and the institutional organization of that production.

Consumption

There are two striking features about the consumption picture over these three decades. One is its sluggish rate of growth and the other is the emergence of new centres at the expense of old ones. In 1956 the capitalist world consumed 154,000 tons of metal; by 1979 it reached 172,000 tons or an annual rate of growth of 0.5 per cent.[40] In 1956 the distribution of that consumption was broadly similar to that established before the war, with 80 per cent going to the United States and Western Europe; by 1979 that had declined to 60 per cent and would continue to drop. While the United States was still the world's largest consumer, it was closely followed by Japan. Within the capitalist world, consumption was much more evenly divided between those whose industrial development was shaped before the war and those who came afterwards.

The history of tinplate is a primary feature of the explanation of both features. In 1956 the United States and Western Europe produced 97 per cent of the capitalist world's requirements; by 1990 that share had dropped to 60 per cent.[41] In 1956 the shift from hot-dipping to electrolytic deposition of the tin on steel was still underway but was virtually complete by 1985 and that reduced the average amount of tin required per ton of plate by two thirds.[42] A threefold increase in tinplate simply preserved what remained the most important outlet.

This technical revolution was accelerated by constraints imposed during the war and the immediate postwar period but it kept tinplate competitive in the overall packaging market. The processes of substitution that were already evident before the war were accelerated and tinplate lost market share, especially to aluminium, and that explains the decline in production in the advanced countries. But without the reduction in cost made possible by the shift, tinplate would have been confined to a very marginal role.[43]

Consumption also remained steady for most other uses. That was less a function of technical innovation in the process of manufacturing but rather in the use of the products. Solder was still the second most important outlet. As automobile radiators were increasingly made of aluminium, rather than copper or brass, solder became redundant but that was compensated by the growth of the electronics industry. The actual quantity used in each item was reduced considerably as a result of new manufacturing techniques but the exponential growth in electronics served to preserve the total demand.

The stagnation in consumption masks the fact that tin had continued to adjust to the major shifts in the world economy. The three decades of the Council are ones which saw rapid growth in the Asian economies. As the whole world became the market for products such as canned fruit from Thailand and electronics from Japan, tin retained a secure foundation.

Production

Production had to adjust to another major shift, the growth of nationalist sentiment which would reduce, if not eliminate, foreign control over mining and smelting.

Nationalization of the major Bolivian mines allowed the government to realize a long standing aspiration for a domestic smelter. With a new German technology which overcame the limitations imposed by altitude and lack of traditional fuels, a smelter was constructed. It began to function from 1971 and treated the high grade concentrates that had been sent to Williams, Harvey. The Texas smelter was sold to private interests in 1956 and continued to treat low grade concentrates. In 1980 Bolivia completed a second plant and then treated virtually the whole of its production.[44]

Williams, Harvey also found itself squeezed from the Nigerian end. As the colony moved towards independence in 1960, a Portuguese firm built a smelter which began to take some of the concentrates previously shipped to Britain. CTS immediately countered with another smelter at Makeri and by 1963 it treated the whole of Nigerian production.[45] Deprived of its main sources of feed, Williams, Harvey was liquidated in the 1980s, leaving Capper Pass as the remaining element of the once dominant British smelting industry.

Return to state management of Banka in 1953 was a prelude to a complete nationalization of GMB's operations on Billiton and Singkep and the whole of Indonesian production was then in the hands of a parastatal, PNT Timah. By 1967 a major smelter was in operation and while some production still went to Singapore, none was available for Arnhem which eventually closed in 1971.

Elsewhere, the concern was less about control than about location. In 1965 Thailand encouraged an American conglomerate, Union Carbide, to build a large smelter in Phuket to treat the whole of her production. However, tin proved a poor fit with its overall plans and its interest was acquired by Billiton in 1971 which then returned to Southeast Asia. Since virtually all other producing countries also maintained or developed local smelters, the international trade in concentrates was sharply reduced and actual smelting capacity increased to over twice the amount actually needed.[46]

Foreign equity interest in mining was reduced through a variety of measures. The most extreme form was in Nigeria which resulted in the absorption of the whole of this interest into the Nigerian Tin Mining Company in 1984. Thailand adopted a more modest objective of a transfer of 60 per cent to local capitalists, though the companies themselves remained more or less intact. The case of Malaya, now Malaysia, was more complicated. Many companies simply changed domicile and included Malaysians on their Boards. The Selangor state government went further and used its control over leases to secure equity interests in both existing companies and new ones.[47] But the most significant development resulted from takeovers of both the Anglo-Oriental and Tronoh groups whose new owners came to an arrangement with the Malaysian government to transfer their interests. These became the basis of Malaysia Mining Corporation.[48]

On the extraction side, the most dramatic change was one in the geographical location of production. Of the six producer members of the Council only Thailand saw significant growth between 1956 and 1980, while both Nigeria and the Belgian Congo, subsequently renamed Zaire,[49] saw precipitous declines. Old mines were reopened and expanded in Cornwall and Australia.

Opening up Amazonia resulted in the location of new deposits which proved to be both extensive and high grade and they would eventually make Brazil into a major producer but one which followed a distinctive path. Where effective administration was possible, foreign capital with previous experience in tin was encouraged;[50] where it was not, deposits were exploited by wildcat miners, garimperos, who proved very difficult to control.

Technological innovation proceeded slowly but it resulted in two major shifts in the pattern of production in the three major Southeast Asian producers. An increasing proportion of production came from offshore deposits extracted by a new generation of very expensive dredges. However, onshore they began to concede to gravel pumps. Overall the proportion of the total output from these three that came from dredging declined from 54 per cent in 1956 to under 15 per cent by 1980. The technical flexibility of gravel pumps, coupled with much lower capital costs, made them more economically efficient than dredges in working many deposits and more socially desirable because of their labour intensity. However, over these three decades there was little sign of improvement in labour productivity.

These shifts in consumption and production led to a change in marketing. With the closure of European smelters and reduction in European demand, trading on the LME became less reflective of the market for physical tin and more driven by speculative forces. That made Penang, where Japan acquired most of its tin, more autonomous.

By 1985 a few features of the world industry that had evolved prior to World War II remained. It was still a cottage industry with a large number of small scale marginal producers. It was also one in which the fate of virtually all producers was bound up with a truly global market. But it had become much more national in the composition of both capital and labour and the policies designed to control its vagaries were no longer vulnerable to the criticism that their primary beneficiaries were wealthy individuals enjoying the very best of European life.

The international tin agreements, 1956–1986

The Council regulated six separate agreements and each followed the procedure laid down in the Havana charter. All interested countries were invited to a conference under the auspices of the United Nations, normally a year before the expiry of the existing agreement. When negotiations led to an initial impasse, it could be prolonged to allow a resolution. One important result was that the market was never disturbed by the possibility of a sudden termination. Apart from the obligatory symmetry between producing and consuming members, that was the only enduring feature from Havana.

Any commitment to universality was compromised by the way in which the United States made it clear that its benevolence would be withdrawn if the People's Republic of China were permitted to join. Later, West Germany insisted on scoring a similar political point when it made the exclusion of East Germany a condition of its participation.[51]

Each of the six agreements operated within a different configuration of political and economic forces and their experience can be briefly summarized.

First agreement (1956–1961)

The problem of excess capacity appeared to be easing as the Council began its work. Consumption was rising and production was falling as result of the new regimes in Bolivia and Indonesia. The buffer stock would absorb most of any remaining excess. Agreements were reached with both Canada and the United Kingdom for an orderly liquidation of their modest stockpiles.[52] However, in 1957 consumption declined and was not offset by the continuing fall in production.

An unforeseen development suddenly exacerbated the problem when the USSR began to sell large quantities of Chinese metal in Western Europe. Export controls, reinforced by increases in the buffer stock, were required and lasted until 1960 when the Soviets succumbed to diplomatic and economic pressure to reduce their disruption.[53] Some smuggling from Thailand indicated a weakness in the control regulations but otherwise they were free of the rancour that plagued the administration of restriction in the prewar agreements. Apart from a brief moment, the floor price had been protected and the 'agreement had survived severe testing'.[54]

However, that success must be qualified. Within the first year of control, at least three members found themselves in some difficulty. Since control only applied in the first instance to exports, minehead stocks continued to accumulate. In the face of a serious decline in revenue, Bolivia, the Belgian Congo and Thailand were all anxious to dispose of these stocks and found the United States prepared to barter surplus food and tobacco. Sympathy with the pressing needs of Bolivia led the Council to approve these arrangements, in spite of the fact that the nearly 8,000 tons that were taken into the stockpile would eventually return to the market. Production control evidently rested on an insecure basis.[55]

Second agreement (1961–1966)

The 1960s began on an entirely new note, with changes in production, trade and stockpile policy. Relaxation of controls revealed a chronic problem: a deficit in overall Western production which only met 82 per cent of the demand on Western markets. Fortuitously, the relationship between China and the USSR began to weaken. China began to sell to Western markets and the USSR started to draw on them. Until 1964 East-West trade showed a net surplus and that closed much of the production gap.

In March 1962 Symington, now Chairman of the Senate Armed Services Committee, announced that of the 348,000 tons held in the stockpile, 164,000 were surplus to strategic requirements. Congress then authorized the sale of 50,000 tons. Thanks to its distinctive reading of the history of tin in which price 'gouging' by a cartel in the 1930s and 1950–1951 played a prominent role, Congress evinced little sympathy with any problems such liquidation would cause for the producers. The State Department had to return to a familiar role as it attempted to ensure that liquidation was consistent with its overall posture of benevolent neutrality.[56] It was not an easy one to play, especially since the mere announcement of this policy caused a price drop and Bolivia started complaining about 'economic aggression'. Since the Council had failed to issue its own protest, it, too, was a target since that reflected the 'blind egoism on the part of the consumers'.[57]

Congressional sentiment prevented any formal agreement with the Council, with the result that there was always uncertainty about the prospect of further declarations of surplus, liquidation authorizations and actual sales.[58] Sales began in 1962 and rapidly increased until 1965. Although this closed the remainder of the production gap, it had a destabilizing effect. At first, it was mitigated by the buffer stock which kept prices just above the midpoint of the range. But this was exhausted in 1963 and then speculation drove prices well above the ceiling, to the point where they were higher, even in real terms, than the level reached at the height of the Korean War boom.

The Council now faced a problem of determining a price policy that was appropriate to these new circumstances of a chronic production shortfall and large American sales.[59] Soon after the commencement of the second agreement, the range was increased by 8 per cent but, once the ceiling had been broken, it was raised by a similar amount and then by another 18 per cent at the beginning of 1965. With the market completely out of the control of the Council, the producers were able to argue for a floor that was sufficiently high to preserve existing capacity in the face of increased costs and encourage fresh investment. The focus of the Council had expanded from market regulation to long term planning.

Where the agreement had started with Bolivian criticism of stockpile sales and the weakness of the Council in combating them, it ended with a much more forceful condemnation in Malaysia. Many companies felt that the relationship between the Council and the GSA amounted to a conspiracy to prevent producers from taking full advantage of the new market conditions. Not only did they discount the prospect of a recurring surplus but they considered it would only arise after their reserves had been exhausted. Although the bulk of Malaysian mining opinion was against them, the government followed their advice and announced that it would not sign a third agreement, at least not without the participation of the United States.[60] Considerable pressure resulted in a change of mind but the episode revealed another source of frailty.

During the course of the second agreement, a major shift in thinking about primary commodities occurred. As the world capitalist economy underwent a major expansion in the 1960s, the same kind of structural asymmetry that had been evident in the 1930s, the deterioration in the terms of trade between primary commodities and manufacturers, was exposed. One response was the expansion of the interest of the United Nations in commodity questions with the formation of the UN Conference on Trade and Development (UNCTAD) in 1964. Equipped with Prebisch's compelling challenge to neo-classical theories of the distribution of benefits from international trade, UNCTAD provided the forum in which countries dependent on primary commodity exports, the Group of 77, could articulate their vision of a New International Economic Order (NIEO). The Havana model was now obsolete.

Third agreement (1966–1971)

Central to the NIEO was the principle of transfer of resources from the developed to the developing world and commodity agreements were seen as an important vehicle to accomplish this objective. Considered the most successful of commodity agreements, tin was now erected as a model to be emulated.

The preamble to the third agreement reflected this new environment. Gone were the Havana exceptions of 'burdensome surplus' and 'widespread unemployment' and in their place was a forthright justification of commodity agreements which 'by helping to secure short-term stabilization of prices and steady long-term development of primary commodity markets, can significantly assist economic growth, especially in developing producing countries'.[61] With the jettisoning of the Havana model, a number of socialist countries joined the Council.

Two trends started during the third agreement and were continued into the fourth. The most encouraging was the expansion of production which had three main sources: (1) continuation of a trend already evident in Malaya and Thailand; (2) recovery in both Bolivia and Indonesia. (3) from consumer members of the Council, of which the most important was Australia. Consumption also rose but at a much slower rate so that while Western sources were generally insufficient to meet demand on Western markets, they were only 5 per cent short. Again, sales of Chinese tin and the American stockpile covered the shortfall but the quantities in both cases were considerably reduced. The Council regained control over the market. On two occasions, modest level of export control were implemented but otherwise the buffer stock served to keep most prices within its range.

Fourth agreement (1971–1976)

The fourth agreement began with the removal of one support to international financial stability, as the dollar was devalued in 1971 and the principle of fixed exchange rates came to an end. As sterling became particularly unstable, the Council decided to replace the LME with Penang as the 'reference point for its own operations'.[62] From 1972 the buffer stock range was expressed in Malaysian ringgits.

Devaluation was soon followed by much a more severe shock, the decision of OPEC in 1973 to raise oil prices which ended the cheap supplies that had sustained the economic growth of Western Europe. Both these events served to advance the UNCTAD agenda which was then endorsed by the General Assembly in 1974 with far reaching consequences.

The dramatic and sudden increase in oil prices had an immediate and general inflationary effect which was particularly severe in the case of tin. In spite of heavy sales from the stockpile and China, from 1971 to 1974 prices rose by 240 per cent, measured in dollars, 187 per cent in ringgits. Thanks to speculation, the rate was much faster than that of any of the other non-ferrous metals. Over the same period, direct costs of the four largest producers rose by 150 per cent, while the net income to their states rose by 280 per cent and that to miners by 370 per cent.[63]

This transfer of resources from the consumers to the producers generated by the overall boom in commodities created several problems. Inflation was doing serious damage to economies of the former, not least because of the pressure on the balance of payments. It produced a recession in 1975 and, as speculators moved out of the market, prices declined. Modest export controls managed to contain its severity. Since the recession proved to be short, by the end of the agreement prices were again on the rise.

Over the course of the fourth agreement, the Council responded to this inflationary pressure by raising the buffer stock range. That was done on five occasions and, at the end, the floor was 160 per cent higher than at the beginning which largely reflected the increase in producers' costs. The ceiling was also raised but, since the goal of price stability required a limit to the size of the range, that was generally below the market price. Apart from the recession of 1975, the Council was again in the position of being unable to control the market.

The UNCTAD framework affected the Council in three other respects. Consumer representation was expanded to include West Germany[64] and the USSR. France and the Netherlands made voluntary contributions to the buffer stock as part of their overall programmes of aid to developing countries. Producer representation was also increased since Australia had become a net exporter and remained on the Council as a producer rather than as a consumer.

Fifth agreement (1976–1982)

The fifth agreement built on two features of the fourth. Producers' costs were explicitly incorporated into the factors determining the floor of the buffer stock range. The expectation that consumers would make voluntary contributions to the buffer stock was reinforced by an article allowing renegotiation of the agreement if they were not forthcoming.

Neither was sufficient to satisfy Bolivia which wanted more specific commitments, both on the part of consumers' buffer stock contributions and on raising the floor to cover her production costs. Oil had made her less dependent on tin and her bargaining position was thereby increased. Bolivia was only prepared to join the agreement on an interim basis and her participation was only finalized in 1977 when the Council succumbed to the pressure she was able to exercise with the support of other producers. An Economic and Price Review Panel was struck which then reported at six-monthly intervals on costs and market trends on the basis of which recommendations were made for adjustments in the floor and ceiling. As far as the consumers' contribution was concerned, the agreement saw some progress, since five other members, including the United Kingdom and Japan, joined France and the Netherlands.

Although these developments suggested an increase in producer power, they were countered by the most important change, the decision of the United States to join the agreement. In spite of her opposition to the 1974 resolution endorsing the NIEO, the United States could not easily remain indifferent to the claim that agreements served a useful function in moderating the commodity cycle. Congress was more convinced by the argument that participation would prevent any further drift towards a scheme controlled by the producers, especially if they became more enamoured of the OPEC model and sought further transfer of resources by raising prices.[65] At the same time, the stockpile objective was reduced which raised the surplus to 174,000 tons. Membership simply formalized the earlier arrangement for consultation about the liquidation of the amounts authorized by Congress.

Bolivia, with considerable support from other producers, and the United States represented different conceptions of the tin agreement, with the former wanting to push it more in the direction envisaged by UNCTAD and the latter wanting to pull it back towards Havana. That tension was exacerbated by the decision of the American delegation to break the norm of consensual decision making and force most questions to a vote.[66]

The fifth agreement began with a recovery in consumption though not to the level reached before the 1974 oil shock and it then started to decline. Meanwhile production continued to rise so that by 1979 the two trends had crossed. Until then production deficits were not covered by sales from either China or the GSA, so that prices continued their upward trend. Producer response to this incentive was limited. Of the original producer members, only Indonesia and Thailand grew and this was marked by an ominous development. A simple adaptation of the gravel pump technology allowed a flotilla of suction boats to extract cassiterite from the seabed. High taxes encouraged most to operate illegally and smuggle their concentrates to Singapore.

By 1980 the second oil shock, prompted by the Iranian revolution, had produced a recession in the industrial world, one that would prove to be more severe and prolonged than in 1975. The declining trend in consumption was accelerated and the Council now faced a serious production surplus, one that the industry had not had to confront since 1939.

Sixth agreement (1982–1990)

The recession coincided with the opening of negotiations for a new agreement and inevitably raised the question of whether it required a revision in price policy. As actual prices overshot the ceiling, it was possible to justify setting the floor on the basis of producer costs in order to preserve capacity and encourage investment. That rationale no longer had any currency among consumers, especially since the costs claimed by producers included taxes imposed by governments.

The obvious cleavage emerged. Bolivia led the producers in insisting on a continuation of the previous policy; the United States proposed a radical revision of the entire agreement. Since export control was seen as the basis of resource transfer, that would be eliminated. Floors and ceilings would be preserved in the short term by a much larger buffer stock but ultimately they would be determined by the market. Negotiations settled one outstanding issue, joint financing of the buffer stock, but since the broader question remained unresolved, the existing agreement was simply extended for a further year.

Extension, however, did not mean preservation of the balance of forces. As the recession began to bite, the ideological support for commodity agreements began to wane and the new

Reagan administration decided not to continue membership.[67] Nor did it return to benevolent neutrality, since it also rejected the 1966 agreement which suspended stockpile sales when the buffer stock was buying. From 1981 the buffer stock had to absorb all such disposals.[68]

Nor were the producers constrained by the agreement. When the first request for adjustment in the buffer stock range was rejected in July 1981, a syndicate, acting on behalf of the Malaysian government, drove prices up by extensive forward dealings on the LME. By the end of year, dealers became sceptical of its ability to sustain its position and started selling short. They miscalculated and by February a severe shortage of cash metal outside the control of the syndicate prevented them from covering their positions. In order to minimize the extent of defaults and bankruptcies, the LME intervened and imposed a limit on their losses leaving the Malaysians outraged at this new interpretation of the rules.[69] The episode not only indicated the desire of the producers to control the absolute price, at the expense of the stabilizing function of the buffer stock, but it undermined any confidence they had in the integrity of the LME. Since the syndicate had acquired 60,000 tonnes, its liquidation put considerable pressure on the market and most went into the buffer stock.

One immediate effect of this manoeuvre was to force consumers to concede a revision of the price range, by around 7 per cent. Since the new ceiling barely covered Bolivia's direct costs, she was not prepared to endorse it by remaining in the agreement. Henceforth, her co-operation with the other producers would be through a new organization, the Association of Tin Producing Countries (ATPC), launched in 1983.

The final agreement retained all the essential features of its predecessor, with one important modification. Consumers made the long overdue concession of joint responsibility for financing the buffer stock. However, as they reviewed the agreement with a view to determining whether to ratify, there were considerable misgivings about its adequacy in these new circumstances. They were set aside on political grounds. The Malaysians were already annoyed at American withdrawal and the British feared that the more aggressive nationalist stance taken by the new Prime Minister would produce a severe backlash should the agreement not be ratified.[70]

British support was sufficient to encourage enough other consumers to ratify and the sixth agreement went into effect from July 1982. But with only 52 per cent of world consumption represented, the agreement remained provisional with the result that the purchasing power of the buffer stock was reduced from 50,000 tonnes to 39,666 tonnes of which half were dependent on external finance.

The Council faced an entirely new and formidable task of preserving an increased price range in the face of declining demand, excess stocks and growing production outside its control. Production cuts were implemented from April 1982, at first at 36 per cent and then increased to 38.6 per cent. While they were probably adequate to deal with the first two aspects, they could not absorb the increase in outside production.[71]

Price increases were directly responsible for a major increase in Cornwall and in illegal production in Southeast Asia. But the main source came from a production momentum already in place in both Peru and Brazil. Compounding the problem was the refusal of the GSA to return to its earlier policy of not undermining the buffer stock and the only concession it made in response to pressure from the Asian producers was to set an annual limit of 3,000 tonnes.

With production control limited, the floor could only be protected through the buffer stock. Although the immediate resources available to the BSM were quite inadequate, they were augmented by bank loans secured against the metal and the right to engage in forward dealings on the LME which provided an additional set of tools.[72] Forward trading had the advantage of immobilizing a stock of metal without including it within the buffer stock limits and the contracts could be simply rolled over as they matured. What enabled it to be continued for so long and then broke it so quickly were major shifts in exchange rates.

Between 1980 and 1984, the American dollar appreciated very quickly, especially against sterling which lost 45 per cent of its value. By contrast, the ringgit only lost 7 per cent of its value, so that as prices dropped towards the floor in Malaysia, they rose in London.[73] A rising market simply expanded the BSM's purchasing power. More loans could be contracted and the profits on forward trades covered all its expenses. As it acquired or controlled more metal, it became the dominant force on the LME. Had the BSM been a commercial trader, the brokers would have taken steps to protect themselves against default. But it was trading on behalf of an international governmental organization whose creditworthiness was not in question. When sterling began to recover its external value in 1985, the processes that had enabled the expansion went into reverse. As the metal declined in sterling, banks wanted more collateral and losses were now made on forward trades. By June, many brokers sensed that the BSM had reached the limit of its resources and could not hold the floor. Expecting further price declines, they became short on their forward positions and as the BSM attempted to squeeze them, the LME again imposed a limit on their losses and its gains. In September, the producers promised to raise a further £60 million to prop up the position but the funds failed to materialize, leaving the BSM with no option but to confess its inability to meet its obligations.[74] All tin markets were then suspended.

The extent of the liabilities was large enough to jeopardize the very survival of the LME. Attempts were made to raise the funds required to settle them and liquidate the metal in an orderly fashion. In the absence of an overriding commitment to the maintenance of the agreement in any form, many Council members were unwilling to meet their moral obligation to stand behind the commitments that the BSM had made on their behalf.[75] The immediate effect of this failure was to force the LME to determine the level at which members were to settle their obligations to each other and, in March 1986, a ringout price of £6,250/tonne was imposed.

Those who suffered losses from this discount of 27 per cent on the final market price then took their cases against brokers, the LME, the Council and its members to whatever court around the world would hear them. While many failed, they all produced embarrassing judicial comments.[76] The costs of litigation were sufficient to encourage all parties to negotiate a final settlement and in March 1990 the creditors received payment for around a third of their claims.[77] The following July, the Council conducted the formalities required to lay itself to rest.

Aftermath

As the buffer stock metal passed into private hands, the industry faced the most serious problem of excess stocks since 1921 and that meant an immediate collapse in price. The ATPC tried to force the absorption of these stocks by continuing export control on the part of its members, with some support from Brazil and China. Governments assisted producers through currency depreciation and the virtual elimination of taxes and royalties. Neither measure was sufficient to prevent widespread closures and virtually all who survived did so at a loss.

The 1980s and 1990s were troubled times for all primary commodities and real prices dropped to levels not seen since the early 1930s. The specific problems of tin were compounded by an increase in sales from the stockpile and by the growth patterns of the two peripheral supporters of ATPC's supply reduction scheme. By the late 1980s Brazil had become the world's largest producer and when she began to decline, that place was taken by China. The ATPC continued export control but its effectiveness was clearly limited and in 1999 the organization was wound up.[78]

Commencement of the sixth millennium of the industry found the metal recovering much of its lost glamour. As an environmentally friendly product, tin was replacing lead in a wide range of products from solder to bullets and its chemical properties were being much more systematically exploited. These more than compensated for the stagnation in tinplate and drove demand to record levels. On the production side, the growth of China and Indonesia shifted the balance in favour of large, well-financed organizations. On the marketing side, tin became a commodity like any other, subject to extreme fluctuations in price. Whether producers have become strong enough to conduct their own form of commodity control is uncertain; if not, they will have to decide whether to capitulate to the market or learn from the experience of those forms that have been explored here.

19 Conclusion

The collapse of the International Tin Council moved a Malaysian delegate, Redzwan Sumum, to write a requiem which concluded:

> In thy demise I feel the stress.
> And tears of regret my soul takes –
> Man learns not from his success,
> And less still from his mistakes.[1]

This could equally have been an epitaph inscribed over the grave of its predecessor. The experience of both organizations offers some salutary lessons and they can be arranged around three sets of questions, about the extent of success, the sources of mistakes and of the failure to learn.

The only overall evaluation of the Council conducted after its demise established three main conclusions. It brought about greater stability in both price and producers' income and it managed to raise income by around 5 per cent for members, 18 per cent for outsiders.[2] The results are not dramatic but sufficient to justify this experiment in tin control, especially if stability is the primary objective.

Diagnosis of the ultimate source of the failure of the Council has pointed in the direction of price policy. Prices were set too high, reducing consumption and stimulating outside production. The only outside production that can be directly attributed to price is that which reflected two fundamental weaknesses in the agreement. Smuggling could only become a major problem because of a failure in administration and the growth of production in a consumer member indicates the limitations of consumer commitment. Price did not undermine the agreement on the production side but exposed its fragility.

Much of the commentary on the 1985 debacle came from those monitoring the British market, where prices rose much faster than elsewhere and where consumption also fell much faster. Before even that is considered a prima facie case for the effect of price, it must be recognized that it may be the continuation of a distinctive trend already evident in the 1970s. Although American consumption registered a decline in the early 1980s, that in Europe and Asia held up so price could well be consistent with the long term consumption trend. Support for this assessment came from a commentator who argued that the floor price of the agreement was about the level required to even out the fluctuations of the commodity cycle.[3]

Without access to primary sources, it is impossible to consider an alternative explanation. From the perspective of the ITC, two aspects are quite puzzling. One is the failure to implement a much more severe set of production cuts. Was the industry so weak that it could not absorb the cuts that were routinely imposed in the 1930s? The second is the failure to develop a plan to deal with an overflowing buffer stock. Was the leadership so weak that it was unable to define and enforce a clear set of options? Some might even have extended the life of the Council; at least one would have given it a dignified funeral.

Some much more fundamental questions can be raised about the overall record of postwar tin control. It provides ample confirmation of Caine's prescient warning against abandoning tried schemes in favour of untested ones.

Given its role in shaping the whole history of postwar tin control, the most important issue is the stockpile policy adopted by the Americans. Had the ITC been allowed to continue in 1946, the stockpile would have been built under conditions similar to those that applied in 1940–1941 and the crisis of 1951–1952 would have been avoided. Guarantees about rates of liquidation would have provided the security required to engage in serious discussion about the incentives required to sustain the industry in relation to long term demand. Although the full dimensions of this problem were not evident for some time, by 1948 the producers had already become alarmed. But by then they were committed to seeking a solution along Havana lines which held out the promise of strengthening, or at least legitimating, commodity control by inclusion of the consumers.

When Clauson argued against Caine, he welcomed effective consumer participation since it would serve to overcome a major weakness of the ITC: the inability to control the parasites. With consumers committed to the agreement, they could enforce its production policy by shutting off their markets to outsiders. Such a commitment would be strengthened by a reciprocal system of supply allocation to protect the ceiling.[4] Without specifying the mutual set of obligations and the common interest that bound them together, the Havana model proved to be inferior to the one that it replaced.

Apart from a common interest in moderating the commodity cycle, consumers and producers shared two others, providing adequate finance for the ITRI and the buffer stock Their refusal to contribute to the former and their reluctance to participate in the latter, indicates a willingness to take a free ride on the sacrifice of the producers. That is also evident in the composition of the Council. The only countries that maintained long term membership were those in Western Europe together with Canada, Japan and India. The absence of the USA and the USSR has already been noted but no less significant was that of newly industrializing countries such as South Korea, Argentina and Mexico.

More information is needed about the actual composition of government delegations to determine whether they gave paramountcy to the needs of industrial users on the basis of which a genuine mutuality of interests could be fostered. If they did, then the Council had the right to expect better data on stocks and consumption trends than it apparently received. Otherwise, such delegations may have been more concerned with the role of tin in their overall macroeconomic and political policies and added little of value to the decisions the Council had to make.

Consumer representation resulted in three further limitations. One was in the role of the Chairman. Since he could not be identified with either side, he had no resources of his own with which to press a decision.[5] Since decisions on major issues required a majority of both constituencies, they were often well behind the market.[6] Perhaps the most important was the way in which consumer participation allowed the Council to become defined as a model commodity agreement during the heady days of the NIEO which made it difficult to adjust to the harsh economic climate of the 1980s.

The agreements weakened the producers' position in other respects. They did not even contain a symbolic threat to freeriders. Defining a producer as a net exporter enabled the

United Kingdom to avoid the controls that were imposed on much smaller producers, Zaire and Nigeria. The strictness of the principle of producer symmetry prevented the recruitment of new members by offering the kind of concession that the ITC had made to secure an important degree of control over the Belgian Congo.[7] Unable to adapt to the changing structure of the industry, the Council was doomed to ultimate failure.

That failure was inevitable is, of course, the judgement of those who see the market as the ultimate arbiter of economic power, a judgement that is well expressed in Alan Clark's statement with which this study began. The most important mistake would be the failure to learn from the actual way in which the Council was structured and the decisions it made. A similar failure to learn was the basis on which the experience of the ITC was set aside in 1946. What then are its enduring lessons which permit a more effective analysis of commodity cartels?

At the core of a commodity agreement is production policy. For the ITC, this was conceptually very simple, since the objective was to match three sets of physical quantities: final demand, commercial stocks and production. Once stocks had been reduced to normal levels, it became clear that stability could only be reached through control over supplementary stocks to cushion the time intervals between the adjustment of these three variables, together with any misjudgements about their magnitude. Since final demand is exogenous and normal stocks are a function of final demand, the only question to settle is the balance between the three instruments under the control of the ITC: production released to the market, producers' stocks and the buffer stock.

The outbreak of war eliminated any opportunity for the ITC to actually determine this balance, so there is no empirical evidence with which to clarify the problems in making these decisions. Although conceptually simple, they are technically quite complicated and they constitute the first level of analysis of the cartel.

In principle, there is no reason why the policies adopted should ever be undermined by independent market forces, as long as effective control over production is maintained. Since the conditions for that control lay outside the ITC itself, they constitute a second level of analysis. Here, three dimensions can be distinguished: (1) conditions for the maintenance of an international market; (2) domestic enforcement of quotas; (3) principles of international and domestic allocation of quotas. They are all inherently political and moral.

The international character of the tin market placed Britain in the dominant position. But that was always open to contest by other major economic powers, United States and Germany, who, for their own political reasons, were prepared to give up the advantage of an international market and strike a side deal with one of the major producers. Since Bolivia had its own internal political reasons to be resentful of British power, such a deal was a latent possibility. That it never came to pass during peacetime was simply a function of the particular configuration of political forces.

Setting production quotas low runs the risk that they will not be adequately enforced. Apart from the teething difficulties in Malaya, that problem mainly arose in the Belgian Congo and Bolivia and both were a result of a structural weakness of the state. In the case of the Belgian Congo, this was a result of granting companies considerable autonomy in determining their policies. In the case of Bolivia, this was a combination of a poorly conceived state policy and an unwillingness to press hard enough on the small miners. The actual amounts of excess exports were not large enough to jeopardize the reputation of the ITC and moral sanctions helped curb the extent of the problem.

The principles of allocation of production quotas raise a much more important set of questions about the operation of commodity cartels. Although the formal market properties, inelastic demand curve and concentration of production, establish the incentives to co-operation, the participants remain competing actors and the conditions of that competition are constantly changing as a result of depletion and technical innovation. All cartelization accomplishes is the displacement but not replacement, of one level of competition to another.

At the core of this process of displacement was the commitment to stabilize an otherwise inherently unstable industry. But the stability that was sought was not simply economic, it was also ideological. The results had to be seen to be 'fair', and that applied to both the relationship between consumers and producers and among the producers. The initial negotiations were conducted within a specific configuration of governments and companies that had been shaped by a particular historical process and 'fairness' covered the crucial sets of compromises.

Since these negotiations did not include the Cornish-Malayan group, it was not restrained from conjecturing another configuration and redefining 'fairness'. In doing so, the Cornish could draw on an extensive repertoire of moral categories: British, pioneers, low-cost and producers of a superior brand of metal. Once the ideological issue had been posed, a profound cleavage then emerged which pitted the Cornish against the ITC. Specific policy proposals simply reinforced the cleavage. Two vicious cycles then emerged. An attempt to stabilize the industry around one ideological conception simply produced another which exacerbated the level of instability. As the configuration conjectured by the Cornish became increasingly obsolete, their ideological commitment to it intensified.

A more extreme version of this problem emerged in the negotiations with Siam in 1936. From the perspective of the ITC, Siam had been granted generous terms in 1931 and 1934, so that her decision to expand productive capacity and demand further concessions was simply immoral. From Siam's perspective, the earlier agreements had no moral authority and she saw herself as making up for the delayed pattern of her earlier development. No concept of fairness could be stretched to cover the actual results and in its absence there was never any assurance that the agreement would be renewed.

It is the dynamic dimension of the industry that makes the search for fairness fruitless. The Nigerian government attempted to take this into account in its domestic allocation of quotas, only to find that the very attempt encouraged the articulation of polarized conceptions of fairness. Nor is this a mistake that is easily avoided, as the FMS government discovered. Its formula originally granted production rights based on a reasonably objective criterion that was universally accepted, established capacity by 1930. But within a decade, the two had become separated, with rights held by lessors whose capacity was exhausted which were then sold to those with capacity but insufficient rights. The government was simply lucky that the war intervened before it was forced to open the Pandora's box of fairness.

One of the more misleading assessments of the ITC is that it simply 'froze the structure and organisation of the greater part of the whole industry.'[8] The problem with the metaphor is that it cannot see beyond any surface results to the underlying processes of congelation and their limits. Three such processes can be identified. One was the need to preserve a level of agreement among the four core members in order to try and rein in Siam. The weaker their commitment to the established standard tonnages, the less likely Siam could be held to its. As long as restriction was implemented, no alternative basis for the allocation of the tonnages among the four core members could be found. A period of economic warfare would have provided such a basis but whenever it was entertained, it was quickly recognized that the costs that the losers could impose on the winners were simply too high. However, when production was unrestricted as in 1937 and 1941, the standards among the core members could be adjusted. The third process derives from the need to preserve options. If a country lets its capacity decline, not only does it undermine its claim to the existing standard but also reduces its ability punish those who threaten warfare. On the other hand, if it augments capacity without being prepared to win such a war, it has to increase the rate of domestic restriction.

The claim about stagnation is simply falsified by the extent of organizational and technical change. One of the factors that drove them was prospect of non-renewal of the ITC which meant that all companies had to be prepared to fight for their existence in a period of economic warfare. As they did so, they could see ever more clearly the advantages of continued co-operation. Thanks to the intensity of the ideological conflicts that surrounded the ITC, it was never defined as a permanent institution. The indirect result was to strengthen the economic imperative that perpetuated its temporary condition.

Resolving the tension created between these two poles of ideological conflict and economic advantage was never easy, especially since it was centered on the largest producer of all which was normally expected to play a leadership role in this form of economic co-operation. This had its origin in the internal organization of the Malayan industry and its peculiar relationship with the colonial state. An entirely different form of internal organization and relationship existed in the NEI which managed to reduce the intensity of its own ideological concerns. Once these internal features are recognized,[9] tin emerges as a less straightforward case of cartelization than the formal economic properties of the industry would suggest. The two largest producers were unable to contain their internal politics to become reliable international actors and it was only because Campbell could rely on the third, with the least at stake, that any real stability to the ITC was found.

One of the striking features about the politics that surrounded the ITC was the frequent recourse to moral epithets. £230 was 'commercial robbery'. Stock control meant 'dictatorship'. Depletion was 'prodigal'. Strategic negotiation was 'blackmail', and concession was 'bribery'. The most severe were reserved for the ITC which was 'vicious' and 'wicked'. Such moralism is perhaps the inevitable result of an attempt to grasp a complex situation from a partisan position.

Scholars, however, are granted the luxury of unravelling that complexity. But such has been the power of the moralism engendered by the politics of the 1930s that it has infected many of the commentaries offered by those writing many years after the demise of the ITC. One of the most bizarre observations is to be found in a study of the conditions of life and thought of Bolivian miners where the animosity towards the tin barons of the 1930s continued unabated forty years later. June Nash simply extrapolated that animosity and concluded that nothing associated with Patiño, Hochschild and Aramayo could have generated any benefits to Bolivia. She therefore felt free to state: 'Their empires in Bolivia were consolidated with world tin magnates after the agreement of 1930 establishing the International Tin Council, which controlled prices in the interests of the tin *consumers*.'[10]

No less enduring was the animosity towards the ITC in Malaya. So convinced was Yip of the rectitude of the Cornish position that he took it at face value and looked for external factors to explain why Malayan interests were 'unfairly penalized'.[11] He found them in an extraordinarily speculative assessment of the way in which the British government formulated policy. Concluding that the continuation of a free market would substantially reduce supplies from Bolivia and Nigeria, he assumed that metropolitan interests in maintaining employment in Williams, Harvey were sufficient to induce the Colonial Office to force Malaya's participation in the ITC against her will.

Those whose predilection is to look for villains can easily substitute Anglo-Oriental for the British government as they try to explain the apparent predicament in which Malaya was placed. Allen and Donnithorne supposed that the Malayan government operated at the behest of Anglo-Oriental and that the group was prepared to subordinate its interests in Malaya for the sake of those that it held in other countries.[12] The first of these claims is even more puzzling than that offered by Yip. The second simply derives from prejudice against international conglomerations of capital without any appreciation of just how Anglo-Oriental balanced its various interests. Unfortunately, these prejudices resonate sufficiently strongly that the judgements they induce are repeated by those with different research agendas and for whom this is largely background coloration.[13] Initial mistakes then become received wisdom.

Much more serious was the damage done to the formation of public policy, especially in the United States. Although Congressional hearings are notorious for their partisanship, they add a level of authority and publicity to their judgements and conclusions which renders mute any informed criticism. This problem is compounded when the partisanship is masked by neo-classical economics, since that produces an apparently objective academic analysis. While the State Department may have had little difficulty in recognizing the more extreme positions that emerged in the postwar hearings, it was always hampered in developing a sober assessment of the problem facing the producers thanks to its uncritical adoption of Knorr as the primary guide to their role in shaping the ITC. Just as academics 'take in each others washing'[14] so do politicians and policy makers. In the case of tin and the ITC, most of the laundry moved around so frequently that a single, dominant story was shaped.

Mineral industries are complex institutions, in large part because they are made up of different elements, working different ore bodies, at various stages of depletion, within different political environments, all of which are very conscious of their distinctive histories. Superimposed on these disparate forces are the special issues that emerge when the various producers and governments seek a solution to their respective problems through co-operation and the formation of a cartel.

Once these features are recognized, it is possible to overcome Redzwan Sumum's pessimism. If commodity cartels are examined from some Archimedean point of optimality, their record will always find critics who cannot see successes and exaggerate mistakes. But if they are examined simply on their own terms, as institutions that struggle to deal with an elaborate pattern of economic, political and ideological forces, their experience is well worth commemorating. The result should lead to better scholarship and better public policy.

Appendix: Archival sources

The International Tin Committee

Since the ITC was administered out of the Colonial Office, its files (National Archives, Kew) provide the primary source for this study. They are extremely comprehensive with considerable informal commentary and are unmatched by any other source. Unfortunately, the records of the ITC itself have disappeared, but it is unlikely that they contained any information not communicated to the Colonial Office. A very miscellaneous collection of documents relevant to the ITC was deposited by Fox in the London School of Economics. McKenna administered the International Tin Pool and his papers at the Hong Kong and Shanghai Bank (London) permit a comprehensive account of its formation and operation.

Governments and the International Tin Committee

The relationship between governments and the ITC should be examined from the standpoint of both the position they formulated internally and the way in which it was represented to the ITC. This is can be done only for two of the main producing countries, Nigeria (National Archives of Nigeria, Kaduna) and the NEI (Algemeen Rijksarchief, The Hague) and five of the minor ones, South Africa (National Archives of South Africa, Pretoria), Australia (National Archives of Australia, Canberra), Burma (India Office and Records, British Library, London), Cornwall (Ministry of Power, National Archives, Kew) and Portugal (Arquivos Nacionais, Lisbon). The destruction of most of the FMS central government files during World War II prevents any analysis of the way in which Malaya handled the conflicting pressures to which it was subjected. Nothing of relevance was found in the Archives d'Outre Mer (Aix-en-Provence) concerning French Indo-China and the Archivo Nacional de Bolivia (Sucre) and Archives Générales du Royaume of Belgium (Brussels) revealed little more concerning the policies of their respective governments. The National Archives of Thailand (Bangkok) contain relevant material in English but sources in Thai were not consulted. However, the political files of the Foreign Office (National Archives, Kew) were particularly useful in examining the relationship with Siam, Bolivia and China. Antenor Patiño wrote regular reports on ITC meetings, copies of which have been preserved in the private archives of the Asociación de Industriales Mineros and the Familia Patiño (Casa Cultural Portales, Cochabamba).

As far as other governments are concerned, the most valuable source is provided by the US Department of State (National Archives and Records Administration, College Park). Fragmentary material was located at College Park in the records of the Department of

Justice, Bureau of Foreign and Domestic Commerce and the Bureau of Mines and in the Roosevelt Presidential Library (Hyde Park). Two State Department officials, Cordell Hull and Herbert Feis, deposited their papers in the Library of Congress. Materials assembled by the McReynolds committee are in the National Archives in Washington. Of the British government departments, only the Treasury appears to have maintained a continued interest but its major file has been lost. Otherwise, there is relevant material in the records of the Board of Trade, Cabinet, Department of Scientific and Industrial Research, the Dominions Office and the Export Credit Guarantee Committee, all of which are at Kew.

Domestic administration

The question of domestic administration can be studied in detail for Nigeria (National Archives of Nigeria, Kaduna) and for the following districts and states in Malaya: Batu Gajah, Ulu Selangor, Trengganu, Perlis and Johore (Arkib Negara Malaysia, Kuala Lumpur and Johor). While government sources are silent for Bolivia, there is considerable material in the archives of Asociación de Industriales Mineros and the Familia Patiño. No attempt was made to locate relevant material for Siam.

Industry associations

The only industry associations which have deposited their records are the Nigerian Chamber of Mines (Guildhall, London) and the Asociación de Industriales Mineros. Both shed light on the internal politics of producers. Unfortunately, there is nothing from either the Malayan or FMS Chambers.

Companies

Corporate records are particularly scanty. The singular exception is a comprehensive set covering the activities of all of the members of Billiton group for which there is a superb guide compiled by M. Gruythuysen and R. Kramer, *Inventaris van het direktie-archief van de N. V. Billiton-maatschappij, 1852–1970*, The Hague: Algemeen Rijksarchief, 1990. They are supplemented by the Groothoof papers which are also deposited in the Algemeen Rikjsarchief. The destruction of the Patiño records in Paris prevents any analysis of the precise role he played at the international level. The material preserved in Archivo Familia Patiño deals primarily with issues of domestic administration and development of the mines under his control. A similar set of records concerning Aramayo are in the Carlos Navarro papers (Archivo Historico de La Paz). Apart from some terse minute books, few internal documents from Anglo-Oriental remain (Malaysia Mining Corporation, Kuala Lumpur). However, the archives of the Bank of England (London) contain much material concerning the phase of its reorganization. Similarly terse minute books, together with some technical information, have been preserved for three members of the Cornish group, Tekka-Taiping, (Guildhall, London) Tronoh and Ayer Hitam (Malaysia Mining Corporation). Other companies who have left fragmentary records include Ipoh Tin Dredging, Camp Bird (Guildhall, London), Renong Tin Dredging (School of Oriental and African Studies, London), Central Mining (Rhodes House, Oxford) and Union Minière (Manchester Science Museum).

Documents provided by British companies, including shareholder lists, to meet the registration requirements of the Board of Trade are preserved in the National Archives, Kew. Registration documents for Australian companies are in the Sydney Stock Exchange and the

Victoria Public Record Office (Melbourne). Annual reports of companies with a listing on the London Stock Exchange have been deposited at the Guildhall.

***

Codes for major depositories and collections

AHLP Archivo Historico, La Paz
- CN Carlos Navarro

ANL Arquivos Nacionais, Lisbon

ANMKL Arkib Negara Malaysia, Kuala Lumpur
- BG Batu Gajah
- TRUS Ulu Selangor

ARA Algemeen Rijksarchief, The Hague
- BM Billiton Maatschappij
- MK Ministerie van Koloniën

Archives Générales du Royaume, Brussels
- AE Affaires Etrangères

BE Bank of England, London

Casa Cultural, Cochabamba
- AIM Asociación de Industriales Mineros
- AFP Archivo Familia Patiño

GH Guildhall, London

Hong Kong and Shanghai Bank, London
- MTF McKenna Tin Files

IORL India Office Records and Library, London

MMC Malaysian Mining Corporation, Kuala Lumpur

NAK National Archives, Kew
- BT Board of Trade
- CAB Cabinet
- CO Colonial Office
- FO Foreign Office
- POWE Ministry of Power
- T Treasury

NANK National Archives of Nigeria, Kaduna
- SNP Secretariat of the Northern Provinces

NARAII National Archives and Records Administration, College Park
- SD State Department
- ATD Anti-Trust Division, Department of Justice

Notes

1 Introduction

1 Ralph Kestenbaum, *The Tin Men. A Chronicle of Crisis*, London: Metal Bulletin, 1991, pp. 1–2. This was the largest commercial default since the South Sea Bubble, Michael Prest, 'The Collapse of the International Tin Agreement', Round Table, no. 302, 1987.

2 Christopher Gilbert, 'International Commodity Agreements: An Obituary Notice?', *World Development,* vol. 24, No. 1, January 1996, p. 6. For an early obituary, see W. Keith Buck, 'Study Groups Versus Commodity Agreements: An Appraisal', *Natural Resources Forum*, November 1987, p. 364.

3 House of Commons, *Parliamentary Debates,* vol. 237, col. 767, 21 July 1988. The lesson was indelible and was repeated by another junior minister, Jeremy Hanley, vol. 265, col. 18, 30 October 1995.

4 Sugar control lasted from 1931 to 1935 and was renewed in 1937. Tea control lasted from 1933 to 1939. Wheat control was short-lived, 1933–1934. Rubber control lasted from 1934 to 1944. For overviews see Joseph Davis, 'Experience under Intergovernmental Commodity Agreements, 1902–1945', *Journal of Political Economy*, vol. 54, no. 3, June 1946; J. W. F. Rowe, *Primary Commodities in International Trade,* Cambridge: Cambridge University Press, 1965, Chapter 12; *Kabir-ur-Rahman Kahn, The Law and Organisation of International Commodity Agreements,* The Hague: Nijhoff, 1982. Recent studies of specific commodities in the 1930s include John Drabble, Malayan Rubber: the Interwar Years, London: Macmillan, 1991 and Bishrupriya Gupta, 'The International Tea Cartel during the Great Depression, 1929–1933', *Journal of Economic History,* vol. 61, no. 1, March 2001.

5 One of the striking features of the major markets in primary commodities over the period from *c.* 1880 to 1940 is the way in which they were subject to detailed routine commentary in a large number of different newspapers, magazines and yearbooks. The information they contain is very extensive and the editorials often provided sophisticated assessments.

6 J. K. Eastham 'Rationalisation in the Tin Industry', *Review of Economic Studies*, vol. 4, no. 1, October 1936; C. A. Myers, 'The International Tin Control Scheme', *Journal of Business of the University of Chicago*, vol. 10, no. 2, April 1937; Klaus E. Knorr, *Tin Under Control*, Stanford: Food Research Institute, 1945; the study with the firmest grasp of the economic issues is by Puey Ungphakorn, 'The Economics of Tin Control', PhD dissertation, University of London, 1949.

7 Josef Wollnik, *Zinn, Wandlungen in der Erzeugung und Verwendung des Zinns nach dem Weltkrieg,* Leipzig, 1936; M. Schut, *Tinrestrictie en Tinprijs*, Haarlem, 1940; Mieczysław Epstein, 'La réglementation internationale d'un marché de matière première. L'exemple du cartel international de l'étain', PhD dissertation, University of Fribourg, 1943.

2 Tin: the foundations of an industry

1 James Muhly, *Copper and Tin. The Distribution of Mineral Resources and the Nature of the Metals Trade in the Bronze Age*, Hampden: Archon, 1973, 1976, p. 171. The field of archaeometallurgy is largely indebted to this work which transcends both disciplinary and regional specialisms.

2 R. J. Forbes, *Studies in Ancient Technology*, Leiden: Brill, 1957, pp. 25–6.

3 E. R. Eaton and Hugh McKerrell, 'Near Eastern Alloying and Some Textual Evidence for the Early Use of Arsenical Copper', *World Archaeology,* vol. 8, no. 2, October 1976, pp. 177–8. James Muhly claims that arsenical copper was a natural alloy, 'Sources of Tin and the Beginnings of Bronze Metallurgy', *American Journal of Archaeology*, vol. 89, no. 2, April 1985, p. 278. If this is the case, then the long transition from arsenical copper to true tin bronzes may be as much a function of the technical advantages discovered by controlling the amount of tin as of the distinctive social properties associated with tin bronzes.

4 This represents an 'extraordinary leap of ingenuity in turning rock into metal', James Muhly, 'The Beginnings of Metallurgy in the Old World', in Robert Maddin, ed., *The Beginning of the Use of Metals and Alloys*, Cambridge, MA: MIT Press, 1988, p. 15.

5 William Baldwin, *The World Tin Market. Political Pricing and Economic Competition*, Durham, NC: Duke University Press, 1983, p. 1.

6 Peter Roddy, *The International Tin Trade*, London: Woodhead, 1995, p. xiii. Tin has also baffled explanations of its mineralogical formation and location; 'Where it is, there it is' was the confession of failure of one mining engineer, R. D. Penhallurick, *Tin in Antiquity: Its Mining and Trade Throughout the Ancient World with Particular Reference to Cornwall*, London: Institute of Metals, 1986, p. 3. There is only one firm geological principle: no granite, no tin.

7 Forbes, op. cit., p. 132. This position is contested by Kristian Kristiansen who argued that bronze axes were at best only marginally superior to those from flint or stone, 'From stone to bronze – the evolution of social complexity in Northern Europe', in Elizabeth Brumfiel and Timothy Earle, *Specialization, Exchange, and Complex Societies*, Cambridge: Cambridge University Press, 1987, p. 46. Metal weapons were as much symbols of social position as functional instruments, Graham Philip, *Metal Weapons of the Early and Middle Bronze Ages in Syria-Palestine*, Oxford: BAR, 1989, vol. I, pp. 151–8.

8 A comprehensive inventory is Jean Deshayes, *Les outils de bronze de l'Indus au Danube (1Ve au IIe Millénaire)*, Paris, 1960.

9 William Fox, *Tin. The Working of a Commodity Agreement*, London: Mining Journal, 1974, p. 7.

10 Georgius Agricola, *De Re Metallica*, tr. Herbert and Lou Hoover, New York: Dover, 1950, p. 413.

11 The implication is that the metal is 'white tin' and this may be a residue of the early difficulty in distinguishing between tin and lead, which led the Romans to refer to tin as 'plumbum candidum'. Unfortunately, in French, silver is sometimes called 'métal blanc' and that led to a confusion between the two metals, John Hatcher, *English Tin Production and Trade before 1550*, Oxford: Clarendon Press, 1973, p. 134. The Chinese also confused lead (black tin) and tin (white lead).

12 Erosion has resulted in an intermediary form, 'eluvial'.

13 Pebbles as large as 32 lbs have been found in Australia, Philip Scalisi, *Classic Mineral Localities of the World, Asia and Australia*, New York: Van Nostrand, 1983, p. 178. Many virgin alluvial deposits can be worked extraordinarily easily; it took only three men three days to extract 600 kg of cassiterite in 1878 in Brittany, Jacques Ramin, *Le problème des Cassitérides: et les sources de l'étain occidental depuis les temps protohistoriques jusqu'au début de notre ère*, Paris: Picard, 1965, p. 23.

14 Bryan Earl, '"Melting tin" in the West of England. A study of an old art', *Journal of the Historical Metallurgy Society,* vol. 19, no. 2, 1985, demonstrates the complexity of an apparently simple process. The metal itself melts at 232ºC.

15 However, early smelting techniques failed to capture a high proportion of the metal, J. P. Northover and C. Gillis, 'Questions in the Analysis of Ancient Tin', in Suzanne M. Young, *et al.* eds, *Metals in Antiquity*, Oxford: Archaeopress, 1999, p. 80.

16 It is unfortunate that the search for a non–Eurocentric view of world history has led towards the conceptualization of a 'World System' with the same structural properties operating over several millennia. This would treat these stages as simply phases in the development of that system, at the expense of a recognition of their qualitative differences.

17 Bronze was simply one form whereby such communication occurred. A useful discussion of the development of bronze in the context of a long succession of prestige goods and their relationship to the emergence of social hierarchies is Li Liu, 'Products of Minds as Well as of Hands: Production of Prestige Goods in the Neolithic and Early State Periods of China', *Asian Perspectives,* vol. 42, no. 1, Spring, 2002.

18 Percival Price, *Bells and Man*, Oxford: Oxford University Press, 1983.

19 V. Gordon Childe, 'Archaeological Ages as Technological Stages', *Journal of the Royal Anthropological Institute* vol. 74, no. 1/2, 1944, pp. 9–10. It must always be remembered that until comparatively recent times all metals were scarce and costly. Muhly cites an estimate of

world copper production of 4 million tons from 5000 BC to 1800 AD, op. cit., p. 173. If production in the eighteenth century is subtracted that gives an annual production of around 650 tons; if all of this were used for bronze, that would result in an annual tin output of 50 tons.

20 The trinity of Stone, Bronze and Iron Ages originated with Lucretius and became an integral part of nineteenth century evolutionary archaeology, V. Gordon Childe, *The Bronze Age*, New York: Biblo, 1963, p. 2. The assumption that this evolution, especially from bronze to iron, was driven by technological innovation, rather than by market forces, is challenged by David A. Warburton, *Macroeconomics from the Beginning: The* General Theory, *Ancient Markets, and the Rate of Interest*, Neuchâtel, 2003, pp. 254–7.

21 Heather Lechtman, 'Arsenic Bronze: Dirty Copper or Chosen Alloy? A View from the Americas', *Journal of Field Archaeology*, vol. 23, no. 4, Winter 1996.

22 Dorothy Hosler, *The Sounds and Color of Power: The Sacred Metallurgical Technology of Ancient West Mexico*, Cambridge, MA: MIT Press, 1994, p. 5. For a fuller description of the role of tin bronze in the process of incorporation of local elites within the Inca empire, see Cathy Lynne Costin and Timothy Earle, 'Status Distinction and Legitimation of Power as Reflected in Changing Patterns of Consumption in Late Prehispanic Peru', *American Antiquity*, vol. 54, no. 4, October 1989.

23 Hosler, op. cit., pp. 191, 231–2.

24 Ibid., p. 243.

25 Tin metal has 'magical' properties since varying degrees of oxidation can produce different colours, from pure yellow to pure white, Carole Gillis, 'The Economic Value and Symbolism of Tin', in Young, op. cit., p. 143. Gillis combines this point with the above analysis to explain the tinfoil coating of Aegean burial vases.

26 This limited role has led to a major confusion, since the many artefacts from another copper alloy, brass, are often called 'bronzes'.

27 Stanley Alpern, 'Did They or Didn't They Invent It? Iron in Sub-Saharan Africa', *History in Africa*, vol. 32, 2005, p. 66. Although the cassiterite deposits lie on the surface as easily collected nuggets, they were forgotten until their rediscovery in the 1940s, Danilo Grébenart, 'Les métallurgies du cuivre et du fer autour d'Agadez (Niger), des origines au début de la période médiévale', in Nicole Échard, ed., *Métallurgies Africaines*, Paris, 1983, p. 116. Niger may also have supplied the bronze smiths at the Phoenician settlement in Carthage.

28 Several tin artefacts were uncovered at some depth during mining operations in the early twentieth century but cannot now be dated, Bernard Fagg, 'Archaeological Notes from Northern Nigeria', *Man*, vol. 46, May-June 1946, pp. 51–2. J. W. Taylor suggests that the Nok not only produced tin but traded it with Carthage, 'A Nigerian Tin Trade in Antiquity?' *Oxford Journal of Archaeology*, vol. 1, no. 3, November 1982, p. 322.

29 It is possible that such bronzes were made on a more extensive basis for export to the Christian states of the Upper Nile, J. E. G. Sutton, 'Igbo-Ukwu and the Nile', *African Archaeological Review*, vol. 18, no. 1, 2001. At least the question of the availability of both the copper and tin within reasonable proximity has been settled by Paul Craddock, *et al.* 'Metal Sources and the Bronzes from Igbo-Ukwu, Nigeria', *Journal of Field Archaeology*, vol. 24, no. 4, Winter 1997.

30 Nicola Sutherland-Harris, 'Trade and the Rozwi Mambo' in Richard Gray and David Birmingham, eds, *Pre-Colonial African Trade*, New York: Oxford University Press, 1970, p. 255. This argues forcefully against the suggestion that the metal was imported from Rooiberg in South Africa. One archaeologist considers that the Rooiberg deposits may have been exploited as early as the third millennium, *South African Digest*, May 22, 1970, p. 6.

31 Early reports suggested an independent and early, *c.* 3600 BC, discovery of bronze. A critical discussion is by H. H. E. Loofs-Wissowa, 'The Development and Spread of Metallurgy in Southeast Asia: A Review of the Present Evidence', *Journal of Southeast Asian Studies*, vol. 13, no. 1, March 1982. The acrimonious tone of both responses and rejoinder reveal the fragile basis of archaeological research in this field.

32 Whether the cassiterite deposits of Upper Burma were exploited at this time is uncertain but those of Vietnam and Yunnan probably were. The Han empire identified tin mines in Yunnan following the extension of its influence in the first century BC, Hans Vogel, 'Bergbau auf Gold, Silver, Kupfer und Zinn in Yunnan bis zum Ende der Mongolenzeit', in Albert Lutz, ed., *Der Goldschatz der Drei Pagoden*, Zurich, 1991, p. 75.

33 Jan Wisseman Christie, 'Trade and State Formation in the Malay Peninsula and Sumatra, 300 B.C.– A.D.700', in J. Kathirithamby-Wells and John Villiers, eds, *The Southeast Asian Port and Polity. Rise and Demise*, Singapore: Singapore University Press, 1990, pp. 42, 51.

34 J. Kathirithamby-Wells, Introduction to ibid.
35 Wisseman Christie, op. cit., p. 45.
36 Expansion would not only increase the benefits but also reduce the costs by eliminating potential sources of piracy. An old German expression captures the structural interrelationship: 'Krieg, Handel, und Piratie, dreieinig sind sie, nicht zu trennen.'[War, Trade and Piracy are an indissoluble trinity.]
37 Louis Malleret, *L'Archéologie du Delta du Mékong*, vol. 2, *La civilisation matérelle d'Oc-Èo*, Paris: EFEO, 1960.
38 Ursula Franklin, 'The Beginnings of Metallurgy in China: A Comparative Approach', in G. Kuwayama, *The Great Bronze Age of China*, Los Angeles, 1983, p. 95. Since the cultures to the west and south used bronze for personal ornamentation, Franklin argued for the independent discovery of bronze in China. The balance of evidence now suggests that the basic techniques of tin bronze manufacture were diffused from Central Asia, Katheryn Linduff, *Metallurgy in Ancient Eastern Eurasia*, Queenston: Mellen, 2004, p. 7.
39 Li, op. cit., p. 22.
40 Lothar von Falkenhausen, *Suspended Music. Chime Bells in the Culture of Bronze Age China*, Berkeley, CA: University of California Press, 1993, pp. 32, 38.
41 Franklin, op. cit., pp. 96–7. The technology permitted the construction of huge cauldrons, one of the largest of which weighed 1,400 kg.
42 K. C. Chang, 'The Importance of Bronzes in Ancient China', in Thomas Chase, ed., *Ancient Chinese Bronze Art*, New York: China Institute in America, 1991, p. 17.
43 Peter Golas, *Mining, Science and Civilisation in China*, Cambridge: Cambridge University Press, 1999, vol. 5, pt.13, p. 76.
44 Jessica Rawson 'The Ancestry of Chinese Bronze Vessels' in S. Lubar and W. D. Kingery, eds, *History from Things*, Washington, DC: Smithsonian Institution, 1993, pp. 59–65. The Yuan and Ming dynasties are an exception but the dings were collected by wealthy merchants. Chime-bells, however, could not serve this function. They became both musically and socially obsolete with the Zhou aristocracy, von Falkenhausen, op. cit., pp. 322–3.
45 The comprehensive report is by Hermann Parzinger and Nikolaus Boroffka, *Das Zinn der Bronzezeit in Mittelasien*, Mainz: von Zabern, 2003, vol. 1. The mines are dated to the early second millennium, although the possibility that they were exploited even earlier is not precluded. Particularly significant is the fact that they are extensive, and therefore expensive, workings of lode veins.
46 Gerd Weisgerber and Jan Cierny, 'Tin for Ancient Anatolia ?' *Der Anschnitt, Anatolian Metal* II, Jan Cierney and Gerd Weisgerber, 'Bronze Age Tin Mines in Central Asia', in Alessandra Giumlia-Mair and Fulvia Lo Schiavo, eds, *Le problème de l'étain à l'origine de la métallurgie*, Oxford: Archaeopress, 2003. This makes an even more forceful case, though at the price of discounting the claims for Afghanistan on the dubious grounds that there is no evidence of exploitation of its alluvial deposits, p. 23.
47 Henri Francfort, *Fouilles de Shortughai*, Paris: Boccard, 1989, vol. 1, p. 403.
48 D. K. Chakrabarti, 'The Problem of Tin in Early India', *Man and Environment*, vol. 3, 1979.
49 Evgeny Chernykh, *et al*, 'Metallurgy of the Circumpontic Area', *Der Anschnitt, Anatolian Metal II*, p. 91.
50 One measure of scarcity is revealed by the data from Mari, a major distribution centre on the upper Euphrates, which imported 485 kg in a year when trade routes to the East permitted purchases at a low price, F. Joannès, 'L'étain, de l'Elam à Mari', in L. De Meyer and H. Gasche, eds, *Mésopotamie et Elam*, Ghent, 1991, p. 71.
51 The Eaton and McKerrell data suggest that the two forms are evenly balanced and show a decline in the Aegean and the Troad over these two periods; the Chernykh data suggest a predominance of arsenical copper and an increase in both forms in the Aegean.
52 The initial suggestions that Kestel was an important source of tin met with considerable scepticism. The charges were met by K. Aslihan Yener and Pamela B. Vandiver, 'Tin Processing at Göltepe, an Early Bronze Age Site in Anatolia', *American Journal of Archaeology,* vol. 97, no. 2, April 1993. In spite of the evidence, one sceptic still continues to claim that Kestel was mined for its gold and was not large enough, Nikolaus Boroffka, *et al.*, 'Bronze Age Tin from Central Asia', in Katie Boyle, *et al.* eds, *Ancient Interactions: East and West in Eurasia*, Oxford: Oxbow Books, 2002, p. 137.
53 K. Aslihan Yener, *The Domestication of Metals: The Rise of Complex Metal Industries in Anatolia*, Leiden: Brill, 2000, p. 98.

54 Vasif Şahoğlu, 'The Anatolian Trade Network and the Izmir Region', *Oxford Journal of Archaeology*, vol. 24, no. 4, 2005, p. 354.

55 Warburton reverses this relationship and speculates that tin production in Anatolia was already inserted into a market system and that the mines were unable to meet the competition from the east. The point plays such an important role in his overall polemic against the Polyani school and his demonstration of the centrality of markets in antiquity that it is repeated frequently, op. cit., pp. 53, 127, 136, 167, 179–180, 189, 193, 232, 273.

56 This is a controversial claim but evidence of actual mine workings is provided by Martin Bartelheim, *et al.*, 'Research into prehistoric metallurgy in the Bohemian/Saxon Erzgebirge', in Bernard Hänsel, ed., *Mensch und Umwelt in der Bronzezeit Europas*, Kiel: Oetker-Voges Verlag, 1998.

57 Colin Shell, 'The Early Exploitation of Tin Deposits in South West England', in Michael Ryan, ed., *The Origins of Metallurgy in Atlantic Europe*, Dublin: Stationery Office, 1978, p. 261.

58 Penhallurick, op. cit., Chapters 19, 24, Sandy Gerrard, *The Early British Tin Industry*, Stroud: Tempus, 2000, Chapter 2. Cornwall also supplied the Southwest of Ireland, W. O'Brien, *Mount Gabriel: Bronze Age Metallurgy in Ireland*, Galway: Galway University Press, 1994, pp. 247–9.

59 There have been many speculations about the location of the Cassiterides. Ramin, op. cit., provides the best argument for the Loire estuary.

60 While the archaeological record offers the promise of a comprehensive survey of the role of tin in prestige goods economies, the documentary record cannot perform the same function for this stage. Since it was constructed by only some of the actors (more often than not winners in these struggles) to serve their own needs, it contains a profound methodological bias.

61 Shipments to Ugarit were to private merchants, A. Malamat, 'Syro-Palestinian Destinations in a Mari Tin Inventory', *Israel Exploration Journal*, vol. 21, no. 1, 1971, p. 37.

62 Joannès, op. cit., p. 74.

63 Ibid., p. 69. One transaction for gold was at the ratio of 107:1. The current ratios are 50 and 2,340 respectively, a devaluation of tin in relation to silver by 80 per cent, in relation to gold by 95 per cent.

64 P. Garelli, *Les assyriens en Cappadoce*, Paris: Maisonneuve, 1963, p. 281. Some have considered the quantities too small to warrant much importance. However, even at current silver values, the annual profit would have been substantial, around $570,000. If Warburton is right in claiming that eastern sources made local tin production uncompetitive, then its cost must have been even higher than this.

65 Cemal Pulak, 'The Copper and Tin Ingots from the Late Bronze Age Shipwreck at Uluburum', in Ünsal Yalçin, ed., *Der Anschnitt, Anatolian Metal I,* vol. 13, Bochum, 2000, pp. 152–3. Lead isotope analysis suggests the eastern origin.

66 These ratios make tin, at best, equivalent to copper, so they do not reflect a scarcity of silver. The sources of Late Bronze Age tin pose a far greater puzzle than do those of the Early Bronze Age on which most attention has been concentrated.

67 Maria Eugenia Aubet, *The Phoenicians and the West: Politics, Colonies and Trade*, trans. Mary Turton, Cambridge: Cambridge University Press, 1993. Aubet claims that the Phoenicians at first traded in tin from Anatolia and Armenia, pp. 60, 72.

68 Alfredo González-Ruibal, 'Facing Two Seas: Mediterranean and Atlantic Contacts in the North-west of Iberia in the First Millennium BC', *Oxford Journal of Archaeology*, vol. 23, no. 3, 2004, p. 292.

69 As the British established themselves as a major seafaring power, it was tempting to suppose that the Phoenicians not only entered the Atlantic but sailed around the Bay of Biscay to acquire tin from Cornwall. That myth was laid to rest by Penhallurick, op. cit., Chapter 21.

70 Muhly, 'Sources of tin', op. cit., pp. 276–7. Calculated on the assumption that a drachma contained 65.4 grains of pure silver. Since the price of tin was nearly seven times that of copper, the copper/silver ratio approximates that established in the late second millennium.

71 R. J. Forbes, 'Hydraulic Engineering and Sanitation', in Charles Singer, ed. *History of Technology*, Oxford: Oxford University Press, 1954, vol. 2, p. 665.

72 R. Knox McElderry, 'Vespasian's Reconstruction of Spain', *Journal of Roman Studies,* vol. 8, 1918, p. 98. Details of tin imports around the western Indian Ocean are recorded in *The Periplus of the Erythraean Sea,* translated and edited by G. W. B. Huntingford, London: Hakluyt Society, 1980, pp. 23, 35–6, 47, 51, 122. There is no consensus about the date of this text. The earliest date proposed would refer to Iberian tin, the latest to English.

73 John Hatcher and T. C. Barker, *A History of British Pewter*, London: Longman, 1974, pp. 20–1.

74 Ibid., p. 8. However, these artefacts were comparatively rare until the third century.

75 Pliny cited by Neil Beagrie, 'The Romano-British Pewter Industry', *Britannia*, vol. 20, 1989, p. 171. Calculated on the assumption that the denarius contained 50 grains of silver and the Roman pound contained 328 grams. The lead/tin ratio was 11:1.

76 Hatcher, *Pewter,* op. cit., pp. 9–10.

77 See Alain Courbin, *Village Bells. Sound and Meaning in the Nineteenth-century French Countryside*, New York: Columbia University Press, 1988, for a detailed analysis of the multiple symbols and functions, both religious and secular, communicated by bells.

78 Price, op. cit., Chs. 4 and 6. One of the largest would have contained around six tons of tin, p. 105. Large bells also served a secular purpose, especially for cities trying to establish their institutional autonomy.

79 A major switch to pewter occurred in response to the devaluation of the Anglo-Saxon silver coinage in the late tenth century, S. R. H. Jones, 'Devaluation and the Balance of Payments in Eleventh-Century England: An Exercise in Dark Age Economics', *Economic History Review*, vol. 44, no. 4., November 1991, pp. 604–5.

80 Carlo Cipolla, *Guns, Sails and Empires: Technological Innovation and the Early Phases of European Expansion 1400–1700*, New York: Pantheon, 1965, pp. 23, 25, 33. The bells themselves could be recycled into cannons, Bert S. Hall, *Weapons and Warfare in Renaissance Europe. Gunpowder, Technology and Tactics*, Baltimore, MD: Johns Hopkins University Press, 1997, pp. 91, 93. Adapting the technology to cope with the stresses produced on firing required considerable skill, John F. Guilmartin, *Gunpowder and Galleys. Changing Technology & Mediterranean Warfare at Sea in the 16th Century*, London: Conway, 2003, Appendix 3.

81 By the mid-seventeenth century cast iron cannons had become reasonably efficient but they only supplied half the total demand, Cipolla, op. cit., p. 73. Cipolla's estimate suggests that at least a fifth of European tin production was consumed in bronze cannons; it does not include the amount consumed by Ottoman cannons.

82 William Manchester, *The Arms of Krupp*, Boston, MA: Little, Brown, 1968, p. 67.

83 This was especially the case with pewter plates which wear easily.

84 John Craig, *The Mint. A History of the London Mint from* A.D. *287 to 1948*, Cambridge: Cambridge University Press, 1953, pp. 182, 416. This set a floor to the price but the experiment was abandoned in 1693 since the coins were well below their intrinsic value and tin was replaced by the more expensive copper.

85 Tin smelting in Portugal is noted by Agricola, op. cit., p. 419. By the sixteenth century, Spanish production had become very marginal, J. Sánchez Gomez, *De Minería, Metalúrgica y Comercio de Metales, 1450–1610*, Salamanca, 1989, pp. 693–4.

86 A brief overview of the history of the German tin industry is by E. Reyer, *Zinn. Eine geologish-montanistisch-historische Monografie*, Berlin, 1881. A study of labour conditions and struggles at its zenith is by Hansjoachim Schonherr, *Anfange arbeitsrechtlicher Regelungen des Kapitalismus im erzgebirgischen Zinnbergbau des 15./16. Jahrhunderts*, Leipzig, 1983.

87 The definitive guide to the whole body of mining law as it applied to tin is Robert Pennington, *Stannary Law. A History of the Mining Law of Cornwall and Devon*, Newton Abbot: David & Charles, 1973. The law was never formally codified and many crucial issues remained subject to customary interpretation.

88 Hatcher, *English Tin*, op. cit., p. 63.

89 These courts adopted the general principles governing secular law and evolved in the same direction. Stewards' courts followed the common law; Vice-Wardens' courts that of equity. The latter continued until 1896, Pennington, op. cit., p. 70.

90 Each county had its own Parliament or Convocation. Devon's was last held in 1600 and Cornwall's in 1752.

91 Coinage was a fixed amount, first set at a very high rate of £4/10/– per ton on Cornish tin, three times the level on Devon tin, Hatcher, *Pewter*, op. cit., p. 6. By the end of the thirteenth century it was reduced to £1/10/– (*c.* 20 per cent of the value); by the eighteenth century it was £4/9/6 (*c.* 6 per cent).

92 One such transaction generated a profit of over 130 per cent, Hatcher, *English Tin*, op. cit., pp. 92–3.

93 One of the largest such capitalists employed 300 men in seven mines in 1357, ibid., p. 62.

94 Ibid., pp. 56–7.

95 The data on the size of blocks presented for coinage conflate two processes, the proportion presented by tinners able to finance their own production and the distribution among the financiers.

The overall trend to greater equality is provided by a comparison between 1300/01 and 1462/63, years in which the quantities presented were very similar. The Gini coefficients are 0.63 and 0.49 respectively, calculated from ibid., pp. 72–3.

96 Ibid., p. 83.

97 Both production and export estimates fail to take into account the quantities that were smuggled to avoid coinage and customs dues.

98 A comparison between the English and French prices suggests that the trade was very lucrative. English prices are in George Lewis, *The Stannaries: A Study of the English Tin Miner*, Boston, MA: Houghton, Mifflin, 1908, pp. 275–7; French prices are in Georges d'Avenel, *Histoire économique de la propriété, des salaires, des denrées et de tous les prix en général, depuis l'an 1200 jusqu'en l'an 1800*, New York: Franklin, 1968–1969, pp. 421–5.

99 The journey lasted one month and cost 3 per cent of the value of the tin, Jean Favier, *Gold and Spices. The Rise of Commerce in the Middle Ages*, New York: Holmes & Meier, 1998, pp. 43, 36.

100 Guilmartin provides a detailed account of the battle, the role played by bronze artillery and the extent of Ottoman losses, op. cit., Chapter 5.

101 S. A. Skilliter, *William Harborne and the Trade with Turkey*, Oxford: Oxford University Press, 1977. The Ottomans relied extensively on bronze cannons, Gábor Ágoston, *Guns for the Sultan: Military Power and the Weapons Industry in the Ottoman Empire*, Cambridge: Cambridge University Press, 2005, pp. 83, 95, 165, 171, 198.

102 R. W. Ferrier, 'Terms and Conditions under which English Trade was Transacted with Safavid Persia', *Bulletin of the School of Oriental and African Studies,* vol. 49, no. 1, 1986, pp. 63–4. The contract stipulated the delivery of 80 tons over the three years of 1632 to 1634, at the equivalent price of £168/ton for metal that was probably bought for around £65.

103 John F. Richards, *The Mughal Empire*, Cambridge: Cambridge University Press, 1995, p. 198.

104 H. V. Bowen. 'Sinews of trade and empire: the supply of commodity exports to the East India Company during the late eighteenth century', *Economic History Review,* vol. 55, no. 3, 2002,

105 Lewis, op. cit., p. 20. This is a good illustration of the technical conservatism of the industry. It took over a hundred years for this technology to be adapted. While the English were well aware of the value of these particles, they had recovered them from the thatch in the roofs of the blowing houses by simply burning them down!

106 Ibid., pp. 19–21.

107 Ernest Hedges, *Tin in Social and Economic History*, London: Arnold, 1964, p. 95. Pewter was systematically developed from the fourteenth century.

108 L. Carrington Goodrich, 'Early Cannon in China', *Isis,* vol. 55, no. 2, 1964, p. 193.

109 Ágoston, op. cit., p. 194.

110 R. B. Mason and M. S. Tite, 'The beginnings of tin opacification of pottery glazes', *Archaeometry*, vol. 39, no. 1, 1997. The technology had its origin in the Mediterranean in the fourth century.

111 Hedges, op. cit., pp. 119–20.

112 Ibid., pp. 94–5.

113 In the early seventeenth century 80 per cent of Chinese consumption was met from Kwangsi (Guangxi), Robert Marks, *Tigers, Rice, Silk, and Silt: Environment and Economy in Late Imperial South China*, Cambridge: Cambridge University Press, 1998, p. 103. By the mid- eighteenth century, the centre of Chinese production had returned to Yunnan with an annual production of over 1,000 tons, Peter J. Golas, 'Tin Mining in 19th and 20th Century China: High Production, Low Technology', in Yang Tsui-hua and Huang Yi-lung, eds, *Science and Technology in Modern China*, Taipei, 1991, p. 267. Kwangsi is a neighbour of Yunnan.

114 The reference cited by Marks above notes an elaborate aqueduct system which implies a level of capital investment comparable to that in England.

115 Barbara Andaya, *Perak. The Abode of Grace. A Study of an Eighteenth-Century Malay State*, Kuala Lumpur: Oxford University Press, 1979, p. 338.

116 The King of Siam ordered Ligor to supply 25 katties (*c.* 32 lbs) of tin for each inhabitant. This was produced by a special class of commoners who were then relieved of their corvée obligations, George Vinal. Smith, *The Dutch in Seventeenth-century Thailand*, DeKalb, IL: Northern Illinois University Press, 1977, p. 75.

117 Gabriel Ferrand, trans., *Voyage du marchand arabe Sulayman en Inde et en Chine redigé en 851*, Paris, 1922, p. 96. The port was called Kalah and Paul Wheatley argued that it was probably Tenasserim, *The Golden Khersonese: Studies in the Historical Geography of the Malay Peninsula before A.D. 1500,* Westport, CT: Greenwood Press, 1973.

118 Holden Furber, *Rival Empires of Trade in the Orient, 1600–1800*, Oxford: Oxford University Press, 1976, pp. 251–2.
119 Trade drove this expansion since it prevented neighbouring states from supporting piracy.
120 M. A. P. Meilink-Roelofsz, *Asian Trade and European Influence in the Indonesian Archipelago between 1500 and about 1630*, The Hague: Nijhoff, 1962, pp. 29, 364 note 13.
121 In 1433 a Chinese text published a brief description of production in the Malaccan hinterland, Ma Huan, *Ying-yai sheng-lan. The Overall Survey of the Ocean's Shores,* ed. and trans. Feng Ch'eng-Chün, Cambridge: Cambridge University Press, 1970, p. 111. Blocks were produced in units of under 2 lbs, to facilitate trade.
122 Victor Lieberman, *Strange Parallels: Southeast Asia in Global Context, c. 800–1830*, Cambridge: Cambridge University Press, 2003, p. 269.
123 By this point, China had withdrawn its support for the great fleet of Zheng Ho and was no longer interested in trade with the South.
124 Meilink-Roelofsz, op. cit., p.165. Tin also came as tribute and from areas under the direct control of the royal court, Ronald Provencher, *Mainland Southeast Asia: An Anthropological Approach*, Pacific Palisades, CA: Goodyear, 1975, p. 39.
125 John Coggin Brown and A. K. Dey, *India's Mineral Wealth*, Bombay: Oxford University Press, 1955, p. 168.
126 Supaporn Ariyasajsiskul, 'The So-called Tin Monopoly in Ligor. The Limits of VOC Power vis-à-vis a Southern Thai Trading Polity', *Itinerario*, vol. 28, no. 3, 2003, p. 96.
127 Wil Dijk, *Seventeenth-century Burma and the Dutch East India Company, 1634–1680*, Singapore: Singapore University Press, 2006. Although this was a lucrative trade in which tin was the most important item, it ended in 1680. In the eighteenth century, Burma prohibited the export of all metals.
128 In late seventeenth-century Ligor, the VOC paid 80 per cent of a normal market price, 50 per cent of the black market price but 20 per cent more than that paid by the crown, Supaporn, op. cit., p. 96. Some of the tin was shipped to Europe where it sold at a discount of 30 per cent against Cornish tin, Lewis, op. cit., pp. 55–6.
129 Supaporn, op. cit., p. 95.
130 Ibid., p. 97.
131 At the time these were known under their Malay names, Ujong Salang, later anglicized to Junk Ceylon and Bangery respectively.
132 G. E. Gerini, 'Historical Retrospect of the Junk Ceylon Island', *Journal of the Siam Society*, 1902, pp. 8–9.
133 Ibid., p. 32. This suggests that the governors could only impose excessive duties on the sale of tin but could not compel miners to work.
134 Bugis are also ethnic Malays but with a different language. As they were dispersed from their homeland on Celebes in the seventeenth century, they became traders and fighters and represent the classic trinity of war, trade and piracy.
135 Gerini, op. cit., p. 54. The EIC monopoly was confined to trade between England and Asia; British country ships were allowed to trade within Asia, D. K. Bassett, 'British "Country" Trade and Local Trade Networks in the Thai and Malay States, *c.* 1680–1770', *Modern Asian Studies,* vol. 23, no. 4, 1989.
136 This follows Graham Irwin, 'The Dutch and the Tin Trade of Malaya in the Seventeenth Century', in Jerome Ch'en and Nicolas Tarling, eds, *Studies in the Social History of China and South-east Asia*, Cambridge: Cambridge University Press, 1970 and S. Arasaratnam, 'Some Notes on the Dutch in Malacca and the Indo-Malayan Trade, 1641–1700', *Journal of Southeast Asian History,* vol. 10, no. 3, 1969.
137 Mary Somers Heidhues, *Bangka Tin and Mentok Pepper: Chinese Settlement on an Indonesian Island*, Singapore: Institute of Southeast Asian Studies, 1992, p. 6.
138 In the late eighteenth century the VOC paid the equivalent of £47/ton but the sultan and others could make illegal sales to passing ships at £67/ton.
139 Jurrien van Goor, 'A Hybrid State. The Dutch Economic and Political Network in Asia', in Claude Guillot, *et al.*, eds, *From the Mediterranean to the China Sea*, Wiesbaden: Harrasowitz, 1998, p. 205. The sultan paid for the costs incurred by the VOC in this incident over the next 60 years.
140 Heidhues, op. cit., pp. 11–12. The technology of both mining and smelting is described in James Jackson, 'Mining in 18th Century Bangka: the Pre-European Exploitation of a 'Tin Island'', *Pacific Viewpoint,* vol. 10, no. 2, 1969, pp. 42–6.

141 On occasion, the VOC refused to buy metal in excess of immediate demand, presumably because of lack of finance. How the credit chain dealt with such interruptions is unclear.
142 Jackson, op. cit., p. 34.
143 Ibid., p. 35. This includes estimates of smuggled tin. The VOC received around half of this total.
144 In 1779 the VOC bought 1,773 tons of Banka tin while British country captains bought 1,786 tons, mainly at Riau, in exchange for opium and textiles, Bassset, op. cit., p. 643. By the second half of the eighteenth century, British traders were better organized than their Indian counterparts and their profits were drawn less from the sale of tin than the purchase of opium against which it was bartered.
145 Andaya, op. cit., pp. 337–40.
146 Coincidentally, the Burmese had devastated Phuket in their attack on Siam.
147 In return, the sultan expected British military protection against Burma and Siam. However, since such support could jeopardize EIC interests in both countries, it was refused. Once established, the EIC could rely on its own military superiority to repulse an attempt to reclaim the island. In 1800 the sultan's successor negotiated another agreement in which territory on the mainland (Province Wellesley) was granted in anticipation of protection. It, too, was fraught with misunderstandings. R. Bonney, *Kedah 1771–1821: The Search for Security and Independence*, Kuala Lumpur: Oxford University Press, 1971.
148 Banka tin could only be sold at a discount on Cornish and at 75 per cent of the price in Canton. It was therefore only shipped to Europe because of its role as ballast in reducing the transport costs on other items.
149 D. Bradford Barton, *A History of Tin Mining and Smelting in Cornwall*, Truro, 1965, pp. 28–9. The arrangement continued until 1813 but the amounts declined considerably. There were two other incentives: the price of tin in China had risen to £101/ton and support for the domestic tin industry would ease the distress created by large scale unemployment in the copper mines, which were unable to meet the competition from Anglesey. Copper not only exacerbated the problem but it also pointed out the solution, since the companies formed a production cartel. The British Government also assisted by waiving the £3/6/– per ton export duty.
150 W. R. Grant, 'Iron in ancient tin from Rooiberg', *Journal of Archaeological Science*, vol. 21, 1994, p. 456, 'The Sourcing of Southern African Tin Artefacts',*Journal of Archaeological Science*, vol. 26, 1999, p. 1117.
151 Alan Smith, 'Delagoa Bay and the Trade of South-Eastern Africa' in Richard Gray and David Birmingham, eds, *Pre-Colonial African Trade*: *Essays on Trade in Central and Eastern Africa before 1900*, London: Oxford University Press, 1970, p. 284.
152 David Randall-Maciver, 'The Rhodesia Ruins: Their Probable Origin and Significance', *Geographical Journal,* vol. 27, no. 4, April 1906, p. 330.
153 A. O. Anjorin, 'Tin Mining in Northern Nigeria During the Nineteenth and Early Part of the Twentieth Centuries', *Odu: A Journal of West African Studies*, vol. 5, 1971, p. 57.
154 Roger Burt, 'The transformation of the non-ferrous metals industries in the seventeenth and eighteenth centuries', *Economic History Review*, vol. 48, no. 1, 1995.
155 One major impurity in many Cornish lodes was arsenic which had to be removed by roasting or calcination. This was captured and sold. It was particularly valuable in the latter part of the nineteenth century, Barton, op. cit., p. 178.
156 Ibid., p. 131.
157 In 1726, Trevenen declared a profit of £7,000, ibid., p. 36. This must have been one of the most technically advanced, since it worked veins over 1,000 feet deep with five engines.
158 This would prevent the entry of competitors.
159 This was extremely small. By the 1840s there were 60 workers who smelted the product of 9,000 miners, ibid., p. 93. This ratio would not change.
160 Pennington, op. cit., pp. 153–4.
161 Lynne Meyers, *Balmaidens*, Penzance: Hypatia, 2004.
162 Barton, op. cit., p. 112. Only careful supervision could obviate this problem.
163 Hedges, op. cit., p. 161
164 Archibald Kidd, *History of the Tin-Plate Workers*, London, 1949, pp. 8–9.
165 Hedges, op. cit., p. 162.

3 Tin and industrial capitalism, 1815–1918

1 In the 1820s, world copper production was around twice that of tin; by the 1910s, it was over seven times, calculated from Christopher J. Schmitz, *World Non-ferrous Metal Production and Prices, 1700–1976*, London: Cass, 1979.

2 Ernest Hedges, *Tin in Social and Economic History*, London: Arnold, 1964, pp. 135–44. This innovation rested on the adapting a formula which alloyed tin with antimony and copper to produce Britannia ware. This was a harder form of pewter and was displacing it from many uses. Babbitt was manufacturing Britannia ware from 1824.

3 This draws on the standard history by Walter Minchinton, *The British Tinplate Industry*, Oxford: Clarendon Press, 1957. There were other areas where Britain lost market share, as in the case of tin toys whose production centre moved from Wolverhampton to Nuremberg, Kenneth Brown, *The British Toy Business: A History Since 1700*, London: Hambledon Press, 1996, pp. 68, 74.

4 Minchinton, op. cit., pp. 26–9.

5 In the 1850s around eight pounds of tin were required for each box of tinplate which sold at 28 shillings. By the 1880s the box needed just over two pounds of tin and sold at 14 shillings. Many other factors were responsible for both the reduction in quantity of tin and the cost of the plate.

6 The first food cans were developed for the Dutch navy, Stuart Thorne, *The History of Food Preservation*, Kirby Lonsdale, 1986, p. 25.

7 Carl Thompson, 'The Heroic Age of the Tin Can. Technology and Ideology in British Arctic Exploration, 1818–1835', in D. Killingray, *et al.*, eds, *Maritime Empires*, Woodbridge: Boydell and Brewer, 2004. These cans were expensive and difficult to open but the fact that they stacked easily and did not break ensured that the early experimentation with the much cheaper glass containers was dropped.

8 Simon Naylor, 'Spacing the Can: Empire, Modernity, and the Globalisation of Food', *Environment and Planning,* vol. 32, no. 9, September 2000, p. 1632. This connection between preserved foods and a 'civilizing' project is discussed more fully by Mona Domosh, 'Pickles and Purity: Discourses of Food, Empire and Work in Turn-of-the-Century USA', *Social and Cultural Geography,* vol. 4, no. 1, March 2003.

9 John Jackson Manley, *The Age of Tin*, London, 1872, part I. This is an example of the vent for surplus form of economic growth.

10 In some lines the canned product was better than the natural one, as for example corn and condensed milk. The latter also made a modest contribution to the increase in efficiency of transoceanic passenger travel since it eliminated the need to carry cows.

11 David Griffith, *Decorative Printed Tins: The Golden Age of Printed Tin Packaging*, New York: Studio Vista, 1979.

12 M. J. Franklin, *British biscuit tins 1868–1939: An Aspect of Decorative Packaging*, London, New Cavendish Books, 1979. The intense competition between biscuit manufacturers led to fresh designs each season and there was no limit to the number of such tins that could be used. Some tins became so elaborate that they had to be sold in a cardboard container.

13 This process of tariff reduction is summarized by Wong Lin Ken, *The Malayan Tin Industry to 1914*, Tucson, AZ: University of Arizona Press, 1965 but in ways that are seriously misleading, pp. 8–9. The 1838 cut is overlooked. The 20 per cent duty on imported ore imposed in 1833 is misrepresented as £400/ton and its subsequent reduction in 1842 is calculated as though it were metal, i.e. on the basis of hundredweights rather than tons. This duty was finally abolished in 1845 and its rationale was never very clear.

14 Daniel Headrick, *The Tools of Empire. Technology and European Imperialism in the Nineteenth Century*, New York: Oxford University Press, 1981, p. 161. However, Phuket was not connected until the early twentieth century, Jennifer Cushman, *Family and State: the Formation of a Sino-Thai Tin-Mining Dynasty, 1797–1932*, Singapore: Oxford University Press, 1991, p. 3.

15 This was approximately the time required to ship and treat the concentrates.

16 A simultaneous pair of contracts for the purchase and sale of spot and future metal balanced losses on one with gains on the other. If supplies of metal for immediate delivery are limited, the spot price will be higher than the future one, a condition known as 'backwardation'. Otherwise, the future price is higher than the spot price by the cost of warehousing and financing, 'contango'.

17 The old adage – He who sells what isn't his'en must buy it back or go to prison – held until 1982.

18 Alex Skelton, 'Copper',in W. Y. Elliott, ed., *International Control in the Non-Ferrous Metals*, New York: Macmillan, 1937, p. 395. Secretan was the Head of the Société Industrielle et Comerciale des Métaux in Paris.

19 Peter Roddy, *The International Tin Trade*, London: Woodhead, p. 12. The syndicate that financed Secretan lost £800,000; their losses on a simultaneous attempt to corner the copper market were around six times as great.

20 Particular needs reflected both the conservatism of many consumers and the technical requirements of their various production processes which could be affected by the different kinds of impurities remaining in these different brands. Neither was sufficiently strong to prevent consumers from moving to other brands if the premium demanded became too large.

21 The largest failure was Wheal Vor which lost £207,000 before folding in 1860, D. Bradford Barton, *A History of Tin Mining and Smelting in Cornwall*, Exeter: Wheaton, 1989. p.107. These speculations rested on the same basis as lotteries, as first noted by Isaac Newton, 'Adventuring for Tyn, which as a lottery hath a great many blanks to one prize', ibid., p. 6.

22 T. R. Harris, *Dolcoath: Queen of Cornish Mines*, Penzance, 1974, Chapter 3.

23 Barton, op. cit., pp. 146–7.

24 Just as those who buy lottery tickets are drawn disproportionately from the poor, so were those who succumbed to the puffery of the mining prospectus, clergymen and poor widows, ibid., p. 163.

25 Philip Flower, *A History of the Trade in Tin*, London, 1880, Appendix.

26 Barton, op. cit., p. 129. Enthoven's interests already included smelting lead from the Isle of Man and the firm sent Cornishmen to Bolivia to demonstrate how the particularly dirty ores could be brought up to a standard required for smelting in Cornwall. Bolivian smelters could only produce an inferior metal.

27 The alluvial deposits were sufficiently rich that they could be exploited with very simple methods. In Queensland it was claimed that the equipment required only cost £5 and two men who earned at least £2 each per week, Walter Hume, *Report on the Queensland Tin Field*, Brisbane, 1874, p. 18. Details of the multitude of workings in New South Wales are provided by Helen Brown, *Tin at Tingha*, Armidale: Brown, 1982, Chapter 3 and for Northeast Tasmania by A. W. Loone, *Tasmania's Northeast*, Launceston, 1981, reprint of 1929 edition. Northwest Tasmania was the site of the one of the richest and most profitable of all lode mines which generated profits of £44/ton of metal, all of which were returned to shareholders who enjoyed an annual dividend of 230 per cent for 20 years, Tasmania Department of Mines, *A Century of Tin Mining at Mount Bischoff, 1871–1971*, Hobart, 1972.

28 Barton, op. cit., p. 173.

29 Ibid., p. 175.

30 For the latter, they were also in competition with two smelters in Hamburg.

31 Anthony Webster, *Gentlemen Capitalists: British Imperialism in South East Asia, 1770–1890*, London: Tauris, 1998, Chapter 7.

32 This required deposing one claimant to the sultanate and investing another. Unfortunately, the first Resident was insensitive to the difficulties in imposing the Victorian principles of government and was promptly murdered by the losing faction. His successor soon found the requisite balance of diplomacy and authority.

33 Wong, op. cit., pp. 53–5. This legal framework was complete by 1895.

34 Ibid., p. 146.

35 Ibid., p. 148.

36 For an analysis of the dominance of the free-standing company in the tin industry, see Jean-François Hennart, 'Transaction Costs and the Multinational Enterprise: The Case of Tin', *Business and Economic History*, vol. 16, 1987. The theory behind this analysis is further developed in his 'International Financial Capital Transfers: A Transaction Cost Framework', *Business History*, vol. 36, no. 1, January 1994. A comprehensive review of the significance of this institution is Mira Wilkins and H. Schröter, eds., *The Free-standing Company in the World Economy, 1830–1960*, Oxford: Oxford University Press, 1998.

37 The directors gained through additional fees and there were other additional management expenses. Many services were provided to the whole group and they included the secretarial work of maintaining share registers and attending to a wide range of administrative and legal obligations together with the more technical assistance required from professional mining engineers.

38 J. S. D. Rawlins, 'French Enterprise in Malaya', *Journal of the Malayan Branch of the Royal Asiatic Society*, vol. 39, December 1966, pp. 75–6.

39 Chai Hon-Chan, *The Development of British Malaya, 1896–1909*, Kuala Lumpur: Oxford University Press, 1964, pp. 28–31.

40 Wong, op. cit., pp. 127–41.

41 Randall Rhohe, 'The Chinese and Hydraulic Mining in the Far West', *Mining History Journal*, 1994.
42 Robert Peele, ed., *Mining Engineers' Handbook*, New York: Wiley, 1941, vol. 1, pp. 10–550.
43 Wong, op. cit., pp. 150–2, *Tin*, January 1936, p. 6.
44 It was suggested that the Cornish connection was a result of the application of the technology of hydraulicing adopted by the china clay industry, G. C. Allen and Audrey Donnithorne, *Western Enterprise in Indonesia and Malaya*, London: Allen & Unwin, 1962, p. 157. This was a groundless speculation which has received much currency.
45 Over the first 20 years the shareholders received an annual return of over 27 per cent on their investment, A. G. Glenister, 'The Gopeng Consolidated Tin Mines', *Mining Magazine*, August 1926, p. 73.
46 Wong, op. cit., pp. 152–3. K. Tregonning states they were neither, *Straits Tin*, Singapore, n.d. p. 9.
47 The group was often called the 'Gopeng group' after its leading company.
48 A brief history of the formation of this influential partnership which long outlasted the deaths of both Osborne and Chappel is in *Straits Times*, 25 June 1923.
49 Many of these companies also acquired properties originally owned by Foo, Ralph Stokes, *Malay tin-fields*, Singapore, 1906, p. 30. Foo sold his interests by 1911. In 1908, Tronoh moved its head office from Redruth to London, which may reflect a dilution in the extent of specifically Cornish equity interests.
50 *West Briton*, 17 August 1943.
51 John Thoburn, *Tin in the World Economy*, Edinburgh: Edinburgh University Press, 1994, p. 53. The relationship between these various techniques is more evident in Dutch where the same root term is used for bucket dredge, 'baggermolen', or 'emmerbagger', gravel pump, 'pompbagger' and suction cutter, 'spuitbagger'.
52 Tin-bearing alluvium was dislodged by high pressure jets of water and the resulting slurry was then pumped to the surface where the particles of cassiterite were separated over long wooden tables, known as palongs. A Bolivian visitor commented: 'This is not mining but a laundry.'
53 Herman Muhlinghaus was with Brandt & Co, one of the many German merchant houses that had reduced the profitability of the Chinese merchants; James Sword was with Gilfillan, Wood, one of whose partners was impressed by a smelter in Sydney, Tregonning, op. cit. p. 8.
54 Capital was raised from the European merchant community, W. G. Huff, *The Economic Growth of Singapore*, Cambridge: Cambridge University Press, 1994, pp. 62–3. As a result, STC became enormously influential within Singapore.
55 Cushman, op. cit., pp. 75–6.
56 Ibid., pp. 76–9.
57 Ibid., pp. 96–7. Frederick Penny was appointed as Managing Director. On his return to England he became part of the 'gentlemanly capitalist' network, as a Conservative MP, serving in various critical positions from Whip to Party Treasurer, Nicolas White, 'Gentlemanly Capitalism and Empire in the Twentieth Century: The Forgotten Case of Malaya, 1914–1965',in Raymond Dummett, ed., *Gentlemanly Capitalism and British Imperialism. The New Debate on Empire*, Harlow: Longman, 1999, p. 181.
58 One small smelter persisted in Kuala Lumpur producing around 300 tons which had to be refined in Singapore before being sold within Asia, *Straits Times*, 14 May 1932. Local smelting ceased in Burma in 1921 and in Siam in 1924. However, it was resumed in Burma by 1936 for export to India.
59 It also ensured that the Malay Civil Service would not be subservient to the Straits Civil Service based in Singapore, Philip Loh, *The Malay States, 1877–1886*, Kuala Lumpur: Oxford University Press, 1969.
60 Johore was eventually persuaded to accept a Resident, termed 'General Adviser', in 1914, Keith Sinclair, 'Hobson and Lenin in Johore: Colonial Office Policy towards British Investors and Concessionaires, 1878–1907', *Modern Asian Studies*, vol. 1, no. 4, 1967.
61 Attempts to find a real coherence to these various units have always been controversial. The term 'Malayan' was used to refer to those who claimed to represent the interests of that ambiguous unity, in spite of the fact they were overwhelmingly of European or Chinese extraction, few of whom intended to become permanent settlers.
62 It was assumed that direct control over the metal would strengthen the company's position in its negotiations with the tinplate manufacturers, *Mineral Industry for 1914*, New York, 1915. Bill Freund claims that Williams, Harvey was behind this initiative, *Capital and Labour in the Nigerian Tin Mines*, Harlow: Longman, 1981, p. 117. If so, this would explain some of the rivalry

that later emerged between the two smelters. Unfortunately, no support is provided for this intriguing suggestion.

63 Shipping and handling costs are much higher on bags of concentrates than on blocks of metal. Labour costs would also be higher.

64 Competition was preserved by exempting exports to Australia and the United Kingdom. In fact, STC became so competitive that it was able to attract around a third of Australian production, N. M. Penzer, *The Tin Resources of the British Empire*, London, 1921, p. 277.

65 Like many such sentiments it was quite irrational, since the Americans always remained free to negotiate with Billiton and the Siamese for an adequate supply. Nor was there any obstacle to the establishment of a smelter in the Straits under American control.

66 STC AGM, *Times,* 13 July 1933.

67 Webster, op. cit., Chapter 4. Mary Somers Heidhues, *Bangka Tin and Mentok Pepper*, Singapore: Institute of Southeast Asian Studies, 1992 speculates that the decision may have had a more particular objective, the prevention of competition with Cornwall, p. 33. It is not clear how Cornwall would be served by leaving a large and very low cost producer outside the sphere of British influence.

68 Malacca had been returned to the Dutch in 1818. In 1819 Raffles attempted to reverse this setback to the EIC and travelled to Singapore where a succession dispute for the Sultanate of Johore enabled him to secure the island, Webster, op. cit., Chapters 3 and 4. Unaware of its tin deposits, Raffles also pressed for retaining Billiton and developing the island as an entrepôt to complement Singapore, Billiton Maatschappij, *Gedenkboek Billiton, 1852–1927*, The Hague: Nijhoff, 1927, vol. 1, pp. 1–2.

69 Malacca was therefore also transferred to the EIC.

70 Heidhues, op. cit., p. 59.

71 Ibid., pp. 67–8.

72 Ibid.

73 Ibid., p. 117.

74 *Gedenkboek*, op. cit. vol. I, p. 7.

75 Much was wasted thanks to the employment of an alcoholic Cornish mining engineer, ibid., p. 15.

76 Wong, op. cit., p. 35.

77 Heidhues' suggestion that this reflected the failure of private capital is mistaken, op. cit., p. 81

78 Abandonment was even considered when prices sagged in the 1890s, J. C. Mollema, *De ontwikkeling van het eiland Billiton en van de Billiton-maatschappij,* The Hague: Nijhoff, 1922, p. 40.

79 *Gedenkboek,* op. cit., pp. 52–4.

80 Ibid., pp. 57–8.

81 Ibid., p. 59. At £32,000 it was considered to be a major investment and the retired chief administrator expected it to result in complete failure.

82 Ibid., p. 63.

83 The little information about Singkep is summarized by A. F. Kamp, *De standvastige tinnen soldaat. 1869–1960,* The Hague: Billiton, 1960, Chapter 13.

84 Cushman, op. cit., p. 9. Cushman provides a detailed account of the formation of the lineage and the way in which it sustained a web of political connections. Unfortunately, her sources are largely silent on the economic implications both for the family and the region during the period of its formation in the nineteenth century.

85 G. E. Gerini, 'Historical Retrospect of the Junk Ceylon Island', *Journal of the Siam Society*, 1902, p. 55.

86 Ruth S. Kerr, 'The Advent of Tin Dredges in Eastern Australia', *The Australian Journal of Historical Archaeology*, vol. 7, 1989.

87 A contrast between a successful and a failed bucket dredge is 'Dredging for Tin at Stanthorpe', *Queensland Government Mining Journal,* 15 January 1904. These early dredges were extremely small. In 1919 the mean cost of the 62 dredges operating in New South Wales was £4,300, with a mean production of 17 tons of concentrate, Edward Kenny, *Tin*, Sydney, 1922, p. 17.

88 F. D. Birch, 'Tropical Milestones: Australian Gold and Tin Mining Investments in Malaya and Thailand, 1880–1930', MA Thesis, University of Melbourne, 1976, pp. 93, 125.

89 John Hillman, 'Australian Capital and Southeast Asian Tin Mining', *Australian Economic History Review*, vol. 45, no. 2, July 2005.

90 Vendors' shares made up around 40 per cent of the capital of dredging companies.

91 As long as those who invested in the equipment were content with a normal rate of return, the whole of the differential rents would be captured by the vendors. Vendors generally received a cash payment as well to reimburse them for their actual costs in developing the property.
92 The only exception to this pattern of British sponsorship of freestanding dredging companies in the prewar period is the Danish East Asiatic Company, a trading conglomerate operating in Bangkok, which introduced a dredge at Pong.
93 Peter Golas, 'Tin Mining in 19th and 20th Century China: High Production, Low Technology', in Yang Tsui-hua and Huang Yi-lung, eds, *Science and Technology in Modern China*, Taipei, 1991.
94 Dr Jarland, 'Au pays de l'étain. Kokiu (Yunnan)', *Revue Indochinoise*, 1921, p. 379.
95 M. A. Leclere, 'Etude géologique et minière des provinces chinoises voisines du Tonkin', *Annales des Mines*, series 9, vol. 20, 1901, p. 440. The price of rice in Kochiu was twice that in Mong-tse, p. 349.
96 Around 50 per cent of the value of the metal before export, ibid., pp. 474–5.
97 Ibid., p. 447.
98 Louis Pichon, *Un Voyage au Yunnan*, Bangkok: White Lotus, 1999, original publication 1893, p. 60.
99 M. de Kergaradec, 'Mines d'étain de Ko-Kieou (Chine)', *Annales des Mines*, vol. 12, 1877, pp. 540–1.
100 L. A de Launay, *La géologie et richesses minerales de l'Asie*, Paris, 1911, p. 152. It is unclear where the balance of power lay between Kochiu and Hong Kong.
101 *Mineral Industry for 1909*, p. 672. This also asserted the principle of central government ownership of the mineral resources and ended free mining.
102 This was designed to preserve French influence in Southern China and was extremely expensive to both build and maintain, Martin Stuart-Fox, 'The French in Laos, 1887–1945', *Modern Asian Studies*, vol. 29, no. 1, February 1995, p. 126. This influence was being threatened by British proposals to build a railway from Bhamo in Burma across Yunnan to the Yangtze. The French railway reached Yunnanfu in 1910.
103 This was an extremely ambitious project. In addition to applying scientific methods to smelting, it included a washing plant which would treat ores through the long dry season and raise the recovery rate. *Decennial Reports on the Trade and Industries, 1902–1911*, Shanghai, 1912, p. 280.
104 William Wong, *Mineral Wealth of China*, Shanghai, 1927, pp. 98–9. It was also difficult to secure a source of feed, André Duboscq, 'Les mines d'étain de Ko-Kicou', *Revue Politique et Parlementaire*, vol. 19, no. 2, June 1921, p. 441.
105 A comprehensive description is Golas, op. cit.
106 W. R. Jones, *Tinfields of the World*, London: Mining Publications, 1925, p. 241.
107 In 1899 transport costs were two-thirds of exploitation costs at Llallagua, Roberto Querejazu Calvo, *Llallagua, Historia de una Montaña*, La Paz: Los Amigos del Libro, 1984, p. 79.
108 The standard biography of the family is Alfonso Crespo, *Los Aramayo de Chichas*, Barcelona: Blume, 1981.
109 Fresh capital was raised in 1909 but a dip in tin prices forced the Aramayos to cover part of the share issue. In 1916 the company decided to move to Geneva to avoid British war taxation and became the Compagnie Aramayo des Mines en Bolivie. By that point, members of the family controlled around 37 per cent of the £600,000 capital.
110 Aramayo also controlled two other companies incorporated in London: Royal Silver Mines and Porco Tin Mines. Neither was successful.
111 Charles Geddes, *Patiño: The Tin King*, London: Hale, 1972, p. 109.
112 The total cost of these four purchases was around £600,000 in cash, much of which was borrowed from the Anglo-South American Bank.
113 The bottleneck created by the shortage of labour following the simultaneous expansion of mining and railway construction created a particular incentive for mechanization. Labour shortages led to a rapid increase in wages and a consequent reduction in the number of shifts miners were prepared to work.
114 Patiño's genius also lay in his mastery of all these issues and his relentless nagging of his administrators.
115 Oploca retained much of the original Bolivian interest, unlike Llallagua where it was soon extinguished.
116 Compañía Minera y Agricola de Oploca, Annual Report for 1908, p. 6.
117 Freund, op. cit., pp. 32–3.

118 The brutality that accompanied the establishment of indirect colonial rule is described in ibid., pp. 43–5.

119 Ibid., p. 33.

120 Many other links between Cornish capital and Nigerian tin companies included Wickett of the Redruth group and Thomas of Tronoh.

121 Minerals Proclamation, in Albert Calvert, *Nigeria and its Tin Fields*, New York: Arno Press, 1977, p. 160, reprint of 1910 edition. Lessees had to show 'sufficient working capital to ensure the proper development and working of the mine'.

122 This account of stock market manipulation follows Ian Phimister, 'Corners and Company-Mongering: Nigerian Tin and the City of London, 1909–1912', *Journal of Imperial and Commonwealth History,* vol. 28, no. 2, May 2000. Phimister also notes that the group had a prior successful foray into the world of tin by manipulating the market in 1909–1911.

123 Olasiji Oshin, 'Developing Infrastructure of Exploitation: The Example of Colonial Transport on the Bauchi Tin Fields, 1902–1914', *Transafrican Journal of History*, vol. 17, 1988. The government had its own incentive since it could use railway tariffs to compensate for the low level of royalties it otherwise received under the terms of the agreement with the Niger Company.

124 The gullibility of the investors is quite remarkable in light of the even more blatant frauds perpetrated in 1909–1910, J. Forbes Munro, 'Monopolies and Speculators: British Investment in West African Rubber, 1905–1914', *Journal of African History*, vol. 22, 1981.

125 A sketch of this remarkable miner is in R. Mackilligin, *The Bisichi Story*, London: Bisichi, 1994, p. 33. Oliver Hoare served as one of the crucial links between Latilla and CGF, p. 34

126 Anthony Kirk-Greene, 'Kitty Cooke Looks Back: A Personal Mining Memoir', *The Nigerian Field,* vol. 26, no. 3, 1961, p. 122.

127 Ibid. Broadbridge is another gentleman capitalist who became Lord Mayor of London and was also active in the Conservative Party.

128 Bernard Fagg, 'Archaeological Notes from Northern Nigeria', *Man*, vol. 46, May-June 1946, p. 52. Both owners and operators received life pensions from the Niger Company.

129 This follows Gail Nattrass, 'The tin mining industry in the Transvaal 1905–1914: Some social and economic implications and perspectives', *South African Journal of Economic History,* vol. 6, no. 1, March 1991. The table on p. 94 exaggerates the significance of South Africa by confusing metal with concentrates and short with long tons.

130 *Mineral Industry for 1914,* p. 738.

131 By 1904 four other companies were formed, Jonathan Crush, 'Settler-Estate Production, Monopoly Control and the Imperial Response: The Case of the Swaziland Corporation, Ltd.', *African Economic History*, no. 8, Fall 1979. One of these survived for a brief period in the early 1920s.

132 D. H. Lenthall, *Tin Production from the Bushveld Complex*, Johannesburg: University of the Witwatersrand, 1974. There were also several small companies in the Cape.

133 This follows John Hillman, 'Chartered Companies and the Development of the Tin Industry in Belgian Africa, 1900–1939', *African Economic History*, vol. 25, 1997.

134 Portuguese-American Tin was a subsidiary of Yuba Consolidated Goldfields, the parent company of Yuba Manufacturing which designed the best dredges working in North America.

135 Pierre Deloncle, 'La mise en valeur du Laos' in Jean Renaud, ed., *Laos: dieux, bronzes, et montagnes*, Paris: Redier, 1930, pp. 149–150.

136 *Mineral Industry for 1909*, p. 684.

137 Goldschmidt established an American subsidiary which became Metal and Thermit during World War I, Walther Däbritz, *Fünfzig Jahre Metallgesellschaft, 1881–1931*, Frankfurt am Main: Osterrieth, 1931, p. 113.

138 In 1912 it reached 27 per cent of American consumption, *Mineral Industry for 1912*, p. 821. In the interwar period this was around 18 per cent.

139 Only 25 per cent of all secondary tin recovered in the USA was in the form of metal, as opposed to alloys or chemicals, United States Congress, House, Subcommittee, Committee on Foreign Affairs, *Investigation on U.S. Dependence on Foreign Tin*, 74th Cong., 1st sess., 1935, p. 27.

140 W. R. Ingalls (American Bureau of Metal Statistics), 'Secondary Tin', *Tin*, February 1935. *Tin Investigation*, op. cit., pp. 686–8.

141 The value of the steel was twice that of the tin. Goldschmidt complained that a cartel of German steel manufacturers was imposing low prices, which made it difficult to compete with English detinners who sold into a more competitive market, *Mineral Industry for 1910*, p. 684.

142 There has been no attention given to the position of the large number of small manufacturers and it is impossible to determine the prices they paid for secondary tins. If the discounts were sufficiently large that they made fresh processes and products economically feasible, there would be no competition at all.

143 Tin was the largest of metal imports. While the USA was also short in nickel and manganese, it could draw on nearby reliable sources from Canada and Cuba, Bruce Netschert, *Mineral Foreign Trade of the United State in the Twentieth Century: A Study in Mineral Economics*, New York: Arno Press, 1977, reprint of PhD Dissertation, Cornell University, 1949.

144 India led the way in 1893, followed by Siam in 1902 and Malaya in 1906, where the Straits dollar (S$) was fixed at 2/4 (£0.12) sterling and which lasted until 1967.

145 A comprehensive list with estimates of capacity is in *Mineral Industry for 1916*, p. 713. Several adjustments must be made to take account of prewar conditions.

146 Over the short term, large producers could exercise the considerable power of strategic negotiation, since they could always finance their operations with credits secured against their stocks of concentrates. Smelters, by contrast, had no flexibility.

147 James Pope testified that a small smelter with an annual capacity of 600–800 tons of metal cost $40,000. One of 8,000 tons would cost much less than $500,000 and could be erected in three months, *Tin Investigation,* op. cit., p. 966.

148 A group of Australian miners managed to use this threat to get the Straits smelters to reduce their returning charges in 1918, Stuart Rosewarne, 'Capital Accumulation and the Export of Mining Capital before World War II' in E. L. Wheelwright and Ken Buckley, eds, *Essays in the Political Economy of Australian Capitalism*, Sydney: Australia and New Zealand Book Co., 1983, p. 195.

149 Although copper is more geographically dispersed than tin, the eight leading firms controlled 48 per cent of world production in 1912, Christopher Schmitz, 'The Rise of Big Business in the World Copper Industry 1870–1930', *Economic History Review,* vol. 39, no. 3, August 1986, p. 395. The largest copper company produced three times the value of Banka.

150 U.S. Bureau of the Census, *Historical Statistics of the United States*, Washington, 1960, pp. 416, 462.

151 Manuel Contreras, 'La minería estañifera en la Primera Guerra Mundial' in *Minería y Economía en Bolivia*, La Paz, 1984, p. 18. In Britain tin alloys remained at about the same level as in peacetime but a shortage of steel reduced demand for tin for tinplate.

152 In many mines, production costs exceeded receipts from tin and they were only viable by virtue of wartime demand for the byproduct of wolfram, Jones, op. cit., p. 27.

153 President of the FMS Chamber of Mines, March 1918, cited in *Mineral Industry for 1918*, p. 693.

154 Barton, op. cit., pp. 264–5. London Tin Smelting misjudged its ability to secure high grade concentrates and was forced to rely on medium and low grade ones from which it could only produce common tin. It was a casualty of the postwar depression, Cornwall Record Office, Truro, Bolitho files, DDX 104/15/21.

155 Budd was concerned that the American smelters would undermine the competitiveness of the British ones in the postwar period and in 1915 encouraged the British government to impose a discriminatory duty on all Nigerian concentrates to destinations outside the UK. That would deny the advantage gained by 'sweetening' Bolivian concentrates with the purer ones from Nigeria. It was dropped in 1939.

156 They turned to a member of the Pearce family, which managed Williams, Harvey, to supervise the smelter, *Engineering and Mining Journal (EMJ)*, 11 March 1915.

157 DDH211/1, cable, Williams, Harvey to National Lead, 29 March 1915.

158 DDH211/2, Pearce to Harvey, 30 May 1916.

159 DDH211/2, Thomas to Vivian, 17 April 1916.

160 DDH211/2, Patiño to Pearce, 15 August 1917.

161 One important shareholder expressed his nervousness about Patiño taking such a large share, C. V. Thomas to Frank Harvey, 25 August 1917, DDH211/2. Thomas would become a major critic of the way in which the ITC operated and the forces, such as Patiño, that ensured its success.

162 The plant cost $800,000 and recorded a surplus of $334,000 in 1919, DDH211/3, statements to shareholders, 1, 8 January 1921.

163 Antonio Mitre, *El enigma de los hornos. La economía política de la fundición de estaño: El proceso boliviano a la luz de otras experiencias*, La Paz, 1993, p. 70.

164 *Mineral Industry for 1917*, p. 678.

165 Bernard Baruch, *American Industry in the War. A Report of the War Industries Board, 1921*, New York: Prentice-Hall, 1941, p. 154. The Board had been pressing the British government for such an agreement since May 1918.

166 Contreras, op. cit., p. 19. In order to ensure that the stocks were sold without loss the US government imposed an embargo on metal imports causing chaos in the immediate post-war market.

4 The problem with tin, 1919–1929

1 60 per cent of world trade was in primary commodities; in 1929, 58 per cent of British exports went to countries dependent on such commodities, Derek Aldcroft, *From Versailles to Wall Street*, London: Allen Lane, 1977, pp. 219, 233.

2 Britain operated the Stevenson scheme (1922–1928) to benefit rubber estates in Ceylon and Malaya. Cuba was similarly able to control the world price of sugar (1926–1929). São Paulo operated a valorization scheme in coffee in 1925. The Webb-Pomerene Act of 1918 exempted collusion among domestic producers in export markets from American anti-trust legislation and permitted the formation of the Copper Export Association (1919–1923) and Copper Exporters, Inc. (1926–1932).

3 These issues were given their first formal clarification by Harold Hotelling, 'The Economics of Exhaustible Resources', *Journal of Political Economy,* vol. 39, no. 2, April 1931, from which a large body of subsequent theorizing derives. The significance of introducing the owner of the mineral in situ as an economic actor and entitled to a rate of return will become evident in Chapter 16 but the results of this body of theory are not relevant to the particular forms of ownership in most of the tin industry.

4 This argument was first expounded in a BBC broadcast by J. W. F. Rowe, 'Keeping up the Price of Tin', *The Listener*, vol. 13, 1936, pp. 327–8 and repeated in his *Markets and Men*, Cambridge: Cambridge University Press, 1936, pp. 158–9, 172. J. K. Eastham came to the same judgement, 'Rationalisation in the Tin Industry', *Review of Economic Studies*, vol. 4, no. 1, October 1936, p. 18. It was then endorsed by Klaus E. Knorr, *Tin Under Control*, Stanford, CA: Food Research Institute, 1945, pp. 79, 87 and in another study that also shaped postwar thinking, P. Lamartine Yates, *Commodity Control*, London: Cape, 1943, p. 145. The argument has continued to circulate, Aldcroft, op. cit., p. 230; Yip Yat Hoong, *The Development of the Tin Mining Industry in Malaya*, Kuala Lumpur: University of Malaya Press, 1966, p. 156; Alton Law, *International Commodity Agreements*, Lexington, KY: Heath, 1975, p. 57; W. Robertson, *Tin. Its Production and Marketing*, Westport, CT: Greenwood, 1982, p. 133; Peter Roddy, *The International Tin Trade*, London: Woodhead, 1995, pp. 18–19.

5 A full account is provided in John Hillman, 'The Bandoeng Tin Pool, 1920–1924', *Malaysian Journal of Economic Studies,* vol. 32, no. 2, 1995. Unlike some later tin pools, the metal was not acquired in the market.

6 The most important is Yip, op. cit., p. 156, repeated in William Baldwin, *The World Tin Market. Political Pricing and Economic Competition*, Durham, NC: Duke University Press, 1983, p. 66, Robertson, op. cit., p. 133, Christopher D. Rogers, 'Consumer Participation in the International Tin Agreements', *Malayan Economic Review*, vol. 14, 1969, p. 115 and John Drabble, *An Economic History of Malaysia, c. 1800–1900*, Basingstoke: Macmillan, 2000, p. 130.

7 Eastham noted that the gross profit to the FMS Government was £105,000 on an initial investment of £2.2 million, while these transaction costs amounted to £408,000, op. cit., p. 25.

8 *Straits Times*, 31 December 1923.

9 Nutt had conducted a successful speculation in 1919 but was warned not to repeat the attempt.

10 Henceforth, the Singapore market operated on the basis of a closed bid system in which both miners and brokers submitted schedules of quantities and prices at which they were prepared to deal and the smelter simply selected the price which would clear the market, Baldwin, op. cit., pp. 197–8. Cash was the basis of both sides of the transaction and the buyers received their metal within the 60 days required to produce it.

11 The memory of the Bandoeng Pool has also lived on the cartel literature, where it obviously does not belong. The first to include it in a comprehensive review of cartels was James M. Griffin, 'Previous Cartel Experience: Any Lessons for OPEC?' in Lawrence R. Klein and Jaime Marquez, eds, *Economics in Theory and Practice*, Dordrecht: Kluwer, 1989. This error has been repeated

by Jaime Marquez, 'Life Expectancy of International Cartels: An Empirical Analysis', *Review of Industrial Organization,* vol. 4, no. 3, 1994 and Valerie Suslow, 'Cartel contract duration: empirical evidence from interwar international cartels', *Industrial and Corporate Change*, vol. 14, no. 5, 2005.

12 F. E. Mair at Tekka AGM, 12 October 1923, cited in Puey Ungphakorn, 'The Economics of Tin Control', PhD dissertation, University of London, 1949, p. 303. Puey provides several other examples of similar assessments from the financial and mining press.

13 Calculated from *Comercio y Minas*, no. 122, March 1930. The reduction was largely confined to Germany, Austria and Belgium which did not recover even the absolute level of demand of 1913.

14 The 'revolution of declining expectations' in food preparation which led to 'soaring sales of canned foods in the 1920s' is documented in H. A. Levenstein, *Revolution at the Table: Transformation of the American Diet*, New York: Oxford University Press, 1988, p. 163.

15 *Metal Industry*, vol. 27, no. 1, 1929, p. 6.

16 Ibid., J. Umhau, 'Consumption of Tin in the United States during 1928', US Bureau of Mines, IC 6165, August 1929, pp. 3–4.

17 London and Cambridge Economic Service in J. M. Keynes, *Collected Works*, London: Macmillan, 1983, vol. 12.

18 Speech at Royal Colonial Institute by Lord Askwith, cited in *Bolivia,* vol. 2, no. 3, January-February 1929.

19 C. G. Moor, *Tin Mining*, London: Pitman, 1928, p. 114.

20 J. W. Furness, 'The Tin Situation from a Domestic Viewpoint', US Bureau of Mines, IC 6018, January 1927, p. 8. Similar warnings were offered about lead supplies in 1923–1925. These were equally misleading, *EMJ*, 28 April 1928.

21 National Archives, Kew (NAK), Board of Trade files, BT329/9, Non-Ferrous Mining Committee, evidence of Budd, 11 November 1919.

22 BT329/11, evidence of Chappel, 13 November 1919.

23 BT329/17, evidence of Pearce, 3 December 1919.

24 W. R. Jones, *Tinfields of the World*, London: Mining Publications, 1925, pp. vi, 231. An even more dramatic claim was made in the same year, which predicted that Bolivia would outperform Malaya by 1925, *Mining Journal* 16 August 1924. Jones remained firmly committed to his prognosis and 25 years later he persuaded the Panel on Import Substitution to grant South Crofty a large subsidy so that the potentiality of Cornwall could be realized, NAK, Treasury files, T228/319, Committee on Industrial Productivity, 17 November 1949. Cf his *Minerals in Industry*, Harmondsworth: Penguin, 1963, pp. 261–2.

25 Arkib Negara Malaysia, Kuala Lumpur (ANMKL), Geologist, 55/27.

26 *Financial News,* 22 February 1928. In this speech Askwith argued for the merging and trustification of British tin producers and smelters.

27 *EMJ,* 23 October 1926.

28 *Industrial Australian and Mining Standard (IAMS),* 23 December 1926.

29 *IAMS*, 4 March 1926.

30 *EMJ*, 25 September 1926.

31 Reginald Pawle, 'The Empire's Tin Resources', Second Empire Mining and Metallurgical Congress *Proceedings*, pt. I, Montreal, 1928, pp. 212, 207.

32 Thomas Kennedy, 'The Future of Tin', *Bolivia*, vol. 2, no. 2, November-December 1928, p. 13.

33 Theodore Hoover, *The Economics of Mining*, Stanford, CA: Stanford University Press, 1933, pp. 6, 196.

34 This simply resonated with the dominant theme of economic discourse when 'nothing big in business could happen without the word rationalization. It was introduced in every company Chairman's speech; it was given prominence in every company prospectus', *Tin*, December 1933, p. 1.

35 Elizabeth S. May, 'The International Tin Cartel', in W. Y. Elliott, ed., *International Control in the Non-Ferrous Metals*, New York: Macmillan, 1937, p. 315.

36 Ibid., pp. 277, 314.

37 Ibid., p. 314. This is presented as an impression for which figures could not be cited but it was certainly widely shared. For a more recent version see Jean-Jacques van Helten and Geoffrey Jones, 'British Business in Malaya & Singapore since the 1870s', in R.P.T. Davenport-Hines and Geoffrey Jones, eds, *British Business in Asia Since 1860*, Cambridge: Cambridge University Press, 1989, p. 169.

38 Eastham, op. cit., p. 20.

39 J. W. F. Rowe, 'The Artificial Control of Raw Material Supplies',*Economic Journal*, vol. 40, no. 156, September 1930. Rowe's subsequent assessments of tin control will be considered in Chapter 16.

40 This term was used to encompass not only capital emanating from Europe but also from Australia and the USA.

41 BT58/81/COS582, Biographical notes.

42 Ernsthausen started in jute in 1874 and by 1909 was one of the largest producers with six mills, D. R. Wallace, *The Romance of Jute*, Calcutta, 1909, pp. 56–7.

43 *Mining World and Engineering Record (MWER)*, 17 March 1928.

44 Ibid. Howeson had 'a profound distaste to disclosing the history of his life to strangers', Mawby & Barrie to Board of Trade, 14 January 1930, BT58/81/COS429, with the result that published references are very misleading. They note, correctly, that he was educated at Uppingham but go on to add a career at Oxford where he earned blues in cricket and chess, for example, *Fortune,* May 1932, p. 83. This describes his brother Adolph, older by four years, with whom John may have been confused, since Adolph died in 1928 just as John was becoming a public figure. Howeson continued to shun publicity and when confronted by a knowledgeable journalist is reported to have said: 'I would pay £2,000 for news of your death.' Patricia Coburn, *The Years of THE WEEK*, London: Comedia, 1985, p. 100.

45 *Sunday Express*, 10 February 1935. This was the last occasion on which anything positive about Howeson would appear in print.

46 The success of a jute mill was dependent on strategic purchases of the raw product in a very volatile market and on the existence of the Indian Jute Mills Association which attempted to control production. The industry was extremely profitable, especially during World War I and Howeson's firm could have easily financed its expansion into tin. An overview of the industry is provided by Amiya Kumar Bagchi, *Private Investment in India 1900–1930*, Cambridge: Cambridge University Press, 1972, Chapter 8. A more detailed analysis of the jute market is H. Sinha, 'Jute Futures in Calcutta', *Economica*, November 1929.

47 Oscar became a naturalized British subject in 1856, serving in the repression of the Indian Mutiny and joining a Masonic lodge in 1857. The children anglicized their name in 1916.

48 William Henry was General Manager, Alliance Bank of Simla and Lt-Colonel in the Indian Defence Force. Godfrey Fell was Army Secretary in India.

49 Malaysia Mining Corporation, Kuala Lumpur (MMC), AOGIT Minute Book No 1. Following the standard pattern of British corporate minutes, these are very brief but they do indicate a developing relationship with Williams, Henry, a produce and share dealing firm jointly owned by Howeson and Garabed Bishirgian. The importance of this relationship will become evident in Chapter 10.

50 Alfred Marcus, *Die grossen Eisen- und Metallkonzerne*, Leipzig: Hirzel, 1929.

51 E. T. McCarthy, *Incidents in the Life of a Mining Engineer*, London: Routledge, 1918; *Further Incidents in the Life of a Mining Engineer*, London: Routledge, 1920.

52 MMC, AOGIT Minute Book, no. 1, 6 October 1927. Norman Cleaveland, *Bang! Bang! in Ampang*, San Pedro: Symcon, 1973, p. 44. Yuba already had some experience in Malaya since Yukon Gold brought Yuba dredges from the goldfields of the Yukon and California.

53 McKenna was something of a maverick in the City since he did not share its commitment to the return of sterling to gold, Philip Williamson, 'Financiers, the Gold Standard and British Politics, 1925–1931', in John Turner, ed., *Businessmen and Politics. Studies of Business Activities of British Politicians, 1900–1945*, London: 1984, p. 107.

54 India was the basis of the Cunliffe-Owen connection, Howard Cox, *The Global Cigarette. Origins and Evolution of British American Tobacco, 1880–1945*, Oxford: Oxford University Press, 2000, pp. 228–9.

55 The Bourse established higher standards than the London Stock Exchange for its listings.

56 Oliver Lyttleton, *The Memoirs of Lord Chandos*, London: Bodley Head, 1962, p. 140. For the overall importance of BMC and Lyttleton see Simon Ball, 'The German Octopus: The British Metal Corporation and the Next War, 1914–1939', *Enterprise & Society*, vol. 5, no. 3, September 2004 and the entry in David J. Jeremy, ed., *Dictionary of Business Biography*, London: Butterworths, 1985, vol. 3.

57 Cited by Howeson at the TST AGM, *Financial Times,* 30 January 1929. CGF also opened up another important connection to the world of non-ferrous metal mining in Sir Edmund Davis.

58 *Investors' Chronicle*, 3 March 1928.

59 Hoare was part of the same network of Conservatives prepared to challenge the old orthodoxies of free trade that included Cunliffe-Lister, John Campbell, *F. E. Smith*, London: Cape, 1983,

p. 362. Other links to the tin network include a close political association to Lyttleton's father when he was Colonial Secretary and to Brabourne whose son served as Hoare's Parliamentary Private Secretary, J. A. Cross, *Sir Samuel Hoare, a Political Biography*, London: Cape, 1977. The expression 'Colonial Secretary' is used throughout, rather than the official, but cumbersome, 'Secretary of State for the Colonies'.

60 J. A. Cross, *Lord Swinton*, Oxford: Clarendon Press, 1982, p. 39. Cunliffe-Lister became Viscount Swinton in 1935. At the Board of Trade, Cunliffe-Lister expressed much interest in the German model of cartelization, BT55/49. When he left office in 1929, Oliver Hoare secured a place on the board of CGF for Cunliffe-Lister. Another CGF Director, Douglas Christopherson was a close friend of McKenna and served as Deputy Chairman of the Midland Bank. Howeson was also on the Board of CGF.

61 The purchase price was £52,500 for the business and £192,000 for the stocks which would have been around 6,500 tons of tin as metal and concentrate, MMC, LTS, Board Meeting, 7 October 1926.

62 Marcus, op. cit., p. 37. Pope attributed the remarkable statements made at this opening 'to the fact that they had very good dinner', United States Congress-House, Subcommittee, Committee on Foreign Affairs, *Investigation on U.S. Dependence on Foreign Tin*, Washington, DC: GPO, 1935, p. 962.

63 *Liverpool Post*, 14 July 1928.

64 R. J. Denyer and K. Heath, 'Mining and Milling Tin-Tungsten Ore at Mawchi Mines, Burma', Institution of Mining and Metallurgy, *Transactions (TIMM)*, vol. 49, 1939-40.

65 H. L. Chibber, *The Mineral Resources of Burma*, London: Macmillan, 1934, p. 181.

66 Burma imposed a very low export duty of 2.5 per cent on all mineral exports, so the contribution of tin to public finance would have been insignificant, Gilbert Stone, *Mining Laws of the British Empire and of Foreign Countries*, London: HMSO, 1924, vol. 6, p. 148. In order to protect the oil industry, Burma restricted all mineral leases to British subjects.

67 J. Coggin Brown and A. Heron, 'Geology and Ore Deposits of the Tavoy District', Geological Society of India, *Memoirs*, vol. 44, pt. 2, 1923, p. 326, *Capital* (Calcutta) 26 June 1920.

68 Tavoy Tin Dredging AGM, 1926.

69 Rhodes House, Oxford, Central Mining 68/1. CTMB took over 17 mines already operated by a Rangoon company, Burma Finance and Mining.

70 The property was eventually sold off to Tavoy Tin in 1933 for the equivalent of £10,000. Those who originally sold the property would have received around £50,000 and the difference is easily explained by the decline in the original valuation at £250 and the level of £220 prevailing at the time of sale.

71 Since equity is the total capital entitled to a share of the net profits, its measure provides a basis for comparing shareholder expectations; since the bulk of the cash was used for equipment and working capital, its measure provides a basis for comparing mining operations.

72 *Tin*, September 1930, p. 9.

73 NAK, Foreign Office files, FO371/18493, Dormer to FO, 2 February 1934.

74 FO371/13264, Minute, 30 April 1928; FO371/13981, Wingfield to FO, 4 July, Coultas to Whitfield, 18 June 1929.

75 Malayan and General was closely associated with a stockbroker, Ridsel. The experience of these and other companies in the group offers support to the claim that many flotations were designed to generate profits for promoters rather than investors.

76 An 'amazing' and difficult character, according to F. K. Excell, *Siamese Tapestry*, London: Hale, 1963, p. 121.

77 Pring and Docker, *Guide to Eastern Tin Stocks for Australian Investors*, Sydney, 1927, p. 25.

78 Yip, op. cit., p. 136.

79 FO371/13264, Waterlow report, 30 April 1928; *Economic Conditions in Siam to the Third Quarter, 1928*, London: HMSO, 1928, p. 22.

80 F. D. Birch. 'Tropical Milestones: Australian Gold and Tin Mining Investments in Malaya and Thailand, 1880–1930', MA Thesis, University of Melbourne, 1976, p. 232. A scandal eventually erupted which exposed the extent to which Khaw Joo Tok was taking advantage of Australian investors and new regulations were implemented which granted control over leases to the Ministry of Agriculture in Bangkok, Paul Battersby, 'Diggers and Diplomats: Australian Mining Entrepreneurs and the Evolution of the Australia-Thailand Bilateral Relationship, 1901–1941', in M. Hayes and S. Smith, eds, *Australia-Thailand Relations in the Twentieth Century*, Bangkok: Australian Studies Centre, Kasetsart University, 2000.

81 Pring and Docker, op. cit., p. 119, Tinsonghkla, Annual Report for 1927–1928, Kopah Tin, Annual Report for 1928–1929. If negotiations failed, the government could exercise powers of expropriation but that would add to the delay and cost.

82 F. K. Excell, *In Siamese Service*, London: Cassell, 1967, p. 125. Excell adds: 'It was surprising that tin mining in Siam had ever gone ahead.' Cf Jennifer Cushman, *Family and State. The Formation of a Sino-Thai Tin-Mining Dynasty*, Singapore: Oxford University Press, 1991, p. 105.

83 In addition there may have been a push factor. As will be seen in Chapter 5, the FMS government severely limited the issuance of new mining leases and that would have left Siam as the main outlet for fresh investment in the region.

84 Sompop Manarungsan, *Economic Development of Thailand 1850–1950*, Bangkok: Institute of Asian Studies, 1989, p. 252.

85 J. C. Ingram, *Economic Change in Thailand, 1850–1970*, Stanford, CA: Stanford University Press, 1971, p. 329, *Statistical Yearbook of the Kingdom of Siam, 1938,* p. 488. This was just under 11 per cent of gross revenues before smelting and shipping costs.

86 FO371/2755, Siam Chamber of Mines to Acting Consul, Senggora, 4 April 1928.

87 FO371/14776, Consul at Senggora to British Legation (Bangkok), 22 May 1930.

88 Shakila Yacob, *The United States and the Malaysian Economy*, Abingdon: Routledge, 2007, Chapter 4.

89 J. B. Newsom, 'Examination of Tin Deposits in the Federated Malay States', *EMJ,* 17, 24 March 1923.

90 *EMJ*, 1 June 1929.

91 Not only was the dredge enormous but so was the property which held an estimated 39,000 tons of tin. This was capitalized at £625,300 with another £150,000 representing the value of the dredge and other assets. Lower Perak was therefore the largest single dredging company.

92 NAK, Colonial Office files, CO717/53/18237, Birch to Pratten, 26 October 1925.

93 Although the cleavage between Anglo-Oriental and the Cornish groups was grounded in the particular rhythm of growth of the tin industry, the vehemence with which it came to be expressed may have reflected something more basic in the Anglo-Malayan business culture. The Rubber Growers Association was also badly factionalized and the leader of one of the factions was 'universally disliked throughout the industry', Nicholas White, *Business, Government and the End of Empire: Malaya, 1942–1957*, Kuala Lumpur: Oxford University Press, 1996, p. 39.

94 Continued dilution of equity interest held by Cornish residents did not affect the pattern of control. Thomas remained based in Camborne and although Mair lived in Kent he kept in close contact with Redruth. The relationship between them was sufficiently strong that they are often treated as having a single policy. When that was the case in this study, they will be called 'Cornish-Malayan' or just 'Cornish'.

95 Yip, op. cit., p. 164. These figures include dulang washers but exclude the amounts produced on land subleased from European companies. The Department of Mines first published such statistics in 1929 when they amounted to 11 per cent of total FMS production. The dulang was a pan that Chinese women used to wash the alluvium they extracted and recover the heavier tin-bearing particles.

96 For an overview of the issues, see Ian Brown, *Economic Change in Southeast Asia, c. 1830–1980*, Kuala Lumpur: Oxford University Press, 1997, pp. 194–200.

97 Takuapa Valley had 40 per cent Chinese and Siamese shareholders, Birch, op. cit., p. 246. Cf. Yip, op. cit., p. 365.

98 Rajeswary Brown, *Capital and Entrepreneurship in South-East Asia*, New York: St. Martin's Press, 1994, p. 85.

99 For example, Loke Yew left a fortune equivalent to nearly £9 million in 1917, ibid., p. 86. The dredge cost is taken from Ambrose Pratt, *Magical Malaya*, Melbourne: Robertson & Mullens, 1931, p. 217.

100 The viability of any explanation needs to be tested against at least three contrary cases: (1) Sew Hoy, a pioneer of gold dredging in New Zealand with four dredges in the 1880s, J. H. M. Salmon, *A History of Gold Mining in New Zealand*, Wellington, 1963, p. 232; (2) Tan Pheck Huad, who operated a Siamese dredge from 1926, Phuwadol Songprasert, 'The Development of Chinese Capital in Southern Siam, 1868–1932', PhD dissertation, Monash University, 1986, p. 365; (3) the anonymous Chinese miner who acquired a dredge from Ipoh Tin Dredging and operated it in Siam from 1936.

101 Yip, op. cit., p. 164.

102 *Straits Times,* 15, 21 March 1930.

103 Lim Chong-Yah, *The Economic Development of Modern Malaya*, Kuala Lumpur: Oxford University Press, 1967, p. 350.

104 *Malayan Tin and Rubber Journal (MTRJ)*, 31 July 1929.

105 The seasonal character to the rainfall on the Jos plateau meant that a consistent supply of power required the construction of a large dam. The one at Kwall Falls cost £250,000.

106 Nigerian Tin and Power took a major equity interest in NESCO and received around £225,000 in compensation for its own expenses. Latilla then joined the Board of NESCO.

107 *Economist*, 31 March 1928. Unlike the arrangement with Alluvial Tin, the companies lost their identity. ATMN invested a further £313,000 in developing this group of mines. Broadbridge retained one company and kept his position on the Council of the Nigerian Chamber of Mines

108 The shares were first exchanged with LTS. The mining assets were transferred to LTMN and Kwall Falls hydroelectric project was transferred to London Nigerian Power Company which provided additional capital to enlarge the plant. Both were wholly owned subsidiaries.

109 These opportunities could be very lucrative. One who made a fortune during the 1930s was Basil Griliopulos, *The Man who Twisted the Lion's Tail. The story of one man's war against the British government*, Hicksville: Exposition, 1974. By 1928 two Africans were among the 63 independent miners.

110 Total government income from all mineral industries in 1928 was £102,000, Penelope Bower, 'The Mining Industry', in Margery Perham, ed., *The Economics of a Tropical Dependency*, London: Faber, 1948, vol. 2, p. 12.

111 Margery Perham, *West African Passage: A Journey Through Nigeria, Chad, and The Cameroons, 1931–1932*, London: Peter Owen, 1983, p. 51.

112 *Economist*, 12 February 1927. Only after the onset of the depression did it become clear that this archaic system of property rights was a serious obstacle, A. K. Hamilton Jenkin, *Natiionalisation of West County Minerals,* London: New Fabian Research Bureau, 1931.

113 Captain Moreing, MP, often spoke of the large quantity of tin in Cornwall, *EMJ,* 14 February 1925.

114 D. Bradford Barton, *A History of Tin Mining and Smelting in Cornwall*, Exeter: Wheaton, 1989, p. 269.

115 At the time, the mining press greeted them with some enthusiasm, *Mineral Industry for 1926*, p. 664, *Mineral Industry for 1927*, pp. 572–3.

116 Camp Bird was originally a gold mine in Colorado but had become a finance company with close links to CGF and would soon be absorbed by it. National Mining Corporation served as one of the links and participated in this and other flotations. The Camp Bird records are in the Guildhall. London. Selection Trust also took out an option on a Cornish tin mine, *EMJ*, 19 December 1925.

117 Some were abandoned long before. Killifreth went bankrupt in 1926, Kingsdown in 1928. In spite of fresh investment of £137,000, Dolcoath never returned to produce even modest quantities and closed in 1928, T. R. Harris, *Dolcoath: Queen of Cornish Mines*, Penzance: Trevithick Society, 1974, Chapter 5.

118 Their histories have been written with a primary focus on recording their technical challenges and accomplishments, Philip Heffer, *East Pool & Agar. A Cornish Mining Legend*, Redruth: Dyllansow Truran, 1985, J. A. Buckley, *A History of South Crofty Mine*, Redruth: Dyllansow Truran, n.d. The story of yet another established producer which also failed to survive the depression, is told by John Corin, *Levant. A Champion Cornish Mine*, Redruth: Trevithick Society, 1992.

119 Helen Brown, *Tin at Tingha*, Armidale: Brown, 1982, p. 54.

120 *Investor's Chronicle*, 13 December 1924. The issue was well oversubscribed.

121 However, that would overlook the plan to compensate shareholders with a free gift of shares in a new Nigerian property being developed by Anglo-Oriental, *Investors' Chronicle*, 11 August 1928. Since the company was never floated, this was an empty gesture.

122 Mining by individuals continued for a long time and one wrote an autobiography, Ion Idriess, *The Tin Scratchers*, Sydney: Angus & Robertson, 1959; the biography of another is Christobel Mattingley, *King of the Wilderness, the Life of Deny King*, Melbourne: Text Publishing, 2001.

123 This was an old mine which had stopped in 1921 and was bought by Hans Hunter who formed a Japanese company, Toyo Kazan Kabushiki Kaisha in 1926. This is another case of gold capital moving into tin, Report on Tin Mining in Japan, FO371/16249.

124 This follows John Hillman, 'The Emergence of the Tin Industry in Bolivia', *Journal of Latin American Studies,* vol. 16, pt. 2, November 1984 and 'Promise and Performance in the Bolivian Tin Mining Industry, 1923–1929', *Canadian Journal of Latin American and Caribbean Studies,* vol. 23, no. 45, 1998.

125 Herbert S. Klein, 'The Creation of the Patiño Tin Empire', *Inter-American Economic Affairs*, vol. 19, no. 2, Autumn 1965, pp. 15–16. Manuel Contreras, 'En Torno a 'La Formación del Imperio del Estaño de Patiño' de Herbert S. Klein', *Historia Boliviana*, vol. 4, no. 2, 1984, provides an important corrective to the details of Klein's account, pp. 206–8.

126 The new company was domiciled in Delaware and headed by an American. The Catavi mill became so efficient that it cut operating costs even in the face of a substantial decline in the grade of the ores, R. R. Beard, 'Property and Operations of Patiño Mines and Enterprises', *EMJ*, 9 August 1930, pp. 170, 237.

127 Manuel Contreras, *Tecnología moderna en los Andes. Minería e ingeniería en Bolivia en el siglo XX*, La Paz: Biblioteca Minera Boliviana, 1994, p. 21.

128 With a total loss of some US$13 million. In addition to the mine, the Guggenheims also operated a dredge at Arofillia which was much more successful.

129 One of the consequences of the system of property rights, which granted concessions in perpetuity, subject only to a modest rent, with no obligation to actually exploit them, was that owners could demand high prices for their properties at least during the boom. Anglo-Oriental formed the London-Bolivian Tin Syndicate to explore the possibility of moving into Bolivia but found nothing at a reasonable price.

130 Helmut Waszkis, *Dr. Moritz (Don Mauricio) Hochschild 1881–1965*, Frankfurt: Vervuert, 2001.

131 Walter Gomez d'Angelo, *La minería en el desarrollo económico de Bolivia, 1900–1970*, La Paz: Los Amigos del Libro, 1978, p. 210. This figure does not include taxes on imports to the mining sector.

132 Such a tax also required surveillance of corporate accounts to prevent the artificial inflating of capital.

133 Manuel Contreras, 'Debt, Taxes, and War: The Political Economy of Bolivia, *c.* 1920–1935', *Journal of Latin American Studies*, vol. 22, pt. 2, May 1990.

134 J. F. Van Wickel, 'Machinery Markets of Netherlands East Indies', US Bureau of Foreign and Domestic Commerce, Trade Information Bulletin no. 395, 1926, p. 18; *IAMS* predicted that there were only 10–11 years left, 4 March 1926.

135 Puey, op. cit., p. 231. Since the government held a majority equity interest in the mines, it drew its revenue from the profits rather than royalties. Without an *ad valorem* royalty, the cut off grade of ore was determined by its direct mining costs, which would result in a wider range of economically viable deposits; hence the importance of developing a comprehensive strategy for extracting the entire deposit.

136 Although the NEI government appointed its own Directors to GMB, it was effectively managed by BM and there was never any tension between these two components.

137 This simply sharpened the competition for Bolivian ores, Hollandsche Metallurgische Bedrijven, Jaarsverslaag, 1929, p. 6. Billiton then considered taking an equity interest in various Bolivian mines and in Dolcoath, Algemeen Rijksarchief, The Hague (ARA) Groothoof papers, file 230.

138 ARA, Ministerie van Koloniën files, MKGA347/F20, Largas to Governor-General, 4 July 1930.

139 Willem Hoogendoorn, *Der Zinnpreis. Eine Untersuchung uber die Zinnpreisbildung und ihre monopolistiche Beeinflussung*, Kiel, 1934, p. 100. Banka was responsible for over 80 per cent of all revenues from tin.

140 *Mineral Industry for 1924,* p. 690.

141 'Yunnan Tin Industry', *Chinese Economic Monthly*, vol. 3, no. 4, 1926, pp. 154–6.

142 Boris Torgasheff, 'Tin in the Far East', *Internationale Bergwirtschaft*, vol. 3, no. 12, December 1928, p. 219. A more optimistic prognosis of the 'Banka of China' is Han Tiauw Tjong, *De Industrialisatie van China*, The Hague: Nijhoff, 1922, p. 254.

143 Although YTC was under the control of the provincial government, Kochiu remained 'practically an independent republic of tin smelters', 'Yunnan Tin Industry', op. cit., p. 156. They made it difficult for YTC to buy the ores required to supplement those from Malaga and make the smelter viable.

144 See Marshall Draper, 'Tin Industry of Yunnan, China', *Mining and Metallurgy,* April-May 1931, for detailed comments on the problems he encountered and the improvements he managed to make.

145 Robert Slessor, 'Chinese Non-Ferrous Metals', Australasian Institute of Mining and Metallurgy, *Proceedings*, New Series, no. 65, 31 March 1927, p. 103.

146 *EMJ,* 3 April 1928.

147 *Société générale métallurgique de Hoboken, 1908–1958*, Brussels: Cuypers, 1957, p. 75. This produced a high quality metal at 99.95 per cent.

148 *EMJ*, 24 March 1928.
149 E. Willard Miller, 'The Mineral Resources of Indo-China', *Economic Geography,* vol. 22, no. 4, October, 1946, p. 269.
150 Pierre Deloncle, 'La mise en valeur du Laos' in Jean Renaud, ed., *Laos: dieux, bronzes, et montagnes*, Paris: Redier, 1930, p. 153.
151 Torgasheff, op. cit., p. 220.
152 *Straits Times*, 15 January 1930.
153 The inherent technical conservatism involved in smelting on a small scale is noted by C. L. Mantell, *Tin. Its Mining, Production, Technology, and Applications*, New York: Hafner, 1970, pp. 118–119. Just as large tin mines were small by comparison with other non-ferrous metals mines, so tin smelters were small by comparison with their counterparts.
154 Ibid., p. 131. Both Penpoll and Williams, Harvey saw major advances which allowed for the operation of much larger furnaces, *Mineral Industry for 1928,* p. 598.
155 Even wartime smelters in Chile and South Africa, which were built much closer to sources of supply, also folded.
156 Josef Wollnik, *Zinn, Wandlungen in der Erzeugung und Verwendung des Zinns nach dem Weltkrieg*, Leipzig, 1936, pp. 100–4. In 1929 Patiño and National Lead each held 25 per cent, with the remainder held by the London metal trading firm, Brandeis, Goldschmidt. In addition to smelting concentrates, the company also treated car radiators shipped by National Lead from the USA.
157 Bryan Little, *Capper Pass*, London: Newman Neame, 1963, pp. 28–9. Capper Pass occupied several niches in the tin market. It treated a variety of different types of concentrates, reprocessed old metals and produced specialized items such as printer's types and solders.
158 This position became very familiar. For example, Margaret Marsh, *Bankers in Bolivia*, New York: Vanguard, 1928, p. 37, Fleming MacLiesh and Cushman Reynolds, *Strategy of the Americas*, New York: Duell, 1941, pp. 28–9; *New Republic*, 17 January 1944, Yacob, op. cit., p. 64.
159 What made it particularly acute was the fact that many commentators were also investors which makes it difficult to read their commentary.
160 *Economist*, 9 April 1927.
161 *Economist,* 17 September 1927.
162 *Economist,* 12 November 1927.
163 J. A. Gallard and Murray Stuart, *Tin: Salient Facts and Opinions*, London: Mining Publications, 1929, p. 45. At the Rambutan AGM, Chappel expressed similar bewilderment at the fact that the price had dropped by £40 while the visible supply had declined over the same twelve months, *MWER*, 21 January 1928.
164 *Economist,* 9 June 1928. Six months later, *Investor's Chronicle* still thought consumption was lagging behind production, 12 January 1929. However, three weeks later Henry corrected this impression and pointed out that consumption would need to rise by at least 10,000 tons to absorb production, 2 February 1929.
165 *Economist,* 16 June 1928. It was to provide a comprehensive statistical picture that Howeson sponsored the publication of an annual, *Tin World Statistics*, beginning in 1928.
166 Moor, op. cit., p. 114.
167 *Investor's Chronicle*, 20 August 1927; *IAMS,* 27 October 1927. Rae was Chairman of Rahman Hydraulic and a representative of the FMS Chamber of Mines in the FMS Legislative Council.
168 *Investor's Chronicle*, 10 September. This letter discounted Dutch criticism of Rae which suggested that the Chinese could continue to produce at £175, 3 September 1927.
169 *IAMS*, 15 December 1927.
170 Kerry estimated that the Chinese would start falling off at £230, *IAMS,* 31 May 1928.
171 A. Parsons, 'Is 45 Cent Tin Likely?', *EMJ,* 28 April, 9 June 1928. 45 cents/lb – £200/ton. This corrected the impression given by an earlier article which was excessively optimistic, *EMJ*, 11 February 1928.
172 *EMJ,* 23 June 1928.
173 Keynes, op. cit., p. 512. This was published on the Continent but not in England. Keynes was also an investor in tin, with shares in Anglo-Oriental and an open position in the metal, though it is unclear just when this occurred.
174 *New York Times (NYT)* , 19 October 1927. The Tin Statistical Bureau also offered its support since current levels of US consumption suggested an exhaustion of supplies within 15 years, *NYT*, 16 October, 18 November 1927.
175 *NYT*, 4 December 1927.

176 *IAMS,* 27 October 1927.
177 *Mineral Industry for 1927,* p. 562.
178 *Mineral Industry for 1928*, p. 587.
179 Eastham, op. cit., p. 20.
180 May, op. cit., p. 316.
181 A. P. L Gordon, *Rationalisation for Tin*, London: Saint Catherine, 1929, p. 18. Similar comments were made by Herbert Oakley, *The Future of Tin*, London: Saint Catherine, 1928, p. 3 and Ernest Harold Davenport, *The Price of Tin. The Problem of Stabilisation*, London: London General Press, 1928, pp. 14–15.
182 President, FMS Chamber of Mines, quoted in *IAMS,* 5 July 1928. A similar diagnosis was offered by *Investors' Chronicle,* 26 May 1928.
183 Some suggest that Patiño participated, Gill Burke, 'The Rise and Fall of the International Tin Agreements', in Jomo, ed., *Undermining Tin. The Decline of Malaysian Pre-eminence*, Sydney, 1990, p. 49, Serge Calabre, *L'étain*, Paris: Economica, 1991, p. 83. Pope suggested that the Midland Bank provided loans against the security of the metal already acquired, *Tin Investigation*, op. cit., p. 953. There is no independent evidence to corroborate these speculations.
184 100 tons could move the market by £9 within the course of a single day, Mieczysław Epstein, 'La réglementation internationale d'un marché de matière première', PhD dissertation, University of Fribourg, 1943, p. 48.
185 MKGA338/T3, Report on meeting, 18 February 1930. AOGIT provided some of the finance and this was done without consulting the Chairman, Sir William Henry.
186 However, as will be seen in Chapter 10, Howeson was prepared to take high risks that called his integrity into question.
187 *Mineral Industry for 1928*, p. 587.

5 The depression: initial responses, 1928–1930

1 FMS, Mines Department, Annual Report for 1928, p. 13.
2 CO717/66/19, FMS to CO, 25 November 1929. Rights under existing EPLs were not affected. The announcement was made without alerting the Colonial Office, which was naturally irritated when it first learnt of it through the press. When pressed for an explanation, the High Commissioner noted that it was considered 'undesirable to introduce a policy of artificial restriction', CO717/69/11, Clementi to CO, 28 January 1930.
3 FMS, Mines Department, Annual Report for 1929, p. 23.
4 FMS, Mines Department, Annual Report for 1928, p. 19. For further comments on the apparent price perversity of the Chinese, see *Economist,* 21 December 1929, *Straits Times,* 14 March 1930.
5 *Bangkok Daily Mail*, 18 August 1930 in MKGA350/T23.
6 *Mining Journal,* 27 September 1930.
7 CO583/155/16, Latilla to Chief Inspector of Mines (CIM), Jos, 2 July 1928; CO583/171/7, Local Council, Nigerian Chamber of Mines, 16 May 1929.
8 CO583/166/5, Note on the Tinfields of Northern Nigeria, 1929, p. 6.
9 CO583/717/7, Flood minute, 11 February. In another minute, Flood commented: 'If the whole tin industry 'goes west' tomorrow, I do not think there would be much harm done to Nigeria', CO717/69/11, 8 February 1930.
10 CO583/717/7, Thompson (Governor of Nigeria) to CO, 5 June 1930.
11 CO583/717/7, Secretariat of the Northern Provinces (SNP) to Lagos, 14 September 1929. In January, rents were cut to 1 shilling, labour obligations were suspended and the royalty was cut to 2 per cent as long as the price was under £180, *Mining Journal*, January 25, 1930. The Chief Inspector of Mines, Langslow Cook, was sympathetic to restriction and one source suggests that he was advocating it as early as 1927, E. Larner, 'A Tin Mining Anniversary at Jos', *The Nigerian Field,* vol. 29, no. 3, July 1954, p. 100.
12 CO583/717/7, Campbell minute, 22 March 1930.
13 CO583/717/7, SNP to Lagos, 14 November 1929.
14 A Decree-Law of April 23, 1928 cut these duties by 50 per cent as long as the London price remained under £280, *Bolivia*, vol. 1, no. 12, July 1928, p. 6.
15 Antonio Mitre, *Bajo un cielo de estaño*, La Paz: Biblioteca Minera Boliviana, 1993, p. 150.
16 Ibid., p. 149.

17 49 per cent of Central Mining was held in France, Jean-Jacques van-Helten, 'La France et L'Or des Boers: Some Aspects of French Investment in South Africa between 1890 and 1914', *African Affairs,* vol. 84, no. 335, April 1985, p. 261.

18 At the same time, CMO consolidated itself by incorporating its affiliated companies, one of which, Colquiri, would become very important. The terms of the share exchange upset Patiño, who held a minority interest, providing another basis to the rivalry with Hochschild that would soon break out.

19 These are the only mines located close to important urban centres and that would give Hochschild a particular advantage in the ensuing conflict with Patiño.

20 Alfred James, 'Tin Industry Faces Facts', *EMJ*, 23 August 1930.

21 Alfred James, 'Tin Prices and Production', *EMJ*, 8 September 1930.

22 These figures were not well received in Bolivia since they undercut the long standing argument that Bolivian producers were suffering from the abnormally high costs imposed by labour and fiscal legislation. They were subject to detailed criticism in *Revista de Economía y Finanzas,* vol. 2, no. 6, January 1932. The critic focussed on a much smaller sample of Malayan companies as the competition that had to be met.

23 James felt that Howeson's target of £250 was underestimated, 'Reorganizing the Tin Industry by Co-operation and Research', *EMJ*, 25 September 1930. James also argued for the necessity of ending the commodity cycle in tin and considered that a buffer stock of 20,000 tons would be required. However, he was unable to translate a sound diagnosis into the elements of an effective solution which rested on a vague valorization scheme.

24 John Hillman, 'Dredging for Tin in Southeast Asia', in Vth International Mining History Conference, *Proceedings*, Milos, 2001, p. 238. Where the Anglo-American commentators defined Anglo-Oriental as a high cost enterprise in 1930, the Germans singled it out as the best example of one with the lowest costs, Horst Wagenführ, 'England in internationalen Kartellen', *Kartellrundschau*, 1940.

25 *EMJ*, 20 July 1929.

26 This was augmented by the sale of the majority of his shares in the Bolivian Banco Mercantil to a consortium of British banks.

27 This was for £60,000 of unissued shares. Since the transaction occurred in January 1930, following the legal formation of CTS, it may have been to give Patiño undisputed majority control of CTS.

28 Oliver Lyttleton, *The Memoirs of Lord Chandos*, London: Bodley Head, 1962, pp. 137–8.

29 CTS was created through a share exchange and the companies retained their separate corporate identities. The whole process was completed between May and September 1930. The remaining smelters in Cornwall, Penpoll and Cornish Tin Smelting, closed in 1931, D. Bradford Barton, *A History of Tin Mining and Smelting in Cornwall*, Exeter: Wheaton, 1989, p. 278. Just how Howeson joined is unclear but there would have been a substantial incentive for him to do so, since it would permit a stable relationship between his growing interests in Malaya and Eastern Smelting. National Lead held a small interest in CTS, United States Congress, House, Subcommittee, Committee on Foreign Affairs, *Investigation on U.S. Dependence on Foreign Tin*, 74th Cong., 1st sess., 1935, p. 791.

30 CTS would soon acquire a controlling interest in BMC. BMC, which was first sponsored by the Treasury in 1918 to prevent the return of German control over Australian lead and zinc concentrates, then passed into Bolivian 'control'. In the 1970s the wheel would turn full circle when CTS, with BMC, was acquired by German metal conglomerate, Preussag, Helmut Waszkis and Peter Waszkis, *The Story of Metal Trading*, London: Metal Bulletin, 2003, pp. 231–2.

31 Since Penny now joined the CTS Board, this established an important political connection.

32 Groothoof/64, Patiño to BM, 5 March 1930, proposing a holding company between the two to acquire shares in Malayan companies, land, smelting and consolidating small units. This would place them in direct competition with Anglo-Oriental; Bagnall to BM, August 27, 1930, proposing a holding company between various British mining companies, STC and BM. STC had made earlier overtures to BM, ARA, Billiton Maatschappij files, BM20, Board Minutes, 29 November 1928.

33 Details of the collection are in CO717/73/17. While Patiño's identity remained hidden it was clear that someone 'with an invincible belief in the future of the tin industry' was aiming to exercise some control over the policies of the companies, *NYT,* 24 November 1929.

34 CO717/73/17, Patiño to Sungei Way, 2 February 1930. The financial editor of the *Sunday Express* stated that Patiño held 80 per cent of some companies, 23 March 1930. Although this was a gross exaggeration, it was reproduced uncritically in Charles Geddes, *Patiño. The Tin King*, London:

Hale, 1972, p. 228, presumably because it increased the apparent power held by Patiño. Patiño's expansion into Malaya would also have significant consequences for his domestic position. Within Bolivia, it fed an increasingly bitter criticism of his role, since it pointed to the way in which the resources of the country, which, after all, were the ultimate sources of all his wealth, were being used to build up her competitors. The fact that simple transfers of equity had no such result did not affect the virulence with which this charge continued to be made.

35 Thomas to Pearce, 26 October 1929, appended to Jean Michel Lecaut, 'S. I. Patiño et le problème de l'étain bolivien, 1928–1947', Memoire de Maîtrisse, Sorbonne, 1976. The correspondence is quite brief but can be interpreted to suggest that the initiative came from Thomas; it was provided in an interview with Patiño's son, Antenor.

36 CO717/73/17, Trade Commissioner (Singapore) to Department of Overseas Trade (DOT), 23 April; FO371/14750, record of meeting with Simms, 23 April 1930. However, a former Chancellor of the Exchequer, Sir Robert Horne, welcomed the investment of American capital in British tin companies, *Manchester Guardian*, 3 April 1930.

37 CO717/73/17, Clementi to CO, 18 May 1930.

38 CO717/73/17, Caine minute, 2 July 1930.

39 CO717/73/17, Cunliffe-Lister to Wilson, 7 January 1931.

40 National Archives and Records Administration, College Park, Maryland (NARAII), RG151, Bureau of Foreign and Domestic Commerce, 272.6, National Lead to Department of Commerce, 14 February 1931.

41 CO717/79/10, CO to FMS, 27 April 1931. In an interdepartmental conference, the Board of Trade representative took the threat of an American smelter seriously.

6 The formation of the International Tin Committee, 1927–1931

1 *Financial Times,* 18 October 1927.

2 Waugh had proposed a restriction scheme in 1922. Qualified support for the principle of intervention to secure stabilization came from an important but more independent voice, Birch letter to *Financial Times*, 28 November 1927. The qualifications included a revision of the target price to the current £265 and a dismissal of the famine thesis.

3 MKGA312/X6, Report on London meeting, 28th GMB Board Meeting, 30 March 1928.

4 The NEI government endorsed the position taken by Billiton, MKGA324/D3, de Iongh to MK, 22 October 1928.

5 This was the only intervention made by Don Simón. The destruction of the Patiño records in Paris makes it impossible to determine what role Don Simón played behind the scenes in tin politics. Since all discussions were conducted in English, it was left to Antenor to play the visible role.

6 MKGA324/D3, Report to 32nd GMB Board Meeting, 8 November 1928. Pearce supported van den Broek.

7 Groothoof/53, *Overzicht van de aan de totstandkoming der tinrestrictie-overeenkomst voorafgaande onderhandelingen*, p. 7.

8 *MWER,* 13 July 1929. Davis' connection with Howeson was noted in Chapter 4. While Howeson may have needed someone to front the initial meeting of the TPA, it is not clear why he would have used someone with such a shady reputation. Davis' biography is in David J. Jeremy, ed. *Dictionary of Business Biography,* London: Butterworths, 1984, vol. 2. This was the only occasion on which Davis played a public role in tin politics.

9 *Times*, 6 June 1929. Although Howeson had assembled a wide range of signatories, they did not include any associated with the Cornish interests.

10 *Times,* 11 June 1929. It was to meet this challenge that A. P. L. Gordon published his plans for tin control in *Rationalisation for Tin,* London: Saint Catherine, 1929.

11 A full list of members is in MKGA336/N, Peat to GMB, 24 December 1929. They include a few non-British companies such as UMHK.

12 Peat was a well regarded accountant and had some experience with the Iron & Steel Federation, *Directory of Business Biography,* op. cit., vol. 4. Both these names added considerable respectability to the TPA.

13 It is tempting to treat subsequent disputes as deriving from a more consistent position than was actually the case and to suppose that they were already in evidence at this stage. Klaus E. Knorr's characterization of the Redruth position in late 1929 could not be more misleading, *Tin Under Control*, Stanford, CA: Food Research Institute, 1945, p. 95.

14 Even this was considered in some quarters as unwise, given the still anticipated famine, *Mining Magazine,* August, 1929, p. 69.
15 NAK, Department of Scientific and Industrial Research files, DSIR17/156, Heckstall-Smith (TPA) to Tizard (DSIR), 30 July 1929.
16 For the first nine months of 1929 deliveries were up by over 26 per cent.
17 Baliol Scott considered that manipulation of the statistical position was largely responsible for sustaining these prices, since invisible stocks increased by 10–12,000 tons, *Mineral Industry for 1929,* pp. 601, 606.
18 Percy at Renong AGM, *Mining Magazine*, 10 October 1929; Henry's prognosis delivered at the Kampar Malaya AGM was even stronger, *Mining Magazine*, 24 October 1929.
19 Thomas at Tronoh AGM, *Financial Times,* 11 July 1929.
20 Stephens at Tronoh AGM, ibid.
21 *Financial Times*, 21 October 1929.
22 *Financial Times,* 28 November 1929.
23 London Tin Market Report, 4 December, in *Straits Times,* 3 January 1930. At this point, British-American Tin was simply a shell without resources to which only CGF and Howeson were committed. Any such withholding scheme would have required additional financing beyond the £1 million.
24 It reflected Howeson's overall bias towards schemes of financial manipulation, though it may simply have been designed to respond to those who felt like Tarbutt who endorsed smelter withholding but who was still 'averse to restriction of primary output', Jantar AGM, *Mining Magazine,* November, 1929. In any case, the scheme was still-born since it was rejected by the smelters, Josef Wollnik, *Zinn, Wandlungen in der Erzeugung und Verwendung des Zinns nach dem Weltkrieg*, Leipzig, 1936, p. 162.
25 *Economist*, 21 December 1929.
26 MKGA336/N, notes on meeting between Groothoof, Fokker and Antenor Patiño in Paris, 2 December 1929.
27 Many years later, Antenor offered an entirely different version. In his interview with Jean Michel Lecaut, he commented both on the strength of the alliance between Howeson and his father and on the importance in the consolidation of smelters in the plans for an agreement among producers, 'S. I. Patiño et le problème de l'étain bolivien, 1928–1947', Memoire de Maîtrisse, Sorbonne, 1976, pp. 11, 19. It is now impossible to determine the precise state of these always contingent alliances.
28 Groothoof/64, Patiño to BM, 5 March 1930. Billiton showed considerable interest, though it is unclear whether anything concrete emerged, BM20, Board Meetings, 3 February, 9 June 1931, 26 January 1932.
29 This was stressed by Howeson in a letter to *Financial News*, 7 January 1931.
30 London Tin Market Report, 11 December 1929, in *Straits Times,* 6 January 1930.
31 Alfred Lanyon letter to *Financial Times*, 3 December 1929.
32 A. P. L. Gordon letter to *Financial Times*, 12 December 1929.
33 That doubled the theoretical maximum to 40 per cent. Australian companies moved particularly quickly to endorse a slightly modified version, *Sydney Morning Herald*, 10 December 1929.
34 *Mining Journal*, 15 February 1930.
35 At the same time, Howeson revealed that British-American Tin held 1,000 and his 'group' another 4,500 tons of metal, of which some was sold forward, MKGA336/N, notes on the meeting of 21 December 1929.
36 BM20, Board Meeting, 17 January, report on GMB Board Meeting of 16 January 1930. Fokker said that Howeson appeared as a speculator and the sooner he disappeared the better.
37 P. Hövig, 'Nederlandsch-Indië en de Tin-Restrictie-Plannen', *ESB*, 11 December 1929. Contracts with the Chinese coolies prevented any reduction in the wage bill, adding considerably to the increase in overall cost of restriction.
38 Archivo Histórico La Paz (ALP), Carlos Navarro files, CNC1D1, Roberts to Navarro, 12 December 1929.
39 Roberts commented rather bitterly that this was one step in Patiño's limitless ambition and the next would be 'the replacement of George V by Simon the Just'. CNC1D1, Roberts to Navarro, 30 January 1930. Fearful of the implications of American anti-trust legislation, PME itself was not a member of the TPA.
40 MKGA338/O4, Resume van de besprekingen, 12 February 1930; *Overzicht,* p. 10. The 5 per cent cut was not very different from a return to the average of 1927–1929.

41 MKGA338/T3, Report on meeting held on 18 February 1930. This admission by Cunliffe-Lister is particularly interesting in light of the uncritically accepted propaganda that TPA could do it alone.
42 MKGA341/O10, 46th GMB Board Meeting, 11 April 1930, emphasis added.
43 Groothoof/198, memo, 1 June 1930.
44 However, the amount actually sold was only 16,928 tonnes, Banka Tinwinning, Annual Report, 1930.
45 MKGA339/F8, de Iongh report, 23 February 1930.
46 London Tin Market Report, 18 December, in *Straits Times*, 6 January 1930.
47 London Tin Market Report, 19 March, in *Straits Times,* 4 April 1930.
48 *Mining Journal*, 15 March 1930.
49 MKGA341/O10, 46th GMB Meeting, 11 April 1930.
50 London Tin Market Report, 12 March, in *Straits Times*, 1 April 1930.
51 London Tin Market Report, 7 May, in *Straits Times*, 2 June 1930.
52 Mair speech to the TPA, *Financial Times,* 17 April 1930. While Mair may have endorsed the principle of restriction, his thinking about its mechanism could be very confused. At the Rambutan AGM he proposed that the TPA should aim at a price range of £215–£225. £225 was selected as the limit beyond which fresh production would be encouraged and when the price was sustained at this level for 20 days, unrestricted mining operations would be authorized. When the price fell to £215 restriction would be reimposed.
53 *Overzicht*, op. cit., p. 13.
54 The FMS Chamber of Mines supported a mandatory Sunday stoppage and this was pressed by Glenister in the FMS Legislative Council but without success, *Proceedings*, 26 June 1930, pp. B36, B55.
55 NARAII, Record Group 59, State Department, Central Decimal Files, SDCDF841.6354, US Consul General (Sydney) to State Department, 10 July 1930, *Times*, 5 June 1930.
56 *Evening News*, 9 July 1930.
57 General Meeting, 14 June 1930, FMS Chamber of Mines, *Yearbook for 1930*, Kuala Lumpur, 1931, p. 46. It was the smaller companies who were sceptical about the new proposals. The Malayan Chamber of Mines also demanded a two month cut, *Financial Times*, 26 June 1930.
58 *Times,* 21 July 1930.
59 MKGA345/L15, de Iongh to Lagas (NEI government), 1 July 1930.
60 SDCDF800.6354, Johnson (US Embassy, The Hague) to State Department, 11 November 1930.
61 MKGA345/L15, Notes on meeting held on 17–18 June 1930.
62 'The Lovett Tin Valorisation Scheme', *Mining Journal*, 25 October 1930. It was first developed in June.
63 Most valorization schemes are designed to absorb existing stocks and fail because of the difficulty of liquidating them. Lovett's scheme would have addressed the source of the problem.
64 A press release issued by the Local Council of the TPA stated that: 'Members of the industry consider that the scheme provides an ideal solution.' *Mining Journal*, 25 October 1930. At least an editorial in the same issue offered a detailed critique.
65 A. P. L. Gordon, 'J'accuse: A Review of Tin Production Policy', *Mining Journal*, 21 June 1930. In response to a supporter of the TPA, Gordon went even further, 'half measures are worse than no measures', *Mining Journal,* 2 August 1930.
66 *Overzicht,* p. 18. Peat had a quite different impression of this meeting, claiming that it was here that the compulsory scheme was initiated,CO323/1154/9, Peat to CO, 22 January 1931.
67 Groothoof/53, van den Broek memo, 19 October 1930.
68 *Overzicht,* pp. 17–18.
69 SDCDF800.6354, report of conversation with van den Broek, communicated by Johnson to State Department, 11 November 1930. Dutch pessimism had already been communicated to Clementi following a meeting with de Graeff who stated that 'as regards the rubber and tin industries the Government of Netherlands East Indies would adopt an attitude of non intervention and leave economic laws to take their course', CO273/565/23, Clementi to CO, 9 September 1930.
70 Knorr, op. cit., p. 97.
71 This concern was raised by *IAMS*, 19 June 1930.
72 Australian Archives, Canberra (AA) A786/R64/8, Soames to Gepp, 10 October 1930.
73 Guildhall (London) GH22449/2, Tekka-Taiping, Board Minutes, 12 December 1929. Although Mair was on the Executive Committee, companies in the Redruth group did not actually become

members of the TPA until three months after its formation. This 'sitting on the fence' was deplored in many circles, *Truth*, 8 January, in *Straits Times*, 8 February 1930.

74 GH22449/2, Tekka-Taiping, Board Minutes, 14 January 1930. Production at Tekka-Taiping in fact increased. Over the period from November to December 1929, the two dredges were producing at a monthly rate of 264 pikuls each; from January to March 1930 this rose to 292 piculs.

75 MMC, Tronoh Board Minutes, 7 July 1930.

76 *Mining Journal*, 19 July 1930.

77 A shareholder criticized its dilatoriness in a letter to the *Financial News*, reprinted in *Straits Times*, 26 April 1930.

78 *Financial Times*, 30 May 1930.

79 *Financial Times*, 4 June 1930.

80 Idris AGM, *MWER*, 19 July 1930. For a discussion of the moral code cultivated and represented by cricket in this period, see Jack Williams, *Cricket and England. A Cultural and Social History of the Inter-war Years*, London: Cass, 1999. Mair added his own moral blackmail at the Pengkalen AGM: 'It would be unjust and unfair to allow any company to stand aside and profit by the sacrifice of others.' *MWER*, 19 July 1930.

81 *Straits Times*, 2 February 1930. Bishrupriya Gupta notes the effectiveness of social sanctions in enforcing voluntary compliance with tea restriction, 'International Tea Cartel during the Great Depression, 1929–1937', *Journal of Economic History*, vol. 61, no. 1, March 2001, p. 154.

82 *Straits Times*, 14 January 1930. One argument against was similar to that made by Banka: it was the Europeans that had expanded and caused the problem.

83 *Straits Times*, 27 February 1930.

84 ANMKL, Mines Department files, Batu Gajah, BG14/30.

85 *Times*, 6 June 1930.

86 Yukon Gold, Annual Report for 1930.

87 Siamese Tin Syndicate AGM, 11 June 1930, emphasis added.

88 Three Palfreyman and both of the Kerry companies exempted themselves; of these, four were in Siam, *IAMS*, 7 August 1930. F. D. Birch exaggerates this lack of compliance by stating that the Australian groups stood completely outside',Tropical Milestones: Australian Gold and Tin Mining Investments in Malaya and Thailand, 1880–1930', MA Thesis, University of Melbourne, 1976, p. 236, cf. Stuart Rosewarne, 'Capital Accumulation and the Export of Mining Capital before World War II', in E. L. Wheelwright and Ken Buckley, eds, *Essays in the Political Economy of Australian Capitalism*, Sydney: Australia and New Zealand Book Co., 1983, p. 202.

89 GH16545/2, Nigerian Chamber of Mines Council Meeting, 1 July 1930.

90 *Straits Times*, 18 January 1930. This news item also reported the much more modest commitment of the Redruth and Tronoh groups to a 24 hour suspension.

91 CO717/69/11, Cunliffe-Lister to Passfield, 23 January 1930.

92 CO717/69/11, FMS to CO, 31 August 1930, Cunliffe-Lister to CO, 22 September, 12 December 1930.

93 CO717/69/11, Nigeria to CO, 4 September 1930. This set a maximum of £1,000 as the government's contribution. In fact, the amount was reduced to a modest £312, since the United Africa Company agreed to split the governmental contribution. As it was entitled to half the royalties, it was prepared to accept its share of the cost.

94 CO717/80/5, report of meetings held on 20 December 1930 and 22 January 1931. A pikul is 133 lbs.

95 CO717/69/11, Campbell minute, 4 September 1930.

96 Campbell's subsequent career would be entirely devoted to administering restriction agreements. In addition to becoming Chairman of the ITC, he also served as Chairman of the International Rubber Regulation Committee (IRRC), the International Sugar Council, and was a member of the International Tea Committee. Campbell also had an India connection since he had previously served with the Indian Civil Service. His formal training was in chemistry.

97 CO323/1108/8, Grindle minute, November 6, 1930; Flood added 'coal, lead, copper, wheat, barley, and coffee'! At what point the Colonial Office became supportive is not clear since comments on the TPA voluntary programme did not indicate any enthusiasm for compulsory restriction, Gent minute, April 10, 1930. This section draws on John Hillman, 'Malaya and the International Tin Cartel', *Modern Asian Studies*, vol. 22, pt. 2, May 1988.

98 CO to FMS and Nigeria, 7 November 1930, George Maxwell, *The Report of a Committee appointed ... by the Nigerian Chamber of Mines on Thursday, the 15th day of December, 1932*, London, 1933, Appendix I.

99 M. Hochschild, 'The Bolivian Mining Industry and the Present Crisis', *Mining Journal*, 22 November 1930.

100 The former is from *Malaya Tribune*, cited in *British Malaya,* February 1931, the latter from Yip Yat Hoong, *The Development of the Tin Mining Industry in Malaya*, Kuala Lumpur: University of Malaya Press, 1966, p. 201.

101 According to the new Governor-General, de Iongh could be very abrasive, S. L. van de Wal, *Herinnergingen van Jhr. Mr. B. C. de Jonge*, Groningen: Walters-Noordhof, 1968, p. 110.

102 CO323/1108/8, Cunliffe-Lister to CO, 27 November 1930.

103 John Keay, *Last Post: The End of Empire in the East*, London: John Murray, 1997, pp. 134–5.

104 The editor of the *Times of Malaya* had long been agitating for restriction and suggested that 1929 would be an acceptable base, CO717/76/13, Jennings to Clementi, 14 September 1930.

105 *Overzicht,* pp. 20–1. The figure for the FMS was soon revised to include the UFMS, making Malaya 37.14 per cent.

106 This erroneous impression was perpetuated in official publications. Even sources which recognize the adjustments have failed to take them into account in their statistical analyses, for example, Yip, op. cit., pp. 190, 196.

107 Even in its communication of the details of the distribution of restriction among the other producers to Malaya and Nigeria, the Colonial Office used the actual figures of the permissible production for 1931 rather than comparative percentages, CO323/1108/8, CO to FMS, Nigeria, 4 December 1930. Since Banka's published production data was not available on a calendar year basis, it was impossible to determine the extent to which it was prepared to restrict.

108 CO323/1108/8, Howeson to Wilson, 22 November 1930. These details were then endorsed by the TPA which then informed its members in a confidential circular which soon appeared in the press, *Financial News*, 27 December 1930.

109 CO323/1108/9, Cunliffe-Lister to Wilson, 4 January 1931.

110 Manuel Carrasco, *Simón Patiño. Un Procer Industrial*, Cochabamba: Editorial Canelas, 1964, p. 129; Wenceslao Molins, *El Estaño. Fundamento Vital de Bolivia*, Buenos Aires, 1937, pp. 176–8; Roberto Quereajzu Calvo, *Llallagua. Historia de una Montaña*, La Paz: Los Amigos del Libro, 1984, p. 168; Charles Geddes, *Patiño. The Tin King*, London: Hale, 1972, pp. 231–2; Jean-Francois Hennart, 'The Tin Industry', in M. Casson, ed., *Mulitinationals and World Trade*, London: Allen & Unwin, 1986, p. 233; Roberto Arce, *Desarrollo económico e histórico de la minería en Bolivia*, La Paz: Plural Editores, 2003, pp. 307–8.

111 It first became evident in Wollnik who published much of the *Overzicht,* op. cit., pp. 157–78. Although Knorr drew extensively on Wollnik, he ignored this crucial section and since his own account became the standard reference, it proved to be very misleading, op. cit.

112 Burton C. Hallowell, 'Administration of Tin Control in Bolivia, 1931–1939', *Inter-American Economic Affairs*, vol. 3, no. 2, Autumn 1949.

113 CO323/1108/8, CO to FMS, 4 December 1930.

114 CO323/1108/8, Grindle minute, 29 November 1930.

115 Peat (TPA) to Marks (Nigerian Committee (London) of TPA), 19 November 1930, Maxwell Report, op. cit., Appendix III. The cable to the TPA representative in Nigeria was extraordinarily pollyanish. It anticipated that restriction would soon bring the price back to £200 and then allow for increased production, ibid., Appendix II. If these naive sentiments were also expressed in the Nigerian Chamber, it is not surprising that the proposals were approved without dissent.

116 GH16545/4, Council Meeting, 21 November 1930.

117 National Archives of Nigeria, Kaduna (NANK) Secretariat of the Northern Provinces files, SNP14034, CIM, memo, 28 November 1930.

118 CO323/1108/8, minute, 4 December 1930.

119 CO323/1108/8, Nigeria to CO, 6 December 1930; reply CO to Nigeria, CO323/1108/9, 29 December 1930: 'quota for Nigeria based on 1929 10,412 = 7,750 'reasonable'. ' This reply indicates something of the rather casual attention to detail on the part of the Colonial Office since Nigeria had already pointed out that its 1929 claim based on exports was 10,590 tons, as opposed to the figure of 10,412 tons which arrived in Liverpool; production was even higher at 10,734 tons. Adjustments would be made retroactively.

120 MKGA355/G1, GMB to STC, 3 January 1931: 'we are informed STC strongly opposing International Scheme; we consider this attitude strongly against our interests ... you promised us in London remain strictly neutral.' Billiton did not accept Bagnall's profession of neutrality and explored the prospects of getting Banka to bring pressure, presumably by threatening ore supplies, GMB to Weyermann, 15 January 1931. GMB was STC's largest supplier.

121 Mair at Gopeng AGM, *Times*, 19 March 1931. Mair stressed that he remained a supporter of restriction as long as the burden was shared equally and pointed to the dishonest propaganda which had disguised the preferential position of the Dutch.

122 CO323/1108/9, Wilson to Howeson, 2 January 1931. Clementi had provided his personal assurance at the Lygon Place Conference that a retroactive start of January 1 would be no problem in Malaya, Hongkong & Shanghai Bank (London), McKenna Tin Files, MTF192/036, Howeson memo, 25 January 1932. The ability of the government to act effectively may well have been compromised by the fact that many of its officials held shares in STC, CO323/1108/8, minute, 10 April 1930. Such investments were naturally frowned upon in London.

123 MKGA355/U1, de Graeff to MK, 22 January, Groothoof to TPA, 25 January 1931.

124 FMS Legislative Council, *Proceedings*, 19 January 1931, pp. B31–2.

125 BG102/31, 'Proposed Tin Restriction Scheme', 19 January 1931.

126 CO323/1154/8, Howeson to Wilson, 5 January 1931.

127. GH22449/3, Tekka-Taiping Board meeting, 21 January 1931 with details of cable sent to Glenister on behalf of the Redruth, Tronoh, Kinta and Malayan Tin Dredging groups, 13 January 1931.

128 Mair letter to *Financial News*, 6 January 1931. This position reflected the views of Lazarus who considered the 22 per cent cut quite inadequate to restore an equilibrium between production and consumption and hence would only benefit the holders of hidden stocks, memo, CO323/1154/8, 12 January 1931. In a direct communication to the CO, Mair argued for a renegotiation, adding 'The present delay may prove irksome to certain enthusiasts who regard restriction as a sort of conjuring trick and to those anxious to close speculative accounts on the Metal Exchange.' 16 January 1931.

129 Mair at Rambutan AGM, *Mining Journal,* 24 January 1931.

130 *Mining Journal*, 10 January, 7 February 1931. However, they drew criticism from shareholders in some of the Cornish companies, Gallagher to Lahat Mines, 16 December 1930, CO323/1154/8.

131 *Times of Malaya*, 7, 8 January 1931. This was reproduced by Lazarus and sent to his clients, CO323/1154/9. Elizabeth May saw the same report but through a set of lenses that could only read the negative remarks, 'The International Tin Cartel', in W. Y. Elliott, ed., *International Control in the Non-Ferrous Metals*, New York: Macmillan, 1937, pp. 323–5.

132 CO323/1154/8, FMS to CO, 13, 19 January 1931. The responses to the circular indicated that 68 per cent (based on 1929 production) were in favour, 12 per cent opposed (including those who qualified their support) and 20 per cent did not reply, *Times*, 20 March 1931.

133 FMS Legislative Council, *Proceedings,* 19 January 1931, p. B30. Although the charge of 'irresponsibility' drew a sharp rebuke from Antenor Patiño, *Mining Journal*, 28 February 1931, it proved to be a prescient observation.

134 The FMS government claimed that the Cornish were unaware of the scheme until the end of December while the others were planning it from early October and used that as part of the argument for delaying the start until 1 March, CO323/1154/8, FMS to CO, 12 January. It was refuted in considerable detail by Peat, CO323/1154/9, 22 January 1931.

135 GH22449/3, Tekka-Taiping Board Minutes, 21 January 1931.

136 While Banka stocks of metal in the NEI could not have been exported after March 1 without affecting production quotas, Malayan holders of metal in the Straits would not have been similarly disadvantaged, since restriction affected exports from the FMS not from the Straits.

137 NAK, Cabinet files, CAB27/447, Tin Restriction Committee, Report. The committee met on 13, 17 February 1931. Clementi's role in securing the support of the British government was naturally not recognized at the time. That would have prevented J. K. Eastham from asserting that the government imposed restriction on a reluctant Malaya as a result of Howeson's special pressure, 'Rationalisation in the Tin Industry', *Review of Economic Studies*, vol. 4, no. 1, October 1936, p. 21.

138 CAB23/66, Cabinet Meeting, 18 February 1931.

139 House of Commons, *Parliamentary Debates*, vol. 248, col. 1,635, 20 February 1931.

140 CO323/1108/9, memo from Strauss, 26 December 1930. This is the same firm that sold the Penpoll smelter to Howeson in 1928.

141 CO323/1154/3, International Tin Conference minutes, 27 February 1931.

142 FMS Legislative Council, *Proceedings,* 13 April 1931, p. B49. Yip claims that many Malayan producers were surprised to learn that the ratification in April meant a retroactive implementation to 1 March, op. cit., p. 203. In fact, the Mines Department had started the process of assessment of each producer as early as January, though by the end of March, the Warden of Mines was getting

concerned that in spite of public warnings, few Chinese were making any attempt to regulate their production, Laird Circular, 30 March 1931, BG54/31.

143 At the FMS Chamber of Mines AGM, 30 May 30 1931, the Senior Warden of Mines made a fine speech in which he supported the results of negotiations leading to the ITC and asked for the sinking of 'petty differences', *Yearbook for 1931*, pp. 23, 36–7. He was referring to Rich (Tronoh) who noted the 'deplorable atmosphere of secrecy' that surrounded these negotiations and to Hutton (Anglo-Oriental) who could find no 'instance of unnecessary secrecy'. Since Rich's charge has remained, the Warden's speech was ineffective.

144 Sam Ah Wing, FMS Legislative Council, *Proceedings,* 13 April 1931, p. B50.

145 *Times,* 27 May 1931. Campbell was irritated to learn of this important change of attitude from the press.

146 The Dutch immediately wrote to Campbell who noted that it would 'wipe out the attitude of hostility growing up in NEI owing to the constant harping by certain sectors in Malaya that Malaya had not got a 'fair' deal'. CO323/1154/11, minute, 27 May 1931.

147 CO323/1154/11, Campbell minute, 26 May 1931.

7 Constructing the machinery of control

1 The Lygon Place meeting had agreed to a proposal from the Ministerie van Koloniën which was designed to prevent an unreasonable rise in price. Increases in quota would be triggered when the visible stocks fell below 25,000 tons and the price rose over £180. MKGA413, van der Ploeg memo.

2 Cited in Klaus E. Knorr, *Tin Under Control*, Stanford, CA: Food Research Institute, 1945, p 105. The full text of the agreement was never published officially and this statement was not included among the other essential details which appeared in the press, *Handelsblad*, 1 March, *Times,* 2 March 1931.

3 Ibid., p. 106.

4 CO323/1154/11, Campbell minute, 4 March 1931.

5 Some even confuse the two, Elizabeth S. May, 'The International Tin Cartel', in W. Y. Elliott, ed., *International Control in the Non-Ferrous Metals*, New York: Macmillan, 1937, pp. 330, 345; Mary Somers Heidhues, *Bangka Tin and Mentok Pepper*, Singapore: Institute of Southeast Asian Studies, 1992, p. 131; Bill Freund, *Capital and Labour in the Nigerian Tin Mines*, Harlow: Longman, 1981, p. 136.

6 CO323/1154/10, Martin minute, 18 March, Campbell minute, 18 March 1931.

7 Groothoof served as the Chairman of the NEI delegation. He had been head engineer for the Department of Government Industries in the 1920s and was the government nominee to GMB. Banka had no independent representation in the Netherlands and Groothoof must have been expected to be particularly sensitive to its interests.

8 Casa Cultural Portales, Cochabamba, Asociación de Industriales Mineros files (AIM) Box 3, Actas 1926–1934, meeting, 19 December 1930. The Bolivian Legation in London had a very different proposal which would have kept Patiño out altogether, noting that: 'Official and commercial circles prefer delegates as nationals of each country if possible occupying official position.' It suggested as delegates the Chargé d'Affaires, Urriolagoitia, and Romero Ovando, with Pearce (CTS) and Roberts (Aramayo) as technical advisors. Files of the Ministerio de Relaciones Exteriores (MRE), La Paz, Bolivian Legation (London) to MRE, 3 March 1931. Romero Ovando was appointed to the delegation; Pearce and Roberts became technical advisors. However, given his position as head of CTS, Pearce became far more important than any other such advisor. Rogers (Central Mining) replaced Roberts following his fatal illness in 1932.

9 CO323/1154/9, Martin minute, 21 March 1931.

10 Maxwell had been forced into retirement following a dispute with the previous High Commissioner, Robert Heussler, *British Rule in Malaya: Malayan Civil Service, 1867–1942*, Westport, CT: Greenwood, 1981, pp. 38–41.

11 CO323/1154/9, FMS to CO, 12 January, Clementi to Calder, 28 February, 1931. Thompson was regarded as far less suitable than Maxwell in Kuala Lumpur. Simms was nominated to allay the 'grave suspicion of leading miners of the bona fides of the TPA in their engineering of this scheme'.

12 CO323/1155/2, Tavoy Tin to Deputy Commissioner, Tavoy, 6 June 1931. In addition to the standard argument about prolonging the life of a limited reserve Tavoy Tin noted that, since most

workers were now Indian and quite mobile, there would be no unemployment problem to worry about! It was also prepared to exempt those who were dependent on the wolfram content of their mixed tin-wolfram concentrates.

13 India Office Records and Library, British Library, London, IORL/E/9/671/102/4, Anglo-Burma Tin Co. to Indian Trade Commission, 23 February 1931.

14 IORL/E/9/671/102/4, Campbell to Dixon (India Office, IO), 5 March 1931.

15 IORL/E/9/671/102/4, Memorandum by Tipper (Minerals Advisor, Indian High Commission), 4 March 1931.

16 CO323/1155/2, Tavoy Tin to Deputy Commissioner, Tavoy, 6 June 1931.

17 IORL/E/9/671/102/4, Government of India (Industries and Labour Dept) to IO, 22 June 1931.

18 National Archives of South Africa (Pretoria), Department of Trade and Industry files, DTI804, ITC to South African High Commissioner (London), 13 April 1931.

19 DTI804, memo, 13 May 1931.

20 DTI804, memo, Sekretaris van Mynwese, 20 June 1931.

21 CO323/1154/10, Campbell to Damras, 13 March 1931.

22 This estimate from *Straits Echo* and published in the *Bangkok Times*, 14 June, was very close to the initial assessment. Those working under a mining lease were assessed at 12,868 tons with an additional 500 tons for dulang washers, *Bangkok Times,* 5 October 1931.

23 CO323/1155/1, Campbell to Damras, 1 June 1931. This position was the result of a discussion between Craig, Mining Advisor to the Siamese government, and Campbell.

24 CO323/1155/1, British Legation (Bangkok) to FO, 14 May , Siam Chamber of Mines, Minutes of 4th AGM, 28 April 1931.

25 FO126/65, Siam Chamber of Mines to Ministry of Commerce, 15 May 1931. This reflected a great deal of nervousness on the part of managers. Debenture holders were often directors and in any case it would hardly be in their interest to force the liquidation of capital assets on such a weak market.

26 FO126/65, Notes on meeting between Edwards and Minister of Commerce, Bangkok, 11 June 1931.

27 Hence there was no 'hard bargaining' as supposed by Yip Yat Hoong, *The Development of the Tin Mining Industry in Malaya*, Kuala Lumpur: University of Malaya Press, 1966, p. 193. A very different account is offered by Paul Battersby, *To the Islands: White Australians and the Malay Archipelago since 1788*, Langham, MD: Lexington Books, 2007, pp. 124–5. This is based on other official sources and suggests that the Siamese supposed that their industry was so dependent on British smelters and merchants that they faced reprisals if they failed to join the ITC. Such fears were groundless in 1931.

28 In preparing the way for Siam's participation, Craig had established a strict separation between the two schemes, since 'the fewer herrings that are drawn across the scent, the better the hounds run', Craig to Campbell, 1 April 1931, CO323/1154/11.

29 CO323/1155/1, Siamese Legation (London) to Campbell, confirming the overall agreement, 8 July 1931.

30 CO323/1155/1, CO to FMS, 4 June 1931.

31 The source of this figure was the imports from Siam to the Straits smelters, *Economist*, 18 July 1931 and did not include production en route; this error has remained in all ITC publications. Official data from *The Record,* nos. 33 and 34, report the total production as 14,610 tons of concentrate, which at 72 per cent is 10,519 tons of metal, and this figure is noted in W. Y. Elliott, ed., *International Control in the Non-Ferrous Metals*, New York: Macmillan, 1937, p. 34

32 One who did was the market correspondent of the *Straits Times* who noted that 'the quota of 10,000 tons was regarded as involving considerable sacrifice as it was thought that the Siamese authorities could have held out for a much larger figure'. London Tin Market Report, 28 June, published 24 July 1933.

33 Windeatt (Osborne & Chappel) at FMS Legislative Council, *Proceedings,* 13 July 1931. Since a cricket team has 11 men, that underestimated the advantage Siam would enjoy.

34 This sense of grievance is particularly evident in Yip who uses the metaphor of bribery to describe the results of these negotiations with Siam, op. cit., pp. 193–4. It is quite inappropriate since nothing was given to Siam.

35 CO323/1155/2, Minutes of the 5th ITC Meeting, 21 August; 7th Meeting, 22 October 1931.

36 Arquivos Nacionais, Lisbon (ANL), Direccão Geral do Comércio e Indústria,172/31, 29 December 1931. Although production had declined after 1927, it had begun to recover in 1929

and even increased in 1930, largely as a result of the Portuguese-American Tin dredge. Portugal was therefore not prepared to consider 1929 as a base.

37 The structure of control in the UFMS will be considered in Chapter 13. It was of little significance in the early years of restriction.

38 Miners who had produced in the past, but who no longer had the capacity to do so, were still granted assessments. These were made dormant until such time as new equipment was installed.

39 In Batu Gajah the initial rules were: (1) each mine must be able to produce its full assessment separately; (2) group limited to six mines unless under same management; (3) interstate grouping not permitted unless under same owner/manager. BG197/31.

40 Grouping which allowed for a one way transfer from dredging companies to kongis was eventually allowed since that would also maintain employment, ANMKL, Mines Department files, Ulu Selangor, TRUS28/35, Macdonald memo, 29 September 1935.

41 *Times*, 20 May, BG54/31, 7 May 1931.

42 CO323/1155/3, Appendix on FMS excess production, 6th ITC Meeting, 22 September 1931. At the 4th Meeting in July, an even more optimistic estimate was given which suggested the need to cut by only 20 per cent so the 25 per cent imposed would have provided a considerable margin for adjustment. The 20 per cent meant that Malaya was expecting a real cut at a lower level than her partners!

43 CO323/1155/2, Campbell to Calder, 24 July 1931.

44 CO323/1155/2, FMS to CO, 14 August 1931. That final figure of 76,000 was revised to 82,000 for the next quota period, which at least softened the rate at which the overexports would be liquidated. Howeson complained to Campbell that the local TPA people had not been consulted in the establishment of this administrative system and noted that the overexports came solely from the European sector, 20 July 1931. The Mines Department then issued an appeal to the main European groups not to produce at all in August and perhaps to defer the amount applicable to that month to the following year, cable to Mair, CO 323/1155/2, 7 July 1931.

45 Yip is a good expression of this resentment, op. cit., p. 201.

46 Nigeria adopted a rather confusing terminology. The expression 'quota' was used rather than 'assessment' and the Quota Committee defined its principles of administration as 'terms of reference'. The Maxwell report stressed the importance of modifying this terminology, George Maxwell, *The Report of a Committee appointed … by the Nigerian Chamber of Mines on Thursday, the 15th day of December, 1932,* London, 1933, p. 17. A more comprehensive discussion of the Nigerian experience is John Hillman, 'The Politics of Equity: Administering Tin Restriction in Nigeria, 1931–1933', *Mining History Journal,* 1997.

47 This meant that those who had taken all the legal and financial steps necessary to start mining operations after the cut-off date found themselves without any right to do so, SNP14839, Barton to SNP, 19 March 1931.

48 SNP14034, Petition to Chief Secretary, 13 December 1930.

49 SNP14034, Byrde to Chief Secretary, 20 December 1930.

50 SNP14034A, CIM to SNP, June 19, 1931. Anglo-Oriental had envisaged this problem and was prepared to meet it by also making a concession from its allotment.

51 SNP14304, Hyde to Chief Secretary, 1 January 1931.

52 This policy was reinforced by continuing to give out new land for mining, without imposing the Malayan condition of its indispensability to the viability of the enterprise. This was particularly 'owing to the inconvenience it might cause more especially to the small producers'. SNP14034, SNP to Lagos, 11 February 1931. The absurdity of this policy was stressed in the Maxwell report, op. cit., p. 2.

53 SNP14034, CIM to SNP, 28 November 1930.

54 SNP14034A, CIM to SNP, 19 June 1931.

55 SNP14034A, CIM to SNP, 22 July 1931.

56 SNP14939 has full details of one such case and its cost.

57 SNP14034, Bisichi Tin to Chief Secretary, 16 March 1931; Graham at Bisichi AGM, *Times,* 24 August 1931.

58 SNP14034A, Mathews to Chief Secretary, 29 June 1931.

59 GH16545/2, Sir Edwin Speed (Naraguta) at Nigerian Chamber of Mines Council Meeting, 30 April 1931.

60 SNP14034A, CIM to SNP, 29 July 1931.

61 SNP14034, Tarbutt to Chief Secretary, 3 July 1931; SNP14034, Chief Secretary to SNP, 15 July 1931.

62 SNP14034A, CIM to SNP, 3 July, CIM to SNP, 8 July, CIM to SNP, 21 July 1931.
63 SNP14034A, Resident, Plateau Province, to SNP, 20 June 1931.
64 This hardship bonus was confined to the amount that was retroactive to the beginning of restriction.
65 It would another three lean years before this could generate any real advantage.
66 SNP14034A, Summarized Statement Expressing the Three Groups of Opinion, 5 August 1931.
67 CO323/1242/8, Langslow-Cook memo to Maxwell Committee, 8 January 1933.
68 GH16545/2, Nigerian Chamber of Mines Council Meeting, 8 September 1931.
69 *El Diario*, 29 March 1931.
70 Ibid.
71 Etchenique to Patiño, 11 May 1931, in Roberto Quereajzu Calvo, *Llallagua. Historia de una Montaña*, La Paz: Los Amigos del Libro, 1984, p. 169.
72 FO371/15802, Annual Report on Bolivia for 1931.
73 CO323/1156/1, Cable, TPA to Bolivian Gvt, 26 March; MRE, Bolivian Legation (London) to La Paz, 31 March 1931.
74 Kabir-ur-Rahman Kahn, 'The Development of the Tin Buffer Stock System: A Diplomatic and Organisational Study of the pre-1945 Phase', *Indian Journal of International Law*, vol. 17, 1977, pp. 302–3.
75 CO323/1156/1, Clementi to CO, 17 July 1931.
76 CAB23/66, 25 March 1931.
77 MRE, Bolivian Legation to La Paz, 19 June 1931. The memo was later published in *El Diario*, 6 September 1931.
78 AIM12, Bolivian Delegation to ITC, 8 December 1931. The actual commitment was made on 22 July 1931, AIM2, Circulares, Ministerio de Hacienda (MH) to Bolivian Delegation.
79 CO323/1156/1, Nigeria to CO, 30 May. The position was repeated on 12 June 1931.
80 CO323/1156/1, Anglo-Oriental petition to CO, supported by seven other companies, Nigeria to CO, 18 June 1931.
81 CO323/1156/1, Campbell minute, 2 June 1931.
82 CO323/1156/1, Campbell memo, 19 June, on Paris meeting of 17 June 1931. The Bolivians were offered an additional concession of 'contributing' tin in the ground, that is of underproducing their quota.
83 CO323/1156/1, notes on telephone call between Howeson and Houwert, 23 June 1931.
84 CO323/1156/1, Campbell minute, 1 July 1931, emphasis in original. The Dutch thought Patiño was in poor financial shape as a result of the failure of the Linoleum Syndicate, MKGA364/U15, minutes of 1st ITP Meeting, 1 August 1931.
85 AIM12, Antenor Patiño to MH, 31 October 1931. All communications between Bolivian ministries and the ITC were through Antenor.
86 CO323/1156/1, Campbell to Martin, 9 July 1931. This episode calls into question any coronation of Patiño as a Tin 'King'.
87 Offering some confidence to the promoters were other withholding arrangements. The Guggenheims held 1,000 tons and an Australian group 500 tons. It was also expected that some prominent American consumers would withhold a further 5,000 tons, BM35, BM Board Meeting, 18 August 1931.
88 MKGA364/U15, minutes of 1st ITP meeting, 1 August 1931. These amounts were part of the original 25,000 ton scheme, presumably they were reduced *pari passu* as the total was reduced to 21,000 tons.
89 Of the initial 1,000 tons, 685 had been bought on the LME and the remainder acquired from GMB, BM35, Board Meetings, 30 November 1931, 16 September 1932.
90 MTF192/038. The source of the pool was subject to much misleading speculation. J. K. Eastham supposed that Anglo-Oriental still held 5,000 tons which it transferred, 'Rationalisation in the Tin Industry', *Review of Economic Studies*, vol. 4, no. 1, October 1936, p. 2. The editor of *Mineral Industry for 1931* thought that little was actually bought on the market, p. 534, while Serge Calabre, *L'étain*, Paris: Economica, 1991, thought it was all bought, p. 84.
91 CO323/1155/2, Campbell minute, 28 July 1931.
92 CO717/80/5, DSIR to Campbell, 5 May 1931.
93 One unanticipated consequence was a further weakening of the TPA since many of the Cornish-Malayan companies no longer saw a rationale for their continued membership. Petaling and Tekka-Taiping resigned in 1933 followed by Tronoh in 1935.

94 The original ITRDC agreement was for three years to 31 December 1934. Its first renewal was for a further four years. Separating the timetables of the two agreements ensured that should the ITC break down, the parties would still have a regular forum at which to discuss tin politics.

8 Rescuing the tin industry, 1931–1933

1 *Mining Journal*, 7 March 1931.
2 CO323/1156/1, Minutes of the 2nd ITC Meeting, 16 May 1931. Some observers were even more cautious and were therefore surprised at the extent of the cut, *Economist*, 23 May 1931.
3 The UFMS had underproduced so the total for Malaya as a whole was 4,777 tons; for the FMS, the excess was 22 per cent of her quota.
4 CO323/1155/3, Campbell to Calder, reporting on conversations with Lazarus and Howeson, 17 September 1931.
5 CO323/1155/2, Campbell to Calder, 24 July 1931.
6 CO323/1155/2, Clementi to CO, 5 August 1931.
7 CO323/1155/3, 30 September, 1, 2 October 1931. The decentralization proposals were designed to establish a coherent political geography in Malaya by developing a framework that would appeal to the sultans of the UFMS and were central to Clementi's domestic agenda.
8 CO323/1155/3, Campbell to Calder, 17 September 1931.
9 MTF192/037, Howeson to McKenna, 11 August 1931.
10 CO323/1155/2, Howeson to van den Broek, 14 July 1931.
11 MKGA365/F17, de Jonge to MK, 18 September 1931, CO323/1155/3, Minutes of the 6th ITC Meeting, 2 September 1931.
12 CO323/1155/4, Minutes of the 7th ITC Meeting, 22 October 1931. Both the Dutch and the Bolivians expressed their complete satisfaction with the arrangements for regularizing the Malayan situation.
13 CO323/1155/1, Campbell to Martin, 15 July 1931. Campbell also advised Martin that he should pay scant attention to the views of some of those cited by Howeson with a curt: 'This Govt: delegate does not.'
14 CO323/1154/4, CO to FMS and Nigeria, 29 October 1931.
15 CO323/1155/5, Nigeria to CO, 9 November, Minutes of the 8th ITC meeting, 24 November 1931.
16 CO323/1155/5, FMS to CO, 27 November; reply, CO to FMS, 2 December 1931.
17 CO323/1154/4, Campbell minute, November 11, 1931.
18 Consumption estimates vary considerably. The figure used here is taken from the ITRDC, *Statistical Yearbook 1938*, The Hague, 1938, p. 68; the Metallgesellschaft estimate was 10,600 tons lower, *Statistische Zusammenstellungen, 1928–1937,* Frankfurt, 1938, p. 19.
19 J. W. F. Rowe, *Markets and Men*, Cambridge: Cambridge University Press, 1936, p. 164.
20 J. A. Cross, *Lord Swinton*, Oxford: Clarendon Press, 1982, Chapter 3.
21 *Mining Journal*, 12 December 1931.
22 One metal broker, Henry Bath, publicly criticized Mair's reading of the statistics in *Mining Journal*, 19 December 1931. In turn, he was criticized in Lazarus to CO, 17 March 1932, CO323/1197/4.
23 Remarks at Tekka AGM, *Mining Journal*, 28 November 1931. In addition, Mair stressed the disproportionate increase in costs borne by Malayan producers in relation to the effects of restriction on price. One of the more curious assertions made on this occasion was that Malaya had cut by 42 per cent over the first 9 months of the scheme, whereas the actual figure was closer to 35 per cent.
24 *Mining Journal*, editorial, 28 November 1931.
25 CO323/1154/5, Minutes of the 9th ITC meeting, 18 December 1931.
26 CO323/1196/12, Mair to Howeson in Howeson to Campbell, 15 February 1932. Emphasis in original.
27 CO323/1196/12, Campbell to Calder, 17 February. Howeson quickly had second thoughts and agreed with Campbell that nothing would buy Mair off: 'He is not a reasonable animal!' Campbell minute, 19 February 1932.
28 CO323/1156/2, Minutes of the 8th ITC Meeting, 24 November 1931. The new schedule was 10 per cent above the old one, so that 5 per cent would be released when the price was maintained over £165 for a month, rising by 5 percentage points for each additional £11.
29 AIM12, Patiño to MH, 5, 24 December 1931.
30 CO323/1197/5, Campbell to Calder, 4 January 1932.

31 Cf. William Fox, *Tin. The Working of a Commodity Agreement*, London: Mining Journal, 1974, p. 146. The price 'fillip' noted by Fox was largely due to the devaluation of sterling.
32 MTF192/038, record of purchases and sales compared with daily prices from *Quin's Metal Statistics and Handbook, 1931*.
33 It appears that the final contribution of the Dutch Syndicate came from moving 1,175 tons from their invisible stocks in Singapore, AIM12, Patiño to MH, 26 February 1932. All reductions in the size of the invisible stocks would ultimately help but if all they did was to simultaneously increase those of the visible, the effect could only be to depress the price.
34 MTF192/036, Howeson to Antenor Patiño, 2 February 1932. The Patiño contribution of 2,500 would leverage another 2,500 from both British and Dutch sources and bring the ITP up to 28,500 tons.
35 MKGA 374/P6, Groothoof to Governor-General, 29 March 1932.
36 MTF192/036, Howeson to McKenna, 27 January 1932.
37 CO323/1197/5, Clementi to CO, 21 February 1932.
38 CO323/1197/5, Campbell to Howeson, 24 March 1932.
39 MKGA374/P6, Minutes of the 8th ITP Meeting, 21 March 1932.
40 MTF192/034, Minutes of the 18th ITP Meeting, 24 January 1933.
41 Elizabeth May, 'The International Tin Cartel', in W. Y. Elliott, ed., *International Control in the Non-Ferrous Metals*, New York: Macmillan, 1937, pp. 332, 337.
42 *Financial Times,* 9 December 1932, 26 July 1933.
43 AIM12, Patiño to MH, 26 January 1932. In presenting this case, Patiño noted that the mine workers would not be laid off but redeployed in development and other work.
44 AIM12, Patiño to MH, 26 February 1932.
45 CO323/1197/5, Clementi to CO, cable, 14 April 1932.
46 Ibid.
47 SNP14034CI, CO to Nigeria, 2 March 1932.
48 CO323/1196/4, FMS to CO, 16 March 1932. The financial incentive was strong since the cost of holding the stocks would be more than compensated by an increase in royalties from the higher price anticipated. But such arguments had more force in London than in Kuala Lumpur which would have to carry the risk.
49 AIM12, Patiño to MH, 26 February 1932.
50 CO323/1197/5, Minutes of the 12th ITC meeting, 22 March 1932. Patiño offered to arrange finance for those who shipped to CTS and hoped the government would finance the small residue of Bolivian production, AIM12, Patiño to MH, 24 March 1932.
51 AIM12, Patiño to MH, 24 March 1932.
52 CO323/1196/4, Clementi to CO, 10 May 1932.
53 The only occasion on which Clementi spoke to London from the East was when he visited Batavia in November 1930 and Passfield gave him a courtesy call, CO273/565/23. Unfortunately, it did not stimulate him to install similar facilities in Singapore.
54 CO323/1196/4, Howeson to Campbell, 11 May. It was formally endorsed by the Malayan and Nigerian Chambers of Mines and by the TPA, *Times*, 4, 13 May 1932. One consequence of the greater initiative taken by Malayan producers was a demand on the part of the FMS Chamber of Mines to nominate Glenister to join the Malayan delegation to the ITC. As long as this was considered a recommendation, it was acceptable, CO323/1196/4, FMS to CO, 10 June 1932.
55 CO323/1197/4, Campbell to Caine, 11 April 1932.
56 *MTRJ,* 15 July 1932.
57 Details of these features of the scheme as discussed by the ITC are presented by Fox, op. cit., pp. 146–7. He suggested that this was a 'bare-faced subordination of the control scheme to the price interests of the pool'. It was probably the reverse. Raising the release price from £165 to £200 meant reducing the rate at which the ITP would feed the market. Unless the ITC were prepared to extend the cut until the ITP was liquidated, there was no assurance that it would actually realize the higher price and any such extension was considered quite unfeasible. That sense of subordination was widespread in Malaya and is reflected in similarly erroneous assertions by Yip Yat Hoong, *The Development of the Tin Mining Industry in Malaya*, Kuala Lumpur: University of Malaya Press, 1966, p. 326, William Baldwin, *The World Tin Market. Political Pricing and Economic Competition*, Durham, NC: Duke University Press, 1983, pp. 72–3.
58 Yip, op. cit., p. 197. This makes it difficult to assess the extent of compliance. The only actual suspension of shipments by producers during July and August was from the NEI to Singapore;

shortage of storage space within the FMS prevented withholding exports, though not sale, to the Straits.

59 CO323/1196/14, FMS to CO, 8 June 1932. Since the international quota meant a domestic one of 25 per cent, Malaya went to some trouble to devise a scheme that would allow the small Chinese miners to produce at an effective rate of 37 per cent. The excess would be financed by the government and wiped off when the international quota reached 40 per cent. The fact that very few took advantage of this provision is a tribute to the resilience of the Chinese miners, Yip, op. cit., pp. 205–6.

60 According to Howeson, this was 'forced upon the Malayan Government by a small but influential section of the local mining industry', *Times*, 20 May 1933.

61 AIM12, Lewis Lazarus, *Etain, 1931–32*, March-April 1932.

62 It should be recalled that dealers on the LME act as principals, not as agents. A fuller account of this speculation which suggests that other dealers were also involved and with their own funds is 'De Toestand op de Tinmarkt', *ESB,* 15 June 1932.

63 *Straits Times*, 7, 8 June, 22 August 1932.

64 STS worked at 60 per cent of its normal output, over twice the level permissible in the FMS. However, STS did support the scheme by withholding sales of its July–August production and earned 'a handsome profit'. STS AGM, *Times,* 22 June 1933.

65 CO323/1197/3, Campbell minute, 26 November 1932.

66 Willem Hoogendoorn, *Der Zinnpreis. Eine Untersuchung uber die Zinnpreisbildung und ihre monopolistiche Beeinflussung*, Kiel, 1934, p. 88.

67 CO323/1242/4, CO to Nigeria, 5 April, 23 May 1933.

68 MTF192/036, memo, 23 June 1932.

69 MKGA382/T18, Hague meeting, 8 July 1932.

70 MKGA378/T12 Groothoof to Governor-General, 26 June; MKGA378/L15, Groothoof to Governor-General, 24 July; MKGA381/U16, Minutes of the 69th GMB Board Meeting, 1 July 1932.

71 MTF192/036, Howeson to Cunliffe-Owen, 25 July 1932. It is possible that others were involved as well. Howeson asked McKenna for a contribution of £100,000. However, the relevant minute of the Advances and General Purposes Committee of Lloyds Bank reports that BMC was approached by Billiton who provided the £600,000, Book no. 2210, 22 June 1932.

72 Patiño to Salamanca, July 27, 1932, in Charles Geddes, *Patiño. The Tin King*, London: Hale, 1972, pp. 233–4. This letter is also cited in Fox, op. cit., p.152, who fails to note that the second pool was quite independent of the first. It is not clear just what difficulties Patiño had in raising these funds since in January he had offered to finance virtually the whole of one month's Bolivian production which would have required around £300,000.

73 Roberto Quereajzu Calvo, *Llallagua. Historia de una Montaña*, La Paz: Los Amigos del Libro, 1984, p. 170.

74 In any case the objective could not have been £180, since the first pool could start liquidating at £165 and the second could only operate while that remained frozen. Nor is the figure of £110 accurate, since the last time that was reached was in April, three months before the formation of the second pool.

75 Puey Ungphakorn, 'The Economics of Tin Control', PhD dissertation, University of London, 1949, p. 379. If this is the full extent of the syndicate's investment, it would have cost just under £1 million. If this amount had been sold in equal amounts from January to April 1933, it would have generated a modest annualized return of around 6 per cent. Hoogendoorn suggests that the second pool was formed with 5,000 tons from the failure of Lazarus, op. cit., p. 89. If so, it could have picked this up at a bargain price.

76 London Tin Market Reports, *Straits Times,* 13 February, 9 March 1933.

77 MTF192/036, memo, 27 February 1933.

78 Patiño to Salamanca, 13 April 1933 in Querezaju Calvo, op. cit., p. 170.

79 Geddes, op. cit., p. 211. BTIC would eventually become a subsidiary of CTS and shared in its fate.

80 Oliver Lyttleton, *The Memoirs of Lord Chandos*, London: Bodley Head, 1962, p. 141. Lyttleton claimed that BTIC also acquired a large block of shares in Australian companies operating in Malaya. If so, these shares were probably those acquired by Howeson in 1929.

81 AIM12, Patiño to MH. From mid-1932, the pressure on the Bolivian government for foreign exchange with which to fight the war with Paraguay over the Gran Chaco must have made it quite unsympathetic to any such pleadings.

82 CO323/1242/4, Pearce to Baddeley, 27 March, Howeson to Campbell, 12 January 1933. These figures are of the assay values of the concentrates and deduction would have to be made for losses in smelting. At ESC these were very small, 0.25 per cent, but considerably larger at STC.
83 US consumption in 1932 exceeded deliveries by 4,280 tons, reducing invisible stocks, Strauss Tin Review, *British Malaya,* May 1933. The figure of 13,146 tons for the decline in invisible stocks is the residual from the other figures; it corresponds to the estimate of 14,000 tons cited in *Minerals Yearbook, 1934*, p. 459.
84 Pope estimated that actual consumption in the USA was not over 4,500 tons in August while deliveries were 8,200 tons, United States Congress, House, Subcommittee, Committee on Foreign Affairs, *Investigation on U.S. Dependence on Foreign Tin*, 74th Cong., 1st sess., 1935, p. 959.
85 *Metal Bulletin*, 23 June 1933.
86 The president of United States Steel stated that this was the cause of his moving away from Straits but he also added that English refined was now a little more pure than Straits, *Tin Investigation*, op. cit., pp. 1012–1013, 1015.
87 *Minerals Yearbook, 1934,* p. 452.
88 Reprinted in *Daily Express*, 22 March and *Mining World*, 25 March 1933.
89 Groothoof/53/4, Houwert memo, 10 April 1933. Mair's reading also flew in the face of the very dramatic decline in overall American demand in 1932.
90 In 1932 STC abandoned its policy of publishing daily sales and it was widely suspected that the company was speculating in tin, with a view to working against the success of restriction, CO323/1197/3, CO to Clementi, 11 November 1932.
91 Strauss Tin Review, *British Malaya*, October 1933.
92 Uncoupling the link between the rate of restriction and releases from the ITP on the termination of the 12 months of the Byrne scheme was undertaken on the suggestion of Nigeria, SNP14034CIII, Nigeria to CO, 3 March 1933.
93 London School of Economics, Fox Papers, Campbell notes, 2 February 1934.
94 *Financial News*, 2 May 1933, endorsed by Yip, op. cit., p. 198.
95 Cf. Baldwin, op. cit., pp. 72–3.
96 Fox, op. cit., p. 154.
97 Klaus E. Knorr, *Tin Under Control,* Stanford, CA: Food Research Institute, 1945, p. 125; May, op. cit., p.338; Yip, op. cit., p.199, Fox, op. cit., p. 154.
98 In the case of metal contributed by the NEI and acquired before September 1931, the return is unlikely to have compensated for the devaluation of sterling.
99 Tin Holdings, Annual Reports, BT31/33325/263565.
100 Such risks are particularly evident in the failure of the bulls associated with Lazarus who had exactly the same market information. The fact that BMC and CGF held the same number of preference shares, even though they held very different numbers of ordinary shares, suggests that they were simply sharing the residual cost of meeting the margin call, rather than expecting any particular profit. It is doubtful whether Tin Holdings could have actually sold any of its assets to meet the margin call; to have done so would have brought down the whole edifice erected around the ITP.
101 Editorials in *El Diario*, 13, 28 March; Juan Muñoz Reyes speech, 17 March; Soria Galvano speech, 19 August 1931.
102 CNC1D3, memo to MH, 20 December 1932, *El Diario*, 12 December 1932.
103 AIM 12, Patiño to MM, 27 July 1932. Patiño's constant reminders of the moral aspects of this problem were reinforced by the AIM and an editorial in *El Diario*, 29 December 1932.
104 Burton C. Hallowell, 'Administration of Tin Control in Bolivia, 1931–1939', *Inter-American Economic Affairs*, vol. 3, no. 2, Autumn 1949, p. 4.
105 CO323/1242/6, as reported by Campbell to Calder, 14 September 1933. Patiño might have added that the fiscal needs of the government with which to fight the Chaco war which broke out in mid-1932 made it quite impossible to impose a further cut. The failure to include the name of Hochschild, who was now three times the size of Aramayo, suggests that he was supporting the small miners. One of the other dealers, Philipp Brothers, formally protested against restriction since the small miners would find it difficult to repay its advances, CO323/1197/3, Derby to FO, 20 July 1932
106 Salamanca to Patiño, 31 May 1933 in Geddes, op. cit., p. 234. Cf. Manuel Carrasco, *Simón Patiño. Un Procer Industrial*, Cochabamba: Editorial Canelas, 1960, p. 130. The Potosí request for 300 tons was made by Davila to Patiño, 12 March 1933, in Querajzu Calvo, op. cit., p. 170. Whether

this full amount was granted is unclear; if any substantial quantity was involved, it must have been included within the exports credited to Patiño.

107 SNP14034, Meeting of the Quota Committee, 9 December 1931. The new system followed the general lines adopted in Malaya. Free transfer simplified the task of administration and permitted more efficient production, at least in the short run.

108 Granting compensating allowances meant that the 1933 quota for small miners was reduced to 42% of their original admitted claims, while that for all other producers was cut to 24%, SNP14034A.

109 SNP14304CI, Resident, Plateau Province, to SNP, 4 April 1932; SNP160964, SNP to Chief Secretary, 4 May 1932.

110 SNP14034, Medium miners to Chief Secretary, 6 June 1932.

111 SNP14034, Rumbold to Local Council, 15 June 1932.

112 The general retrenchment of the Nigerian Civil Service in 1931 had severely reduced the establishment of the Mines Department in spite of the extra work generated by tin restriction, SNP14270/II.

113 SNP14034, minute, 18 June 1932.

114 *Nigerian Daily Times*, 8 November 1932.

115 *Nigerian Daily Times,* 22 November, 1 December 1932.

116 GH16545/2, Nigerian Chamber of Mines, Special meetings of all Nigerian tin producers, 15 December 1932, 25 April 1933.

117 Maxwell Report, op. cit., p. 22. It therefore refused to accept the request of the CIM to settle the issue itself, Langslow-Cook memo, 8 January 1933, CO323/1242/8.

118 Maxwell Report, op. cit., p. 20.

119 Ibid., pp. 23–4.

120 CO323/1242/8, Baddeley to Howeson, 5 May 1933.

121 SNP14034, CIM to SNP, 10 July 1933.

122 SNP14034C, Resident, Plateau Province, to SNP, 12 July 1933.

123 Pahang Consolidated, *Sixty Years of Tin Mining. A History of the Pahang Consolidated Company, 1906–1966,* London, 1967, p. 35. The case was finally disposed of on 15 December 1932.

124 House of Commons, *Parliamentary Debates*, vol. 270, col. 264, 8 November 1932. Chamberlain supposed that the next obvious candidate was the far more complicated meat trade. He was considered a strong supporter of cartels, Clemens Wurm, *Business, Politics, and International Relations: Steel, Cotton, and International Cartels in British Politics, 1924–1939*, Cambridge: Cambridge University Press, 1993, p. 54. Even as late as 1935, Chamberlain's comments on tin were still being used, Rotterdamsche Bankvereenigung, *Monthly Review*, August–September 1935, p. 193.

125 Patricia Clavin, *The Failure of Economic Diplomacy. Britain, France, Germany, and the United States,* Basingstoke: Macmillan, 1996, Chs 6–7.

126 Sydney Caine, *Prices for Primary Producers*, London: Institute for Economic Affairs, 1963; Kabir-ur-Rahman Kahn, *The Law and Organisation of International Commodity Agreements*, The Hague: Nijhoff, 1982, pp. 57–65; Ronald Brand, *et al.*, 'Legal Aspects of International Commodity Agreements', in Osman Suliman and Mahdi El-Baghdadi, eds, *Global Commodity Price Stabilization,* Westport, CT: Quorum, 1995, pp. 7–8.

127 Cunliffe-Lister speech, 19 June 1933, League of Nations, Conf. M.E./C.E.17.

128 United Kingdom delegation memo, ibid. The case for maintaining a pivotal price at around £200 was not made and the whole strategy was subject to immediate criticism by the *Financial Times*, 13 July 1933.

129 NAK, Ministry of Power files, POWE22/60, Nott-Bower notes on the first meeting of the tin subcommittee, 15 July 1933.

130 League of Nations, Conf. M.E./C.E./102(1), Report by the Sub-committee on Tin, 19 July; *Journal of the World Monetary and Economic Conference*, 21 July 1933, p. 204.

131 Official statement delivered to the Subcommittee on Tin, 20 July 1933, League of Nations, Conf. M.E./C.E./109. W. Y. Elliott, ed., *International Control in the Non-Ferrous Metals*, New York: Macmillan, 1937, claimed that it was not included in the Conference Proceedings and therefore reproduced it, pp. 52–3. The terminological confusion between the pool as a specific form of stock control and the more comprehensive agreement governing production was not uncommon. One of the American delegates was a member of the House Foreign Affairs Committee, Sam McReynolds. Considered a close friend of Hull, their relationship ensured co-operation between the House and the State Department on tin policy. As the Americans were sounding a warning

note about tin, they were also active in persuading the conference to endorse their plans for an international agreement that would raise the price of silver for the benefit of their miners, Joseph Davis, 'Experience under Intergovernmental Commodity Agreements, 1902–1945', *Journal of Political Economy*, vol. 54, no. 3, June 1946, p. 213.

132 Some discounting is necessary to take account of the overall inflation in the USA. However, over the four months from April to July, prices of staples rose by 52 per cent, while tin rose by 80 per cent, Charles Kindleberger, *The World in Depression, 1929–1939*, Berkeley, CA: University of California Press, 1986, pp. 222–4.

133 Groothoof/53/4.

9 Renewing the second agreement, 1933–1934

1 *Papers Relating to the International Tin Control Scheme*, London: HMSO, 1935, Cmd 4825, p. 5.

2 Ibid., p. 9.

3 CO323/1242/4, Houwert memo, 12 April 1933.

4 CO323/1242/5, Minutes of Singapore Conference, 26–27 April 1933. This position simply reflected the views of the FMS Chamber of Mines and the Selangor Miners Association, *Straits Times,* 4 March 1933. The Dutch also expressed their interest in the formation of a buffer stock, of 15,000 tons with equal contributions from the governments of Malaya, NEI and Bolivia. Although Clementi and other official Malayan representatives were in favour, there was much opposition and the matter was dropped.

5 CO323/1242/5, Minutes of Singapore Conference, 26–27 April 1933.

6 *Times of Malaya* editorial, cited in Charles Geddes, *Patiño. The Tin King*, London: Hale, 1972, p. 236. No date but context suggests early June 1933. As will be seen in Chapter 13, few could have long survived at this price.

7 FMS Legislative Council, *Proceedings,* 26 June 1933, p. B97.

8 Idris AGM, *Financial News,* 11 May 1933.

9 Pengkalen and Tekka-Taiping AGMs, *Financial News*, 19 May 1933.

10 Simms at Sungei Besi AGM, *Financial News*, 15 June 1933.

11 At the Tronoh AGM, Thomas explicitly distanced himself from Simms, *Financial News,* 12 July 1933.

12 FMS Legislative Council, *Proceedings,* 18 December 1933, Memorandum on the Negotiations Leading to the Signing of the Agreement for the Renewal of Tin Restriction, No. 40, Appendix D, 21 August 1933, p. C439.

13 Calder commented: 'Malayan delegation has been seriously handicapped because the Dutch know that Clementi has accepted Mr Houwert's conclusions.' That elicited the following response from Campbell, 'she has already all she can fairly claim: perhaps more'. 12 June 1933, CO323/1242/8.

14 Victor Lowinger was the new head of the Malayan Information Agency and was considered the High Commissioner's representative on the Malayan delegation.

15 CO323/1242/8, Clementi to Lowinger, 15 May 1933.

16 CO323/1242/5, Cameron to CO, 15 June 1933.

17 AIM2, Patiño to MH, 20 March 1933.

18 AIM2, Patiño to MH, 1 August 1933.

19 AIM2, Patiño to MH, 28 June 1933. The concession meant that Bolivia gave up 3,066 tons over the life of the first agreement.

20 In addition, it was defined as a purely temporary agreement for which temporary concessions had to be made, the full cost of which was quite unanticipated. Bolivia made this point clear when it indicated that the extension of the agreement beyond the original termination date rendered those concessions inoperative, Groothoof/199, Bolivian Delegation to Dutch Delegation, 4 April 1933.

21 MKGA406/C23, Campbell to Groothoof, with details of Patiño cable, 25 September 1933.

22 A compromise proposal presented to the ITC by the Dutch delegation applied the same scale of refunds to Bolivia, Nigeria and Billiton. In any event, Billiton was not prepared to renew the concession it had made to Singkep, Groothoof/53, memo, 13 February 1933.

23 CO323/1242/6, Campbell to Calder, 14 September 1933.

24 MKGA405, de Jonge to MK, 12 September 1933.

25 CO323/1242/6/11730/5, Campbell to Calder, 14 September 1933. Bolivia was also quite prepared to walk away from a new agreement if the Dutch did not concede, AIM2, Patiño to MH, 19 October 1933.
26 CO323/1242/6/11730/5, Campbell to Calder, 14 September 1933.
27 Colijn had considerable experience in the NEI and was also Colonial Minister, H. J. Langeveld, *Hendrikus Colijn 1869–1944*, Amsterdam: Balans, 1998. Cunliffe-Lister had established a good working relationship with him during the initial negotiations in 1930 and together they shaped the way in which the World Monetary Conference addressed the question of commodities, Swinton, *I Remember*, London: Hutchinson, 1948, pp. 75–6; *Sixty Years of Power*, London: Hutchinson, 1966, p. 104.
28 As this issue was explored, the Colonial Office did not find much support from the Foreign Office through which any such pressure would have to have been exerted, FO371/17327, Nicholls memo, 19 September 1933.
29 CO852/4/6, Campbell to Thomas, 12 July 1935. Campbell used this to dissuade Thomas from attempting to reopen the ratio between Malaya and the NEI in the next round of negotiations.
30 MKGA406/C23, Campbell to Patiño, 21 September 1933.
31 MKGA406/C23, Campbell to van den Broek, 25 September 1933.
32 Groothoof/199, Howeson to van den Broek, 27 September, Bolivian Delegation to Dutch Delegation, 28 September 1933.
33 MKGA406/C23, Telephone conversation, 29 September 1933.
34 MKGA406/C23, Colijn to MK, September, 28, 1933. This followed yet another meeting with Campbell and Cunliffe-Lister and it reflected the activities of Colonel Williams who was actively promoting plans to establish a smelter in the USA drawing on Bolivian concentrates, CO323/1242/4, UK Legation (La Paz) to FO, 21 February, 6 April 1933. Such a proposal received much support in both Bolivia and the United States and would eventually be realized in 1940.
35 MKGA406/C23, Telephone conversation, 29 September 1933.
36 CO323/1242/5, Clementi to CO, 2 May 1933.
37 De Iongh was still hoping for common selling arrangement with the smelters: 'Punten … over de te volgen gedragslijn in zake tin.' 18 August 1932, Groothoof/53. The buffer stock was therefore a considerable concession from this extreme position.
38 CO323/1242/4, Minutes of the 26th ITC Meeting, 23 May 1933.
39 National Archives of Thailand, Bangkok, Financial Advisor's files, 0301/1/11/6, Craig report, 27 May 1933.
40 Division among the forces that had ended the absolute monarchy in 1932 was resolved by a coup in which Colonel Phibun played a major role, one that was consolidated by the role he played in crushing an attempted royalist counter-coup in October 1933. Phibun cultivated the support of the Japanese to counter the presumed support of the British for the ousted conservative forces. That relationship would grow as Phibun increased his power, Edward Flood, 'Japan's Relations with Thailand: 1928–1941', PhD dissertation, University of Washington, 1967, Chapter 2.
41 CO323/1242/6, Minutes of Special ITC Meeting, 10 October 1933.
42 Since this crucial qualification was never publicized, it was easy to claim that Siam had actually lost ground in the second agreement, remarks of the representative from Phuket to the People's Assembly, *Bangkok Times*, 15 February 1935.
43 There is some variation in the statistics on true assay values. Those that were used for Malaya and the NEI were 74.7 per cent and 72.6 per cent respectively. However, Scott states that they were really 75.3 per cent and 73.4 per cent, *Mining Journal,* 17 February 1934. If so, then Malaya gave up 578 tons and the NEI 400 tons.
44 *Times*, 6 December 1933.
45 *IAMS,* 1 February 1934.
46 CO323/1242/7, FO to Calder, 12 December 1933.
47 United Kingdom, Mines Department, Report of the Advisory Committee for the Metalliferous Mining and Quarrying Industry, London: HMSO, 1932. This committee still laboured under the illusion of the imminent exhaustion of the Eastern alluvial deposits, p. 10.
48 POWE22/60, memo, 12 July 1933.
49 POWE22/60, Nott-Bower minute, 20 July 1933.
50 POWE22/60, Campbell to Nott-Bower, 24 July 1933.
51 This decision was publicly endorsed by Herbert Thomas, *Cornish Post*, 12 August 1933.
52 POWE22/60, Arthur Thomas to Nott-Bower, 18 September 1933.
53 *Daily Express*, 26 September 1933.

54 CO323/1242/6, Minutes of Special ITC meeting, 10 October 1933.
55 Charles Thomas at the Tehidy Minerals AGM, *West Briton and Cornwall Advertiser*, 12 March 1934. That would mean a flat rate of 3,250 tons. Cornwall only barely exceeded 2,000 tons on two occasions in the 1930s. Thomas prefaced his position by noting that: 'Tin restriction has, undoubtedly, been of great benefit.'
56 POWE22/60, Arthur Thomas to Nott-Bower, 11 July 1934.
57 ANL, Assumptos Financieros, 6 Comercio, Portuguese Embassy (London) to Ministerio do Naciones Exteriores, 16 October 1933.
58 CO323/1233/10, Cunliffe-Lister minute, 1 August 1933.
59 France was considered the 'foremost advocate of international industrial agreements', and 'seized every opportunity to obtain the unqualified support of other Governments for international cartels'. Board of Trade memo, October 12, 1932 in FO371/16422. Laurence Ballande identified 140 international cartel agreements, 67 of which had French participation, the highest rate of all, *Essai d'étude monographique et statistique sur les ententes économiques internationales*, Paris: Librairie technique et éonomique, 1936, p. 311.
60 Archives Générales du Royaume, Brussels, Affaires Etrangères files, AE/II 2607bisI, Colonies to Foreign Affairs, 31 July 1933. This section follows John Hillman, 'Chartered Companies and the Development of the Tin Industry in the Belgian Congo, 1900–1939', *African Economic History*, vol. 25, 1997.
61 AE/II2607bis, Colonies to Foreign Affairs, 11 October 1933.
62 *Financial News*, 8 January 1934.
63 *Mineral Industry for 1933*, p. 566.
64 CO323/1242/7, Minutes of the 31st ITC Meeting, 7 December 1933. See also William Fox, *Tin. The Working of a Commodity Agreement*, London: Mining Journal, 1974, p. 157. This argument suggests an uncharacteristic willingness on the part of the state to adopt its own mining policy and to force it on the producers. However, the Governor was certainly anxious to see the development of mining as a means of absorbing unemployment, P. Ryckmans, 'Le pays du Grands Lacs', *Le Miroir du Congo Belge*, vol. 1, Brussels, 1929, pp. 272–3.
65 Had the Belgians taken the 10,500 standard tonnage they would have gained 2,300 tons. There was to be no carry forward of unused quota.
66 They were described as 'absurd' in a *Straits Times* editorial, 14 March 1934.
67 CO323/1301/3, Baddeley memo, February 2. This opposition was not supported by the Nigerian government as long as the Congo was considered an outsider for the purposes of article 20, which, of course, it was, SNP14034C, Nigeria to CO, February 24. The Colonial Office made it clear that while it was asking for instructions for the delegations, the ITC would make the actual decision, if necessary by majority vote, SNP14034C, CO to Nigeria, February 19, 1934.
68 CO323/1301/3, Campbell minute, 13 February 1934. These terms did limit the development of some fresh deposits and encouraged Somico to concentrate more on gold, *La Belgique Financière, Coloniale et Maritime*, February 1935, p. 19.
69 *Mining Journal*, 14 July 1934. The communiqué also announced the signing of a Buffer Stock agreement which effectively completed the process of establishing the machinery of control for the second agreement. A full account of its formation is provided in Chapter 10.
70 NAK, Dominions Office files, DO35/246/8933/5, Campbell to Caine, 28 August 1933.
71 DO35/246/8933/6, Dickson to Eales (High Commissioner to South Africa), 24 October 1933.
72 DO35/246/8933/6, Campbell minute, 23 November 1933.
73 DO35/246/8933/6, minute, 22 May 1934.
74 DO35/246/8933/22, McCreedy Tin Mine, 26 February, Dickson to Eales, 16 April 1934. The final offer by Swaziland was a standard tonnage of 375 and a flat rate of 250, though its preference remained to be outside as long as South Africa was not a member, letter to ITC, 22 June 1934.
75 DO35/246/8933/21, Campbell minute, 6 June 1934.
76 Such an appeal was made by Maxwell at the TPA AGM, *Times*, 31 May 1933; there were no takers.
77 DTI804/24, South African Tin Producers, Draft Agreement, 14 June 1934. Zaaiplaats publicly expressed its support at its Annual General Meeting, 16 December 1933. However, a sceptical note was sounded by the *South African Mining and Engineering Journal (SAMEJ),* 13 January 1934.
78 DTI809, Trade Commissioner (London) to Department of Commerce (Pretoria), 1 March 1934.

79 DTI809, Mines Inspectorate to Department of Commerce, Pretoria, 19 September 1934. Anglo-French may have feared that permitting its South African subsidiary, Rooiberg, to participate would compromise its position in relation to Anglo-Burma.
80 CO323/1242/4, Campbell memo, 1 May 1933.
81 Fox, op. cit., p. 157; CO323/1301/5, Minutes of the 37th ITC Meeting, 8 August 1934.
82 FO371/1849, Harding (Consul-General, Yunnanfu) to UK Legation (Peking), 28 September 1934.
83 It was subsequently alleged that the Chinese authorities were deliberating minimizing production figures 'in order not to alarm the international trade'. *Manchester Guardian*, 13 January 1936.
84 IORL/E/9/671/102/4, Turner memo 10 July, cable to India, 20 July 1933.
85 IORL/E/9/671/102/4, Burma to Secretary of State for India, 7 October, 1933.
86 IORL/E/9/671/102/4, Lindsay (India Trade Commission, London) to Department of Industry (New Delhi), 20 October 1933.
87 IORL/E/9/671/102/4, Government of Burma Revenue Department to Department of Industry, 19 November. This offer was then formally made to Campbell on 14 December 1933.
88 IORL/E/9/671/102/4, Campbell to Turner, 15 December 1933, Tipper to Turner, 4 January 1934.
89 IORL/E/9/671/102/4, Blenkinsop to India Office, 3 February 1934.
90 IORL/E/9/671/102/4, Government of India to India Office, 13 August 1934. Fox mistakenly supposed that the Tavoy Chamber's formal assent was something more than a grudging support of the principle and confused the role of the government of India with its representative in London, op. cit., p. 157.
91 *Times*, 3 December 1934.
92 National Archives of Australia, Canberra, AA-A981/Conferences/117/2, Bruce to Lyons, 30 May 1933.
93 AA-A601/1/763/44/1, Memo, 19 September 1933.

10 Stabilizing the tin market, 1934–1936

1 A very comprehensive *Statistical Bulletin* was published monthly from August 1933. By 1937 it had a circulation of 600 copies.
2 For example, the American Bureau of Metal Statistics estimated actual consumption in the United States for 1935 at 61,940 tons, while the comparable figure from the Bureau of Mines was 55,928 tons.
3 CO323/1242/7, Minutes of the 31st ITC Meeting, 7 December 1933. The consumption figure was simply an extrapolation of the estimated consumption rate for November. In preparing this position, Lowinger ignored many warnings from Howeson, 25 October, and only Baddeley was prepared to give him support.
4 CO323/1242/7, Minutes of the 31st ITC Meeting, 7 December 1933.
5 *Economist*, 4 November 1933.
6 CO323/1242/7, Campbell to Calder, 14 November 1933.
7 MKGA442/E10, Director of Economic Affairs to Governor-General, 12 April, Governor-General to MK, 16 April 1934. The Dutch also contemplated a higher maximum of 12,000 tons.
8 Even Yip Yat Hoong, whose position is otherwise very supportive of the Malayan critics of the ITC, recognized the value of such a buffer stock, *The Development of the Tin Mining Industry in Malaya*, Kuala Lumpur: University of Malaya Press, 1966, p. 223.
9 SNP18460, minute, 2 March 1934. Cf. Yip, op. cit., p. 224.
10 CO323/1242/7, CO to FMS and Nigeria, 22 December 1933.
11 CO323/1301/8, Campbell minute, 20 March, Martin minute, 21 March, 1934.
12 However, they had the support of Austral-Malay, one of the largest independent groups in Malaya.
13 CO323/1301/3, LME to CO, 25 January 1934. The *Daily Express* offered its own inimitable comment: 'The whole pool idea is fantastic and should not receive a moment's consideration from any body of intelligent businessmen.' 19 December 1933.
14 CO323/1242/7, Howeson to Hoare, 11 October 1933. The new member of the Malayan delegation, Lowinger, attempted to get him to change this position but failed, 19 December 1933. Clementi's support was publicly known, Strauss Tin Review, *British Malaya,* March 1934.
15 FMS Legislative Council, *Proceedings,* 19 March 1934, p. B28.
16 Rae's remarks are particularly interesting since he recognized that the ITC had provided 'considerable service to the industry' in the past but that it was now time to start the process of

'releasing this country from the shackles of restriction'. FMS Legislative Council, *Proceedings,* 19 March 1934, p. B21.

17 *Financial News*, 26 January 1934. The Malayan Chamber of Mines, however, voted 4/3 in favour.

18 Maxwell may simply have been reflecting the sentiment of connections made during his term as Chief Secretary of the FMS. He was replaced by someone more reliable, Sir Samuel Wilson, just retired as Permanent Undersecretary at the Colonial Office and also a director of CGF.

19 Yip, op. cit., p. 233; *Financial News*, 12 January 1934. Maxwell was not a member of the Executive Council and was not opposed to the principle of a buffer stock.

20 CO323/1301/8, FMS to CO, 24 March 1934. It was expected that the proposal would lose by 60/40, Strauss Tin Review, *British Malaya,* April 1934.

21 CO323/1301/5, Campbell minute, 6 July 1934.

22 Since there was no mechanism to force producers to release their stocks, they could easily join the inevitable speculation.

23 CO323/1242/7, Baddeley note, 10 December 1933.

24 *Mining Journal*, 14 July 1934, Jos correspondent report.

25 SNP14034, Russell (Acting CIM) minute, 15 January, Thompstone (SNP) minute, 18 January 1934.

26 SNP14034/C, CIM to SNP, 22 January 1934.

27 SNP14034/VI, Cameron to CO, 2 February 1934. When the scheme was implemented, only Bisichi, Amari and a few individuals stood outside. Most of the producers, including Anglo-Oriental, took up the government's offer to finance at £90 per ton at 3 per cent interest, SNP22280/I, Meeting between Tin Producers and Mines Department, 13 June 1934. A full list of contributors is in CO323/1302/2, memo, 17 August 1934.

28 GH16546/2, Nigerian Chamber of Mines, General Meeting Minute Book, vol. 2, minutes of meeting, 16 February 1934.

29 GH16545/4, Nigerian Chamber of Mines, Council Meeting Minute Book, no. 3, Council Meeting, 13 February 1934.

30 AIM2, Patiño to MH, 2 June 1934.

31 CO323/1301/8, Campbell minute, 8 February 1934.

32 CO323/1301/8, CO to Nigeria, 23 March 1934.

33 CO323/1301/8, Cunliffe-Lister to FMS, 28 March 1934.

34 Strauss Tin Review, *British Malaya,* April 1934.

35 CO323/1301/8, McKenna's comments noted in Campbell minute of 22 March 1934.

36 CO323/1302/1, Mair to Campbell, 25 April 1934.

37 CO323/1301/4, Minutes of Special ITC Meeting, 2 May 1934. The announcement provoked an intervention from the Foreign Office which continued to see this form of an international agreement as 'very irregular'. CO323/1302/1, FO to Calder, 2 June 1934.

38 Bolivian reluctance made the Dutch furious: 'They think [Bolivia] wholly untrustworthy and unreliable, watchful for every opportunity to blackmail.' CO323/1302/1, Campbell to Calder, 20 April 1934.

39 CO323/1302/1, Campbell to Calder, 26 April 1934.

40 After Howeson's departure, it became routine for the British and Dutch to discuss tin policy 'without any undue risk of a leakage to Patiño', CO852/625/3, Meeting on Tin, 13 June 1945.

41 CO323/1301/5, Minutes of the 36th ITC Meeting, 10 July 1934.

42 MKGA442/E10, Director of Economic Affairs to Governor General, 12 April 1934, Governor General to MK, 16 April 1934. This figure was later endorsed by a New York metal dealer, C. S. Trench, Trench to Houwert, 30 December 1935, Groothoof/52/3. Trench was one of the few dealers to express some support for the ITC and its policies.

43 *Minerals Yearbook, 1935* states that the buffer stock was outside the statistics of visible supplies, p. 514. However, when a shipment of buffer stock metal was inadvertently placed in a non-reporting warehouse in New York, the ITC promptly issued a statement that this had been corrected.

44 Strauss Tin Review, *British Malaya,* February 1934.

45 Strauss Tin Review, *British Malaya*, May 1934, June 1934. The fact that Howeson was appointed as Nigeria's representative to the buffer stock committee would have eliminated many of the remaining doubts.

46 CO323/1301/8, Mair to Campbell, 9 April 1934.

47 BM21, Board Meeting, 12 September 1934. MKGA445/A5, Campbell to Groothoof, 1 March 1935. Details of their market operations are in the files of the Bank of England, G1/244.

48 TPA, *International Tin Control and Buffer Stocks*, London, 1944, p. 13. The range for tin was –1.1 per cent to +2.6 per cent, for copper, –17.6 per cent to + 3.7 per cent.
49 BM21, Board Meeting, 6 November 1935. These were losses by BM; losses by LTC and GMB would have been more than compensated by the overall price support. It was estimated that otherwise the price would have dropped to £180, CO323/1301/5, Lowinger to FMS, 14 August 1934.
50 Strauss Tin Review, *British Malaya,* November 1934.
51 P. Cunliffe-Lister, 'The Economic Policy of the Colonial Empire', *United Empire*, October 1934, p. 580. When challenged in the House of Commons, Cunliffe-Lister stressed the responsibility held by the government to look after the interests of consumers, to which the United States government in particular attached great importance, *Parliamentary Debates,* vol. 298, col. 1920, 6 March 1935.
52 Strauss Tin Review, *British Malaya*, March 1934.
53 *SAMEJ,* 5 October 1935. Since the buffer stock aimed at stability and relieved consumers of the costs of carrying some stocks, there were no *prima facie* grounds for consumer opposition.
54 CO323/1311/14, Cunliffe-Lister memo, 26 May 1934.
55 CO323/1311/14, Campbell memo, 12 June 1934. The accusation of stupidity was largely based on the failure to retain its largest supplier, Billiton, because it was 'unreasonable and discourteous'.
56 Straits Settlements Legislative Council, *Proceedings*, 1934, p. 41.
57 Ibid.
58 Ibid. The reference is to Psalm 53, verse 5.
59 *Minerals Yearbook, 1935*, p. 509.
60 Pahang Consolidated AGM, *Economist*, 25 November 1933.
61 Pahang Consolidated AGM, *Economist*, 8 December 1934.
62 FO371/18569, US Embassy, London, aide-memoire, 13 April 1934. Rather than forcing the State Department to provide a basis for this allegation, Campbell replied with a memo exonerating the ITC on the grounds that US policy was generally to inflate prices back to their 1926 level and tin had some way to go to reach that target and indeed was still below metals generally. The problem of establishing a basis to determine the fairness of tin prices would recur and will be discussed further in Chapter 16. A more moderate position was argued within the Department which noted: 'A high world price for tin is not a grave detriment to American interests', Livesey memo, 3 March 1934, SDCDF 824.6374.
63 *Economist*, 28 April 1934. In early 1935 Campbell met the editor and told him that his 'pontificating' on the price question was tiresome and encouraged him to prepare an article documenting the claim that £230 was too high and establishing an alternative figure. This was not a challenge to which Sir William Layton was prepared to respond, CO852/4/6, Campbell to Thomas, 12 July 1935.
64 Yip, op. cit., p. 225. Yip defines 1934 as a whole as this turning point but the issues were really only joined around the discussion of the fourth quarter. It does not follow that low cost producers would prefer higher production and a lower price on economic grounds alone. Such preferences are a function of anticipated depletion paths.
65 *Financial Times,* 31 July, cited in preparatory notes for the 37th ITC Meeting. It was also considered 'necessary' and generally expected by Campbell in his minute to Boyd immediately before the August meeting, 10 August 1934, CO323/1301/5.
66 MKGA432/522, cable to Governor-General, 14 August 1934.
67 CO323/1301/5, Minutes of the 37th ITC Meeting, 13 August 1934.
68 The Bolivian government was pressing for 50 per cent in order to maximize revenue to pay for the Chaco war, AIM2, Patiño to MH, 19 July, 10 November 1934. In addition, the return to 40 per cent placed Patiño in a very difficult financial situation. The earlier increase had been the basis on which Patiño had secured a loan of £265,650 from the Anglo-South American Bank, which in turn was advanced against future royalties to the Bolivian government to meet its war needs. The Bank was now reluctant to continue to advance funds against the lower quota, Casa Cultural Portales (Cochabamba), Archivo Familia Patiño files, AFP12, Paris to Oruro, 7 July, 16, 22 August 1934.
69 CO323/1301/5, Cunliffe-Lister minute, 16 August 1934.
70 CO323/1301/5, Lowinger to Chief Secretary, FMS government, 14 August 1934.
71 The State Department claimed that the pressure it exercised was responsible, SDCDF800.6354, Hull to Tower (AISI) 21 May 1938. But that appears to have been wishful thinking. Augmenting the role of consumers was stressed at the World Monetary Conference and this arrangement was considered to have gone 'considerably beyond any obligation assumed by HMG ... at the

Economic Conference'. FO to US Embassy (London), 26 April 1934. It was also noted that the American government was restricting cotton without any provision for the representation of foreign consumers.

72 This has been seized on as confirmation of the claim that the ITC was a producers' organization, Klaus E. Knorr, *Tin Under Control*, Stanford, CA: Food Research Institute, 1945, p. 245. A fuller discussion of the problem of consumer participation is provided in Chapter 16.

73 CO323/1301/6, Minutes of the 38th ITC Meeting, 22 October 1934. That provision was akin to the right to exclude technical advisors from actual decisions and was never implemented.

74 Firth was an active promoter of price stabilization schemes to support the tinplate industry in 1919 and of a more comprehensive scheme for rationalizing the steel industry in 1933, J. C. Carr and W. Taplin, *The History of the British Steel Industry*, Cambridge, MA: Harvard University Press, 1962, pp. 398–490, 503. He also was a member, with Lyttleton, of the Industrial Reorganization League. This was formed by Conservatives in 1934 to promote rationalization of industry along progressive lines, Harold Macmillan, *Winds of Change, 1914–1939,* London: Macmillan, 1966, p. 371. Firth could therefore be expected to have considerable sympathy with the ITC. However, he was not well regarded among other steelmakers, Duncan Burn, *The Steel Industry, 1939–1959*, Cambridge: Cambridge University Press, 1961, pp. 53–6.

75 Limiting the panel to two members had the 'advantage' of ensuring that Germany did not sit around the same table as France.

76 *Mineral Industry, 1934*, p. 553.

77 Cordell Hull, *Memoirs of Cordell Hull*, New York: Macmillan, 1948, vol. 1, p. 457. Like so many of the State Department initiatives on tin, it was poorly thought out. The British government had no tin and failed to see the quid pro quo for making substantial payments on its debt, SDCDF800.51W89 (Great Britain), Feis memo, 4 June 1935.

78 SDCDF811.6354, Veatch to Caldwell, May 1934. In this memo, Veatch relied on the judgement of the *Economist* that a price of £225 could not be justified.

79 Such a policy was consistent with that of other major industrial powers, especially France and Germany, which had also placed embargoes on the export of tin plate scrap to build up their domestic detinning plants.

80 The Republicans objected, seeing it as entailing quite unnecessary expense and queried the jurisdiction of a Foreign Affairs Committee over the matter. Both were prescient sentiments. It took the intervention of Hull to remove the obstacle imposed by the House Rules Committee, SDCDF811.6354, Hull to Bankhead, 15 May 1934; the House debate is in *Congressional Record*, vol. 78, pt. 11, 15 June 1934, pp.11745–55.

81 United States, House, Subcommittee, Committee on Foreign Affairs, *Investigation on U.S. Dependence on Foreign Tin*, 74th Cong., 1st sess., 1935., pp. 1005–6.

82 Ibid., pp. 37–8.

83 Ibid., p. 31.

84 Ibid. Quite uncharacteristically, this suggests that price was not in fact of any concern.

85 Elizabeth Hannan provided a thoughtful survey of the results of the McReynolds investigation in 'Congress Investigates Tin', *International Law and Relations*, vol. 5, no. 4, 13 January 1936. However, she concluded by criticizing the committee for not going nearly far enough in breaking this 'monopoly', pp. 15 ff.

86 This obsession with a domestic smelter derived from the fact that the technical counsel to the committee was Henry Buckman who was actively involved in creating the American smelting industry in the early 1920s, ibid., pp. 1, 7. Hannan indulges in the groundless speculation that the lack of interest on the part of the large consumers may have been a function of stock ownership in British smelters, p. 8. Since the US was able to smelt other metals efficiently, it seemed that only artificial political barriers stood in the way of a tin smelter, Caldwell speech to American Tin Trade Association, 8 March 1935, CO852/4/1.

87 With the passage of the Strategic Minerals Act in June 1939, the Department of the Interior received an annual appropriation of $500,000 to support prospecting, *Minerals Yearbook, 1939*, p. 681. Nothing substantial was ever found.

88 *Tin Investigation,* op. cit., pp. 26, 35–6. NARA, Roosevelt Library (Hyde Park), FDROF931, Percy Barbour, 'An analysis of the Congressional Tin Investigation Report', 6 May 1935, p. 24. What makes the error particularly puzzling is that it is flagrantly inconsistent with the statistical data presented on the facing page, p. 27. Barbour reviewed the conceptual problems and concluded that the committee demonstrated a 'complete inability for the analysis of the situation'.

89 HR 7675, 74th Congress, 1st Session, 23 April 1935.

90 *American Metal Market (AMM),* 4 June 1935, has a full page advertisement by a manufacturer of solder, Lissberger, urging defeat of HR Bill 7675. A metal dealer, Pope, who had no sympathy with restriction, similarly argued for the defeat of the bill, Pope to McReynolds, 10 May 1935, in NARAI (Washington) HR 74A F15.4. The AISI also protested the bill, *AMM,* 8 June 1935.
91 'Tin Consumers Threatened by Radical Legislation', 1 May 1935.
92 FDROF931, Russell Clark to Roosevelt, 24 May 1935. Roosevelt's reply indicates that this was the first he had heard of the proposed legislation. Cf. Barbour, op. cit., p. 30.
93 The bill was never reported out of the House Ways and Means Committee, SDCDF, 811.6354, Veatch memo, 17 May 1935.
94 Strauss Tin Review, *British Malaya,* June 1935.
95 United States, Congress, Senate, S 3381, 16 February 1936. Roosevelt signed the executive order designating the Department of State as the body responsible for issuing such licences the following day.
96 FDROF931, Wilson to Roosevelt, 26 January 1935. Had Baruch's proposal led to the kind of formal agreement that was eventually signed with the ITC in 1940, the direct participation of the US government would have changed the subsequent course of tin politics.
97 FDROF931, Hull to Roosevelt, 29 May 1935. This was still based on the assumption that Britain would provide tin as a credit against war debt. McReynolds floated a different conception of such an exchange which would involve granting the UK credits against its war debt to the extent of the duties foregone on the imports of ore from Malaya and Nigeria, *Tin Investigation,* op. cit., pp. 24–5.
98 *AMM,* 18 August 1942, commented on the way in which the unfounded optimism concerning the possibility of substantial tin deposits being developed in the USA was still being canvassed during the war as a result of the McReynolds report, 'we know of no volume with more misinformation on any subject'. One account claimed that there was an ore body with over 11,000 tons of metal content in North Carolina alone and that British control over smelting blocked the incentive to develop it, Sylvia Pass, 'Tin has been made into a dunce-cap for Uncle Sam', *Ken*, 27 July 1939.
99 Howard Cox, 'Business on Trial: The Tobacco Securities Trust and the 1935 Pepper Debacle', *Business History*, vol. 49, no. 6, November 2007.
100 A. Dauphin Meunier, *La Cité de Londres*, Paris: Gallimard, 1940, p. 82.
101 *Financial News,* 6 February 1935.
102 With supplementary loans from the Midland Bank and further investments from Howeson and Cunliffe-Owen's Dean Finance. For the role of Dean Finance, see Howard Cox, *The Global Cigarette. Origins and Evolution of British American Tobacco, 1880–1945*, Oxford: Oxford University Press, 2000, pp. 319–21.
103 Bishirgian's character and background is sketched out in *NYT,* 22 February 1936.
104 Howeson held two thirds of Williams, Henry and Bishirgian the remainder. During the reorganization, Williams, Henry was transferred to Sumatra Rubber, owned by Howeson and Hardy. Bishirgian also held an interest in AOGIT.
105 Jayandee was owned by Howeson, his wife, Dorothy, and their three children. Dorothy's brother served as a Director of Williams, Henry.
106 The financial press was divided in its assessment of the prospectus. *Investors' Chronicle* reviewed it favourably, 8 September 1934; *Economist* was critical, 6 September 1934. While he did not suspect the exposure to pepper, Lyttleton considered that there was something very wrong with the prospectus and so advised Howeson when he was asked to underwrite the flotation, *The Memoirs of Lord Chandos*, London: Bodley Head, 1962. p. 142. Indeed, the prospectus is quite extraordinary, since investors in the £112,500 of ordinary shares were also required to subscribe a further £300,000 in the preference shares, to be redeemed at the end of 1935 at an annualized profit rate of 12.4 per cent.
107 While it is clear that McKenna was familiar with the shellac position, there are contradictory statements as to whether he was also aware of the more serious pepper position. *The Week* claimed that he was advising friends to buy shares in James & Shakspeare because of the pepper, 20 February 1935, Coburn, Patricia, *The Years of THE WEEK*, London: Comedia, 1985, p. 106. However, he was exonerated by the Official Receiver, a judgement supported by A. R. Holmes and E. Green, *Midland: 150 Years of Banking Business*, London: Batsford, 1986, p. 195. Nonetheless, the suspicion remains that McKenna was too big to fall.
108 CO852/10/17, Howeson to van den Broek, probably January, 1935. Coincidentally, the island of Banka was also the largest producer of pepper.

109 In any case, such deception would have ended their relationship quite independently of the outcome of the pepper pool and with it all the far more important games Howeson was playing in tin.

110 Campbell caught the ambivalence in a minute, 1 April 1935, CO852/10/17, which noted:

> Many indices pointing to his probable knowledge and complicity; but apparently nothing definite and convincing so far. It may be that the whole thing was run by people with whom he had business connections of a close kind and that he was unaware of the action taken till the crisis had almost been precipitated.

111 For the very different treatment accorded Lord Kylsant who also served twelve months for issuing a misleading balance sheet, see P. N. Davies, 'Business Success and the Role of Chance: The Extraordinary Philipps Brothers, *Business History,* vol. 23, no. 2, July 1981, p. 217 and Nicolas Davenport, *Memoirs of a City Radical*, London: Weidenfeld and Nicolson, 1974, p. 22. The implications for the reorganization of the Anglo-Oriental financial empire will be considered in Chapter 13.

112 House of Commons, *Parliamentary Reports,* vol. 298, col. 2,223, 7 March 1935. This debate was preceded by the extensive use of Question Time to embarrass the ITC. Major Nathan alone posed ten such questions after the scandal broke.

113 This would actually entail a slight reduction since the gross 4 per cent bonus for 1934 now fell away.

114 This may have been one consequence of restriction itself, as more metal was sold outside the LME at a slight discount on its prices as part of a strategy of price maintenance.

115 MKGA445/A5, telephone conversation, van Buuren, Houwert and Groothoof, 13 March 1935. Houwert even went so far as to discount the articles in the mining press on the grounds that they were simply paid for by the opposition.

116 AIM2, MH to Patiño, 20 March, Patiño to MH, 20 March 1935.

117 CO852/4/1, Minutes of the 41st ITC Meeting, 14 March 1935.

118 AIM2, Patiño to MH, 24 June 1935. Virtually the whole of the Bolivian share of the buffer stock was pledged as security against a loan from BMC which was due at the end of December 1935, *La Razón,* 24 November 1935.

119 CO852/4/1, Minutes of the 42nd ITC Meeting, 6 June 1935.

120 Although world tinplate exports increased by over 8 per cent from 1933 to 1934, British exports dropped by over 14 per cent which reduced her market share from 62 per cent to 48 per cent. Over the same period, US exports increased by 94 per cent, ITRDC, *Statistical Bulletin*, September 1936, Table E-1.

121 CO852/4/1, Minutes of the 42n ITC Meeting, 6 June 1935. There had been a general expectation that the quota would be raised to 55 per cent but it was thought that the ITC was keeping supplies tight to prevent a bear attack by the dealers, *Economist,* 15 June 1935. Strauss criticized the decision as soon as it was announced and also suggested it should have been 55 per cent, *British Malaya*, July 1935. That, too, was on the low side.

122 CO852/4/6, Campbell to Thomas, 12 July 1935. Firth was then asked for a reasoned statement justifying his position but he failed to deliver on his promise to provide one.

123 CO852/4/1, Elibank to MacDonald, 24 July 1935.

124 CO852/4/1, MacDonald to Elibank, 28 July 1935; Elibank's reply on 31 July held fast to his original position. The only other evidence was a letter from an American chemical company to the TPA complaining specifically about the backwardation and warning about a shift to substitutes.

125 Knorr considered the March decision to have been the crucial mistake and suggested that the existence of the buffer stock encouraged the ITC to set the quota on the conservative side, *Tin Under Control*, Stanford, CA: Food Research Institute, 1945, p. 146. In this he was supported by Lyttleton, *Economist*, 18 January 1936.

126 CO852/4/1, Minutes of the 43rd ITC Meeting, 9 August 1935.

127 CO852/4/1, Minutes of the 44th ITC Meeting, 11 September 1935.

128 CO852/4/1, Minutes of the 45th ITC Meeting, 3 October 1935.

129 CO852/4/1, Minutes of the 46th ITC Meeting, 22 October 1935. Patiño decided not to block the increase to 80 per cent, in return for which Bolivia was given an extension to enable her to produce the increase once her productive capacity had been restored, AIM2, Patiño to Ministerio de Industria, 21 December 1935. Since quotas continued to increase in 1936, the concession was not needed.

130 In part, this was because Hochschild's contribution was not accepted until he had secured the warrants being held by his bank as security for a loan. AIM2, Patiño to MH, 27 February 1935. Eventually some accommodation was reached with Hochschild but Bolivia remained 32 tons

short, AIM2, Patiño to MH, 4 October 1935. Late contributions carried a financial penalty at an annual interest rate of 6 per cent.

131 CO852/4/3, 5th report of the Buffer Stock committee, 3 March 1935. It also aimed at keeping the Straits premium under £3.

132 CO852/4/7, Campbell to Thomas, 17 May, Campbell minute, 22 May 1935.

133 *Times*, 9 May 1935.

134 CO852/4/4, LME memo to Board of Trade, to CO, 21 May 1935. It tried again after Cunliffe-Lister was replaced as Colonial Secretary by Malcolm MacDonald but with no more success, LME to MacDonald, 7 August 1935. Had the dealers passed on their sentiments to their clients, they could easily have encouraged some to switch to less controversial metals. 1935 was the year of the lowest ratio of turnover to total consumption. It recovered slightly from this low of 48 per cent but remained far below the 1926–1933 average of 120 per cent.

135 Strauss refuted the TPA claim that most consumers only bought in the forward market, by noting that 12.5 per cent of the metal he sold was for prompt delivery, *Times*, 3 August 1933.

136 Cf. William Fox, *Tin. The Working of a Commodity Agreement*, London: Mining Journal, 1974, p. 167, for a very different account. Fox claimed that this release was automatic since the market price was exceeding the target price but he must have been thinking of the nominal price for Straits in London which was not relevant.

137 This price was £245 but since there were no sellers, it was artificially high. When the market reopened the following day, the price was a more reasonable £232. Failure to recognize this difference led Knorr to exaggerate the price policy of the ITC, op. cit., pp. 146–7.

138 Fox papers, Memo on Buffer Stocks 1933–1939, Campbell to Midland Bank, 27 July 1935. One consequence of the shortage of standard tin was a greater interest in the use of common tin and the discount against standard dropped from £5 to £3/10/–, *British Malaya*, January 1935. Supplying common tin would have been a serious admission of failure to secure sufficient supplies and could undermine the role of standard, just as standard was undermining the role of Straits.

139 Pope Weekly Market Review, 26 July 1935; Trench Tin Report, 25 July 1935; Strauss Tin Review, *British Malaya*, September, October, 1935.

140 Fox, op. cit., p. 167. Both Puey Ungphakorn, 'The Economics of Tin Control', PhD dissertation, University of London, 1949, p. 384 and Knorr, op. cit., p. 147 claim that it did, but this may be the result of confusing profits made as a result of the market operations of the stock and profits from the difference between cost of production and final sale. Most of the metal in the stock was available for sale in February when the price was £233; even if it was all sold when the market recovered in July, the price realized would have been slightly lower at £232.

141 CO852/139/6, Malayan Chamber of Mines Buffer Stock memo, March 1938, Appendix 5.

142 Consumers drew down their stocks in anticipation of a price slump following the dissolution of the ITC.

143 John Hillman, 'The Mining Industry and the State: The Politics of Tin Restriction in Bolivia, 1936–1939', *Bulletin of Latin American Research,* vol. 21, no.1, January 2002.

144 CO852/4/2, Minutes of the 47th ITC Meeting, 6 December 1935. This decision was taken in spite of the scepticism of both Lowinger and Groothoof on Campbell's pleading that the ITC could not allow itself to again be subject to the severe criticism that it had received.

145 CO852/33/2, Minutes of the 48th ITC Meeting, 19 February 1936.

146 AIM2, Patiño to MH, 7 March 1936. This is a particularly self-serving letter since Patiño grossly misrepresented the debate, supposing that all the others were committed to 90 per cent and the resulting 85 per cent was simply due to his efforts on behalf of Bolivia.

147 With one major difference. Since the local position in the NEI was articulated by the government it did not break out into the public arena and hence did not become evident to the other members of the ITC. However, the local government found exactly the same difficulty as that in Malaya in getting its delegates to follow instructions, MKGA469/C4, van Buuren to Governor-General, 31 January, Governor-General to MK, 6 February 1936,

148 MKGA466/Y, van Buuren to Governor-General, 10 December 1935. On Campbell's request, the NEI had overexported 1,455 tons to help relieve the stock position in the last quarter of 1935. These were probably drawn from local stocks and the NEI therefore did not need to produce at the 90 per cent level during the first quarter since it had this excess to wipe off.

149 MKGA469/C5, telephone conversation between van Buuren, Groothoof, Houwert, van den Broek, 22 February. Again, van Buuren stressed the high marginal costs for Banka with an 85 per cent quota for just one quarter. Houwert overestimated the level of British stocks; they were 600

tons on February 22 and would drop to 400 by the end of March. Only Firth supported van den Broek.

150 *Investors' Chronicle,* 22 February 1936, Strauss Tin Review, *British Malaya,* March 1936. However, a month later a survey of the stock position suggested that the ITC was on the right course, *AMM,* editorial, 28 March 1936.
151 MKGA471/T8, Campbell to van den Broek, 30 March 1936.
152 CO852/33/2, Campbell minutes, 3 March, 4, 6 April 1936.
153 CO852/33/2, Minutes of the 51st ITC Meeting, 25 June 1936. In March 1936, Hitler occupied the Rhineland.
154 MKGA477/P17, Verslag van de ITC vergaderingen, Juni en Juli 1936.
155 Had this gesture been made with the announcement of the 90 per cent quota, the market fall would have been arrested. Campbell, however, thought there was some excuse since the 'Bolivians are now dealing with what will probably prove to be the most difficult gvt. They clearly did not know where they were domestically.' minute, 27 June 1936, CO852/33/1.
156 CO852/33/2, Gutt at the 52nd ITC Meeting, 6 July 1936.
157 CO852/33/1, Campbell minute, 28 September 1936.

11 Renewing the third agreement, 1935–1936

1 CO852/4/6, Houwert to Campbell, 29 April 1935.
2 Brian Montgomery, *Shenton of Singapore: Governor and Prisoner of War*, London: Secker & Warburg, 1984.
3 *Straits Times,* 19 January 1935.
4 Thomas may have been anxious to avoid the ignominious fate of Clementi who upset the business community, Robert Heussler, *Completing a Stewardship: The Malayan Civil Service, 1942–1957*, Westport, CT: Greenwood, 1983, p. 27.
5 CO852/4/6, Howeson to Campbell, October 9, 1935. Only 2.7 per cent of the eligible votes abstained and only 1.6 per cent of those who voted followed the opposition of Pahang and Tronoh. Thomas stressed that Tronoh's opposition was not to the principle of restriction, *Times,* 11 June 1936.
6 CO852/4/6, FMS to CO, 1 October 1935.
7 CO852/4/6, Rowson minute, 14 June 1935.
8 CO852/4/6, Thomas to Campbell, 14, 19 June 1935.
9 The first came soon after this correspondence when Campbell had to explain the elementary principle of price inelasticity of demand which made the desire for a higher production but at a lower price as 'obviously in opposition to the basic plan of the control scheme', CO852/4/6, Campbell to Thomas, 12 July 1935.
10 MKGA381/A17, Campbell memo, 4 February 1936.
11 CO852/33/1. At the 48th ITC Meeting on 19 February 1936, Picard requested a minimum of 2,500 tons. Such a figure was not only unrealistic in light of past performance but it would vitiate the whole point of moving to a standard tonnage.
12 Fox papers, Box 4, Blenkinsop, Notes on the establishment and development of the ITC, 10 November 1936, p. 36.
13 POWE22/89, Campbell to Nott-Bower, 15 December 1936.
14 The ITC wanted Burma in on even more favourable terms, Blenkinsop to Turner, 27 February 1937, IORLM/3/56; a week later it learnt that Burma was abandoning any support, *Times*, 3 March 1937.
15 Fox papers, Box 4, Blenkinsop, Notes.
16 *SAMEJ,* 3 October 1936.
17 MKGA475/W14, Houwert report on Congo visit, 26 May 1936.
18 CO852/33/6, van den Broek to Lowinger, 9 May 1936.
19 AE/II No 532 (2935) memo, 28 July 1936.
20 CO852/33/6. These terms involved a substantial concession from the initial demand of 13,700 tons, which was later justified in light of the Congo's performance during World War II.
21 MKGA477/P17. Van Buuren started to criticize the Dutch delegation in July 1936 and he then felt he found further evidence in the BM Annual Report, van Buuren to Governor-General, 2 February 1937. Unqualified support for van den Broek was provided by Billiton to MK, 28 December 1936,

MKGA480/T. This internal feud continued for many months. Billiton had taken a small share in a new prospecting company in Katanga in February 1933.

22 Edward Flood, 'Japan's Relations with Thailand: 1928–1941', PhD dissertation, University of Washington, 1967, pp. 112–13. This section also draws on John Hillman, 'The Freerider and the Cartel: Siam and the International Tin Restriction Agreements, 1931–1941', *Modern Asian Studies*, vol. 24, no. 2, May 1990.

23 Sir Josiah Crosby, *Siam: The Crossroads*, London: Hollis & Carter, 1945, p. 103. Fuller background is in P. Fistié, *Sous-developpement et utopie en Siam. Le programme de réforme présenté en 1933 par Pridi Phanonmyong*, Paris: Mouton, 1969. Phibun was a strong supporter of Mussolini and Hitler, not least because of their irredentism, Richard J. Aldrich, *The Key to the South. Britain, the United States, and Thailand during the Approach to the Pacific War, 1929–1942*, Kuala Lumpur: Oxford University Press, 1993, pp. 174–5.

24 Flood, op. cit., pp. 165–6.

25 This contrast between the two figures that dominated Siamese politics is qualified in K. Suwannathat, 'Thai Wartime Leadership Reconsidered: Phibun and Pridi', *Journal of Southeast Asian Studies*, vol. 27, no. 1, March 1996.

26 Sompop Manarunsgan, *Economic Development of Thailand, 1850–1950*, Bangkok: Institute of Asian Studies, 1989, p. 108. John Drabble, *Malayan Rubber. The Interwar Years*, London: Macmillan, 1991, p. 181.

27 CO852/4/6, Campbell to Thomas, 29 April 1935. Siam's aggressive position on rubber was based on her earlier success with tin, CO323/1304/2, Campbell minute, 17 December 1934. Rubber interests attempted to stiffen the CO position by pointing out that an increase in the tin standard would be a prelude to another demand for concessions on rubber, *Financial News*, 28 July 1936.

28 A strict conversion of the de facto flat rate of 10,000 tons with increases over 65 per cent would give a standard tonnage of 15,500 tons. The transition from a flat rate to a standard tonnage could generate some confusion among commentators, for example, Klaus E. Knorr, *Tin Under Control*, Stanford, CA: Food Research Institute, 1945, p. 151, Paul Battersby, Paul, *To the Islands: White Australians and the Malay Archipelago since 1788*, Langham, MD: Lexington Books, 2007, p. 127.

29 The existing terms were denounced by Reginald Le May who had been Financial Advisor when they were negotiated in 1931, as the result of 'ignorant handling of the situation on this [UK] side'. *Times*, 23 June 1936.

30 CO852/33/3, Siam statement to 53rd ITC Meeting, 23 September 1936.

31 Puey Ungphakorn, 'The Economics of Tin Control', Ph.D. dissertation, University of London, 1949, p. 406.

32 In 1934, assessed claims were around 17,500 tons of metal, plus a further 350 tons for dulang washers, FO 628/50. That already represented an increase of 25 per cent over the capacity established in 1931.

33 CO852/4/6, Henggeler to Lowinger, 16 August 1935.

34 For example, Bangrin was restricted to 64 per cent of its assessment in 1936 while the producers in the FMS worked on 67 per cent of theirs. By contrast, dulang washers were granted much higher quotas than those in the FMS, Puey, op. cit., p. 406. Of course, companies could always increase actual production by buying quotas from other producers and many of these new producers may have been created with the sole purpose of effecting such sales.

35 Butler at STS AGM, *Times*, 23 June 1936.

36 CO82/4/6, Campbell note on Siam, reporting on conversation with Pradit, 28 October 1935. However, they may have changed their position since the Siam Chamber of Mines passed a resolution supporting renewal and the Chairman lamented: 'Unfortunately, the Siamese Chamber carries no weight with the Government which does not wish to co-operate with it.' *Times*, 16 June 1936.

37 Official Siamese statement, *Bangkok Times*, 14 March 1936.

38 *Sri Krung*, 21 December 1935. It was repeated a few days later by representatives of the Chinese miners in a meeting with the Siamese Department of Mines, CO852/4/6, Boyd to Thomas, 26 December 1935. The Dutch accepted the figure of 18,600 tons and recognized that this would make it extraordinarily difficult to get anything below 18,000 as the standard tonnage, MKGA469/C5, telephone conversation with van Buuren, 22 February 1936.

39 CO852/33/6, Lowinger to Thomas, 9 January 1936.

40 CO852/33/6, Thomas to Lowinger, 21 December 1935.

41 CO852/33/6, Boyd to Thomas, 26 December 1935.

42 CO852/33/6, Thomas to Lowinger, 21 December 1935.

43 Ibid.

44 *Sri Krung*, 21 December 1935. A less irresponsible version was published on 25 December.

45 CO852/33/6, Siamese Legation to Campbell, 23 March 1936; MKGA470/V6 cable to Governor-General, 10 March 1936.

46 A similar figure was offered as representing Siam's capacity by one of the few Western commentators to express any sympathy with her bargaining position, Virginia Thompson, 'Siam: Manoeuvering towards Self-Sufficiency', *Far Eastern Survey,* vol. 7, no. 35, 21 December 1938.

47 CO82/33/6, Crosby to FO, 5, 6 March. Baddeley commented: 'Crosby not a fit person to discuss matters with the Siamese.' 10 March 1936. He was particularly upset that an official of the British government would stoop to such negotiating tactics.

48 Crosby was considered to be an intimate friend of the Thai elite, especially Pradit, Aldrich, op. cit., pp. 153, 173.

49 CO852/33/6, FO to Calder, 31 March 1936.

50 *Financial News*, 28 July 1936.

51 CO852/33/7, Campbell to Calder, 30 October 1936.

52 However, it was still 16 per cent over its value prior to the devaluation of sterling in 1931.

53 MKGA476/O16, van Buuren memo on draft agreement, 22 June 1936.

54 CO852/33/6, Lowinger memo, 27 July 1936.

55 AFP28, Bolivian Delegation to PME, 31 July 1936.

56 CO852/33/3, as reported to the 54th ITC Meeting, 21 October 1936.

57 CO852/33/7, Campbell to Thomas, 13 October, reply, 14 October 1936.

58 The Dutch position was reinforced by the press, which considered a concession to Siam at the expense of Bolivia as 'a sort of premium paid for an obstinate claim for special advantage – something one frequently sees in the case of other cartels'. Rotterdamsche Bankvereenigung, *Monthly Review,* August–September 1936, p. 219.

59 Bank of England, London (BE) G1/245, Bunbury to Norman, 12 November 1936. This may have been the result of pressure from the British government, Bunbury note, 28 October 1936.

60 CO852/33/7, Campbell to Calder, 30 October 1936.

61 In fact, the true assay values over the life of the third agreement ranged between 72.3 per cent and 72.9 per cent, which reduced the extent of these concessions.

62 Pradit was prepared to resign if the agreement was rejected, MKGA483/K28, Dutch Legation (Bangkok) to Ministry of Foreign Affairs (The Hague), 28 October 1936.

63 CO852/33/7, Campbell to Calder, 30 October 1936. While this was written in the darkest hour towards the end of negotiations, Campbell had long come to this conclusion, CO852/4/6, minute, 25 March 1936.

64 MKGA486/T, Aanteekeningen ... van de 56ste vergadering van het ITC, 11 December 1936.

65 CO852/72/13, Minutes of the 57th ITC Meeting 5 January 1937.

66 CO852/33/7, Crosby to FO, 19, 29 November 1936.

67 AIM2, Mair letter, 30 October, in Patiño to Ministerio de Minas (MM), November 21, 1936.

68 AIM2, Patiño to MM, 21 November 1936.

69 AIM2, Patiño to MRE, 5, 22 October 1936. The Dutch records provide some evidence of Patiño's success since they considered that Bolivia might even emerge strengthened from a period of open competition, MKGA483/K28, memo on negotiations for renewal.

70 AIM2, MM cable to Bolivian Delegation, 7 September, Patiño to MM, 5 October 1936.

71 MKGA483/K28, Dutch Legation (Bangkok) to MFA, 28 October 1936.

72 CO967/8, Groothoof to Campbell, 10 November 1936. Crosby defended himself against the most serious allegation but left the impression that he could have done much more to secure what was, after all, a major policy objective of his government. The Foreign Office file on this incident has 'not been selected for preservation'.

73 Cf. William Fox, *Tin. The Working of a Commodity Agreement*, London: Mining Journal, 1974, p. 171 and Yip Yat Hoong, *The Development of the Tin Mining Industry in Malaya*, Kuala Lumpur: University of Malaya Press, 1966, p. 238, who considered that the ITC 'had no alternative but to give in'.

74 Once the agreement had been renewed, the Chamber invited all Siamese miners to become members and moved its headquarters to Bangkok, *Straits Times*, 30 August 1937

75 One factor that may have stiffened the attitude towards the ITC was the determination to re-establish absolute sovereignty by abrogating the treaties which granted extra-territorial judicial

privileges. That was also accomplished in 1936, Vichitvong Na Pombhejara, *Pridi Banomyong and the making of Thailand's modern history*, Bangkok, 2001, Chapter 9.

76 *Financial News*, 4 June, 28 July 1936.

77 CO852/33/7, Campbell to Boyd, 28 October, Campbell minute, 27 November 1936.

78 Campbell himself concluded that the politics would make it 'difficult, probably impossible to renew'. POWE22/89, Campbell to Nott-Bower, 3 July 1937.

79 United Kingdom, Colonial Office, *Papers Relating to the International Tin Control Scheme*, London: HMSO, 1938, Cmd. 5879, Article 13. Members protected by minimum tonnages could not vote on quota changes unless they were directly affected by them.

80 CO852/33/7, Campbell to Crosby, 23 November 1936.

81 CO852/72/13, Campbell minute, 10 May 1937. Cf. Yip, op. cit, pp. 239–40, who indulges in a groundless speculation on the politics of setting quotas based on the assumption that they are determined primarily by cost considerations.

82 The NEI government remained anxious to ensure that Banka's special position was protected by determining the scope of any position in advance, MKGA476/O16, van Buuren memo, 22 June 1936. However, no pre-agreed position could be sustained unless coalitions could be formed with other delegations.

83 G1/248, Campbell to Bunbury, 17 March 1937.

84 *British Malaya,* February 1936.

85 Knorr, op. cit., p. 156. This was over the objection to the principle of consumer representation expressed by France, CO852/33/6, memo, 4 February 1936.

86 The Belgian Congo was exempted from this provision. Shipping difficulties tended to concentrate exports in December and the exemption was consistent with the principle of continued expansion. The Congo voluntarily limited this right to 8.33 per cent of its standard tonnage and little use was actually made of the privilege. CO852/72/13, Gutt to Campbell, 11 December 1936, Appendix III to Minutes of the 57th ITC Meeting.

87 CO852/73/5, FO to CO, 2 June, CO to FO, 16 June 1937.

88 CO852/73/5, Calder minute, 8 June 1937. Such flexibility would be needed to determine the level of quotas for 1938.

89 CO967/3, Maffey report on conversation with Lowinger, 13 September 1935.

90 CO967/3, Thomas to CO, 20 July 1935.

91 CO967/7, Thomas to CO, 11 December 1936.

92 CO967/7, minute, 12 December 1936. Cf. CO967/3, Gent minute, 1 July 1935. Even in the more relaxed atmosphere of the first ITC meeting under the new agreement, on being formally nominated for the Chairmanship, Campbell quipped that he was 'most appreciative but not the least grateful'. CO852/72/13, Minutes of the 57th ITC Meeting, January 5, 1937.

93 CO967/7, CO to Thomas, 17 December 1936.

94 Ibid.

95 CO852/33/10, Calder minute, 11 November 1936.

96 Ibid.

97 This left the Nigerian delegation without anyone 'who is actually acquainted with conditions in the country'. Lyttleton was 'more reasonably considered to belong to the Malayan delegation because his interests are overwhelmingly Malayan'. SNP14034, CIM to SNP, 5 November 1938.

12 Riding the commodity roller-coaster, 1937–1939

1 'The Recovery in Commodities', *Westminster Bank Review*, no. 275, January 1937; Charles Kindleberger, *The World in Depression, 1929–1939*, Berkeley, CA: University of California Press, 1986, pp. 262–3. Demand for tin was particularly high in the USSR which increased its purchases from 9,664 tons in 1936 to 25,125 tons in 1937. Germany also showed a significant increase, P. Hövig, 'De Tinpositie', *ESB*, 27 April 1938. Other factors which contributed to an extraordinarily unstable market include: the commencement of the Sino-Japanese war, intensification of the conflict between Arabs and Jews in Palestine and a gold scare in the United States, Lyttleton at BTIC AGM, *Times,* 2 February 1938.

2 Strauss Tin Review, *British Malaya*, March 1937. The daily average fluctuation in spot tin on the LME over the period from January 1 to April 30 was 1.62 per cent, for copper 2.42 per cent; the ratio between the two metals is the same in the forward markets. The maximum price for spot tin reached in this period was 35 per cent over the price on January 1, whereas for copper it was 58

per cent. Fluctuations were slightly higher in tin for March as a result of the rapid change in the quota.

3 Strauss Tin Review, *British Malaya*, April 1937.

4 *AMM,* 18 February 1937; Strauss Tin Review, *British Malaya*, January 1937. Cables protesting against the prospect of such a cut were received from Wheeling Steel and Cleveland Babbitt Metals, CO852/73/6, 24 February 1937. These are among the very few direct communications between consumers and the ITC.

5 CO852/73/6, Trench to FMS, 20 February 1937.

6 Campbell was anxious lest the Conference end by denouncing control schemes, especially since the USA was 'peeved' about both tin and rubber, POWE22/89, Campbell to Nott-Bower, 3 March 1937.

7 CO852/72/13, Minutes of the 58th ITC Meeting, 5 March 1937.

8 The ITC communique noted that current reduction in stock levels was simply a function of Chinese New Year and that the 100 per cent quota would mean a monthly increase of stocks by 1,600 tons. It also stressed that the level of minehead stocks could be increased under the new agreement, *Times*, 6 March 1937.

9 However, at the same time an unlikely source was complimenting the ITC on its ability to control the market, Strauss Tin Review, *British Malaya,* April 1937. This must have been written before the turbulence that followed the 100 per cent decision.

10 *Wall Street Journal*, 13 March 1937.

11 G1/248, Campbell to Bunbury, 17 March 1937.

12 CO852/72/13, Minutes of the 59th ITC Meeting, 18 March 1937.

13 Just before the 6 March meeting, Strauss commented that a purely statistical assessment could lead the ITC to actually cut the quota, *Times*, 5 March 1937.

14 *AMM*, editorial, 'American Tin Consumers Taken for a Ride', 10 March 1937.

15 *AMM,* editorial, 'Ineffective Tin Control', 25 March 1937. The harshness of this criticism should have been muted in light of an examination of the scope of the speculation in the major non-ferrous metals. This was measured by comparing the turnover in the forward market for the first quarter of 1937 with that for the same period in 1936. Tin increased by 1.88, copper by 3.88 and lead by three times, *AMM*, 2 April 1937.

16 AIM2, Patiño to MM, 26 June 1937.

17 CO852/72/13, Minutes of the 60th ITC Meeting, 11 June 1937. Malaya was very fortunate that the 125 per cent quota was not adopted since she was not in a position to produce it. Uncertainty as to when and how any shortfall would be made up was one source of market instability. The revelation that Malaya could not produce even at 90 per cent (the domestic equivalent of an international rate of 125 per cent) of capacity would have seriously undermined her bargaining position in the subsequent revision of the standard tonnages.

18 CO852/72/13, Minutes of the 61st ITC Meeting, 9 September 1937. Firth was the only one to argue for a higher quota but it was really quite impracticable.

19 CO852/72/14, Minutes of the 62nd ITC Meeting, 25 October 1937. When Firth had to resign from the ITC on being forced out of Richard Thomas, Campbell sent a letter of regret and referred to incidents such as this as 'free and frank expositions of the situation' and expressed his appreciation of 'your vigour in expounding your opinions'. SDCDF800.6354, Campbell to Firth, 6 June. Firth replied: 'on no occasion was there the slightest sign of resentment or ill will, and it is the recollection of this happy atmosphere that increased my regret that circumstances make it necessary for me to resign.' 21 June 1940.

20 *Metallwirtschaft*, 29 October 1937.

21 CO852/72/14, Malayan Chamber of Mines to CO, 16 November 1937.

22 Some consolation could be taken from the fact that approximately half of the total increase was immobilized in the USSR.

23 CO852/139/4, Minutes of the 65th ITC Meeting, 11 June 1938, Appendix VIIIA.

24 The problem revealed in the position of French Indo-China could be discounted. Her entitlement had been reduced on the renewal of the agreement and she had not engaged in any tough bargaining to secure it.

25 Such a warning was issued by Leadenhall, 'Malaya's Case for Higher Tin Assessment', *Financial Times*, 29 March 1938. The author also suggested a new distribution of the standard tonnages which would have granted Malaya only 3 per cent more than she received in the actual redistribution.

26 Arthur Miles letter to *Straits Times*, May 12, 1938.

27 CO852/72/14, Lowinger to Thomas, 17 January 1938.

28 CO852/72/14, Clauson minute, 31 December 1937. Malaya was not informed of this overriding constraint on her freedom of action.
29 AIM12, Patiño to MM, 9 December 1937.
30 *Tin*, February 1938, provides the only intelligible account of the formula but without any explanation as to the thinking behind it. It involved several stages. The members were divided into three categories: those who performed well in 1937, Malaya, NEI and Nigeria; Siam; those who performed badly. For obvious reasons, Siam was to be left alone. Those who performed well were granted an increase of 13.7 per cent on their existing standards, around half of which was met by various reductions from all the others but primarily from Bolivia.
31 Perhaps in order to draw attention away from the change in the standard tonnages, the new quota was presented as a special 80 per cent for Malaya on the existing standard rather than 70 per cent on the modified one. The FMS Department of Mines presented the results in a very misleading fashion in its *Bulletin of Statistics*. The first quarter period is presented as though it were 70 per cent on the old standard whereas it was 70 per cent on the new and the second quarter period is presented as 55 per cent plus a special 7.56 per cent, both on the old. While this is equivalent to 55 per cent on the new, the juxtaposition of the figures for the two quarters, which have quite different referents, gives the impression that Malaya only had a special benefit for the second quarter. The table is reproduced in Yip Yat Hoong, *The Development of the Tin Mining Industry in Malaya*, Kuala Lumpur: University of Malaya Press, 1966, p. 252.
32 AIM12, Patiño to MM, 31 December 1937. The Malayan delegation proposed 1937 performance as the basis for the allocation. Patiño calculated that Malaya had benefited to the extent of about £4 million as a result of the total underperformance.
33 CO852/72/14, Minutes of the 63rd ITC Meeting, 10 December 1937, Appendix IV.
34 As soon as the proposal was floated, the *Straits Times* ran a strong critical editorial, 23 December 1937.
35 CO852/139/4, Minutes of the 64th ITC Meeting, 18 February 1938. Campbell made a serious mistake in the way in which he conceded the demand of the press for an oral version of the communiqué which included both the quota rate and the details of the internal transfers. It resulted in considerable misinterpretation, confusion and criticism of the ITC itself, *Times*, 22, 26 February 1938.
36 AIM12, Patiño to MM, 2 March 1938.
37 Several versions of a buffer stock scheme were proposed, ranging from a tight comprehensive selling agreement to establish a uniform price, as favoured by Gutt, to a loose reserve stock of metal held at the smelters, as favoured by Lyttleton. Lyttleton's proposal addressed the fact that it was much easier to turn off the production tap than to turn it back on but the main advantage was political. Producers would not be asked to trust the market judgement of a professional manager. The actual result was something of a compromise between these two extremes. The various drafts are in CO852/73/3.
38 For a fuller discussion of the administrative and legal issues involved in the operation of this and the preceding buffer stock scheme, see Kabir-ur-Rahman Kahn, 'The Development of the Tin Buffer Stock System: A Diplomatic and Organisational Study of the pre-1945 Phase', *Indian Journal of International Law*, vol. 17, 1977.
39 It is also possible that there was an informal understanding with the LME that market operations would be conducted only to protect the floor and ceiling. Khan suggests that publishing the limits established a 'quasi-right' which could be invoked by third parties if there were 'any digression from that criterion'. ibid., p. 312.
40 League of Nations, *Report of the Committee for the Study of the Problem of Raw Materials*, Geneva, 1937, p. 18.
41 Ibid., p. 19. This paragraph was quoted in the Memorandum on the Advantages of a Buffer Pool, distributed by the FMS Department of Mines, 14 May 1938, TRUS10/38. Annex II to the League's report developed the arguments for stock control at some length. While the report is couched in authoritative, objective prose, one of the authors was van Gelderen of the Ministerie van Koloniën, which, of course, was heavily committed to these schemes. The model advocated was an interesting, although impractical, one: joint producer/consumer control and finance by consumers. Clauson was also on the committee and offered to shape the final report in this direction in order to get Malaya to swallow the scheme but was officially discouraged from such manipulation, Shuckburgh minute, 14 June 1937, CO852/73/3.
42 *Straits Times*, 12, 11 May 1938.
43 CO852/139/6, Campbell to TPA, 24 February 1938.

44 CO852/139/6, Nigerian Chamber of Mines, Resolution, 8 March 1938. The Chamber even suggested using the buffer stock tin as collateral for a loan with which to buy metal.
45 CO852/139/6, Nigeria to CO, 4 April 1938.
46 This prompted Wollnik to provide a generally sympathetic account, stressing that the ITC would learn from the earlier mistakes, 'Der Kampf um dem Zinn-Pufferpool', *Frankfurter Zeitung*, 28 December 1937.
47 CO852/139/6, Malayan Chamber of Mines to Thomas, 9 March 1938. A group composed of Lyttleton, Pawle and Stephens expressed their opposition to the official position. Their absence from this critical meeting had allowed the minority to get its way. They wanted the issue of revision uncoupled from that of the buffer stock, saw the latter as 'vital', and did its best to expose the flaws in the arguments supporting the official position. Lyttleton, *et al.* to Thomas, 19 March 1938.
48 CO852/139/6, FMS Chamber of Mines to Thomas, 28 March 1938.
49 CO852/139/6, letter from unidentified miner to Thomas in Thomas to CO, 24 April 1938. Emphasis in the original.
50 *Straits Times,* 11 May 1938. Bagnall was severely criticized for breaking the neutrality expected of a smelter by Elibank, *Financial Times,* 4 April 1938.
51 CO852/139/6, Clauson minute, 14 May 1938.
52 CO852/139/6, Campbell minute, 13 May 1938. Dutch resentment was compounded by the fact that, in spite of repeated invitations to leading Malayan miners, only Glenister and an official from the Mines Department actually visited Banka or Billiton. Glenister proved to be very suspicious and did not make a good impression.
53 *Straits Times,* 3 March 1938.
54 FMS Chamber of Mines, *Yearbook for 1938*, p. 79. This represented about 98 per cent of the eligible votes. Yip supposes that some of the result can be attributed to the decline in antagonism towards the ITC as a result of the revision of the standard tonnages, p. 246. This did not in fact occur until a month later. Nor was the result affected by any fear of the Dutch invoking clause 22 which was not known outside the ITC.
55 CO852/139/6, Thomas to CO, 24 May 1938. The conditions were: revision of the standard tonnages and acceptability by the American consumers. Both Todd and Firth supported the principle of the buffer stock, which forces some qualification to Knorr's assertion that consumers were opposed, *Tin Under Control,* Stanford, CA: Food Research Institute, 1945. p. 163. Strong support for the principle of the buffer stock came from Leadenhall, 'Merits and Defects of the Tin Buffer Pool Scheme', *Financial Times,* 28 April 1938.
56 CO852/139/6, MacDonald to Thomas, 30 May; CO852/140/1, Campbell to Gent, 8 June 1938.
57 CO852/139/4, Minutes of the 66th ITC Meeting, 20 June 1938. The basis for this is unclear, especially since the current price was determined by heavy accumulated private stocks. It was not shared by one of the most thoughtful Dutch commentators on tin policy who considered a normal level to be within the range of £220–£230, P. Hövig, op. cit. Groothoof later defended this range against an attack made in the official publication, *Economische Weekkroniek,* stressing its appropriateness the light of current circumstances, 12 July 1939, ARA, 3850/21.7.39/2. At the same time, both Calder and Campbell felt that a range of £190–£210 would be more defensible, CO198/8, minutes, 15, 16 March 1939.
58 Given the fiscal constraints to which they were subject, Bolivian companies may have needed the higher range in order to justify financing the buffer stock.
59 *Straits Times,* 6 August 1938.
60 CO852/275/9, Campbell to Clauson, 20 July 1940. Under normal circumstances Ellinger could have commanded a salary of £10,000 but had to settle for £1,400.
61 The scheme was endorsed by the League of Nations Economic Committee in its Report to the Council, 48th session, T188/20. Tin was the only case in which a buffer stock was introduced into a prewar commodity agreement.
62 Tin Buffer Stock Scheme, 1938/41, *Report and Accounts,* London, 1943, p. 3. Over the 15 months during which it operated on the LME, the stock was turned over more than fourfold.
63 William Fox, *Tin. The Working of a Commodity Agreement*, London: Mining Journal, 1974, p. 180. The intervention took the form of a letter to the British government, 17 June, Appendix II to the Minutes of the 66th ITC Meeting, 20 June 1938, CO852/139/4.
64 SDCDF800.6354, memo on meeting between Veatch and Ruimschotel (NEI), 22 April 1938.
65 Had the administration been freed from Congressional budgetary constraints, it should have followed the example of Germany and the USSR and established its own stockpile. The ITC

was subject to severe criticism for reducing production in 1938 in the face of the increasing likelihood of war which deflected attention from the real target which was the weakness of the American political system. That may explain the extent of the severity of the criticism which will be considered in Chapter 16.

66 SDCDF800.6354, Kennedy to State Department, 29 April 1938. The £200 was considered unreasonable in any circumstances, not just those prevailing at that moment.

67 SDCDF800.6354, Feis memo, 3 June 1938. There was no response to Kennedy's request for substantiating evidence.

68 In his regular reports to the Department, the American Consul in Singapore consistently repeated the oppositional rhetoric that had become so pronounced during 1938. His very limited grasp of the way in which the industry operated allowed him to claim that the level of dividends paid by STC on its reduced nominal capital 'confirms the exorbitant price American consumers have had to pay'. SDCDF800.6354, McEnelly to State Department, 21 March 1938.

69 SDCDF800.6354, ITRDC memo, 19 July 1938.

70 SDCDF800.6354, response to ITRDC, August 1938.

71 Ibid.

72 SDCDF800.6354, draft for presentation to UK government, 9 February 1939. This was formally communicated in Kennedy to FO, 7 March, Gordon to Dutch Foreign Ministry, 7 March 1939, FO371/23979. A similar observation was made by Lotte Müller-Ohlson, *Non-Ferrous Metals. Their Role in Industrial Development*, Cambridge: Woodhead, 1981, p. 231. This treated the price at the point of formation of the buffer stock as a norm, rather than as an unacceptable low.

73 FO371/23968/9966, Ashton-Gwatkin minute, 28 June 1939. The FO was not at all pleased at Campbell's refusal to provide any reassurances for Kennedy, 'Sir J. Campbell is always uncompromising and not always right. That is one of his difficulties.'

74 CO852/198/8, Minutes of the 70th ITC Meeting, 27 April 1939, Memorandum on US tin complaints, Appendix I.

75 There is no evidence from the State Department records that there was any consultation with other government agencies, nor any familiarity with the detailed grasp of the complexities of the tin market being published by Trench in *AMM.*

76 Patricia Glover, 'Low Malayan Tin Quota Leads to Chinese Unemployment', *Far Eastern Survey*, vol. 8, no. 13, June 21, 1939. Many were left quite destitute but most were absorbed on public works and in squatter-farming, Francis Loh Kok Wah, *Beyond the Tin Mines. Coolies, Squatters and New Villages in the Kinta Valley, Malaysia, c.1880–1980,* Singapore: Oxford University Press, 1988, p. 37.

77 CO852/139/4, Clauson minute, 18 March 1938.

78 CO852/139/4, Minutes of the 65th ITC Meeting, 2 June 1938.

79 CO852/139/4, Glenister to MacDonald, 4 June 1938.

80 CO852/139/4, Clauson draft of letter from MacDonald to Thomas, June 1938.

81 CO852/143/7, Campbell minute, 6 May 1938, cited in John Drabble, *Malayan Rubber: The Interwar Years*, London: Macmillan, 1991, p. 192.

82 CO852/199/1, Clauson minute, 31 January 1939.

83 CO852/140/1, Lyttleton to Campbell, 7 June 1938.

84 CO852/140/1, CO to Thomas, October 1938. Such attributes do not seem to have been recognized in Malaya, where the government refused even to reply to his various communications, Campbell to Gent, 8 June 1938. The loss of Lowinger would have created a problem for Campbell since he could then no longer rely on an alliance with him to control meetings of the delegation. He would then have to reveal the fact that he held the delegation's vote and could overrule any internal opposition which would generate further antagonism.

85 CO852/140/1, CO to Thomas, 6 July 1938.

86 CO852/140/1, Campbell to Gent, 8 June 1938, Campbell minute, 7 September 1938.

87 CO852/140/1, Thomas to CO, 27 March 1939. Thomas dutifully laid out the CO position in his explanation to Ward, 8 December 1938, CO852/199/1.

88 CO852/1991/1, Campbell minute, 12 April 1939. One ideal candidate would have been Elibank who was always supportive of the principle of restriction but equally prepared to be critical of specific decisions.

89 CO852/140/1, Thomas to Campbell, 15 August 1938.

90 CO852/140/1, Thomas to CO, 15 August 1938.

91 Campbell considered Thomas 'in the pocket of Sir John Bagnall', and the *Straits Times* under his control, CO852/140/1, Campbell to Gent, 8 June 1938. In any case, what needed further

exploration were the implications of the dual role played by Thomas. As Governor of the Straits Settlements and living in Singapore, he operated within a culture firmly committed to free trade, one which was well expressed in the *Straits Times*. Setting this aside would not have been easy when he changed hats and became High Commissioner on leaving for Kuala Lumpur.

92 CO852/139/4, Minutes of the 66th ITC Meeting, 20 June 1938.

93 CO852/139/4, Minutes of the 67th ITC Meeting, 14 September 1938.

94 Tin Buffer Stock Scheme 1938/41, *Report,* op. cit., suggests that was this was entirely responsible for the rise from June to August, p. 3; Strauss considered that it was very minor, *British Malaya*, July 1938.

95 CO852/139/5, Minutes of the 68th ITC Meeting, 22 November 1938. This was the only occasion on which a formal vote was taken and it was not on a question of principle but rather of timing. Expanding the size of the buffer stock was made possible by Siam's ratification of the agreement, though she was still uncertain as to whether to participate.

96 Particular pressure came from veterans of the Chaco war, Dionisio Foianini, *Misión Cumplída*, Santa Cruz, 1991, p. 211.

97 A more comprehensive account of this conflict and its implications for the overall position facing the tin industry in Bolivia is John Hillman, 'The Mining Industry and the State: The Politics of Tin Restriction in Bolivia, 1936–1939', *Bulletin of Latin American Research*, vol. 21, no. 1, January 2002.

98 Manuel Contreras and Napoleón Pacheco, *Medio siglo de minería mediana en Bolivia, 1939–1989*, La Paz: Biblioteca Minera Boliviana, 1989, p. 9. The decree was promulgated on 26 April 1939.

99 AIM12, Patiño to MRE, 27 May 1938.

100 AIM12, Patiño to MRE, 25 July 1938.

101 AIM12, Patiño to MRE, MM, President, AIM, Banco Minero, 9 March. The AIM then reinforced these messages in circulars to its members, AIM2, Circulares 1936–1939, 2, 3 March 1939.

102 AIM12, Patiño to MRE, MM, President, AIM, Banco Minero, 23 March 1939. The terms were very generous. On quotas of 40 and 45 per cent, around two thirds of the increase over the current 35 per cent would be set against arrears. At 40 per cent that meant 20 months before liquidation was complete, at 45 per cent 10 months. Only at 50 per cent was a higher proportion required which would assure a rapid rate of liquidation but in March 1939 the prospects of that were remote.

103 Cf. Fox, op. cit., p. 176.

104 The German Minister in La Paz saw Busch as clearly moving towards a totalitarian state along national-socialist lines, León Bieber, 'El comercio germano-boliviano, 1936–1939', in Nikolaus Böttcher and Bernd Hausberger, eds, *Dinero y negocios en la historía de América Latina*, Frankfurt: Vervuert, 2000, pp. 492–3.

105 Ibid., p. 504. As these proposals were being discussed in Berlin, Busch issued two decrees nationalizing both the Banco Minero and the Banco Central. Unfettered government control of both institutions would be an indispensable condition of any such bilateral agreement.

106 Ibid, p. 505. Bolivia would have run an enormous risk in such a commercial agreement, since transactions would have been in the accounting currency of ASKImarks, with no assurance that the imports would otherwise fairly match the exports. Foianini claimed that only the outbreak of war prevented the agreement, op. cit., pp. 171–2.

107 CO852/199/4, Rawlins (UK Legation, La Paz) to FO, 24 July, FO to CO, 31 July, Campbell to Calder, 4 August 1939. The Treasury attached conditions which would have protected two British ore-dealing firms against Busch's plans to eliminate them.

108 Gosálvez was considered an 'important levelling influence on the often mercurial character and mentality of the president'. Herbert Klein, *Parties and Political Change in Bolivia, 1880–1952* Cambridge: Cambridge University Press, 1969, p. 304. Such an influence was sorely needed if Busch were to shape a coherent economic policy.

109 AFP13, Arce to Simón Patiño, 24 August 1939.

110 Decree of 1 October 1939 in Banco Minero de Bolivia, *Tasas e impuestos sobre la industría minera en Bolivia* La Paz, 1941, p. 141.

111 Some indication of the seriousness with which the new government was prepared to tackle the problem of the small miners is given in the quota distribution for the third quarter of 1939. When the international quota was first raised to 60 per cent following the outbreak of war and made retroactive, the Banco Minero was allotted a monthly quota for the small miners of 196 tonnes or

7.5 per cent of the total; under all earlier distributions they had never received less than 9 per cent, AIM2, Dirección General de Minas y Petroleo to AIM, 9 September 1939.

112 SDCDF841.6354, Johnson (US Embassy, London) to State Department, 26 January 1939.

113 Robert Gibson-Jarvie, *The London Metal Exchange*, London: Woodhead-Faulkner, 1976, p. 59.

114 Strauss Tin Review, *British Malaya,* November 1938.

115 Given the level of invisible stocks, this task would not have been any easier had the ceiling been set at a lower level.

116 SDCDF841.6354, Johnson to State Department, 26 January 1939.

117 SDCDF800.6354, State Department to Kennedy, 2 March 1939.

118 SDCDF800.6354, Trench to Campbell, 3 March 1939.

119 Strauss Tin Reviews, *British Malaya*, February, March, April 1939. The VYB market report estimated that consumption in 1939 would be even lower than in 1938, which argued for keeping the quota low, *Mining Journal,* 18 March 1939. The *Times*' City Editor also argued for 35 per cent to allow the buffer stock to realize the cash needed to protect the floor, 20 March 1939.

120 SDCDF800.6354, memo of conversation between Kennedy, Leith-Ross (FO), Campbell, Clauson and Calder, 17 March 1939.

121 This observation was later used to support the allegation that Campbell regarded the ITC as solely a 'producers' show'. NARAII, RG60, Department of Justice files, Anti-Trust Division (ATD), subject file, Tin, Report on Tin Industry, note 141. Unfortunately, only the footnotes to a very comprehensive review have survived. This charge was unfair in light of three practical considerations: (1) votes were not the basis of decision-making; (2) Stalin was unlikely to be forthcoming about his plans; (3) meetings of the ITC were already very large for the kind of free discussion that allowed a consensus to emerge.

122 CO852/140/1, Leith-Ross to CO, 19 July 1941. This was based on the experience of Professor Noel Hall who had no connection with either of the two main offenders, tin and rubber.

123 CO852/198/8, Minutes of the 69th ITC Meeting, 22 March 1939.

124 Fox, op. cit., p. 183.

125 CO852/140/1, Minutes of the 71st ITC Meeting, 14 June 1939. Todd wanted to raise it to 60 per cent.

13 Development under restriction: the producers

1 Whether that is to the advantage of the participants depends on many factors, especially comparative depletion rates and the extent to which higher gross incomes are offset by higher costs on a smaller production.

2 *Financial News*, 15 June 1933.

3 Calculated from production data in J. B. Were, *Tin Shares as an Investment,* Melbourne, 1938 and *WCM.*

4 BT31/33632/307645, Burmese Hydraulic Tin Ltd. prospectus.

5 *Mineral Industry for 1934,* p. 567. Kwangsi's new capacity was at first estimated to be as high as 20,000 tons annually, *Minerals Yearbook, 1935*, p. 518. Further details are provided in FO371/18153, Wilson-Brand report, 1 June 1934.

6 *Minerals Yearbook, 1939*, p. 691.

7 Dorothy Borg, *The United States and the Far Eastern Crisis of 1933–1938*, Cambridge: Harvard University Press, 1964, pp. 259–60. The head of the American Economic Mission was very anxious to see such a development in order to secure supplies for strategic reasons; some interest was elicited from United States Steel.

8 E. M. de Villa, *The Study of Mines in China, Indochina and Malaya*, Hong Kong, 1935, p. 162.

9 A. J. H. Latham, *The Depression and the Developing World*, London: Croom Helm, 1981, p. 63 China remained on silver until 1935 and tin producers suffered as a result of the rise in the price of silver from 12d/oz to 36d/oz from 1931 to 1935, *British Malaya,* May 1935. The segmentation of the market meant that Chinese producers captured only 55 per cent of the rise in international prices from their low of 1930, calculated from Hsiao Liang-li, *China's Foreign Trade Statistics, 1884–1949*, Cambridge, MA: Harvard University Press, 1974, Tables 3, 9a.

10 Children around the ages of 11–12 constituted the bulk of the labour force. Those who deserted were shot and the rest experienced a mortality rate of 40–50 per cent, FO371/22151, Harding report, 15 October 1934. Fuller descriptions of the abominable conditions under which the children lived and died are in Nicol Smith, *Burma Road*, Garden City, 1942, pp. 160–9 and 'League Expert

Exposes Yunnan Tin Mine Conditions', *Far Eastern Survey*, vol. 6, no. 18, September 1937, pp. 208–9.

11 FDR Box 11, Memorandum on tin in China, 22 January 1945.

12 Smith was told of some with annual incomes of US$250,000, op. cit., p. 165.

13 *Mineral Industry for 1938,* p. 600. This may have been as much for patriotic as for economic reasons.

14 CO852/73/2, Lyttleton to Export Credit Guarantee Department, 21 December 1937.

15 Knuckey, Lyttleton's representative in Yunnan, confirmed that Miao, Economic Council of Yunnan, was committed to production control, Consul-General (Yunnanfu) to FO, 14 September 1937, IORL/M/3/376.

16 CO852/73/2, Campbell to Lyttleton, 30 December 1937.

17 CO852/198/11, Strathallan memo, 21 March 1939. The project foundered on the reluctance of the Export Credit Guarantee Advisory Committee to provide the necessary financial guarantees, NAK, ECG 1/18–19.

18 FO371/22151, Harding to British Embassy (Peking), 7 November 1935. Profits are reported at a rate of 100 per cent but without any indication of the capital invested or whether they are gross or net. A more reliable indicator is the cost of production of actual metal, which is recorded at the equivalent of £167/ton, *Bulletin d'études géologiques et minières coloniales*, 15 July 1939.

19 *Chronique des Mines Coloniales*, May 1936.

20 These can be calculated at £61 per ton of metal content, before allowing for £11 for Japanese overheads, *Tin*, April 1936, p. 26. 1938 costs show a rise of 20 per cent over 1935 and with lower prices, profits were reduced to £21/ton, Annual Report for 1938 in BT31/32868/225182.

21 US Department of the Interior, *Tin Deposits of the Republic of Mexico*, Geological Survey Bulletin 935–C, Washington, DC: GPO, 1942, p. 105.

22 *IAMS,* 2 February 1936.

23 For example, Endurance Tin Mining treated 113,730 cubic yards in 1935–6 from which it produced 80 tons of concentrates. When other sources are added, it sold concentrates containing 65 tons of metal on which it made a profit of £12 per ton during a period when the average international price was around £210, *IAMS*, 1 September 1936. These new marginal ventures included a property at St Helens in Tasmania which was operated by STS from 1934 to 1939.

24 Sue Harlow, *Tin Gods. A Social History of the Men and Women of Maranboy, 1913–1962*, Darwin: Historical Society of the Northern Territory, 1997, p. 26; Helen Brown, *Tin at Tingha*, Armidale: Brown, 1982, p. 54.

25 'Tin Dredging in North Queensland', *Chemical Engineering and Mining Review*, 15 March 1938, *WCM*, 1 May 1937. Anglo-Oriental also explored the prospect of returning to Australia, since it started boring at Bynoe Harbour and Darwin, *IAMS*, 15 July 1937.

26 Senator J. B. Hayes in a question to the Minister of Trade, 12 May 1938, Commonwealth of Australia, *Parliamentary Debates*, 1938, p. 1,125. The response was negative and such a policy would have been difficult to implement in view of the fact that around a third of Australian production was exported.

27 GH22429/5-7, Camp Bird Minute Books.

28 Ministério da Economia, *Elementos para o estudo da Indústria Mineira em Portugal nos anos de 1930 a 1939*, Lisbon, 1941, pp. 88–9.

29 *Mineral Industry for 1935*, p. 581.

30 Scott cites a production of 2,000 tons and a net profit of £200,000, *Mineral Industry for1938*, p. 599. Since the total Argentine production for 1938 is recorded at 1,111 tons metal content, these figures are grossly exaggerated.

31 For example, *Mineral Industry for 1938*, p. 600. In 1938, the industry attempted to get a government subsidy to tide it over the depression but without success, POWE16/149.

32 Moreing at the East Pool and Agar AGM, 3 May 1934; Anderson at the Geevor AGM, 8 June 1935.

33 Philip Heffer, *East Pool and Agar. A Cornish Mining Legend*, Redruth: Dyllansow Truran, 1985, pp. 51–4.

34 J. A. Buckley, *A History of South Crofty Mine*, Redruth: Dyllansow Truran, n.d., pp. 156–169.

35 *Mining Journal*, 3 March 1938.

36 French Cameroons first started to produce in 1933; by 1934 it reached a total of 225 tons, *Metallwirtschaft*, vol. 16, no. 21, 1937, p. 490.

37 This follows John Hillman, 'Chartered Companies and the Development of the Tin Industry in the Belgian Congo, 1900–1939', *African Economic History*, vol. 25, 1997.

38 The committee was made up of representatives of each of the states and was advisory to the High Commissioner. The only formal authority it had over each of the states derived from the always ambiguous power of the High Commissioner.
39 This was actually proposed but the context within which negotiations for the fourth agreement were conducted rendered it moot.
40 J. C. Pasqual 'The Limestone Caves of Perlis', *Singapore Free Press*, 11 August 1921. Most cases paid for the cost of blasting and left just enough to continue boring to the next pot hole, ANMKL, British Advisor files, Perlis, AP432/1349, Petition, 10 October 1928. It was, therefore, a system generating continuous employment but discontinuous output.
41 AP432/1349, responses from Pasqual, 6 January, the Chinese community, 10 January, De Graf, 11 January 1931.
42 AP432/1359, Laird minute, 4 May 1931.
43 AP343/51, Campbell to Loh Ah Tong, 22 April 1931.
44 AP335/51, High Commissioner to British Advisor, Perlis, 13 August 1932.
45 AP335/51, reply, 27 August 1932.
46 AP335/51, Greig memo, 30 August 1932. Protocol required that discussions between Kuala Lumpur and Kangar follow a circuitous route via Singapore and Alor Star. Details of the formula are in Greig to Venables (British Advisor) 3 January 1933, AP630/135.
47 AP508/1351, Venables to Legal Advisor, Kedah; Campbell to Venables, 23 October 1932.
48 AP382/1353, Venables memo.
49 AP313/52, Venables to High Commissioner, 6 December 1933.
50 AP382/1353, MacDonald memo. Dulang permits were again reissued.
51 AP613/57, British Advisor to Warden of Mines, Johore, 15 January 1939.
52 AP679/57, Day minute, 23 July 1939.
53 AP659/57, Bean to British Advisor, 13 September 1939.
54 ANMKL, Commissioner of Lands and Mines files, Trengganu, CLM45/1940, de Mowbray memo, 21 January 1940.
55 *WCM*, 23 October 1930.
56 CLM264/1350.
57 Bundi attempted to transfer its quota to Pahang but could not agree on the terms of compensation to Trengganu, CLM177/52, CLM to British Advisor, 21 November 1933.
58 CLM107/37, Harris to McKerron, 27 January 1937.
59 ANMKL, British Advisor files, Trengganu, BAT911/38, Bean to British Advisor, Trengganu, 11 August, BAT170/37, Bean memo, 19 September 1938.
60 Some background is provided by Yuen Choy Leng, 'Japanese Rubber and Iron Investments in Malaya, 1900–1940', *Journal of South-East Asian Studies,* vol. 5, no. 1, March 1974. ISK had previously operated in Johore but was prevented from developing tin, p. 28.
61 Arkib Negara Malaysia, Johor (ANMJ), Mines Office files, Johore, MOJ17/29, minute, 29 January 1929.
62 It was encouraged by a report by E. S. Willbourn, 'The Geology and Mining Industries of Johore', *Journal of the Malaysian Branch of the Royal Asiatic Society,* vol. 6, pt 4, November 1928, p. 28.
63 MOJ17/29 Warden of Mines to Commissioner of Lands and Mines, 2 February 1930.
64 Two hundred acres came from the state and the remainder from a rubber company after some pressure from the state government.
65 ANMJ, General Advisor, Johore GA/30, Marshall (Warden of Mines) to State Secretary, Johore, 10 October 1931.
66 MOJ168/31, Winstedt to Pretty, 6 August 1932.
67 GA625/33, Harris memo, 4 November 1933. These rules specified that no investment made after 1930 would be eligible for an assessment but since Johore formulated its rules several months later, the relevant date was changed to 1931.
68 CO323/1302/5, Campbell minute, 6 April 1934. He added that since the parent company, Sungei Besi, had 'strongarmed the FMS Government over the purchase of their power station, it is a little unreasonable for it to expect the FMS Gvt to allot part of the FMS quota to Johore for its benefit'. The allegation of bribery may explain the lack of interest by a new Warden of Mines in helping Pelepah get this transfer, MOJ62/32, Memo, 8 November 1932.
69 GA268/34, High Commissioner to Colonial Office, 9 May 1934.
70 While Johore had endorsed the first restriction agreement, it had given no formal undertaking in relation to the second, MOJ129/33, draft letter 8 June 1934.
71 Pelepah Tin Dredging, Annual Report for 1939–1940.

72 CO323/1302/4, Singapore to CO, 29 March 1934.
73 Straits Settlements Legislative Council, *Proceedings,* 1934, p. 39.
74 *Administrative Report for the State of Kelantan, 1931,* Kelantan, 1932, p. 8.
75 *Annual Report on ... the State of Kelantan for 1933*, Kelantan, 1934, p. 11; *Annual Report ... for 1936*, p. 19.
76 *Annual Report ... for 1937*, p. 24.
77 Zaharah binti Haji Mahmud, *Change in a Malay Sultanate: A Historical Geography of Kedah Before 1939*, Kuala Lumpur: Oxford University Press, 1966, pp. 214–15; *Administrative Report for the State of Kedah for 1348*, Alor Star, 1929, p. 10.
78 In 1938 Kampong Kamunting produced 77 per cent of the level reached in 1937, whereas in the FMS its authorized production would have been 58 per cent, *WCM*, 5 August 1939.
79 Quotas for tin concentrates from gold mines were formally established and reviewed by the Central Committee prior to the determination of the operating quota for the other categories, BG64/35, Report on Rulings made by the Central Committee, 15 November 1934.
80 BG62/33. In accordance with a policy of reducing the number of dulang washers every application for transfer was to be refused 'whatever the circumstances'. That did not prevent the issuing of a few new passes, BG33/34. By 1936 the rules were relaxed and transfer of passes to family members had become routine, BG35/36.
81 In 1934 there were 1,638 passes in force in Batu Gajah of which 38 were held by Malays and 10 by Tamils, BG33/34.
82 A. Azmi Abdul Khalid, 'The Social Organization of the Mining Industry during the Depression, 1929–1933, in Malaya', *Journal of the Malaysian Branch of the Royal Asiatic Society,* vol. 65, pt 2, December 1992, p. 97. The cited rate of S$3–S$5 per month is around half of the amounts paid for legal transfers between regular miners.
83 BG137/32, Greig memo, 19 July 1932.
84 BG137/32, Laird minute, 8 August 1932.
85 BG137/32, Greig minute, 22 August 1932. The Chamber of Mines was only able to express its opinion once the ruling was made.
86 However, applicants often found that their claims were routinely denied on the grounds that the land should be reserved for agricultural and other purposes, 'Prospecting for Tin in Malaya', *Chemical Engineering and Mining Review,* 11 July 1938. Cf. Yip Yat Hoong, *The Development of the Tin Mining Industry in Malaya*, Kuala Lumpur: University of Malaya Press, 1966, p. 388.
87 The figure of £150 as normal with no restriction is taken from a memo from Harris, Acting Warden of Mines, Perak, 1 October 1932, BG219/32.
88 *British Malaya*, July 1937, p. 8.
89 In Batu Gajah there were 36 miners with one-pikul assessments in 1937, 25 per cent of all Chinese gravel pump mines and together they produced 1,287 tons of tin concentrates, BG168/38, memo, 21 April 1938.
90 CO852/198/11, Memorial from Perak Miners' Association.
91 CO852/198/11, Campbell minute, 29 September 1939, Thomas to CO, 30 August 1939.
92 Campbell endorsed that position, fearing that 'any radical change' would jeopardize the industry's support for renewal of the agreement, CO852/198/11, minute, 17 October 1939.
93 BG51/39. For the first two quarters of 1937 the balance was reversed but from July 1937 to February 1939, the balance is consistently towards the Chinese. Cf. Yip, op. cit., p. 260.
94 FMS, Department of Mines, *Monthly Bulletin of Statistics*, June 1939, September 1939, Table 36
95 CO852/198/11, Thomas to CO, 8 August 1939.
96 The 43 dredges built between 1928 and 1930 had an average bucket size of just over 10 cubic feet and could dig to 72 feet; the 11 built between 1937 and 1940 had an average bucket size of over 16 cubic feet and could dig to 94 feet.
97 While this income contributes to the profit and loss account, it is a depletion of the real capital held by the company. The cost of acquisition of the land has to be deducted to get a real profit.
98 Since Tronoh had to realize a large proportion of its investments in order to do so, other income was correspondingly reduced.
99 The organizational structure was simplified in 1938–1939. Yukon Gold changed its name to Yukon-Pacific to reflect the fact that it no longer held any interests in gold. Yukon-Pacific then merged with its parent company to become Pacific Tin Consolidated, eliminating one layer in the corporate hierarchy. Malay States and Ampang remained subsidiaries of the new Pacific Tin, Shakila Yacob, *The United States and the Malaysian Economy*, Abingdon: Routledge, 2007, p. 78.

100 David was a stockbroker who had made heavy commitments to several tin mining ventures, financed against the security of large blocks of shares and the prospect of selling leases in Siam. But the inevitable delays in transferring leases in Siam, coupled with plummeting share values, forced him into bankruptcy in early 1934, *Straits Times*, 3 February 1934. That would make him very receptive to an initiative from Anglo-Oriental. The nominal value of the issued shares in the two companies he brought into Southern Kinta was £684,250 for which shares with a nominal value of £207,750 were exchanged. David eventually recovered from this relapse and again developed tin properties, though on a far smaller scale.

101 *Straits Times*, editorial, 17 October 1934.

102 MMC, LTC, Board Meeting, 17 May 1934.

103 On the basis of this assessment in the first 11 months of 1940, the company would have received a quota of 752 tons but it only managed to produce 582, or 77 per cent, *Mining Yearbook*, 1941.

104 Pahang Consolidated AGM, *Economist*, 25 November 1933.

105 Dorothy Thatcher, *Pai Naa*, London: White Lion, 1974, p. 14.

106 *Financial News*, 22 December 1932.

107 Pahang Consolidated AGM, *Economist*, 25 November 1933.

108 *Economist,* 28 April 28, 1934; *Minerals Yearbook, 1939*, p. 691, J. W. F. Rowe, *Markets and Men*, Cambridge: Cambridge University Press, 1936, pp. 166–7; Lionel Robbins, *Economic Planning and International Order*, London: Macmillan, 1937, p. 130; Etienne Dennery, *Le problème des matières premières*, Paris: IICI, 1939, p. 94; *Economische Weekkroniek,* 1 July 1939.

109 The life of the mine is taken as the total of 63,000 tons of concentrate actually extracted between 1933 and 1965. Had production, costs and price remained at the average of 1933–1939, the NPV of the mine would have been £780,000. The NPV of unrestricted production at an annual rate of 3,760 tons, zero costs and a price of £62 is £625,000. This confirms the judgement made by *Investor's Chronicle*, 10 October 1936.

110 CO717/138, Thomas to CO, 6 September 1937.

111 The Department had a dual mandate, to prevent contamination of rivers through silting and to irrigate agricultural land so that it could be used for rice cultivation and thereby reduce Malaya's heavy dependence on imported rice. The first constrained existing mines, the second blocked the development of new ones.

112 The overall case for doing so and the implications for other developmental strategies are presented by Patricia Glover, 'Malayan Tin Producers Clash with Agricultural Interests', *Far Eastern Survey*, vol. 8, no. 14, 5 July 1939.

113 Sir Lewis Fermor, *Report upon the Mining Industry of Malaya*, Kuala Lumpur, 1940, p. 165.

114 Ibid., pp. 169–74. Alternative uses were permitted if they generated a higher return.

115 Ibid., p. 178.

116 CO717/138, Lyttleton commented to Clauson, 'one of the best things of its kind I have ever seen'. 30 January 1940. The ex-Warden of Mines totally disagreed with some of the proposals but endorsed its thrust in 'making tin exploitation paramount', Laird to Gent, 9 October 1939, CO852/198/11.

117 CLM26/38, British Advisor (Trengganu) to Commissioner of Lands and Mines, 11 December 1939.

118 CO717/138, Sultan of Perak at Durbar of Rulers, 25 November 1939. Thomas personally endorsed this position, 24 June 1940.

119 Remarks of Rae, FMS Legislative Council, *Proceedings,* 30 April 1935.

120 Fermor, op. cit., p. 138. This principle would retain the existing burden of restriction on existing mines resulting in the kind of operational inefficiencies which are otherwise decried on p. 136.

121 Ibid., p. 225.

122 Ibid., p. 74.

123 CO717/132/9, Campbell minute, 1 March 1939. Calder provided a point by point criticism, 28 February 1939, though this may have been of a draft which was more contentious than the published version.

124 Fermor, op. cit., pp. 114–15.

125 CO717/138/3, Campbell minute, 3 April 1939.

126 *Report of the Commission of Enquiry ... into the Conduct and Management of the Mines Department*, Kuala Lumpur, 1941. At the same time, a corruption scandal erupted in the Army over the awarding of building contracts, CO717/144/13, 16 September 1940.

127 *Report,* op. cit., p. 8. This was simply a form of 'honest graft'. The official in question had a total salary of S$75,000 since 1931 but had saved S$76,000 and lamely claimed that his local living expenses were covered by his racing winnings, *Straits Times,* 10 October 1941.
128 *Straits Times*, 16 October 1940, 7 January 1941. *Report*, op. cit., pp. 10, 22.
129 *Straits Times,* 18 December 1940. Macdonald had no direct role in the revision of the assessment.
130 *Report*, op. cit., pp. 16–17.
131 FMS Chamber of Mines, *Yearbook,* 1934. The land itself received no assessment but could be added to existing leases. STS acquired sufficient ground to assure itself of a 'long life', *Economist,* 23 June 1934. Malayan producers then began to press for a similar relaxation in the FMS.
132 *Straits Times,* 28 May 1934. A head tax was also imposed on immigrants which added to the expense of labour recruitment.
133 Virginia Thompson, 'Siam Maneuvering towards Self-Sufficiency', *Far Eastern Survey*, 21 December 1938.
134 Puey Ungphakorn, 'The Economics of Tin Control', PhD dissertation, University of London, 1949, p. 406.
135 Rayne Kruger, *The Devil's Discus*, London: Cassel, 1964, pp. 45–6.
136 CO852/73/4, Crosby to FO, 28 January 1937.
137 Kenneth Landon, *Chinese In Thailand*, New York: Institute for Pacific Relations, 1941, pp. 254–5. An important exception was the granting of a lease to Mitsubishi in 1938, William Swan, *Japanese Economic Activity in Siam. From the 1890s until the Outbreak of the Pacific War*, Gaya: Centre for South East Asian Studies, 1986, p. 63.
138 It may have been facilitated by the allocation of 16,000 shares in British American Tin Mines to a member of the Ministry of Defence, BT31/38522/294254.
139 *Tin*, January 1939. The complications derived from the very fine dissemination of the cassiterite throughout the ore body which initially resulted in a recovery rate of 30 per cent.
140 *Metal Industry*, 15 January 1937.
141 The initial capital was largely American, 'a fact which the company does not want made public'. SDCDF892.6354, Porter (US Legation, Bangkok) to State Department, 21 June 1933. Horner also had an ambitious plan to develop a smelter in Siam and Porter claimed that official support wilted in the face of threats of reprisals from the British financial advisor. This would feed the sense of resentment at the arbitrary control over the industry that was emerging in the State Department. The episode is discussed in more detail in Richard J. Aldrich, *The Key to the South. Britain, the United States, and Thailand during the Approach to the Pacific War, 1929–1942*, Kuala Lumpur: Oxford University Press, 1993, pp. 117–20. The economic rationale was never clear, since a smelting monopoly would raise costs to producers and provide no additional leverage against the ITC.
142 BM soon sold its share since van den Broek was not prepared to contemplate moving the legal domicile of the company from London to the FMS, BM21, Board Meeting, 21 March 1939.
143 This was yet another case where Anglo-Oriental marginalized Tronoh, since it held some of the original debentures and Simms had to leave the Board of British American Tin Mines.
144 Billiton's interest in expanding in Siam was prompted by a concern that the proposed fusion with Banka would not materialize, BM21, Board Meeting, 1 April 1937.
145 CO852/73/4, Crosby to FO, 28 January 1937.
146 CO852/73/4, Crosby to FO, 2 February 1937. Crosby reported that the retained quota would be sold at16.8 Bhat (£1/10–)/ton metal. This is a nominal amount which suggests that it was an administrative fee for redistribution. However, if it was sold independently of any redistribution, the rate seems very low, especially by comparison with the £40/ton paid to the Ministry of Defence for the transfer of some of its quota and may be the result of a confusion between tons and pikuls. In subsequent years, the government reserved around 10 per cent for itself.
147 CO852/73/4, Crosby to FO, 28 January 1937.
148 R. Mackilligin, *The Bisichi Story*, London: Bisichi, 1994, p. 36.
149 Bill Freund, *Capital and Labour in the Nigerian Tin Mines*, Harlow: Longman, 1981, pp. 122–3.
150 Ibid., pp. 125–6.
151 Ibid. Columbite sold for $560 per pound, *Minerals Yearbook , 1938,* p. 676.
152 SNP14034/IX, minutes of the Quota Committee, 6 December 1936. Compensating Allowances were also being relinquished as small miners retired or died.
153 SNP14034/IX, minutes of the meetings of the Quota Committee, 14, 21 December 1936. With the conversion, the requirement that entitlements from special claims be worked from specific

properties was dropped. Further extraction and transfers of leases had made the assessment of 1931 obsolete but then so were the assessments of the ordinary claims.

154 However, Nigeria refused to change the terms of either the committee or its principles.

155 SNP14034/IX, Terms of Reference from 1 January 1937. Although these rules allowed for greater flexibility, they still prevented any new producers from entering the industry, except by buying an existing operation.

156 SNP14034/IX, CIM to SNP, 12 January 1937.

157 Ibid.

158 SNP14034/XI, CIM to SNP, 5 November 1938.

159 Delinquents were penalized by cutting up to 50 per cent of the shortfall from their admitted claims; those with authorized excesses were rewarded in the same proportion. The residue was then distributed among all non-delinquents, so that the total domestic allocation remained unchanged. It was feared that reducing this would undermine Nigeria's claim to its standard tonnage.

160 SNP14034/XI, Minutes of the Quota Committee, 28 February, 7 March 1938.

161 NANK, Mines Department, Jos, JosMinesC392, CIM to SNP, January, 1939. By contrast, the productivity of land in Malaya was nearly 10 times greater and the rent charged was less than half that in Nigeria.

162 Ibid. The rents bore particularly heavily on the independent miners since they had to have enough land to support their admitted claim of 20 tons.

163 Ibid. At the same time a Hausa labourer received £18 per annum.

164 JosMinesC392, Adams (Chief Commissioner, Northern Provinces) minute, 25 March 1939.

165 Oliver Lyttleton, *The Memoirs of Lord Chandos*, London: Bodley Head, 1962, pp. 142–3.

166 G1/244, Spens memo, 1 April 1935.

167 G1/244, Committee Investigation, Spens to Bunbury, 5 November 1935.

168 Lyttleton, op. cit., p. 143. Norman was another strong supporter of schemes of industrial rationalization, Philip Williamson, 'Financiers, the Gold Standard, and British Politics, 1925–1931', in John Turner, ed., *Businessmen and Politics. Studies of Business Activities of British politicians, 1900–1945*, London: Heinemann, 1984, pp. 113, 116–17.

169 The major conflict was between the preference and ordinary shareholders in AOMC. Keynes led a successful protest on behalf of the former and sufficient ordinary shares were bought out to allow the merger, *Financial Times*, 18 December 1936.

170 G1/245, Bunbury memo, 11 November 1938.

171 G1/248, Norman to Peak, 26 June 1936.

172 It should be recalled that Howeson was considered to have been indispensable in securing the renewal of the second agreement.

173 Norman commented to Bunbury: 'Howeson must be looked upon as dead and buried as far as London and this country are concerned.' 19 June 1936, G1/248.

174 G1/245, Norman to Bunbury, 19 June 1936.

175 In early 1937, BTIC raised nearly £1 million to buy CGF's tin holdings, including its shares in AOMC. At the same time CGF doubled its investment in BTIC, *Times*, 15 January 1937.

176 G1/245, Bunbury to Norman, 30 November 1938, reporting on conversation with Campbell. This provided the context within which Bunbury commented on the earlier overture from van den Broek.

177 In addition, the Volksraad enjoyed considerably greater power than did the FMS Legislative Council, G. H. C. Hart, *Towards Economic Democracy in the Netherlands Indies*, New York: Institute of Pacific Relations, 1943, pp. 25–9.

178 This follows a report based on the records of the Colonial Ministry by K. Biegman, 'De Tinfusie', July 1944 in Groothoof/50.

179 Not only did Billiton have a large equity interest in Singkep but its concession was due to expire in 1937, A. F. Kamp, *De standvastige tinnen soldat,* The Hague: Billiton, 1960, p. 153.

180 D. Gerritsen 'De Tinfusie', *Koloniale Studiën*, vol. 19, no. 3, June 1935. In turn, this was subject to criticism by the retired head of Banka, P. Hövig, 'De Tinfusie', *Koloniale Studiën,* vol. 21, no. 5, October 1937.

181 This provoked Groothoof to present the argument for fusion in a less technical manner, *De Tinfusie in Nederlandsch-Indië*, The Hague: Nijhoff, 1937, which in turn provoked a rejoinder, D. Gerritsen 'Nogmals de Tinfusie', *Koloniale Studiën*, vol. 22, no. 4, August 1938.

182 When the war was over Billiton assumed responsibility for the management of Banka.

183 A. L. ter Braake, "Rationalisatie der Banka Tinwinning," *Koloniale Studiën,* vol. 25, no. 2, 1941, p. 283.

184 The total cost of this expansion programme was around £1.5 million, all internally financed.
185 Kamp, op. cit., p. 153.
186 Only one other new mine was developed in this period. Sante Fé commenced operations in 1936 and became a respectable medium-sized producer.
187 He also attempted to consolidate Oploca and PME but this foundered on the same kind of problem of evaluation that bedevilled the fusion discussions in the NEI. The only other important development was the consolidation of the mines he owned personally into a new company, Bolivian Tin and Tungsten, in 1936.
188 The size of the resource transfer is a function of two variables: (1) the amount of sterling delivered beyond the amounts needed for local expenses; (2) the difference between the rate at which the Banco Central actually bought sterling and the rate that would have otherwise prevailed on an open market. It was a crude device which often left companies with surplus bolivianos and Patiño considered acquiring fresh properties to absorb them but none proved sufficiently attractive.
189 With twice the volume of production in 1945, these depreciation costs declined to £28 per ton at the official exchange rate.
190 Helmut Waszkis, *Dr. Moritz (Don Mauricio) Hochschild 1881–1965*, Frankfurt: Vervuert, 2001, pp. 126–9.
191 John Hillman, 'Mining Industry and the State: The Politics of Tin Restriction in Bolivia, 1936–1939', *Bulletin of Latin American Research*, vol. 21, no. 1, January 2002, p. 61.
192 Ricardo Martinez Vargas, *Estaño en Bolivia y el Comité Internacional*, La Paz: Echaniz, 1936, p. 31.
193 K. C. Tregonning, *Straits Tin*, Singapore: Straits Times, n.d., pp. 54–5. At 3.3 lbs per cubic yard, Kyerwa was a rich mine but labour shortages meant that it could only produce 8 tons in each of the seven months when weather conditions permitted concentration, C. T. Sweet, 'The Kyerwa Tin Deposits, Tanganyika', *TIMM*, vol. 54, 1940. The Kywera Syndicate acquired the properties first developed in association with Billiton and operated by Bukoba (Tanganyika) Tinfields.
194 *Mineral Industry for 1938*, p. 602.
195 Tin smelting stopped in 1943 on orders of the British government, Tregonning, op. cit., p. 58. Presumably this was to allow greater efficiencies at Williams, Harvey which had lost much of its Bolivian sources.
196 Ibid., pp. 53–4.
197 Ibid., p. 56.
198 *Mineral Industry for 1938*, p. 601.

14 Tin consumption and research

1 Following the devaluation of sterling, German smelters found it difficult to compete for Bolivian concentrates and they turned increasingly to secondary recovery, Willy Neitzel, *Die Wirkungen internationaler Kartelle in Kupfer, Zink und Zinn auf die deutsche Volkswirtschaft*, Cologne, 1936, p.100.
2 The most obvious of these was in collapsible tubes, *Minerals Yearbook, 1936*, p. 403.
3 SNP12848/I, Tavener to ITRDC, 11 November 1932. This irrational nervousness continued, Bureau of Mines, IC 6930, November 1936, p. 10.
4 *Minerals Yearbook, 1932–33, 1940*. Tin reached 81 per cent of its 1928 level, whereas copper and lead remained stagnant at 62 and 66 per cent respectively.
5 Domestic advantages lie in the ease of storage of the can and the processing of its contents.
6 James W. McKie, *Tin Cans and Tin Plate*, Cambridge, MA: Harvard University Press, 1959, p. 138.
7 A full description of the technology, together with a brief history of the improvements since its inception in the seventeenth century is provided by C. L. Mantell, *Tin. Its Mining, Production, Technology, and Applications*, New York: Hafner, 1970, Chapter 14.
8 V. Lowinger, 'Production and Utilisation of Tin', *Crown Colonist*, January 1944, p. 19.
9 The process was first perfected by a German company in 1934, Peter Roddy, *The International Tin Trade*, London: Woodhead, 1995, p. 17. For a description of the technology, see Mantell, op. cit., Ch 10; McKie, op. cit., pp. 41–4.
10 CO852/4/8, Campbell minute, 29 January, 1935.
11 Chen May Yee, *Born and Bred in Pewter Dust. The Royal Selangor Story*, Kuala Lumpur: Archipelago Press, 2003, p. 27.

12 *Minerals Yearbook, 1932–33*, p. 353, *1934,* p. 524, *1935*, p. 588, *1937*, p. 741.
13 FMS Chamber of Mines, *Yearbook,* 1933, Council Meeting, 22 November, 1933.
14 *Proposals for Expanding the Activities of the ITRDC*, New York, 1951, pp. 7–8; Ernest Hedges, *Tin in Social and Economic History*, London: Arnold, 1964, p. 51.
15 *Metal Industry*, 1934, p. 3.
16 Testimony of James Pope, United States Congress, House, Subcommittee, Committee on Foreign Affairs, *Investigation on U.S. Dependence on Foreign Tin*, 74th Cong., 1st sess., 1935, p. 966.
17 The only exciting new product was speculum. This recovered the properties of a tin-copper alloy which had first been exploited in classical times to produce a mirror and a wide range of potential applications in household items from bath taps to trays. It was produced commercially from mid-1939, *Tin and its Uses*, January 1940.
18 CO435/9, Hanson report, 1942. One of the unexplored issues is the extent to which existing industrial users benefited from the technical support offered by the ITRDC. Such benefits would offset any apparent 'exploitation' of consumers.
19 Trevor Tarring, *Trading in Metals*, London: Mining Books, 1983, p. 89. Less positive assessments are offered by William Fox, *Tin. The Working of a Commodity Agreement*, London: Mining Journal, 1974, p. 20 and William Baldwin, *The World Tin Market. Political Pricing and Economic Competition*, Durham, NC: Duke University Press, 1983, pp. 54–5.
20 By 1937 425,000 copies of technical publications had been distributed, ITRDC, Second General Report, London, 1938, p. 51. The scientific and commercial press gave extensive publicity to these publications.
21 CO852/73/7, Campbell minute, 21 January 1937. The annual budget of the ITRDC at this time was £33,100, CO852/140/2.
22 CO852/954/3, Minutes of the 52nd ITRDC Meeting, 1 December 1949.
23 *Minerals Yearbook, 1936*, p. 468. In addition to private expenditure in the search for substitutes, the American government was spending £40,000 per annum, *British Malaya,* June 1935. This incentive should not be overestimated. There was a substantial reversion to tinplate by the petroleum industry, SDCDF824.63, memo on Department of Commerce survey, 10 April 1934.
24 *Minerals Yearbook, 1939*, p. 682.
25 *SAMEJ*, 5 October 1935.

15 The International Tin Committee and World War II, 1939–1942

1 Alvin Barber, 'Tin Control in a Major War', *Far Eastern Survey,* vol. 8, no. 23, 22 November 1939, pp. 270–1.
2 BMC was appointed to manage NFMC for a 'very generous' annual fee of £221,400, John Cassington, *A Metal Man's Wartime Diary*, Eynsham, 1941, p. 90.
3 Leo Fishman, 'Wartime Control over Tin in Great Britain', *Journal of Political Economy*, vol. 54, no. 3, June 1946, p. 420.
4 That relieved the British of the necessity of placing an embargo on shipments of concentrates destined to produce metal for Germany. The closure of the Arnhem smelter did not affect the treatment of NEI concentrates since they were now shipped to Penang, Institut für Weltwirtschaft, *Die Zinnindustrie des südostasiatischen Raumes*, Kiel, 1941, p. 22.
5 *AMM*, 12, 20 July 1939.
6 SDCDF800.6354, Trench to Feis, 30 August 1939. Trench noted that visible supplies in the USA were around five weeks whereas those in the UK were around five months. If the UK then imposed export restrictions, the surplus in the UK could not compensate for the deficit in the USA.
7 CO852/253/1, Campbell to ITC delegates, 28 August 1939. Only one third of the retroactive increase would have been available for fresh production, since the remainder was required to wipe off accumulated excesses. Campbell had the votes necessary to issue an immediate communiqué but this was delayed for several days in an attempt to contact Patiño in Italy.
8 CO852/253/1, Campbell to ITC delegates, 8 September 1939.
9 FO371/23968, Lothian (UK Ambassador, Washington) to FO, 6 November 1939, communicating the complaints of the State Department which also sought to have half the buffer stock kept in the USA. Since the buffer stock was designed to stabilize the London price, it is not easy to see how this demand could have been granted without a considerable expansion in its size.
10 SDCDF800.6354, Kennedy to State Department, 11 September 1939.
11 Trench market letters, 2, 15 September 1939.

12 CO852/253/1, Todd to Campbell, 12, 28 September 1939.
13 CO852/253/1, Kennedy to MacDonald, 12 September 1939.
14 Costs to producers in Malaya rose by around £17/ton, as a result of a new war tax and an increase in the returning charges by nearly £5, as the price war between ESC and STC came to an end. According to Jorge de la Barra, both companies operated the same tariff schedule which raised their charges from 1.4 per cent to 3.6 per cent of the market value of the metal, *Singapore y el estaño en la Malaya*, La Paz, 1941, p. 76. Since the metal was sold in Singapore, consumers had to bear the additional wartime costs of shipping and insurance, which amounted to £8/ton, Walter Montenegro, *Estaño en Malaya*, La Paz: Fenix, 1943, pp. 121–3.
15 A parallel order governed trading in Singapore, where the local price was set at the equivalent of £227, which was designed to result in a strict cif parity between Straits tin in New York and standard tin in London.
16 The LME was closed from 1 to 5 September to allow for the evacuation of children from London and again from 13 to 17 September.
17 William Fox, *Tin. The Working of a Commodity Agreement*, London: Mining Journal, 1974, p. 184.
18 By late September it had risen to 75 cents. This was a 30 cent premium over delivery for late November, *AMM*, 28 September 1939. Serge Calabre, *L'étain*, Paris: Economica, 1991, p. 84 presents this as a sterling price, £397, which could not be more misleading.
19 Since prices in New York were not determined in an organized commodity market, there were no turnover data, which made it impossible to interpret the nature of the problem. In any case, many consumers were able to buy at the fixed price in Singapore, Cassington, op. cit., p.12. Gouging may well have been confined to those consumers who could run down their stocks and return them to the market.
20 CO852/253/1, Clauson minute, 14 September 1939; CO717/140/20, Campbell minute, 19 December 1939.
21 CO852/253/1, Campbell minute, 8 September 1939.
22 CO852/253/2, Lyttelton to Campbell, to van den Broek, 3 October 1939.
23 CO852/253/2, Trench to Campbell, 3 October 1939. The shift to tinplate was based on the assumption that the United States would pick up the export trade lost by Britain and Germany, *NYT*, September, 1939.
24 CO852/253/2, Campbell to van den Broek, 3 October 1939.
25 A daily demand of 2,000 tons and a supply of 240.
26 CO852/253/2, cable from Singapore to CO, 5 October 1939.
27 CO852/253/2, Trench to Campbell, 7 October 1939. This was followed by a particularly sarcastic cable a week later.
28 *Economist,* 14 October 1939. Such optimism should have been tempered by a recognition of the implication of the tin and rubber for timber barter deal struck at the same time between the UK and the USSR. While this involved Cornish tin and hence did not affect ITC supply, it would reduce the level of stocks in the UK and delay the point at which UK exports could again feed the market.
29 CO852/253/2, Malayan Chamber of Mines to CO, 10 October 1939. By this time, the elements sympathetic to the ITC, such as Stephens and Lyttleton, had resigned from the Chamber, leaving it under the control of a Cornish rump.
30 Fishman, op. cit., pp. 420–1, Klaus E. Knorr, *Tin Under Control*, Stanford, CA: Food Research Institute, 1945, p. 173.
31 CO852/253/2, Campbell to Kennar, representing Todd, 25 October 1939.
32 House of Commons, *Parliamentary Debates,* vol. 355, cols. 416–17, Stokes questions, 5 December 1939.
33 CO852/253/2, Campbell to Calder, 23 October 1939.
34 CO852/253/2, Campbell to Kennar, 25 October 1939.
35 SDCDF800.6354, Memo of meeting between Feis, Veatch and Todd, 30 January 1939. This was in response to the concern of the State Department for more effective American representation on the ITC following the failure of the Hughes appointment.
36 ATD, Report on Tin Industry, note 140, citing a report of 21 February 1940, based on interview with Klein. The fear was that the ITC would then use the information to rig the market.
37 CO852/253/2, Trench to Campbell, 24 October 1939.
38 CO852/253/2, Campbell note to ITC delegates, 27 October 1939.

39 CO852/253/2, Clauson minute, 21 November 1939. They had wanted a revision of the maximum from £230 to £240 to take account of the increase in costs, MKGA 562/T37, van den Broek to Lyttelton, 14 September 1939.
40 CO852/253/2, Lyttleton to Campbell, 24 October 1939. Lyttleton attributed this to a leakage from Bagnall who, naturally, had never supported a fixed price.
41 CO852/253/2, FO to Lothian, 24 October 1939. Commitment to this figure conveniently overlooked the fact that the devaluation of sterling from $4.86 to $4.03 meant that the new £230 was equivalent to an old £198, that is below the buffer stock limit!
42 CO852/253/2, Todd to Campbell, 23 November 1939.
43 CO852/253/2, Trench to Campbell, 25 November 1939.
44 CO852/198/8, Minutes of the 72nd ITC Meeting, 1 December 1939.
45 FO371/23968, Wilson-Young minute on meeting held on 29 November 1939. Patiño was unaware of this decision. In his response to Campbell he made Bolivia's support for the new quota contingent on a return to a free market in London and therefore took considerable domestic credit for having achieved this result! AIM12, Comité Internacional 1938, exchange of cables between MH and Patiño, 28 October, 4, 9, 12 December in Patiño to MM, 25 March 1940
46 At the same time, the Ministry of Supply revised the price schedule for the non-ferrous metals under its control and they all rose at a much faster rate. By comparison, tin would remain cheap for the rest of the war.
47 SDCDF811.23 (Raw Materials), Kennedy to State Department, 28 November 1939.
48 SDCDF811.23 (Raw Materials), memo on Campbell interview, 30 November 1939. Viles was the longstanding consumer representative on the IRRC.
49 CO852/312/4, Campbell minute on Trench report, 6 January 1940. In 1938 US exports had been a mere 205 tons. The *AMM* reported that 500 tons shipped from the USA to Sweden were against the wishes of the American government, 2 December 1939.
50 A similar issue affected British companies who were faced with a wartime excess profits tax at 100 per cent, which meant confiscation of part of their capital, Fishman, op. cit., pp. 415–18.
51 CO852/275/8, Lyttelton to Campbell, 7 February 1940.
52 SDCDF800.6354, Johnson (US Embassy, London) to State Department, 20 February 1940.
53 SDCDF800.6354, Gordon (US Embassy, The Hague) to State Department, 6 April 1940.
54 CO852/312/4, Campbell to Lyttelton, 8 February, Campbell to Clauson, 26 February, Minutes of the 73rd ITC Meeting, 26 February 1940. Patiño's assessment is in AIM12, Comité Internacional 1938, Patiño to MM, 25 March 1940.
55 CO852/312/4. There was some concern that adding further stocks would force a decline in price to £150 and there was no particular confidence that the cut would even keep the price above £230, Clauson to Waley, 2 March 1940.
56 CO852/275/8, Clauson's notes on meeting, 23 February 1940. The reference to Lyttleton is to his connections with Patiño through BTIC.
57 Nor it is clear that the accumulation of stocks would have benefited the British war economy, given the cost of warehousing and financing them. Fishman, op. cit., p. 433, is severely critical of British tin policy since it failed to build up stocks to a level that he considered adequate but overlooks the opportunity cost that such diversion of resources would have entailed.
58 *Times*, 15 June 1940.
59 House of Commons, *Parliamentary Debates*, vol. 357, col. 2,091, 2 February, vol. 358, cols. 371–2, 6 March, cols. 1,189–92, 13 March, vol. 362, col. 458, 26 June 1940.
60 CO852/275/8, Clauson to Owen (Treasury), 23 July 1940.
61 One reason was the availability of free sterling in New York which could now be used to import standard tin, eliminating the premium traditionally paid for Straits. In March the Treasury insisted on dollar invoicing for a wide range of exports including tin. With demand for free sterling reduced, the value fell dramatically but it was now irrelevant to the tin market.
62 CO852/275/8, Minutes of the 74th ITC Meeting, May 27, 1940. That, of course, reduced the effect of the 80 per cent quota for the second quarter.
63 J. A. Robinson',The International Tin Restriction Schemes with special emphasis on British Malaya', PhD dissertation, George Washington University, 1945, p. 279. The obstacle lay in Congressional opposition to buying foreign commodities.
64 Knorr, op. cit., p. 174. It was not until August that Congress authorized funds and even then cut Roosevelt's request for $25 million for the whole stockpile programme to $10 million, Robinson, op. cit., p. 279. The total amount acquired by the Army and Navy Munitions Board under this programme was 11,457 tons, none of which was used during the war.

65 CAB102/193, Combined Statistical Summary, p. 133.

66 When Campbell was approached by the Americans for the ITC's co-operation in building a stockpile in 1939, the Dutch were anxious to give a considerable reduction on the prevailing market price, by 7 per cent and another member was prepared to go as far as 16 per cent, CO852/199/4, Campbell to Clauson, 23 March 1939.

67 Fox, op. cit., pp. 185–6. It was, however, 8 per cent higher than the prices prevailing from January to April and some in the USA considered that 45 cents would have been sufficient to bring forth the tin required, *Minerals Yearbook, 1940*, p. 669. The issue proved to be an important point of contention in postwar tin politics.

68 CO852/275/8, Minutes of the 75th ITC Meeting, 8 July 1940. The increase to 130 per cent coincidentally solved a serious problem for Nigeria which would otherwise have had some difficulty in wiping off the excess accumulated as a result of the cleaning out of stocks in 1939.

69 *Minerals Yearbook, 1940*, pp. 681–2.

70 CO852/434/7, Campbell to Burgess, 2 February 1941.

71 Extension for a further year meant a possible total of 132,000 tons for the stockpile, CO852/434/7, Lowinger to Campbell, 16 September 1941.

72 A narrowly strategic focus would have seen a small smelter built immediately to treat low grade Bolivian concentrates, with a decision on enlargement deferred until the shipping situation was clarified. Had consumers' stocks been controlled, the stockpile would have easily covered any transitional period required for enlargement.

73 Around half of the remaining sterling would be needed for realization costs payable in sterling.

74 The assets were held offshore in the Channel Islands, now under German occupation, and formally frozen. A fuller discussion of the evolution of the relationship between Patiño, Britain, and the United States is in John Hillman, 'Bolivia and British Tin Policy, 1939–1945', *Journal of Latin American Studies*, vol. 22, pt. 2, May 1990.

75 NARAII, RG59, State Department, Office of American Republics, Box 21, Duggan memo, 25 June, memos of conversations between Guachalla and Duggan, 30 October 30, 1940, between Guachalla and Feis, 7 February 1941. In addition, Patiño enjoyed a poor reputation in New Deal America. He was portrayed as a tax-evader, pro-Fascist, holding Bolivia in feudal bondage and the source of US tin problems, Edsel Kelly, 'The Battle for Tin', *The Nation*, November 1940, pp. 473–4. The first of these charges was well-founded, the remainder were not.

76 CO852/434/7, van den Broek to Pearce, 3 March 1941. As a result of this competition, van den Broek felt that the terms of the contract were quite unattractive and wanted CTS to assume a 45 per cent share in the venture.

77 Details of the contract are in *EMJ*, December 1940. Normally, smelting charges per ton of concentrate are increased as the proportion of metal declines. This contract reversed, although only apparently, that principle to encourage the shipment of very low grade concentrates containing between 18 per cent and 25 per cent metal. The editor explained to his puzzled readers that this followed the practice adopted in previous contracts with the UK, Germany and Holland. However, he stressed that the whole contract must be seen primarily as a benevolent political gesture towards Bolivia, p. 32. The actual smelting costs far exceeded those paid under the terms of the contract.

78 *Minerals Yearbook, 1940*, pp. 681–2.

79 *Quin's Metal Statistics and Yearbook,* 1941, p. 117. Keeping the price just below parity eliminated the incentive for Patiño to ship to the USA. Later in the year, export restrictions threatened to lower the LME price, forcing the Ministry of Supply to re-enter the market to prop it up.

80 Of the 108,000 tons of stocks in the USA, 50,039 were held by various government agencies, *Minerals Yearbook, 1941*, p. 706.

81 *Tin*, January 1942, p. 3.

82 CO852/275/8, Campbell to Clauson, 19 December 1940. Since the metal produced would be available commercially, it would serve to displace the privileged position enjoyed by the Straits smelters in the US market. Confirmation of this rumour came via Lyttleton with the views of the Chairman of Continental Can who saw the MRC/Bolivia contract as more than 'a war time expedient, but designed to drive a wedge into the ITC … the outcome of long negotiation between consumer interests, the Senate Foreign Affairs Committee and the State Department'. CO852/434/7, Lyttleton to Campbell, 18 February 1941. Once there was a functioning smelter and a relationship with Bolivia, justifying a subsidy in the form of a tariff on imported metal would be much easier.

83 CO852/275/13, Campbell minute, 15 October 1940.

84 CO852/275/8, Campbell minute, 15 November 1940.
85 Patiño had already alerted Campbell to Bolivia's definite unwillingness to contemplate a change in the existing tonnages, CO852/275/13, Campbell minute, 22 November 1940.
86 One usually moderating force was absent. Camus had planned to attend the meeting but was killed in the London blitz.
87 Van den Broek proved to be quite right. The US smelter found it impossible to treat much of the low grade Bolivian concentrates.
88 CO852/275/8, Minutes of the 76th ITC Meeting, 16 December 1940. Campbell added: 'I have, as usual, toned the thing down quite a lot in the minutes. The substance remains; but a good deal of the acerbity has been cut out'. Nonetheless, van den Broek's forceful and 'reasonable' position comes through very clearly.
89 Ibid. Although Thailand was represented at this meeting, her delegate was not pressed for a comment on this formula.
90 CO852/435/5, Minutes of the 77th ITC Meeting, 20 March 1941.
91 This article was tightened on the advice of the Foreign Office, following the model of the Sugar Agreement, FO371/29129, Hart to Campbell, 27 August 1941. The full text of the agreement was published as *Agreement between the Governments of Belgium, Bolivia, the United Kingdom and the Netherlands for the International Control of the Production and Exports of Tin*, London: HMSO, 1942, Cmd 6396.
92 It was not easy for the other delegations to take the Bolivian position seriously since, in spite of the generous terms of the MRC contract, she was only producing at 95 per cent of her standard over the second half of 1940, while the NEI and Nigeria were both producing at 130 per cent of theirs. Patiño had made the position even more difficult by at first refusing to meet in London and proposing that the whole committee meet in Lisbon! SDCDF800.6354, Winant (US Embassy, London) to State Department, 13 March 1941.
93 SDCDF800.6354, Hull to US Embassy (London), 8 March 1941; 811.6354, Guachalla to Feis, 12 March 1941, in which the Bolivian Ambassador expressed his 'deep appreciation for the valuable help given to Bolivia by the Department of State'.
94 SDCDF800.6354, Winant to State Department, 25 March 1941.
95 SDCDF800.6354, Winant to State Department, 3 April 1941.
96 No less absurd was the position taken by Montenegro which assumed that a high proportion of Malayan production was marginal so that her peak years were simply the result of exceptionally high prices and therefore concluded that a normal level for the interwar period would be just under 50,000 tons, op. cit., p. 129.
97 CO852/435/5, FMS to Lowinger, 8 May 1941. A few months earlier Clauson had noted: 'If Lowinger receives the wrong instructions, he is sensible enough to disregard them until we can put the matter right though higher channels.' CO852/175/8, minute, 22 November 1940.98 At the Tronoh AGM, Rich argued against the five year term and at the Sungei Besi AGM Simms went further, arguing for an end to the ITC since the 'new agreement would probably have the effect of creating a form of control which could remove British control from a predominantly British industry'. *Mining World*, 15 November 1941.
99 CO852/435/5, Campbell minute, 9 May 1941. Campbell considered that the quota would have to drop to 60 per cent once the US stockpile had been completed, so that the case for the continuation of control was overwhelming.
100 CO852/435/5, Clauson minute, 15 May 1941.
101 CO852/435/5, CO to FMS, 22 May 1941.
102 CO852/435/5, Calder to Woolley (Nigeria), 19 August 1941. The Selangor Miners Association was still pressing for 1937 as the basis of the new standard tonnages, *Straits Times*, 23 May 1941.
103 CO852/435/5, Thomas to CO, 26 November 1941.
104 CO852/435/5, Clauson minute, 29 December 1941. This is rather harsh since at the Pahang AGM, Ward supported the continuation of the ITC 'in its present form until the expiry of hostilities'. *Times,* 18 December 1940.
105 CO852/435/5, Thomas to CO, 9 January 1942. The original proposal would have seen Lowinger and Wilcoxson dropped and Rich moved to full membership on the delegation. Campbell regretted the prospect of losing Lowinger but did not see Wilcoxson as likely to give any serious trouble. In his estimation, Rich had 'but little ability'. Minute, 19 December 1941.
106 CO852/275/11, Lyttleton to Campbell, 4 November 1940.
107 CO852/275/11 Campbell to Clauson, 18 November 1940.
108. CO852/435/4, Campbell to Calder, 16 April 1941.

109 CO852/435/4, Mooney to Campbell, 10 May 1941.
110 CO852/435/4, Calder minute, 13 May 1941. The issue was finally settled in early 1942 with the appointment of the LTC Chairman, Spens, to the delegation and making Burgess a technical advisor. Mooney was kept on for the sake of the appearance of continuity, even though he was 'really no use'. Campbell minute, 9 January 1942.
111 The name was changed in May 1939 as one of the first initiatives undertaken by Phibun on becoming Premier. 'Siam' was of foreign origin with supposed connotations of submissiveness; 'Thai' meant free and was supposedly the term originally used to refer to the inhabitants, Edward Flood, 'Japan's Relations with Thailand: 1928–1941', PhD dissertation, University of Washington, 1967, p. 221.
112 CO852/435/5, Calder to Woolley, 19 August 1941.
113 CO852/435/5, Minutes of the 79th ITC Meeting, 16 September 1941.
114 CO852/435/5, Campbell minute, 28 October 1941. It is worth noting that Thailand had maintained its commitment to participate in the ITRDC which was also renewed in 1941, Lowinger to CO, 23 April 1942, CO435/9.
115 Moving to a more formal agreement was not something desired by the other governments, nor by the Colonial Office, all of which were concerned about the rigidities that would be created. The need for flexibility in interpretation was one of the lessons of the 1938 readjustments and the experience of the new rubber agreement suggested that it would not be easy to maintain.
116 FO371/29129, Crosby to FO, 7 November 1941, communicating the overall anxiety of the Minister of Foreign Affairs. American diplomats dismissed this as 'the usual difficulty of Bangkok wishing to throw its weight about and to haggle in hope of getting something for nothing'. SDCDF800.6354, Winant to State Department, 1 December 1941.
117 CO852/435/5, Crosby cable to FO, 19 October, FO371/29129, Thai Legation (London) to Campbell, 24 October 1941.
118 CO852/435/5, Thai Legation (London) to Campbell, 20 November 1941. While this was clearly out of the question, the rather poorly drafted letter indicates that Thailand would have accepted her own earlier proportion.
119 Campbell checked with the leading British and Australian companies operating in both Thailand and Malaya, SDCDF800.8354, Winant to State Department, 1 December 1941.
120 CO852/435/1, Joshi Market Report, 24 July 1941. Germany paid at the rate of £400 per ton, while the Straits smelters paid £260. The difference of £140 was too tempting to be overlooked by the Thai officials.
121 William Swan, *Japanese Economic Activity in Siam. From the 1890s until the Outbreak of the Pacific War*, Gaya: Centre for Southeast Asian Studies, 1986. pp. 100–1. Japan ended up with 19 per cent of Thai production in 1941.
122 In addition to direct pressure by making such consignment a condition of renewal of leases, the government doubled the proportion of the overall Thai quota that it reserved for sale to producers. Such sales were now contingent on consignment of the concentrates produced, with the result that at least a fifth of the quota during the base period was subject to conditions that most producers would have found politically unacceptable.
123 Flood, op. cit., pp. 223–4, 619–20. This forced Shell and Standard Vacuum to close their operations in August 1939. For the effect on mining, see FO837/996, Tronoh to FO, 2 April 1941.
124 The very suggestion reflected a naive innocence of the growing relationship between Thailand and Japan.
125 CO852/435/5, Campbell to Calder, November 22, minutes of meeting at Colonial Office, November 28, Minutes of the 80th ITC Meeting, 1 December 1941.
126 SDCDF800.6354, Hull to US Embassy (London), 17 September 1941.
127 SDCDF800.6354, reply, 22 September 1941.
128. SDCDF800.6354, Hull to US Embassy (London), 5 December 1941.
129 SDCDF800.6354, Campbell to Steyne (US Embassy, London), 12 December 1941.
130 CO852/435/7, British Embassy (Washington) to FO, 5 April 1941.
131 CO852/435/9, Campbell minute, 9 April 1941.
132 BE, OV6/244, Ronald (FO) to Somervell (Board of Trade), 17 October 1940.
133 OV19/12, notes on meeting at the Board of Trade, 3 April 1941.
134 CO852/435/7, Campbell minute, 12 May 1941.
135 CO852/434/7, Campbell to Figg and Clauson, 11 July 1941.
136 CO852/434/7, Campbell to Clauson, 13 October 1941. As a result, Lowinger found himself negotiating on behalf of the ITC without plenipotentiary authorization.

137 CO852/434/7, Lowinger to Campbell, 9 September 1941.
138 CO852/435/7, Minutes of the 79th ITC Meeting, 16 September 1941, Appendix I.
139 The crucial test would have come at the point at which rising costs to producers forced them to renegotiate price.
140 CO852/435/9, Campbell minute, 3 November 1941.
141 Kent (FMS) Tin Dredging AGM, 19 November 1941. What makes this particularly intriguing is that it is diametrically opposed to the kind of criticism levelled by his son-in-law in early 1940. Perhaps they were both simply prepared to pick up whatever cudgels came to hand without any thought for principle.
142 CO852/435/2, FMS to CO, 17 December 1941.
143 CO852/435/3, Campbell minute, 18 March 1942.
144 CAB115/116, Archer (British Purchasing Commission, Washington) to Fennelly (Ministry of Supply), 8 January 1942. Negotiations with the Dutch in January must have rested on the extraordinarily naive assumption that they were in a better position to resist the Japanese than the British. The NEI fell in February 1942.
145 President Peñaranda's internal political position was beginning to weaken, not least because of his decision to break relations with the Axis in January 1942 and a generous attitude towards tin would serve to bolster it.
146 CO852/435/3, Clauson minute, 26 March 1942; FO371/31490, Ashton minute, 16 October 1942.
147 CO852/434/11, Minutes of the 83rd ITC Meeting, 28 September 1942.
148. CO852/434/11, Clauson to Nigeria, October 1942. However, that did not mean the suspension of all administrative work at the domestic level. The Nigerian Quota Committee continued to meet in 1945 to authorize the transfer of admitted claims, Mines Department, Annual Report for 1945, p. 8.
149 BT28/219, Morris to Palmer, 14 October 1942.
150 BT28/219, memo, 13 October 1942.
151 *NYT*, 6 October 1942. The report of the 105 per cent quota, with its clause permitting excess production, was also reported in an extraordinarily misleading fashion. It was defined as a concession to consumers, a gesture to preserve the existing machinery for continued control after the war, *NYT*, 28 September 1942. The State Department noted that these articles were written by a freelance journalist who considered that their sale was promoted by including a 'special note of apprehension'. SDCDF800.6254, 5 November 1942. The British press was divided in its response. The *News Chronicle* denounced it as 'old and vicious', 6 October 1942. The *Financial News* gave its unqualified support and noted the serious difficulties in granting equal voting power to consumers, 10 October 1942. The position adopted by the editor of *The Statist* is particularly interesting. He first criticized the agreement for its lack of adequate consumer representation, 3 October. But he modified his position when he saw the text, 17 October and then backed away entirely in response to the TPA, 7 November 1942. He admitted the advantages of the ITC for consumers and 'only wants consultation with consumers'.
152 United Kingdom, *Joint Declaration by the President of the United States of America and Mr Winston Churchill ... .August 14, 1941*, London: HMSO, 1941, Cmd. 6321. This clause was designed to undercut the Fascist claim that Germany and Italy were deprived of access to raw materials, Klaus Knorr, 'Access to Raw Materials in the Postwar World', *Harvard Business Review,* vol. 21, no. 3, Spring 1943. Its operational significance was completely unclear.
153 However, this remained the public position of the British government. Stokes also used the Atlantic Charter as a way of returning to the attack on the ITC in the House of Commons, only to elicit an assurance that the agreement was consistent with the Charter and that it served consumers as much as producers, *Parliamentary Debates,* vol. 388, cols. 1,837–8, 20 October 1942. This was the last occasion on which the government came to the defense of the ITC. As the Colonial Office prepared the official answer, it went even further, arguing that such a regulation agreement was a condition of the general prosperity envisaged by the Charter. That, too, was the last positive comment Clauson would makc.
154 CO852/275/9, Nigeria to CO, 12 December 1939.
155 CO852/275/9, Clauson minute, 2 February 1940. British producers may also have been reluctant to accept further funds subject to EPT.
156 CO852/275/9, Campbell minute, 2 February 1940.
157 FO371/29129, Campbell at a general meeting of British government officials, 11 November 1941. The £1.5 million remaining would have bought around 5,400 tons, which may have been

sufficient to support the market outside the USA during a transition to a quota of only 30–40 per cent required to meet real demand.

158 CO852/522/4, Minutes of the 81st ITC Meeting, 10 April 1942. Earlier reasons for retention had also vanished when sterling became completely inconvertible and when the EPT regulations were modified to deal with the particular problems facing mining companies.

159 Cf. Paul Hexner, *International Cartels*, Chapel Hill, NC: University of North Carolina Press, 1945, p. 124, Fox, op. cit., p. 184. The £230/16/– includes £4 from the interest received on the cash balances. Fox underestimated the size of the profit by the amount that the very formation of the buffer stock itself had boosted prices.

160 *Financial Times,* 10 May 1944.

161 Tin Buffer Stock Scheme, 1938/41, *Report and Accounts*, London, 1943. p. 7.

162 *Tin*, July 1949.

16 The International Tin Committee and its critics

1 American tinplate manufacturers carried much higher stocks than other users. At the end of 1941 they held 65 per cent of consumer stocks but only accounted for 44 per cent of total consumption, *Minerals Yearbook, 1941*, pp. 713, 719.

2 ATD, Nolting memo, 10 January 1945. This cannot be for want of looking since the Department had adopted a very critical position on the ITC from 1942.

3 Paul Hexner enquired into the effectiveness of this form of representation among American consumers and found they considered it 'helpful'. *International Cartels*, Chapel Hill, NC: University of North Carolina Press, 1945. p. 245.

4 This was stressed in an editorial in *Financial News*, 14 October 1942.

5 CO852/431/10, Caine memo, November 1941.

6 CO862/430/15, Rowe to Caine, 12 February 1942, Appendix A.

7 CO862/430/15, Campbell minute, 2 March 1942.

8 CO852/431/10, Campbell to Caine, 11 November 1941.

9 Feis Papers, Library of Congress, Washington, Box 6, Veatch report to Foreign Affairs Committee, 21 August 1934.

10 SDCDF811.6354, Veatch memo, 22 August 1934. It even considered that the ultimate consumer might be directly represented on control committees, 856d.6176, State Department to US Embassy, London, 12 March 1938. Lack of market familiarity would be a serious drawback to effective participation.

11 Kenneth L. Mayall, *International Cartels: Economic and Political Aspects*, Rutland: Tuttle, 1951, p. 93.

12 SDCDF800.6354, Feis to Clayton, 24 December 1941.

13 SDCDF800.6354, Veatch memo on meeting with Hughes, 12 May 1938.

14 Ibid. That fear was well justified, since the role of US Steel as the price leader of the steel industry soon came under attack by the Temporary National Economic Committee, Duncan Burn, *The Steel Industry, 1939–1959*, Cambridge: Cambridge University Press, 1961, Chapter 8.

15 SDCDF800.635, Vcatch memo on meeting with Hughes, 12 May 1938.

16 SDCDF800.6354, Kennedy to State Department, 4 April 1938.

17 SDCDF800.6354, meeting, Todd, Feis and Veatch, 24 October 1938. Todd considered he was learning from Firth's mistake who had 'rubbed members of the committee the wrong way'.

18 SDCDF800.6354, AISI to Todd, 19 April, Gallman memo, 21 June 1943.

19 SDCDF800.6354, Berman (Department of Justice), 8 June 1944.

20 ATD, Economic Warfare Section, Tin file, interview with Feis, 6 October 1944. Since Feis wanted to work within the arrangements for consumer representation in the fourth agreement, it was premature for Fox to dismiss them as 'ludicrous', *Tin. The Working of a Commodity Agreement*, London: Mining Journal, 1974. p. 195. Steyne was interviewed on the same day. Even though he considered the ITC as being 'born in sin', he also advised against the adoption of a 'theological attitude', which would prevent continuing to work with the ITC. Feis argued for the necessity and viability of reformed regulation agreements in 'Raw Material Prices and Controls', Academy of Political Science, *Proceedings,* vol. 21, no. 2, January 1945, pp. 44–7.

21 Fox Papers, Box 4, Blenkinsop notes, 10 November 1936. Stock levels were determined more by the technical requirements of manufacturing, rather than by prices, *Minerals Yearbook, 1935*, p. 514.

22 These inherent difficulties in defining a normal level of stocks are compounded by shifts in the pattern of smelting. The norm for European consumers was considerably reduced when Billiton starting shipping concentrates to Arnhem instead of metal from Singapore.

23 *Mineral Industry for 1941*, p. 547. There is no indication of how serious an inconvenience this proved to be.

24 If 5 per cent is taken as a reasonable margin of error, the ITC underestimated consumption for nine of the 20 quarters from 1934 to 1938 and overestimated for only two.

25 An alternative adopted by Knorr is to compare fluctuations in the 1930s with those in the 1920s. The method is statistically simple and measures the gap between the high and the low for each calendar year and then aggregates these differences for the two decades. It shows that fluctuations under the ITC were just as pronounced as those in the 1920s, *Tin Under Control*, Stanford, CA: Food Research Institute, 1945, pp. 198–9. Apart from its statistical and conceptual shortfalls, this conveniently overlooks the far greater external economic turmoil in 1930s, with devaluations, rearmament and war. Keynes used the same method to demonstrate the volatility of the prices of four commodities from 1928–1937, 'The Policy of Government Storage of Foodstuffs and Raw Materials', *Economic Journal,* vol. 48 , September 1938, pp. 450–1. Tin comes out much better than any that Keynes selected, wheat, lead, rubber and cotton. Ballande also used this method and found that tin was much more stable than copper, lead or zinc from 1931 to 1937 with the exception of 1933, *Essai d'étude monographique et statistique sur les ententes économiques internationales*, Paris: Librairie technique et économique, 1936, p. 15.

26 Lead is similar to tin in that demand is driven by personal consumption and the two prices followed an identical rhythm from 1921–1925, Mieczysław Epstein, 'La réglementation internationale d'un marché de matière première', PhD dissertation, University of Fribourg, 1943, p. 137. Copper is more dependent on capital investment.

27 Since the argument is comparative, it does not rest on this definition of normalcy. Those who wish to explore the implications of other assumptions can obtain the data set from the author. While it only contains monthly data for lead and copper, it contains daily data for both spot and forward prices for tin.

28 For an example of a critical commentary which fails to undertake even an elementary examination of the data before coming to the conclusion that the attempts to eliminate price oscillations failed, see Lotte Müller-Ohlson, *Non-Ferrous Metals. Their Role in Industrial Development*, Cambridge: Woodhead, 1981, p. 232. Puey claims that the buffer stocks had no material effect on price stability, 'The Economics of Tin Control', PhD dissertation, University of London, 1949, p. 419. Perhaps, but the data have to be interrogated to a much greater extent to warrant such a conclusion.

29 For a review of the limitations of this assumption, see Bernard Hodgson, ed., *The Invisible Hand and the Common Good*, Berlin: Springer, 2004.

30 J. K. Eastham, 'Rationalisation in the Tin Industry', *Review of Economic Studies*, vol. 4, no. 1, October 1936, pp. 28–30.

31 Melvin T. Copeland, *A Raw Commodity Revolution*, Cambridge, MA: Harvard University Press, 1938, p. 26.

32 J. W. F. Rowe, 'Some General Economic Aspects', in W. Y. Elliott, ed., *International Control in the Non-Ferrous Metals*, New York: Macmillan, 1937, p. 74. In an article that appeared simultaneously, C. A. Myers reached the same conclusion, 'The International Tin Control Scheme', *Journal of Business of the University of Chicago*, vol. 10, no. 2, April 1937, pp. 122–3. Rowe admitted that he had changed his first position on tin published in 1930 in light of the specific problems posed by a combination of chronic excess capacity and a general trade depression.

33 Rowe, op. cit., pp. 74–9.

34 J. W. F. Rowe, *Primary Commodities in International Trade*, Cambridge: Cambridge University Press, 1965, p. 144.

35 M. Schut, *Tinrestrictie en Tinprijs*, Haarlem, 1940, represents the best attempt by a professional economist to construct a case for tin control. It was subject to mild criticism by Puey, op. cit., p. 103. His strictures on the absence of a demand function for stocks are well taken but there are no data to supply it.

36 Schut, op. cit p. 104.

37 New technologies would reduce these costs but the experience of Caracoles stands as a salutary of reminder of the pitfalls of placing too much confidence in their development.

38 Charles Kindleberger comments: 'Economists are confounded that history is replete with examples of errors of economic policy, as seen by contemporary analysts, which turned out to be useful.' *The World in Depression, 1929–1939*, Berkeley, CA: University of California Press, 1986, p. 284.

39 In the case of Tronoh, those costs amounted to nearly £15/ton metal, adding 20 per cent to working costs. Calculated from Annual Reports, 1930–1933.

40 Yukon Gold, Annual Report for 1937, p. 2. The Tronoh data are consistent with this since costs declined at a faster rate than the decline in production from 1937 to 1938.

41 One commentator who adopted this criterion concluded that: 'Prices remained remarkably moderate at least well into 1936. No substitutes or newcomers were encouraged and one cannot say that the consumers have been hit hard by a policy which involved these prices.' Herman Kranold, *The International Distribution of Raw Materials*, London: Routledge, 1938, p. 120.

42 FO371/18569, memo to US Embassy, 20 April 1934.

43 ITRDC, *Statistical Yearbook, 1938,* The Hague, 1938.

44 CO198/12/15020/B10, Calder minute, 15 March 1939. Calder supposed that a better range would be £190–£210 but provided no grounds for this conclusion.

45 The data refer to a sample of companies listed on the London Stock Exchange and is representative of the more established and more profitable ones. It therefore only permits comparison between the leading companies in each of these categories.

46 For a fuller discussion, see John Hillman, 'The Impact of the International Tin Restriction Schemes on the Return to Equity of Tin Mining Companies, 1927–1939', *Business History,* vol. 39, no. 3, July 1997.

47 There are two possible explanations for this. One is that Patiño's purchases were on such a scale as to prop up the share prices. The second is the presence of the distinction that Eastham later formalized, between the 'solid' Malayan groups, Mair, Thomas and Stephens, and supposedly 'weaker' companies.

48 S. H. Frankel, *Investment and the Return to Equity in the South African Gold Mining Industry, 1887–1965*, Oxford: Blackwell, 1967, p. 81.

49 Cunliffe-Lister stressed that 50–60 per cent of manufactures were taken by primary producers, 'The Economic Policy of the Colonial Empire', *United Empire*, October 1934. p. 579. Demonstration of the negative impact of depressed commodity prices on the demand for industrial goods was first developed by the economist Nicolas Kaldor and the issues are reviewed by Albert Maizels, *Commodities in Crisis: The Commodity Crisis of the 1980s and the Political Economy of International Commodity Policies*, Oxford: Clarendon Press, 1992, pp. 85–6. A more extravagant claim for the ITC and its buffer stock was made by W. A. Lewis who considered that, had such machinery covered most commodities in 1929, there would have been neither slump nor war, *Economic Survey, 1919–1939*, London: Allen & Unwin, 1949, p. 174.

50 Myers was sceptical that raising prices on an inelastic demand curve would have any positive overall multiplier effect, op. cit., p. 122. Of course, many economists had theoretical grounds on which to object to inflationary solutions. For an extreme example, see Lionel Robbins, *The Great Depression*, London: Macmillan, 1934, pp. 131–2, 140–1, 163 and his *Economic Planning, and International Order*, London: Macmillan, 1937, Chapter 6.

51 The 'high' prices of 1939 did not prevent consuming corporations from earning pre-tax profits of 12.3 per cent on their capital, United States, Federal Trade Commission, *Tin can and tinware manufacturing corporations*, Washington: GPO, 1941, p.5.

52 No attempt has been made to calculate the change in the position of the labourers. While the FMS Department of Mines recorded data on wage rates, they are far too crude to allow any analysis.

53 Raúl Prebisch, 'The Economic Development of Latin America and its Principal Problems', *Economic Bulletin for Latin America*, vol. 7, no. 1, February 1962. The argument was first developed in 1950.

54 For comprehensive reviews see Maizels, op. cit, espec. Chapter 3 and John Spraos, *Inequalising Trade?*, Oxford: Clarendon Press, 1983.

55 Prince Casaro to Kennedy, 8 July 1838, cited in Eugene Staley, *Raw Materials in Peace and War*, New York: Council on Foreign Relations, 1937, p. 172. Emphasis added.

56 Palmerston simply dispatched gunboats to seize Sicily's ships and she quickly capitulated.

57 The standard complaint of price gouging reappears not as the result of a successful appeal to a normative tribunal but as indicating the existence of a problem that has to be addressed politically.

58 In order to treat Bolivia as a proxy for other tin producers, the quantities used are those of her entitlement, rather than actual exports. The index of import prices for the NEI is very similar to

that for Bolivia, Gé Prince, 'Economic Policy in Indonesia, 1900–1942', in J. Thomas Lindblad, ed., *New Challenges in the Modern Economic History of Indonesia,* Leiden, 1993, p. 180.

17 The demise of the International Tin Committee, 1945–1946

1 T247/10, Keynes memo, 'International Regulation of Primary Products', December 1942. This was subsequently published in *Journal of International Economics*, vol. 4, 1974. The model was endorsed by Nicolas Kaldor, 'The Role of Commodity Prices in Economic Recovery', *World Development,* vol. 15, no. 5, May 1987.
2 T247/10, Keynes to Padmore, 22 February 1943. The most successful experience was to be found in nickel, controlled by the International Nickel Company, Keynes to Clauson, 9 July 1943. A laudatory assessment of nickel control is offered by Alex Skelton, 'Nickel', in W. Y. Elliott, ed., *International Control in the Non-Ferrous Metals*, New York: Macmillan, 1937, pp. 178–9. In light of the stress placed on consumer representation in commodity agreements, it should be noted that International Nickel was able to serve consumer interests without any direct participation on their part.
3 T247/10, Keynes to Kahn (Ministry of Supply), 7 February 1944.
4 T247/10, minutes of interdepartmental meeting, 24 January 1944.
5 John Drabble, *Malayan Rubber: the Interwar Years*, London: Macmillan, 1991, p. 24. Much of the particular difficulty with rubber can be attributed to the incoherence of US policy both in relation to civilian consumption and the building of a government reserve stock. Since the RFC sponsored a large synthetic rubber programme, the power of natural rubber producers was inevitably weakened.
6 CO852/625/5, Clauson was quite prepared to concede overall responsibility to the Ministry of Supply, minute, 18 December 1945.
7 From 1942 to 1945, Lyttelton served as Minister of Production but there is no sign that his experience in tin was drawn upon.
8 BT64/2162, Warburton memo on Tin Control, 19 December 1944.
9 CO852/659B/1, Carstairs, memo on Post War Commodity Policy, 18 November 1943. General support for the principle of cartelization was provided in a comprehensive review undertaken by the Board of Trade, *Survey of International Cartels and Internal Cartels, 1944, 1946*, London: Department of Industry, 1976, p. 12.
10 Clauson minute, 29 July 1943, cited by Robin Palmer, 'The Nyasaland Tea Industry in the Era of International Tea Restrictions, 1933–1950', *Journal of African History,* vol. 26, no. 2/3, 1985.
11 CO852/659B/1, Clauson minute, 12 January; Caine minute, 1 January 1944. Caine often took a sound but 'lonely, dissenting view' within the Colonial Office, Wm. Roger Louis, *Imperialism at Bay, 1941–1945*, Oxford: Clarendon Press, 1977, p. 403.
12 *Daily Express,* 13 October 1944.
13 The exception was the AIM which continued to press for a renewal of the existing agreement in all respects, noting that the American producers exercised unfettered control over the cotton market, AIM12, AIM to Ministerio de Economía Nacional, 16 September 1946.
14 Tin Producers' Association, *International Tin Control and Buffer Stocks*, London: 1944.
15 For example, *Manchester Guardian Commercial*, 11 March 1944.
16 'Malayan Tin and the Restrictionist Mentality', *Crown Colonist,* 19 January 1945. Thomas died in January 1941 and Mair in August 1943, leaving Simms and Rich as the remaining leaders of the Cornish group. The animus against the ITC evident in this article also resulted in a very selective quotation from the Fermor report and a misplacing of the rapid change in quotas that occurred in the autumn of 1939 to the spring of 1940.
17 *Proposals transmitted by the Secretary of State of the United States of America to His Majesty's Ambassador at Washington, December 6, 1945*, London: HMSO, 1946, Cmd 6709. Although these were presented as American proposals, Clauson claimed authorship of some of the key paragraphs, CO825/625/6, Clauson to Melville, 8 August 1946.
18 British support for American international economic policy was a condition of the American loan granted in 1945, Alex Cairncross, *British Economic Policy since 1945*, Oxford: Blackwell, 1992, p. 52.
19 Others appeared on coffee and tea in 1944 and Knorr subsequently wrote the one on rubber.
20 Norman Cleaveland, *Bang! Bang! in Ampang*, San Pedro: Symcon, 1973, p.134.

21 *AMM*, 28 February 1945. Cleaveland noted that Knorr could easily have learnt a great deal from the retired general manager of Anglo-Oriental in Malaya who was still living in Oakland, just 30 miles from Stanford, op. cit., p.134. *Mining and Metallurgy*, May 1945, published a scathing review which condemned the study as 'an academic treatise abortive of constructive suggestion', since it failed to recognize the fundamental instability caused by the commodity cycle. The author considered that from 1938 the ITC had accomplished its objective and should simply be modified to include better consumer representation.

22 Robert Maddox, *The War within World War II: The United States and International Cartels*, Westport, CT: Praeger, 2001. The National Association of Manufacturers distributed a pamphlet, *N.A.M. Looks at Cartels*, which attempted to extend public concern about private industrial cartels to intergovernmental commodity agreements. Curiously, it does not include tin among the list of such agreements which are denounced, *NYT*, 31 October 1946.

23 ATD, Cartel Memoranda, Box 43, Gorman to Berg, 10 September 1945. One important industry voice against the application of this policy to minerals markets was that of the editor of the *EMJ*, September 1943.

24 ATD, Cartel Memoranda, Box 6, Comer review of Cartel Agreements in International Trade, pp. 105–6. Another way of misreading the position accorded to Bolivia was to compare the notional 130 per cent entitlement of 59,835 tons for 1941 with the notional 105 per cent entitlement of 46,768 tons for 1942 from which it was easy to conclude that the ITC was actually cutting her back! These gross misjudgements overlooked the central feature of the ITC communiqué of 28 September 1942 which explicitly encouraged unlimited production.

25 David Podell before the Subcommittee on the Mining and Minerals Industry of the Special Committee to Study and Survey Problems of Small Business Enterprise, Hearings, 30 March 1943,Washington, DC: GPO, 1943, part 18, pp. 2, 326–7; *Business Week*, 30 January 1943; *New Republic*, 27 March 1944; *Harper's*, June 1945. *Business Week* soon corrected the impression that there was any real problem with tin, 31 July 1943.

26 ATD, Cartel Memoranda, Box 6, Nolting memo, 10 January 1945.

27 ATD, Cartel Memoranda, Box 6, Till memo on the International Tin Cartel, 27 May 1943, p. 33. There are many points at which Till herself also found it difficult to distinguish fact from fiction but the most outrageous is the allegation that 'a large part of the English and Dutch management personnel are now working for the Japanese'. p. 25.

28 Although no names were mentioned, this allegation was publicized by Carleton Beals, *Lands of the Dawning Morrow*, Indianapolis, IN: Bobbs-Merrill, 1948, p. 128.

29 Feis papers, Box 127, draft letter, October 1943. Dogmatism in the Department of Justice was an important factor in delaying the formulation of a coherent policy with respect to oil in 1943, Daniel Yergin, *The Prize. The Epic Quest for Oil, Money, and Power*, New York: Simon & Schuster, 1991, p. 378.

30 ATD, Tin file, Johnson to Wohlforth, 28 November 1944.

31 US, Congress, Senate, Special Committee Investigating the National Defense Program, *Hearings*, 79th Cong., 1st sess., part 32, 21 September 1945, pp. 15680–1.

32 CO852/625/5, Minutes of the 86th ITC Meeting, 19 September 1945. Similar tributes were made during his lifetime, Ricardo Martinez Vargas, *Estaño en Bolivia y el Comité Internacional*, La Paz: Echaniz, 1936, p. 18.

33 Groothoof/65, 10th ITRDC review of the World Position of Tin (1940–1945), pp. ii–iii.

34 This was made much more explicit in an interview between Caine and van den Broek soon after the ITC meeting, CO852/625/5, Caine minute, 23 October 1945.

35 CO852/625/1, Burgess memo on Post War Tin Position, 31 March 1944, p. 13. Clauson commented that Stephens 'is old and tired and a sick man and runs away from decisions', minute, 10 October 1944. Stephens died two months later. It should be recalled that by this time the two represented half the production of Nigeria and just under half of the corporate capacity in Malaya.

36 CO852/625/6, TPA to ITC, 1 March 1946, pressing for the renewal of the agreement. The TPA even made an eleventh hour request, on 30 October 1946.

37 CO852/625/5, Caine minute, 10 October 1945. The name proposed was Spens, Chairman of LTC. Not only was he a member of the Nigerian delegation to the ITC but he had held a series of major administrative positions with the Ministries of Supply and Production during the war.

38 CO852/625/5, Minutes of subcommittee meeting, 5 December 1945.

39 CO852/625/6, Statement from Belgian Producers, 30 January 1946.

40 CO852/625/3, FO to Washington, May 1946. Not only was this politically wise but it would avoid the embarrassment of the ITC, on which the UK held only a minority of votes, formally

recommending its own renewal. It would be even worse if the other governments accepted the recommendation, since the ITC would keep the secretariat and pre-empt the possibility of a Study Group.

41 CO852/652/6, Minutes of the 87th ITC Meeting, 7 March 1946.
42 CO852/625/6, Clauson to Rich, 12 August 1946.
43 CO852/625/6, Lowinger to Clauson, 8 July 1946. As this debate proceeded, Edward Mason published *Controlling World Trade*, New York: McGraw-Hill, 1946, in which he made an exception to his lukewarm support for commodity agreements and argued that the United States should actively co-operate with a remodelled ITC, pp. 142–4, 192.
44 CO852/625/6, Lowinger to Clauson, 10 July 1946.
45 Fox Papers, Box 7, Houwert, Memo on Cmd 6709, p. 2.
46 CO852/625/6, Clauson to Lowinger, 8, 12 July 1946. 'Rubbing the producers' faces in the dust' would prove to be a fair characterization of American price policy on several occasions, not least at that very moment in the negotiations with Bolivia.
47 CO852/625/6, Clauson to van den Broek, 8 August, 7 September, 1946.
48 CO852/625/6, report on Kennedy's views, 22 March, meeting at the FO, 31 May 1946.
49 CO852/625/6, Clauson minute, 12 June 1946.
50 Consistent with the desire to keep as much control over the conference as possible, there was no question of broadening the representation of consumers. Canada, in particular, had become a significant consumer during the war but her presence was discouraged, CO852/625/7, meeting, 15 August 1946. The UK delegation did not include representation for Nigeria and Malaya.
51 The ITSG took over the statistical functions that Billiton had undertaken on behalf of the ITRDC, while its research functions were transferred to a new organization, the International Tin Research Institute (ITRI).
52 BT11/3022, memo on International Tin Conference, 12 October 1946. The ITSG would therefore have a broader mandate than the Rubber Study Group.
53 Ibid.
54 CO852/625/7, Minutes of the 88th ITC Meeting, 12 December 1946. This was a quarter of the cost of operating the IRRC, CO852/522/3, Campbell memo, 26 February 1943.
55 CO852/625/4, Minutes of the 88th ITC Meeting, 12 December 1946.
56 CO852/625/4, Clauson minute, 13 September 1946.
57 CO852/625/4, Minutes of the 88th ITC Meeting, 12 December 1946.
58 Fox commented that the experience of the seven Study Groups suggests that they were formed to 'prevent rather than stimulate further action towards effective international agreements'. *Tin. The Working of a Commodity Agreement*, London: Mining Journal, 1974, p. 208.

18 From the International Tin Committee to the International Tin Council, 1946–1985

1 Negotiations with the Belgians are briefly reviewed by Gerhard Mollin, *Die USA und der Kolonialismus: Amerika als Partner und Nachfolger der belgischen Macht in Afrika, 1939-1965*, Berlin: Akademie Verlag, 1996, pp. 223-5. Any attempt to gain a concession in the form of a commitment to a commodity agreement was blocked by continued Congressional agitation about the ITC. However, something may have been secured in return for abandoning a tin-grain barter deal with the USSR.
2 The Bolivian Ambassador took considerable offence at this negotiating style, Victor Andrade, *My Missions for Revolutionary Bolivia, 1944-1962*, Pittsburgh: University of Pittsburgh Press, 1976, pp. 69-70.
3 The RFC negotiators had not considered this possibility since Argentina only had a small smelter and the British government would not authorize the purchase of additional Bolivian concentrates. What they overlooked was the desire of Argentina to overcome the constraints imposed by the CTC on the expansion of its tinplate industry. Although the Bolivia/Argentine contract was never implemented, it served its immediate purpose which was to force the hand of the RFC.
4 The text is in *Minerals Yearbook, 1945*, pp. 14-17. In view of the insistence on putting Bolivia on a secure enough foundation to withstand competition, it is worth noting that Congressional special interests forced Truman to set aside his opposition to the provision that subsidized domestic producers.

5 Hence the paradox of "socialist" Britain leading the way to restoration of a free market against the wishes of the "capitalist" United States.
6 *Minerals Yearbook, 1947*, p. 1170. Yip Yat Hoong, *The Development of the Tin Mining Industry in Malaya*, Kuala Lumpur: University of Malaya Press, 1966, pp. 297-310.
7 United Kingdom, *United Nations Conference on Trade and Employment*, London: HMSO, 1948, Cmd 7375. The Charter was designed as the basis of an International Trade Organization to complement the International Monetary Fund and the World Bank in establishing the conditions under which postwar capitalism could flourish. As an Anglo-American initiative, it represented a set of compromises between a commitment to an unbridled market and a pragmatic conception of regulated capitalism.
8 Ibid., Article 56, clause 3a.
9 Ibid., Article 62, clauses a, b.
10 Ibid., Article 56, clause 3c.
11 Ibid., Article 63, clause b.
12 Largely as a result of the failure of the State Department to develop the political support required to overcome Congressional indifference and opposition, Susan Aaronson, *Trade and the American Dream*, Lexington: University Press of Kentucky, 1996. However, the Department remained committed to its conception.
13 The Wheat Agreement took effect from July 1949, following ratification by the United States Senate. It was designed to replace the existing bilateral system with a multilateral one. Consumer members would commit themselves to specific quantities to be bought at no less than an agreed floor price, while producer members made a reciprocal commitment to the same quantities to be sold at no more than a ceiling price. The fact that the United States was an enthusiastic promoter of an agreement which covered its largest export may have led those negotiating one for tin to suppose that the antipathy towards this form of regulation could be suspended in their case as well.
14 International Tin Study Group, *Draft International Agreement on Tin, March 1950*, The Hague, 1950. The text elaborated the general principles, with the details, especially those on the price range, to be settled at the tin conference. A strong argument in favour of an agreement, which even stressed the success of the ITC, was made in an editorial in *EMJ*, July 1950.
15 Robert Caro, *The Years of Lyndon Johnson. Master of the Senate*, New York: Vintage, 2003, pp. 308-14.
16 Ibid., p. 317.
17 Preparedness Subcommittee of the Committee on Armed Services, *Sixth Report, Tin,* 82d Cong., 1st sess., March 1951, pp. 31, 48.
18 Excluding the stockpile and the minimal level of stocks required for commercial operations by both consumers and the Texas smelter.
19 Testimony of Donald Cook, Chief Counsel to the Preparedness Subcommittee, *Hearing on Stockpiling of tin and rubber*, 82d Cong., 1st sess., 24 July 1951, p. 13.
20 Truman appointed Symington in May in order to clean up the RFC, then the subject of political scandal for the many dubious business loans it provided. He had a strong business background and may have brought some familiarity with tin from his previous position with the National Security Resources Board.
21 Preparedness Subcommittee, *Hearing,* op. cit., pp. 16-17, 22.
22 Ibid., p. 22.
23 Later it became clear that the tin and rubber bonanza provided the Malayan government with the resources it needed to conduct a very expensive campaign which effectively ended the civil war.
24 FMS Chamber of Mines and the All-Malaya Chinese Mining Association, *Tin. A reply to the Report on Tin by the Preparedness Subcommittee*, Singapore, 1951, p. 1. Much of the report was based on an even more acerbic draft by Norman Cleaveland, written after several attempts to inform the Johnson subcommittee of the facts about tin had been rebuffed, *Bang! Bang! in Ampang*, San Pedro: Symcon, 1973, pp. 122-4.
25 Indonesian Memorandum included in the Hearing record, op. cit., p. 23.
26 This argument would resurface in the 1980s. It neglects the positive externalities of mining and the negative ones of aid.
27 Not only were certain manufacturers unable to develop their products but restrictions on permitted uses often meant higher unit costs that more than offset any savings on price.
28 The net cost on the tin side of the agreement was £143,000, William Fox, *Tin. The Working of a Commodity Agreement,* London: Mining Journal, 1974, p. 238.

29 This may have been coincidental, since at the same time Truman refused to appoint Symington Secretary of Defense. He then conducted a successful campaign for the open Missouri Senate seat.
30 Preparedness Subcommittee, *Supplemental Report on Tin*, 82d Cong., 2d sess., 1952. p. 1.
31 *AMM*, 25 July 1952. The editorial sarcastically commented that 'the merit and significance of the report have been adequately assessed by the reception it received from the American press. It has been virtually ignored.' The committee assumed that the 202 cents was reached as a result of a draft of the first tin report being leaked to the "cartel."
32 James C. Olson, *Stuart Symington: a life*, Columbia: University of Missouri Press, 2003, pp. 227-30.
33 Although the McReynolds committee had a similar fixation, it called several witnesses from the industry to its public hearings. Johnson only called Symington and Larson of the General Services Administration (GSA), the agency responsible for the stockpile. Johnson's skill at political manipulation suggests that this ignorance was also willful, Cleaveland, op. cit., p. 133. For example, the mission to Malaya became very sympathetic to the plight of the miners but its report was initially classified and then not released until November 1952, a year after it had returned.
34 *EMJ,* December 1951, p. 71.
35 Fox, op. cit., pp. 223-4.
36 Malaya and the Belgian Congo both conceded market share to Bolivia and Indonesia, presumably in recognition of the special problems they encountered, ibid., pp. 222-3. Those concessions only operated for the first agreement.
37 Ibid., pp. 224-5.
38 This had been commissioned by the Eisenhower administration to establish the basic principles of American foreign economic policy and left the door only very slightly ajar to participation in commodity agreements.
39 This also revealed the persistence of longstanding cleavages. The Chinese were overwhelmingly in favour; the Europeans were split between those who followed Anglo-Oriental and those followed the Cornish interests. Leadership in the industry, however, had changed. The Chinese were represented by an exceedingly able and influential Anglophile, Colonel H. S. Lee; he was one of the figures around which an independent Malaysia would be constructed. The mantle of Cornish leadership had passed to Simms, who was remained much more firmly opposed than Thomas, his father-in-law, to any form of commodity control.
40 Calculated from International Tin Council, *Statistical Yearbook 1959*, London, 1959, pp. 22-3 and Peter Roddy, *The International Tin Trade*, London: Woodhead, 1995, Appendix V. Confining the analysis to the capitalist world reflects the absence of any data for the USSR. 1979 was chosen for comparison to avoid any specific distortion associated with the recession of the 1980s.
41 Calculated from *Statistical Yearbook*, op. cit., p. 33 and Roddy, op. cit., Appendix II.
42 Ibid., p. 46. It would have dropped further had much of the new tinplate production not been destined for canned fruit with a high acidic content which required a thicker coating. One of the major advantages of electrolytic deposition is that the precise thickness of the coating for the can can be determined for each product line. Since statistics on tinplate are by weight, rather than area, the efficiency with which tin is now used is underestimated by a considerable margin, W. Robertson, *Tin. Its Production and Marketing*, Westport: Greenwood, 1982, p. 89.
43 A full cost analysis would make tinplate much more competitive since its production creates fewer pollutants and uses less energy, Roddy, op. cit., p. 45.
44 Mahmood Ali Ayub and Hideo Hashimoto, *The Economics of Tin Mining in Bolivia*, Washington: World Bank, 1985, Ch. 6. With limited prospects for further forward linkages, this smelter imposed an additional burden on the Bolivian economy, John Thoburn, *Multinationals, Mining and Development. A Study of the Tin Industry*, Farnborough: Gower, 1981, p. 82.
45 Bill Freund, *Capital and Labour in the Nigerian Tin Mines*, Harlow: Longman, 1981, pp. 215-216.
46 Roddy, op. cit., Appendix I.
47 Thoburn, op. cit., p. 86.
48 Ibid, pp. 66, 82-3.
49 This name applied from 1971 to 1997. Independence left Rwanda free to decide her future and she declined to continue membership in an organization considered contaminated by its colonial origins.

50 Ibid., p. 70. They included Billiton, Grace, National Lead and Patiño. Patiño also operated a tin smelter and transferred all its interests to the Canadian conglomerate, Brascan, in 1980.
51 Fox, op. cit., p. 250.
52 Ibid., p. 284.
53 Ibid., pp. 293-9. Quotas were imposed on Russian sales, especially in the UK.
54 John Thoburn, *Tin in the World Economy*, Edinburgh: Edinburgh University Press, 1994 , p. 92. H. Heymann claims that the problem was exacerbated by the decision to raise the floor to £740 and by mismanagement of the buffer stock, 'The International Tin Scheme,' in E. A. G. Robinson, ed., *Problems in Economic Development*, London: Macmillan, 1965.
55 Fox, op. cit., pp. 307-8. The MNR land reform had left Bolivia chronically short of food and she was forced to strike a very bad bargain.
56 Ibid, p. 328. The Department's earlier concerns about Bolivia, Malaysia and Indonesia were augmented by the chaos in the Congo and the new role played by Thailand in supporting American policy in Vietnam.
57 Guillermo Bedregal, *Monopolios contra países pobres: la crisis mundial del estaño*, Mexico: Siglo XXI, 1967, p. 172. What made the Bolivians so angry was the inconsistency with the commitments made in August 1961 at Punta del Este when the new Treasury Secretary announced that the United States would join the tin agreement, ibid., pp. 153, 174. Indonesia was also angry since such sales were a violation of the promise made to the NEI that releases would only be made in a war emergency, Christopher Brown, *The Political and Social Economy of Commodity Control*, London: Macmillan, 1980, p. 13.
58 Declining consumption reduced the target and increased the surplus. The destabilizing consequences of uncertainty are discussed in William Baldwin, *The World Tin Market. Political Pricing and Economic Competition*, Durham: Duke University Press, 1983, pp. 62-3, 177 and Patricia Perkins, *World Metal Markets. The United States Strategic Stockpile and Global Market Influence*, Westport: Praeger, 1997, pp. 84-5.
59 From 1962 to 1966 over half the surplus, 81,000 tons, was disposed of. Selling such an amount required extensive informal discussions with the Council, Fox, op. cit., Ch. 15.
60 Ibid., pp. 355-8. The issues are discussed in more detail by Yip, op. cit., Ch. 15.
61 Baldwin, op. cit., p. 87.
62 Ibid., p. 90.
63 Calculated from Ayub and Hashimoto, op. cit., p. 34 and discounted to reflect the rising dollar value of the ringgit.
64 Although German consumers were supportive, the government considered that the agreement was "cartel-like" and less interested in stabilization than in raising prices. Pressure from other members of the European Economic Community resulted in a change of mind, Werner Gocht, *Der metallische Rohstoff Zinn*, Berlin: Dunker & Humblot, 1969, p. 180.
65 Baldwin, op. cit., pp. 92-3. Corporate interests, however, remained opposed.
66 Ibid., pp. 98-9.
67 The withdrawal of the United States was followed by that of the USSR and most of the other Eastern European members.
68 Since many other stockpile objectives were still deficient, the administration wanted to sell off surpluses to finance new purchases.
69 Eventual losses to the Malaysian government were $150 million and a further $60 million to their associate, Marc Rich, Helmut and Peter Waszkis, *The Story of Metal Trading*, London: Metal Bulletin, 2003, p. 232.
70 Ralph Kestenbaum, *The Tin Men. A Chronicle of Crisis*, London: Metal Bulletin, 1991, pp. 28, 34-5, Michael Prest, 'The Collapse of the International Tin Agreement,' *Round Table*, no. 302, 1987, p. 172.
71 Modest relief came from China which reduced the level of its sales.
72 Many were both complicated and risky and the Council was not properly informed about their use. Some members sensed a potential problem but were overruled by political fears that revelations would bring the agreement to a premature close, Kastenbaum, op. cit., pp. 44-5, 130-1.
73 David Williamson, 'The ITC collapse: background and consequences,' in Robin Amlot, ed., *The Market for Tin*, London: Tin Publications, 1986, suggests that this increase 'accelerated the decline in consumption,' p. 14. However, it would not affect the United States, where prices declined, nor Japan where the yen remained very strong.
74 Kastenbaum, op. cit., p. 46.
75 Indonesia and Thailand both refused to provide the necessary support, Prest, op. cit., p. 173.

76 The courts could not find a basis on which to hold either the Council or its members liable. The issues are reviewed by B. S. Chimni, *International Commodity Agreements: A Legal Study*, London: Croom Helm, 1987, pp. 203-12.

77 Kastenbaum, op. cit., Chs 12, 13.

78 By this time, Brazil and China had both joined, while Thailand and Malaysia had left since they were no longer net exporters. Private companies then provided financial support for the ITRI, the sole institution that remains from the history of tin control.

19 Conclusion

1 Redzwan Sumum, 'The Goose that Laid the Golden Egg,' reproduced in John Thoburn, *Tin in the World Economy*, Edinburgh: Edinburgh University Press, 1994. p. 179.

2 Subramaniam Pillay, 'Performance of the world tin industry: effects of the international tin agreements, 1956-85,' Ph. D dissertation, University of British Columbia, 1990, p. 185.

3 Reg Eccles, 'The economics of tin supply,' in Robin Amlot, ed., *The Market for Tin*, London: Tin Publications, 1986, p. 63. More detailed analysis suggested that of the large producers only Brazil and Indonesia would be profitable at a price under the floor and that the development of entirely new deposits would require a far higher one, Andrew Sabin, 'Tin availability - market economy countries,' in ibid., p. 84.

4 Christopher Rogers, 'A Scheme for Enforcing the Tin Ceiling Price,' *Intereconomics*, November 1970. Supply allocation was authorized in several agreements but there was never any effort to develop a concrete plan, William Fox, comments, *TIMM,* vol. 80, October 1971, p. A156.

5 The lack of the Chairman's authority was particularly evident in the immediate aftermath of the crash.

6 William Fox, comments, *TIMM*, vol. 81, April 1972, p. A109.

7 William Fox fell into the trap of dismissing such concessions as 'bribery.' *Tin. The Working of a Commodity Agreement,* London: Mining Journal, 1974, p. 265.

8 J. W. F. Rowe, *Primary Commodities in International Trade*, Cambridge: Cambridge University Press, 1965, p. 144.

9 This dimension is a major focus of the work of Debora Spar, *The Cooperative Edge. The Internal Politics of International Cartels*, Ithaca: Cornell University Press, 1994.

10 June Nash, *We Eat the Mines and the Mines Eat Us*, New York: Columbia University Press, 1979, p. 246. Emphasis added.

11 Yip Yat Hoong, *The Development of the Tin Mining Industry in Malaya*, Kuala Lumpur: University of Malaya Press, 1966. p. 272. Cf. William Baldwin, *The World Tin Market. Political Pricing and Economic Competition*, Durham: Duke University Press, 1983. pp. 70-1. Important correctives are Lim Chong-Yah, *Economic Development of Modern Malaya*, Kuala Lumpur: Oxford University Press, 1967, pp. 60-1 and Li Dun Jen, *British Malaya. An economic analysis*, Kuala Lumpur: INSAN, 1982. p. 73. Yip's position is the mirror of that from another self-defined victim, the Bolivian official who claimed that British control over smelting forced Bolivia into the ITC to protect British mines in the Far East, Edward Tomlinson, *New Roads to Riches in the Other Americas*, New York: Scribners, 1939, p. 407.

12 G. C.Allen and Audrey Donnithorne, *Western Enterprise in Indonesia and Malaya*, London: Allen & Unwin, 1962, p. 162.

13 Jean-Jacques van Helten and Geoffrey Jones, 'British Business in Malaya and Singapore since the 1870s,' in R.P.T. Davenport-Hines and Geoffrey Jones, eds, *British Business in Asia since 1860*, Cambridge: Cambridge University Press, 1989, p. 170, K. C. Tregonning, *Straits Tin*, Singapore: Straits Times, n.d, p. 49. Tregonning cites not only Allen and Donnithorne but also Fermor who refrained from any such allegation.

14 Norman Cleaveland, *Bang! Bang! in Ampang*, San Pedro: Symcon, 1973, p. 134.

Bibliography

I Official documents

Bolivia, Banco Minero, *Memorias.*
—— *Tasas e impuestos sobre la industría minera en Bolivia*, La Paz, 1941.
Commonwealth of Australia, *Parliamentary Debates.*
Decennial Reports on the Trade and Industries, 1902–1911, Shanghai, 1912
Estaño en Bolivia, La Paz, 1935.
Federated Malay States, Legislative Council, *Proceedings.*
—— Mines Department, *Annual Report.*
—— *Report of the Commission of Enquiry ... into the Conduct and Management of the Mines Department*, Kuala Lumpur, 1941.
Fermor, Lewis Sir, *Report upon the Mining Industry of Malaya*, Kuala Lumpur, 1940.
French Indo-China, Inspection générale des mines et de l'industrie, *L'industrie minérale en Indochine*, Hanoi, 1931.
International Tin Study Group, *Draft International Agreement on Tin, March 1950*, The Hague, 1950.
Kelantan, *Annual Report on the Social and Economic Progress of the People of the State of Kelantan.*
Kedah, *Administrative Report for the State of Kedah.*
League of Nations, *Report of the Committee for the Study of the Problem of Raw Materials*, Geneva, 1937.
Nigeria, Mines Department, *Annual Report.*
Portugal, Ministério da Economia, *Elementos para o estudo da Indústria Mineira em Portugal nos anos de 1930 a 1939*, Lisbon, 1941.
Stone, Gilbert, *Mining Laws of the British Empire and of Foreign Countries*, London: HMSO, 1924.
Straits Settlements, Legislative Council, *Proceedings.*
Tasmania, Department of Mines, *A Century of Tin Mining at Mount Bischoff, 1871–1971*, Hobart, 1972.
United Kingdom, Board of Trade, *Survey of International Cartels and Internal Cartels, 1944, 1946*, London: Department of Industry, 1976.
—— Department of Mines, Report of the Advisory Committee for the Metalliferous Mining and Quarrying Industry, London: HMSO, 1932.
—— House of Commons, *Parliamentary Debates.*
—— *Papers Relating to the International Tin Control Scheme*, London: HMSO, 1935, Cmd 4825.
—— *Papers Relating to the International Tin Control Scheme*, London: HMSO, 1938, Cmd 5879.
—— *Joint Declaration by the President of the United States of America and Mr Winston Churchill ... August 14, 1941*, London: HMSO, 1941, Cmd 6321.
—— *Agreement between the Governments of Belgium, Bolivia, the United Kingdom and the Netherlands for the International Control of the Production and Export of Tin*, London: HMSO, 1942, Cmd 6396.
—— *Proposals transmitted by the Secretary of State of the United States of America to His Majesty's Ambassador at Washington, December 6, 1945*, London: HMSO, 1946, Cmd 6709.
—— *United Nations Conference on Trade and Employment*, London: HMSO, 1948, Cmd 7375.
United States, Bureau of Mines, Information Circulars.

—— Bureau of Foreign and Domestic Commerce, Trade Information Bulletins.
—— Congress, House, Subcommittee, Committee on Foreign Affairs, *Investigation on U.S. Dependence on Foreign Tin*, 74th Cong., 1st sess., 1935.
—— Congress, Senate, Subcommittee on Mining and Minerals Industry of the Special Committee to Study and Survey Problems of American Small Business, *Hearings*, part 18, 30 March 1943.
—— Congress, Senate, Special Committee Investigating the National Defense Program, *Hearings*, 79th Cong., 1st sess., part 32, 21 September 1945.
—— Congress, Senate, Preparedness Subcommittee of the Committee on Armed Services, *Sixth Report, Tin*, 82d Cong., 1st sess., March 1951.
—— Congress, Senate, Preparedness Subcommittee, *Hearing on Stockpiling of tin and rubber*, 82d Cong., 1st sess., 24 July 1951.
—— Congress, Senate, Preparedness Subcommittee, *Supplemental Report on Tin*, 82d Cong., 2d sess., 1952.
—— Department of the Interior, *Tin Deposits of the Republic of Mexico*, Geological Survey Bulletin 935-C, Washington, DC: GPO, 1942.
—— Federal Trade Commission, *Tin can and tinware manufacturing corporations*, Washington, DC: GPO, 1941.

II Books, articles, theses

Aaronson, Susan, *Trade and the American Dream*, Lexington, KY: University Press of Kentucky, 1996.
Adams, F. Gerard and Jere Behrman, *Commodity Exports and Economic Development: the Commodity Problem and Policy in Developing Countries*, Lexington, KY: Lexington Books, 1982.
Ágoston, Gábor, *Guns for the Sultan: Military Power and the Weapons Industry in the Ottoman Empire*, Cambridge: Cambridge University Press, 2005.
Agricola, Georgius, *De Re Metallica*, trans. Herbert and Lou Hoover, New York: Dover, 1950.
Albarracin Millan, Juan., *El poder financiero de la gran minería Boliviana*, La Paz: Ediciones Akapana, 1995.
Aldcroft, Derek, *From Versailles to Wall Street*, London: Allen Lane, 1977.
Aldrich, Richard J. *The Key to the South. Britain, the United States, and Thailand during the Approach to the Pacific War, 1929–1942*, Kuala Lumpur: Oxford University Press, 1993.
Allen, G. C. and Audrey Donnithorne, *Western Enterprise in Indonesia and Malaya*, London: Allen & Unwin, 1962.
Alpern, Stanley, 'Did They or Didn't They Invent It? Iron in Sub-Saharan Africa', *History in Africa*, vol. 32, 2005.
Andaya, Barbara, *Perak. The Abode of Grace. A Study of an Eighteenth-Century Malay State*, Kuala Lumpur: Oxford University Press, 1979.
Andrade, Victor, *My Missions for Revolutionary Bolivia, 1944–1962*, Pittsburgh, PA: University of Pittsburgh Press, 1976.
Anjorin, A., 'Tin Mining in Northern Nigeria During the Nineteenth and Early Part of the Twentieth Centuries', *Odu: A Journal of West African Studies*, vol. 5, 1971.
Arasaratnam, S., 'Some Notes on the Dutch in Malacca and the Indo-Malayan Trade, 1641–1700', *Journal of Southeast Asian History*, vol. 10, no. 3, 1969.
Arce, Roberto *Desarrollo económico e histórico de la minería en Bolivia*, La Paz: Plural Editores, 2003.
Aubet, Maria Eugenia, *The Phoenicians and the West: Politics, Colonies and Trade*, trans. Mary Turton, Cambridge: Cambridge University Press, 1993.
Avieny, Wilhelm, *Strukturwandlungen der Weltmetallwirtschaft*, *Kieler Vorträge*, no. 67, Jena, 1941.
Ayub Mahmood, Ali and Hideo Hashimoto, *The Economics of Tin Mining in Bolivia*, Washington, DC: World Bank, 1985.
Bagchi, Amiya Kumar, *Private Investment in India 1900–1930*, Cambridge: Cambridge University Press, 1972.
Baldwin, William, *The World Tin Market. Political Pricing and Economic Competition*, Durham: Duke University Press, 1983.
Ball, Simon, 'The German Octopus: The British Metal Corporation and the Next War, 1914–1939', *Enterprise & Society*, vol. 5, no. 3, September 2004.

Ballande, Laurence, *Essai d'étude monographique et statistique sur les ententes économiques internationales*, Paris: Librairie technique et économique, 1936.

Barber, Alvin, 'Tin Control in a Major War', *Far Eastern Survey,* vol. 8, no. 23, November 22, 1939.

Bartelheim, Martin, E. Niederschlag and T. Rehren., 'Research into Prehistoric Metallurgy in the Bohemian/Saxon Erzgebirge', in Bernard Hänsel, ed., *Mensch und Umwelt in der Bronzezeit Europas*, Kiel: Oetker-Voges Verlag, 1998.

Barton, D. Bradford, *A History of Tin Mining and Smelting in Cornwall*, Exeter: Wheaton, 1989.

Baruch, Bernard, *American Industry in the War. A Report of the War Industries Board, 1921*, New York: Prentice-Hall, 1941.

Bassett, D. K., 'British 'Country' Trade and Local Trade Networks in the Thai and Malay States, c. 1680–1770', *Modern Asian Studies,* vol. 23, vo. 4, 1989.

Battersby, Paul, 'Diggers and Diplomats: Australian Mining Entrepreneurs and The Evolution of the Australia-Thailand Bilateral Relationship, 1901–1941', in M. Hayes and S. Smith, eds, *Australia-Thailand Relations in the Twentieth Century*, Bangkok: Australian Studies Centre, Kasetsart University, 2000.

—— *To the Islands: White Australians and the Malay Archipelago since 1788*, Langham, MD: Lexington Books, 2007.

Beagrie, Neil, 'The Romano-British Pewter Industry', *Britannia*, vol. 20, 1989.

Beals, Carleton, *Lands of the Dawning Morrow*, Indianapolis, IN: Bobbs-Merrill, 1948.

Beard, R. R., 'Property and Operations of Patiño Mines and Enterprises', *Engineering and Mining Journal*, 9 August 1930.

Bedregal, Guillermo, *Monopolios contra países pobres: la crisis mundial del estaño*, Mexico: Siglo XXI, 1967.

Bieber, León, 'El comercio germano-boliviano, 1936–1939', in Nikolaus Böttcher and Bernd Hausberger, eds, *Dinero y negocios en la historía de América Latina*, Frankfurt: Vervuert, 2000.

Billiton Maatschappij, *Gedenkboek Billiton, 1852–1927*, The Hague: Nijhoff, 1927.

Birch, F. D., 'Tropical Milestones: Australian Gold and Tin Mining Investments in Malaya and Thailand, 1880–1930', MA Thesis, University of Melbourne, 1976.

Bonney, R., *Kedah 1771–1821: The Search for Security and Independence*, Kuala Lumpur: Oxford University Press, 1971.

Borg, Dorothy, *The United States and the Far Eastern Crisis of 1933–1938*, Cambridge, MA: Harvard University Press, 1964.

Boroffka, Nikolaus, J. Cierney and J. Lutz., 'Bronze Age Tin from Central Asia', in Katie Boyle, *et al.* eds, *Ancient Interactions: East and West in Eurasia*, Oxford: Oxbow Books, 2002.

Bowen, H. V., 'Sinews of trade and empire: the supply of commodity exports to the East India Company during the late eighteenth century', *Economic History Review,* vol. 55, no. 3, 2002.

Bower, Penelope, 'The Mining Industry', in Margery Perham, ed., *The Economics of a Tropical Dependency*, London: Faber, 1948, vol. 2.

Brand, Ronald, T. S. Shenkin and E. J. Meehan., 'Legal Aspects of International Commodity Agreements', in Osman Suliman and Mahdi El-Baghdadi, eds, *Global Commodity Price Stabilization*, Westport, CT: Quorum, 1995.

Brown, Christopher, *The Political and Social Economy of Commodity Control*, London: Macmillan, 1980.

Brown, Helen, *Tin at Tingha*, Armidale, Brown, 1982.

Brown, Ian, *Economic Change in Southeast Asia, c. 1830–1980*, Kuala Lumpur: Oxford University Press, 1997.

—— and A. K. Dey, *India's Mineral Wealth*, Bombay: Oxford University Press, 1955.

Brown, J. Coggin, and A. Heron, 'Geology and Ore Deposits of the Tavoy District', Geological Society of India, *Memoirs*, vol. 94, pt. 2, 1923.

Brown, Kenneth, *The British Toy Business: A History Since 1700*, London: Hambledon Press, 1996.

Brown, Rajeswary, *Capital and Entrepreneurship in Southeast Asia*, New York: St. Martin's Press, 1994.

Buck, W. Keith, 'Study Groups versus Commodity Agreements: An Appraisal', *Natural Resources Forum*, November 1987.

Buckley, J. A., *A History of South Crofty Mine*, Redruth: Dyllansow Truran, n.d.

Burke, Gill, 'The Rise and Fall of the International Tin Agreements', in Jomo, ed., *Undermining Tin. The Decline of Malaysian Pre-eminence*, Sydney, 1990.

Burt, Roger, 'The transformation of the non-ferrous metals industries in the seventeenth and eighteenth centuries', *Economic History Review*, vol. 48, no. 1, 1995.

Burn, Duncan, *The Steel Industry, 1939–1959*, Cambridge: Cambridge University Press, 1961.

Caine, Sydney, *Prices for Primary Producers*, London: Institute for Economic Affairs, 1963.

Cairncross, Alex, *British Economic Policy since 1945*, Oxford: Blackwell, 1992.

Calabre, Serge, *L'étain*, Paris: Economica, 1991.

Calvert, Albert, *Nigeria and its Tin Fields*, New York: Arno Press, 1977, reprint of 1910 edition.

Campbell, John, *F. E. Smith*, London: Cape, 1983.

Caro, Robert, *The Years of Lyndon Johnson. Master of the Senate*, New York: Vintage, 2003.

Carr, J. C. and Taplin, W., *The History of the British Steel Industry*, Cambridge: Harvard University Press, 1962.

Carrasco, Manuel, *Simon Patiño. Un Procer Industrial*, Cochabamba: Editorial Canelas, 1960.

Cassington, John, *A Metal Man's Wartime Diary*, Eynsham: Quin, 1941.

Chai Hon-Chan, *The Development of British Malaya, 1896–1909*, Kuala Lumpur: Oxford University Press, 1964.

Chakrabarti, D. K., 'The Problem of Tin in Early India', *Man and Environment*, vol. 3, 1979.

Chang, K. C., 'The Importance of Bronzes in Ancient China', in Thomas Chase, ed., *Ancient Chinese Bronze Art*, New York: China Institute in America, 1991.

Chen May Yee, *Born and Bred in Pewter Dust. The Royal Selangor Story*, Kuala Lumpur: Archipelago Press, 2003.

Chernykh, Evgeny, L. I. Avilova and L. B. Orlovskaya., 'Metallurgy of the Circumpontic Area', *Der Anschnitt, Anatolian Metal II.*

Chibber, H. L., *The Mineral Resources of Burma*, London: Macmillan, 1934.

Childe, V. Gordon, 'Archaeological Ages as Technological Stages', *Journal of the Royal Anthropological Institute,* vol. 74, no. 1/2, 1944.

—— *The Bronze Age*, New York: Biblo, 1963.

Chimni, B. S., *International Commodity Agreements: A Legal Study*, London: Croom Helm, 1987.

Christie, Jan Wisseman, 'Trade and State Formation in the Malay Peninsula and Sumatra, 300 B.C.–A.D. 700', in J. Kathirithamby-Wells and John Villiers, eds, *The Southeast Asian Port and Polity. Rise and Demise*, Singapore: Singapore University Press, 1990.

Cierney, Jan and Gerd Weisgerber, 'Bronze Age Tin Mines in Central Asia', in Alessandra Giumlia-Mair and Fulvia Lo Schiavo, eds, *Le problème de l'étain à l'origine de la métallurgie*, Oxford: Archaeopress, 2003.

Cipolla, Carlo, *Guns, Sails and Empires: Technological Innovation and the Early Phases of European Expansion 1400–1700*, New York: Pantheon, 1965.

Clavin, Patricia, *The Failure of Economic Diplomacy. Britain, France, Germany, and the United States*, Basingstoke: Macmillan, 1996.

Cleaveland, Norman, *Bang! Bang! In Ampang*, San Pedro, CA: Symcon, 1973.

Coburn, Patricia, *The Years of THE WEEK*, London: Comedia, 1985.

Contreras, Manuel, 'En torno a "la formación del imperio del estaño de Patiño' de Herbert S. Klein', *Historia Boliviana*, vol. 4, no. 2, 1984.

—— 'La minería estañifera en la primera guerra mundial', in *Minería y Economía en Bolivia*, La Paz, 1984.

—— 'Debt, Taxes, and War: The Political Economy of Bolivia, c. 1920–1935', *Journal of Latin American Studies*, vol. 22, no. 2, May 1990.

—— *Tecnología moderna en los Andes. Minería e ingeniería en Bolivia en el siglo XX*, La Paz: Biblioteca Minera Boliviana, 1994.

—— and Napoleón Pacheco, *Medio siglo de minería mediana en Bolivia, 1939–1989*, La Paz: Biblioteca Minera Boliviana, 1989.

Copeland, Melvin T., *A Raw Commodity Revolution*, Cambridge, MA: Harvard University Press, 1938.

Corin, John, *Levant. A Champion Cornish Mine*, Redruth: Trevithick Society, 1992.

Costin, Cathy Lynne and Timothy Earle, 'Status Distinction and Legitimation of Power as Reflected in Changing Patterns of Consumption in Late Prehispanic Peru', *American Antiquity*, vol. 54, no. 4, October 1989.

Courbin, Alain, *Village Bells. Sound and Meaning in the Nineteenth-century French Countryside*, New York: Columbia University Press, 1988.

Cox, Howard, *The Global Cigarette. Origins and Evolution of British American Tobacco, 1880–1945*, Oxford: Oxford University Press, 2000.
—— 'Business on Trial: The Tobacco Securities Trust and the 1935 Pepper Debacle', *Business History*, vol. 49, no. 6, November 2007.
Craddock, Paul, J. Ambers, D. Hook, R. Farquhar, V. Chikwendu, A. Umeji and T. Shaw., 'Metal Sources and the Bronzes from Igbo-Ukwu, Nigeria', *Journal of Field Archaeology*, vol. 24, no. 4, Winter 1997.
Craig, John, *The Mint. A History of the London Mint from A. D. 287 to 1948*, Cambridge: Cambridge University Press, 1953.
Crespo, Alfonso, *Los Aramayo de Chichas*, Barcelona: Blume, 1981.
Crush, Jonathan 'Settler-Estate Production, Monopoly Control and the Imperial Response: The Case of the Swaziland Corporation, Ltd', *African Economic History*, no. 8, Fall 1979.
Crosby, Josiah, *Siam: The Crossroads*, London: Hollis & Carter, 1945.
Cross, J. A., *Sir Samuel Hoare, a Political Biography*, London: Cape, 1977.
—— *Lord Swinton*, Oxford: Clarendon Press, 1982.
Cunliffe-Lister, P., 'The Economic Policy of the Colonial Empire', *United Empire*, October 1934.
Cushman, Jennifer, *Family and State. The Formation of a Sino-Thai Tin-Mining Dynasty*, Singapore: Oxford University Press, 1991.
Däbritz, Walther, *Fünfzig Jahre Metallgesellschaft, 1881–1931*, Frankfurt am Main: Osterrieth, 1931.
Davenport, Ernest Harold, *The Price of Tin. The Problem of Stabilisation*, London: London General Press, 1928.
Davenport, Nicolas, *Memoirs of a City Radical*, London: Weidenfeld and Nicolson, 1974.
Davis, Joseph, 'Experience under Intergovernmental Commodity Agreements, 1902–1945', *Journal of Political Economy*, vol. 54, no. 3, June 1946.
Davies, P. N., 'Business Success and the Role of Chance: The Extraordinary Philipps Brothers, *Business History*, vol. 23, no. 2, July 1981.
de Kergaradec, M., 'Mines d'étain de Ko-Kieou (Chine)', *Annales des Mines*, vol. 12, 1877.
de la Barra, Jorge, *Singapore y el estaño en la Malaya*, La Paz: np, 1941.
de Launay, L. A, *La géologie et richesses minerales de l'Asie*, Paris: Béranger, 1911.
Deloncle, Pierre, 'La mise en valeur du Laos' in Jean Renaud, ed., *Laos: dieux, bronzes, et montagnes*, Paris: Redier, 1930.
Dennery, Etienne, *Le problème des matières premières*, Paris: IICI, 1939.
Denyer, R. J. and K. Heath, 'Mining and Milling Tin-Tungsten Ore at Mawchi Mines, Burma', Institution of Mining and Metallurgy, *Transactions*, vol. 49, 1939–40.
Deshayes, Jean, *Les outils de bronze de l'Indus au Danube (IVe au IIe Millénaire)*, Paris: Geuthner, 1960.
de Villa, E. M. *The Study of Mines in China, Indochina, and Malaya*, Hong Kong: Daily Press, 1935.
Dijk, Wil, *Seventeenth-century Burma and the Dutch East India Company, 1634–1680*, Singapore: Singapore University Press, 2006.
Domosh, Mona, 'Pickles and purity: discourses of food, empire and work in turn-of-the-century USA', *Social and Cultural Geography*, vol. 4, no. 1, March 2003.
Drabble, John, *Malayan Rubber: the Interwar Years*, London: Macmillan, 1991.
—— *An Economic History of Malaysia, c. 1800–1900*, Basingstoke: Macmillan, 2000.
Draper, Marshall, 'Tin Industry of Yunnan, China', *Mining and Metallurgy*, April–May 1931.
Duboscq, André, 'Les mines d'étain de Ko-Kicou', *Revue Politique et Parlementaire*, vol. 19, no. 2, June 1921.
Earl, Bryan, ''Melting tin' in the West of England. A study of an old art', *Journal of the Historical Metallurgy Society*, vol. 19, no. 2, 1985.
Eastham, J. K., 'Rationalisation in the Tin Industry', *Review of Economic Studies*, vol. 4, no. 1, October 1936.
Eaton, E. R. and Hugh McKerrell, 'Near Eastern alloying and some textual evidence for the early use of arsenical copper', *World Archaeology*, vol. 8, no. 2, October 1976.
Eccles, Reg. 'The economics of tin supply', in Robin Amlot, ed., *The Market for Tin*, London: Tin Publications, 1986.
Elliott, W. Y., ed., *International Control in the Non-Ferrous Metals*, New York: Macmillan, 1937.
Epstein, Mieczysław, 'La réglementation internationale d'un marché de matière première', PhD dissertation, University of Fribourg, 1943.

Excell, F. K., *Siamese Tapestry*, London: Hale, 1963.

—— *In Siamese Service*, London: Cassell, 1967.

Fagg, Bernard, 'Archaeological Notes from Northern Nigeria', *Man*, vol. 46, May–June 1946.

Falkus, Malcolm, 'Early British business in Thailand', in R. P. T. Davenport-Hines and Geoffrey Jones, eds, *British Business in Asia since 1860*, Cambridge: Cambridge University Press, 1989.

Favier, Jean, *Gold and Spices. The Rise of Commerce in the Middle Ages*, New York: Holmes & Meier, 1998.

Ferrand, Gabriel, trans., *Voyage du marchand arabe Sulayman en Inde et en Chine redigé en 851*, Paris: Brossard, 1922.

Ferrier, R. W., 'Terms and Conditions under which English Trade was Transacted with Safavid Persia', *Bulletin of the School of Oriental and African Studies,* vol. 49, no. 1, 1986.

Feis, Herbert, 'Raw Material Prices and Controls', Academy of Political Science, *Proceedings,* vol. 21, no. 2, January 1945.

Fishman, Leo, 'Wartime Control over Tin in Great Britain', *Journal of Political Economy*, vol. 54, no 5, October 1946.

Fistié, P., *Sous-developpement et utopie en Siam. Le programme de réforme présenté en 1933 par Pridi Phanonmyong*, Paris: Mouton, 1969.

Flood, Edward, 'Japan's Relations with Thailand: 1928–1941', PhD dissertation, University of Washington, DC, 1967.

Flower, Philip, *A History of the Trade in Tin*, London: G. Bell, 1880.

FMS Chamber of Mines and the All-Malaya Chinese Mining Association, *Tin. A reply to the Report on Tin by the Preparedness Subcommittee*, Singapore, 1951.

Foianini, Dionisio, *Misión Cumplída*, Santa Cruz: Sirena, 1991.

Forbes, R. J., 'Hydraulic Engineering and Sanitation', in Charles Singer, ed., *History of Technology*, Oxford: Oxford University Press, 1954, vol. 2.

—— *Studies in Ancient Technology,* Leiden: Brill, 1957.

Fox, William, *Tin. The Working of a Commodity Agreement*, London: Mining Journal, 1974.

Francfort, Henri, *Fouilles de Shortughai*, Paris: Broccard, 1989, vol. 1.

Frankel, S. H., *Investment and the Return to Equity in the South African Gold Mining Industry, 1887–1965*, Oxford: Blackwell, 1967.

Franklin, M. J., *British Biscuit Tins 1868–1939: An Aspect of Decorative Packaging,* London: New Cavendish Books, 1979.

Franklin, Ursula 'The Beginnings of Metallurgy in China: A Comparative Approach', in G. Kuwayama, *The Great Bronze Age of China*, Los Angeles, 1983.

Freund, Bill, *Capital and Labour in the Nigerian Tin Mines*, Harlow: Longman, 1981.

Furber, Holden, *Rival Empires of Trade in the Orient, 1600–1800*, Oxford: Oxford University Press, 1976.

Gallard, J. A. and Murray Stuart, *Tin: Salient Facts and Opinions*, London: Mining Publications, 1929.

Garelli, P., *Les assyriens en Cappadoce*, Paris: Maisonneuve, 1963.

Geddes, Charles, *Patiño. The Tin King*, London: Hale, 1972.

Gerini, G. E., 'Historical Retrospect of the Junk Ceylon Island', *Journal of the Siam Society*, 1902.

Gerrard, Sandy, *The Early British Tin Industry*, Stroud: Tempus, 2000.

Gerritsen, D., 'De Tinfusie', *Koloniale Studiën*, vol. 19, no. 3, June 1935.

—— 'Nogmals de Tinfusie', *Koloniale Studiën*, vol. 22, no. 4, August 1938.

Gibson-Jarvie, Robert, *The London Metal Exchange*, London: Woodhead-Faulkner, 1976.

Gilbert, Christopher, 'International Commodity Agreements: Design and Performance', *World Development,* vol. 15, 1987.

—— 'International Commodity Agreements: An Obituary Notice?', *World Development,* vol. 24, no. 1, January 1996.

Gillis, Carole, 'The Economic Value and Symbolism of Tin', in Suzanne M. Young, A. M. Pollard, P. Budd, and R. A. Iver., eds, *Metals in Antiquity*, Oxford: Archaeopress, 1999.

Glenister, A. G., 'The Gopeng Consolidated Tin Mines', *Mining Magazine*, August 1926.

Glover, Patricia, 'Low Malayan Tin Quota Leads to Chinese Unemployment', *Far Eastern Survey*, vol. 8, no. 13, 21 June 1939.

—— 'Malayan Tin Producers Clash with Agricultural Interests', *Far Eastern Survey*, vol. 8, no. 14, 5 July 1939.

Gocht, Werner, *Der metallische Rohstoff Zinn*, Berlin: Dunker & Humblot, 1969.

Golas, Peter, 'Tin Mining in 19th and 20th Century China: High Production, Low Technology', in Yang Tsui-hua and Huang Yi-lung, eds, *Science and Technology in Modern China*, Taipei, 1991.
—— *Mining, Science and Civilisation in China*, vol. 5, pt. 13, Cambridge: Cambridge University Press, 1999.
Gomez d'Angelo, Walter, *La minería en el desarrollo económico de Bolivia, 1900–1970*, La Paz: Los Amigos del Libro, 1978.
González-Ruibal, Alfredo, 'Facing Two Seas: Mediterranean and Atlantic Contacts in the North-west of Iberia in the First Millennium BC', *Oxford Journal of Archaeology*, vol. 23, no. 3, 2004, p. 292.
Goodrich, L. Carrington, 'Early Cannon in China', *Isis*, vol. 55, no. 2, 1964.
Gordon, A. P. L., *Rationalisation for Tin*, London: Saint Catherine, 1929.
Grant, W. R., 'Iron in ancient tin from Rooiberg', *Journal of Archaeological Science*, vol. 21, 1994.
—— 'The Sourcing of Southern African Tin Artefacts', *Journal of Archaeological Science*, vol. 26, 1999.
Grébenart, Danilo, 'Les métallurgies du cuivre et du fer autour d'Agadez, Niger, des origines au début de la périod médiévale', in Nicole Échard, ed., *Métallurgies Africaines*, Paris, Société des Africanistes, 1983.
Griffith, David, *Decorative Printed Tins: The Golden Age of Printed Tin Packaging,* New York: Studio Vista, 1979.
Griffin, James M., 'Previous Cartel Experience: Any Lessons for OPEC?' in Lawrence R. Klein and Jaime Marquez, eds, *Economics in Theory and Practice*, Dordrecht: Kluwer, 1989.
Griliopulos, Basil, *The Man who Twisted the Lion's Tail. The Story of One Man's War Against the British Government*, Hicksville: Exposition, 1974.
Groothoff, A., *De Tinfusie in Nederlandsch-Indië*, The Hague: M. Nijhoff, 1937.
Guilmartin, John F., *Gunpowder and Galleys. Changing Technology and Mediterranean Warfare at Sea in the 16th Century*, London: Conway, 2003.
Gupta, Bishrupriya, 'International Tea Cartel during the Great Depression, 1929–1937', *Journal of Economic History*, vol. 61, no. 1, March 2001.
Hall, Bert S., *Weapons and Warfare in Renaissance Europe. Gunpowder, Technology and Tactics*, Baltimore: Johns Hopkins University Press, 1997.
Hallowell, Burton C., 'Administration of Tin Control in Bolivia, 1931–1939', *Inter-American Economic Affairs*, vol. 3, no. 2, Autumn 1949.
Han Tiauw Tjong, *De Industrialisatie van China*, The Hague: Nijhoff, 1922.
Hannan, Elizabeth, 'Congress Investigates Tin', *International Law and Relations*, vol. 5, no. 4, January 13, 1936.
Harlow, Sue, *Tin Gods. A Social History of the Men and Women of Maranboy, 1913–1962*, Darwin: Historical Society of the Northern Territory, 1997.
Hart, G. H. C., *Towards Economic Democracy in the Netherlands Indies*, New York: Institute of Pacific Relations, 1943.
Harris, T. R., *Dolcoath: Queen of Cornish Mines*, Penzance: Trevithick Society, 1974.
Hatcher, John, *English Tin Production and Trade before 1550*, Oxford: Clarendon Press, 1973.
—— and T. C. Barker, *A History of British Pewter*, London: Longman, 1974.
Headrick, Daniel, *The Tools of Empire. Technology and European Imperialism in the Nineteenth Century*, New York: Oxford University Press, 1981.
Hedges, Ernest, *Tin in Social and Economic History*, London: Arnold, 1964.
Heffer, Philip, *East Pool and Agar. A Cornish Mining Legend*, Redruth: Dyllansow Truran, 1985.
Heidhues, Mary Somers, *Bangka Tin and Mentok Pepper*, Singapore: Institute of Southeast Asian Studies, 1992.
Hennart, Jean-Francois, 'The Tin Industry', in M. Casson, ed., *Mulitinationals and World Trade*, London: Allen & Unwin, 1986.
—— 'Transaction Costs and the Multinational Enterprise: the Case of Tin', *Business and Economic History*, vol. 16, 1987.
—— 'International Financial Capital Transfers: a Transaction Cost Framework', *Business History*, vol. 36, no. 1, January 1994.
Heussler, Robert, *British Rule in Malaya: The Malayan Civil Service, 1867–1942*, Westport, CT: Greenwood, 1981.
—— *Completing a Stewardship: the Malayan Civil Service, 1942–1957*, Westport, CT: Greenwood, 1983.

Heymann, H., 'The International Tin Scheme', in E. A. G. Robinson, ed., *Problems in Economic Development*, London: Macmillan, 1965.

Hexner, Paul, *International Cartels*, Chapel Hill, NC: University of North Carolina Press, 1945.

Hillman, John, 'The Emergence of the Tin Industry in Bolivia', *Journal of Latin American Studies*, vol. 16, pt. 2, November 1984.

—— 'Bolivia and the International Tin Cartel, 1931–1941', *Journal of Latin American Studies*, vol. 20, no. 1, May 1988.

—— 'Malaya and the International Tin Cartel', *Modern Asian Studies*, vol. 22, pt. 2, May 1988.

—— 'Bolivia and British Tin Policy, 1939–1945', *Journal of Latin American Studies*, vol. 22, no. 2, May 1990.

—— 'The Freerider and the Cartel: Siam and the International Tin Restriction Agreements, 1931–1941', *Modern Asian Studies*, vol. 24, no. 2, May 1990.

—— 'The Bandoeng Tin Pool, 1920–1924', *Malaysian Journal of Economic Studies* vol. 32, no. 2, 1995.

—— 'The Impact of the International Tin Restriction Schemes on the Return to Equity of Tin Mining Companies, 1927–1939', *Business History*, vol. 39, no. 3, July 1997.

—— 'Chartered Companies and the Development of the Tin Industry in the Belgian Congo, 1900–1939', *African Economic History*, vol. 25, 1997.

—— 'The Politics of Equity: Administering Tin Restriction in Nigeria, 1931–1933', *Mining History Journal,* 1997.

—— 'Promise and Performance in the Bolivian Tin Mining Industry, 1923–1929', *Canadian Journal of Latin American and Caribbean Studies*, vol. 23, no. 45, Spring 1998.

—— 'Dredging for Tin in Southeast Asia, 1921–1931', in V[th] International Mining History Conference, *Proceedings*, Milos, 2001.

—— 'Mining Industry and the State: The Politics of Tin Restriction in Bolivia, 1936–1939', *Bulletin of Latin American Research*, vol. 21, no. 1, January 2002.

—— 'Australian Capital and South-East Asian Tin Mining', *Australian Economic History Review*, vol. 45, no. 2, July 2005.

Hochschild, M., 'The Bolivian Mining Industry and the Present Crisis', *Mining Journal*, 22 November 1930.

Hodgson, Bernard, ed., *The Invisible Hand and the Common Good*, Berlin: Springer, 2004.

Hoogendoorn, Willem, *Der Zinnpreis. Eine Untersuchung uber die Zinnpreisbildung und ihre monopolistiche Beeinflussung*, Kiel: np, 1934.

Holmes, A. R. and Green, E., *Midland: 150 Years of Banking Business*, London: Batsford, 1986.

Hoover, Theodore, *The Economics of Mining*, Stanford, CA: Stanford University Press, 1933.

Hosler, Dorothy, *The Sounds and Color of Power: The Sacred Metallurgical Technology of Ancient West Mexico*, Cambridge, MA: MIT Press, 1994.

Hotelling, H., 'The Economics of Exhaustible Resources', *Journal of Political Economy,* vol. 39, no. 2, April 1931.

Hövig, P., 'Nederlandsch-Indië en de Tin-Restrictie-Plannen', *Economisch-Statistische Berichten*, 11 December 1929.

—— 'De Tinfusie', *Koloniale Studiën,* vol. 21, no. 5, October 1937.

Huff, W. G., *The Economic Growth of Singapore*, Cambridge: Cambridge University Press, 1994.

—— 'Monetization and Financial Development in Southeast Asia before the Second World War', *Economic History Review,* vol. 56, no. 2, 2003.

Hull, Cordell, *Memoirs of Cordell Hull*, New York: Macmillan, 1948.

Hume, Walter, *Report on the Queensland Tin Field*, Brisbane, 1874.

Huntingford, G.W.B, trans. and ed., *The Periplus of the Erythraean Sea*, London: Hakluyt Society, 1980.

Idriess, Ion, *The Tin Scratchers*, Sydney: Angus & Robertson, 1959.

Ingram, J. C., *Economic Change in Thailand, 1850–1970*, Stanford, CA: Stanford University Press, 1971.

Institut für Weltwirtschaft, *Die Zinnindustrie des südostasiatischen Raumes*, Kiel, 1941.

Irwin, Graham, 'The Dutch and the Tin Trade of Malaya in the Seventeenth Century', in Jerome Ch'en and Nicolas Tarling, eds, *Studies in the Social History of China and South-east Asia*, Cambridge: Cambridge University Press, 1970.

Jackson, James, ' Mining in 18th Century Bangka: the Pre-European Exploitation of a 'Tin Island', *Pacific Viewpoint*, vol. 10, no. 2, 1969.

James, Alfred, 'Tin Industry Faces Facts', *Engineering and Mining Journal*, 23 August 1930.

—— 'Tin Prices and Production', *Engineering and Mining Journal*, 8 September 1930.

James, C. 'International Control of Tin Ore', *Harvard Business Review*, vol. 11, October 1932.

Jarland, Dr, 'Au pays de l'étain. Kokiu (Yunnan)', *Revue Indochinoise*, 1921.

Jenkin, A. K. Hamilton, *Nationalisation of West Country Minerals*, London: New Fabian Research Bureau, 1932.

Jeremy, David J., ed., *Dictionary of Business Biography*, London: Butterworths, 1984–6.

Joannès, F., 'L'étain, de l'Elam à Mari', in L. De Meyer and H. Gasche, eds, *Mésopotamie et Elam*, Ghent, 1991.

Jones, S. R. H., 'Devaluation and the Balance of Payments in Eleventh-Century England: An Exercise in Dark Age Economics', *Economic History Review*, vol. 44, no. 4. November 1991.

Jones, W. R., *Tinfields of the World*, London: Mining Publications, 1925.

—— *Minerals in Industry*, Harmondsworth: Penguin, 1963.

Kahn, Kabir-ur-Rahman, 'The Development of the Tin Buffer Stock System: A Diplomatic and Organisational Study of the pre-1945 Phase', *Indian Journal of International Law*, vol. 17, 1977.

—— *The Law and Organisation of International Commodity Agreements*, The Hague: Nijhoff, 1982.

Kaldor, Nicolas, 'The Role of Commodity Prices in Economic Recovery', *World Development*, vol. 15, no. 5, May 1987.

Kamp, A. F., *De standvastige tinnen soldat*, The Hague: Billiton, 1960.

Keay, John, *Last Post: The End of Empire in the Far East*, London: John Murray, 1997.

Kennedy, Thomas, 'The Future of Tin', *Bolivia*, vol. 2, no. 2, November–December 1928.

Kenny, Edward, *Tin*, Sydney, 1922.

Kerr, Ruth, 'The Advent of Tin Dredges in Eastern Australia', *The Australian Journal of Historical Archaeology*, vol. 7, 1989.

Kestenbaum, Ralph, *The Tin Men. A Chronicle of Crisis*, London: Metal Bulletin, 1991.

Keynes, J. M., 'The Policy of Government Storage of Foodstuffs and Raw Materials', *Economic Journal*, vol. 48, no. 191, September 1938.

—— 'International Regulation of Primary Products',*Journal of International Economics*, vol. 4, 1974.

Khalid, A. Azmi Abdul, 'The Social Organization of the Mining Industry during the Depression, 1929–1933, in Malaya', *Journal of the Malaysian Branch of the Royal Asiatic Society*, vol. 65, pt. 2, December 1992.

Kidd, Archibald, *History of the Tin-Plate Workers*, London: National Union of Sheet Metal Workers, 1949.

Kindleberger, Charles, *The World in Depression, 1929–1939*, Berkeley, CA: University of California Press, 1986.

Kirk-Greene, Anthony, 'Kitty Cooke Looks Back: A Personal Mining Memoir', *The Nigerian Field*, vol. 26, no. 3, 1961.

Klein, Herbert, 'The Creation of the Patiño Tin Empire', *Inter-American Economic Affairs*, vol. 19, no. 2, Autumn 1965.

—— *Parties and Political Change in Bolivia, 1880–1952*, Cambridge: Cambridge University Press, 1969.

Knorr, Klaus E., 'Access to Raw Materials in the Postwar World', *Harvard Business Review*, vol. 21, no. 3, Spring 1943.

—— *Tin Under Control*, Stanford, CA: Food Research Institute, 1945.

Kranold, Herman, *The International Distribution of Raw Materials*, London: Routledge, 1938.

Kristiansen, Kristian, 'From stone to bronze — the evolution of social complexity in Northern Europe', in Elizabeth Brumfiel and Timothy Earle, eds, *Specialization, Exchange, and Complex Societies*, Cambridge: Cambridge University Press, 1987.

Kruger, Rayne, *The Devil's Discus*, London: Cassel, 1964.

Landon, Kenneth, *Chinese in Thailand*, New York: Institute for Pacific Relations, 1941.

Langeveld, H. J., *Hendrikus Colijn 1869–1944: dit leven van krachtig handelen*, Amsterdam: Balans, 1998.

Larner, E. A., 'Tin Mining Anniversary at Jos', *The Nigerian Field*, vol. 19, no. 3, July 1954.

Latham, A. J. H., *The Depression and the Developing World*, London: Croom Helm, 1981.

Law, Alton, *International Commodity Agreements*, Lexington, KY: Heath, 1975.

Lecaut, Jean Michel, 'S. I. Patiño et le problème de l'étain bolivien, 1928–1947', Memoire de Maîtrisse, Sorbonne, 1976.

Lechtman, Heather, 'Arsenic Bronze: Dirty Copper or Chosen Alloy? A View from the Americas', *Journal of Field Archaeology*, vol. 23, no. 4, Winter 1996.

Leclere, M. A., 'Etude géologique et minière des provinces chinoises voisines du Tonkin', *Annales des Mines* series 9, vol. 20, 1901.

Lenthall, D. H., *Tin Production from the Bushveld Complex*, Johannesburg: University of the Witwatersrand, 1974.

Levenstein, H. A., *Revolution at the Table: Transformation of the American Diet*, New York: Oxford University Press, 1988.

Lewis, George, *The Stannaries: A Study of the English Tin Miner*, Boston, MA: Houghton, Mifflin, 1908.

Lewis, W. A., *Economic Survey, 1919–1939*, London: Allen & Unwin, 1949.

Li Dun Jen, *British Malaya. An Economic Analysis*, Kuala Lumpur: INSAN, 1982.

Li Liu, ''Products of Minds as Well as of Hands': Production of Prestige Goods in the Neolithic and Early State Periods of China', *Asian Perspectives,* vol. 42, no. 1, Spring, 2002.

Lieberman, Victor, *Strange Parallels: Southeast Asia in Global Context, c. 800–1830*, Cambridge: Cambridge University Press, 2003.

Lim Chong-Yah, *Economic Development of Modern Malaya*, Kuala Lumpur: Oxford University Press, 1967.

Linduff, Katheryn, *Metallurgy in Ancient Eastern Eurasia*, Queenston: Mellen, 2004.

Little, Bryan, *Capper Pass*, London: Newman Neame, 1963.

Loh, Philip, *The Malay States, 1877–1886*, Kuala Lumpur: Oxford University Press, 1969.

Loh Kok Wah, Francis, *Beyond the Tin Mines. Coolies, Squatters and New Villages in the Kinta Valley, Malaysia, c.1880–1980*, Singapore: Oxford University Press, 1988.

Loofs-Wissowa, H. H. E., 'The Development and Spread of Metallurgy in Southeast Asia: A Review of the Present Evidence', *Journal of Southeast Asian Studies,* vol. 13, no. 1, March 1982.

Loone, A. W., *Tasmania's Northeast*, Launceston: Typographical Arts, 1981, reprint of 1929 edition.

Louis, Wm. Roger, *Imperialism at Bay, 1941–1945*, Oxford: Clarendon Press, 1977.

Lowinger, V., 'Production and Utilisation of Tin', *Crown Colonist*, January 1944.

Lyttleton, Oliver, *The Memoirs of Lord Chandos*, London: Bodley Head, 1962.

McCarthy, E. T., *Incidents in the Life of a Mining Engineer*, London: Routledge, 1918.

—— *Further Incidents in the Life of a Mining Engineer*, London: Routledge, 1920.

McElderry, R. Knox, 'Vespasian's Reconstruction of Spain', *Journal of Roman Studies,* vol. 8, 1918.

Mackilligin, R., *The Bisichi Story*, London: Bisichi, 1994.

McKie, James W., *Tin Cans and Tin Plate*, Cambridge: Harvard University Press, 1959.

MacLiesh, Fleming, and Cushman Reynolds, *Strategy of the Americas*, New York: Duell, 1941.

Macmillan, Harold, *Winds of Change, 1914–1939*, London: Macmillan, 1966.

Ma Huan, *Ying-yai sheng-lan. The Overall Survey of the Ocean's Shores,* ed. and trans. Feng Ch'eng-Chün, Cambridge: Cambridge University Press, 1970.

Maddox, Robert, *The War within World War II: The United States and International Cartels*, Westport, CT: Praeger, 2001.

Maizels, Albert, *Commodities in Crisis: The Commodity Crisis of the 1980s and the Political Economy of International Commodity Policies*, Oxford: Clarendon Press, 1992.

Malamat, A., 'Syro-Palestinian Destinations in a Mari Tin Inventory', *Israel Exploration Journal*, vol. 21, no. 1, 1971.

Malleret, Louis, *L'Archéologie du Delta du Mékong*, vol. 2, *La civilisation matérelle d'Oc-Èo*, Paris: EFEO, 1960.

Manchester, William, *The Arms of Krupp*, Boston, MA: Little, Brown, 1968.

Manley, John Jackson, *The Age of Tin*, London: Tweedie, 1872.

Marsh, Margaret, *Bankers in Bolivia*, New York: Vanguard, 1928.

Mantell, C. L, *Tin. Its Mining, Production, Technology, and Applications*, New York: Hafner, 1970.

Marcus, Alfred, *Die grossen Eisen- und Metallkonzerne*, Leipzig: Hirzel, 1929.

Marks, Robert, *Tigers, Rice, Silk, and Silt: Environment and Economy in Late Imperial South China*, Cambridge: Cambridge University Press, 1998.

Marquez, Jaime, 'Life Expectancy of International Cartels: An Empirical Analysis', *Review of Industrial Organization,* vol. 4, no. 3, 1994.

Martinez Vargas, Ricardo, *Estaño en Bolivia y el Comité Internacional*, La Paz: Echaniz, 1936.
Mason, Edward, *Controlling World Trade*, New York: McGraw-Hill, 1946.
Mason, R. B. and M. S. Tite, 'The beginnings of tin opacification of pottery glazes', *Archaeometry*, vol. 39, no. 1, 1997.
Mattingley, Christobel, *King of the Wilderness, the Life of Deny King*, Melbourne, 2001.
Maxwell George, *Report of a Committee appointed ... by the Nigerian Chamber of Mines on Thursday, the 15th day of December, 1932*, London: np, 1933.
May, Elizabeth S., 'The International Tin Cartel', in W. Y. Elliott, ed., *International Control in the Non-Ferrous Metals*, New York: Macmillan, 1937.
Mayall, Kenneth L., *International Cartels: Economic and Political Aspects*, Rutland: Tuttle, 1951.
Meilink-Roelofsz, M. A. P., *Asian Trade and European Influence in the Indonesian Archipelago between 1500 and about 1630*, The Hague: Nijhoff, 1962.
Meunier, A. Dauphin, *La Cité de Londres*, Paris: Gallimard, 1940.
Meyers, Lynne, *Balmaidens*, Penzance: Hypatia, 2004.
Miller, E. Willard',The Mineral Resources of Indo-China', *Economic Geography,* vol. 22, no. 4, October, 1946.
Minchinton, Walter, *The British Tinplate Industry*, Oxford: Clarendon Press, 1957.
Mitre, Antonio, *El enigma de los hornos. La economía política de la fundición de estaño*, La Paz: Biblioteca Minera Boliviana, 1993.
—— *Bajo un cielo de estaño. Fulgor y ocaso del metal en Bolivia.*, La Paz: Biblioteca Minera Boliviana, 1993.
Molins, Wenceslao, *El Estaño. Fundamento vital de Bolivia*, Buenos Aires, 1937.
Mollema, J. C., *De ontwikkeling van het eiland Billiton en van de Billiton-maatschappij,* The Hague: Nijhoff, 1922.
Mollin, Gerhard, *Die USA und der Kolonialismus: Amerika als Partner und Nachfolger der belgischen Macht in Afrika, 1939–1965*, Berlin: Akademie Verlag, 1966.
Montenegro, Walter, *Estaño en Malaya*, La Paz: Fenix, 1943.
Montgomery, Brian, *Shenton of Singapore: Governor and Prisoner of War*, London: Secker & Warburg, 1984.
Moor, C. G., *Tin Mining*, London: Pitman, 1928.
Müller-Ohlson, Lotte, *Non-Ferrous Metals. Their Role in Industrial Development*, Cambridge: Woodhead, 1981.
Muhly, James, *Copper and Tin. The Distribution of Mineral Resources and the Nature of the Metals Trade in the Bronze Age*, Hampden: Archon, 1973, 1976.
—— 'Sources of Tin and the Beginnings of Bronze Metallurgy', *American Journal of Archaeology*, vol. 89, no. 2, April 1985.
—— 'The Beginnings of Metallurgy in the Old World', in Robert Maddin, ed., *The Beginning of the Use of Metals and Alloys*, Cambridge: MIT, 1988.
Munro, J. Forbes, 'Monopolies and Speculators: British Investment in West African Rubber, 1905–1914', *Journal of African History*, vol. 22, 1981.
Myers, C. A., 'The International Tin Control Scheme', *Journal of Business of the University of Chicago*, vol. 10, no. 2, April 1937.
Nash, June, *We Eat the Mines and the Mines Eat Us*, New York: Columbia University Press, 1979.
Nattrass, Gail, 'The Tin Mining Industry in the Transvaal 1905–1914: Some Social and Economic Implications and Perspectives', *South African Journal of Economic History,* vol. 6, no. 1, March 1991.
Naylor, Simon, 'Spacing the can: empire, modernity, and the globalisation of food', *Environment and Planning,* vol. 32, no. 9, September 2000.
Neitzel, Willy, *Die Wirkungen internationaler Kartelle in Kupfer, Zink und Zinn auf die deutsche Volkswirtschaft*, Cologne: np, 1936.
Netschert, Bruce, *Mineral Foreign Trade of the United State in the Twentieth Century: A Study in Mineral Economics*, New York: Arno Press, 1977. Reprint of PhD dissertation, Cornell University, 1949.
Newsom, J. B., 'Examination of Tin Deposits in the Federated Malay States', *Engineering and Mining Journal,* 17, 24 March 1923.
Northover, J. P. and C. Gillis, 'Questions in the Analysis of Ancient Tin', in Suzanne M. Young, A. M Pollard, P. Budd and R. A. Ixer, eds, *Metals in Antiquity*, Oxford: Archaeopress, 1999.

Oakley, Herbert, *The Future of Tin*, London: Saint Catherine, 1928.
O'Brien, W., *Mount Gabriel: Bronze Age Metallurgy in Ireland*, Galway: Galway University Press, 1994.
Olson, James C., *Stuart Symington*, Columbia, MA: University of Missouri Press, 2003.
Oshin, Olasiji, 'Developing Infrastructure of Exploitation: The Example of Colonial Transport on the Bauchi Tin Fields, 1902–1914', *Transafrican Journal of History*, vol. 17, 1988.
Pahang Consolidated, *Sixty Years of Tin Mining. A History of the Pahang Consolidated Company, 1906–1966*, London: Pahang Consolidated, 1967.
Palmer, Robin, 'The Nyasaland Tea Industry in the Era of International Tea Restrictions, 1933–1950', *Journal of African History,* vol. 26, no. 2/3, 1985.
Parsons, A., 'Is 45 cent tin likely?', *Engineering and Mining Journal*, 28 April, 9 June 1928.
Parzinger, Hermann and Nikolaus Boroffka, *Das Zinn der Bronzezeit in Mittelasien*, Mainz: von Zabern, 2003, vol. 1.
Pass, Sylvia, 'Tin has Been Made into a Dunce-Cap for Uncle Sam', *Ken*, July 27, 1939.
Pawle, Reginald, 'The Empire's Tin Resources', Second Empire Mining and Metallurgical Congress, *Proceedings,* pt. 1, Montreal, 1928.
Peele, Robert, ed., *Mining Engineers' Handbook*, New York: Wiley, 1941, vol. 1.
Penhallurick, R. D., *Tin in Antiquity: Its Mining and Trade Throughout the Ancient World with Particular Reference to Cornwall*, London: Institute of Metals, 1986.
Pennington, Robert, *Stannary Law. A History of the Mining Law of Cornwall and Devon*, Newton Abbot: David & Charles, 1973.
Penzer, N. M., *The Tin Resources of the British Empire*, London: Rider, 1921.
Perham, Margery, *West African Passage: A Journey through Nigeria, Chad, and the Cameroons, 1931–1932*, London: Peter Owen, 1983.
Perkins, Patricia, *World Metal Markets. The United States Strategic Stockpile and Global Market Influence*, Westport, CT: Praeger, 1997.
Philip, Graham *Metal Weapons of the Early and Middle Bronze Ages in Syria-Palestine*, Oxford: BAR, 1989.
Phimister, Ian, 'Corners and Company-Mongering: Nigerian Tin and the City of London, 1909–1912', *Journal of Imperial and Commonwealth History,* vol. 28, no. 2, May 2000.
Phuwadol Songprasert, 'The Development of Chinese Capital in Southern Siam, 1868–1932', PhD dissertation, Monash University, 1986.
Pichon, Louis, *Un Voyage au Yunnan*, Bangkok: White Lotus, 1999, original publication 1893.
Pillay, Subramanian, 'Performance of the World Tin Industry: Effects of the International Tin Agreements, 1956–1985', PhD dissertation, University of British Columbia, 1990.
Pratt, Ambrose, *Magical Malaya*, Melbourne: Robertson & Mullens, 1931.
Prebisch, Raúl, 'The Economic Development of Latin America and its Principal Problems', *Economic Bulletin for Latin America*, vol. 7, no. 1, February 1962.
Prest, Michael',The Collapse of the International Tin Agreement', *Round Table,* No. 302, 1987.
Price, Percival, *Bells and Man*, Oxford: Oxford University Press, 1983.
Prince, Gé, 'Economic Policy in Indonesia, 1900–1942', in J. Thomas Lindblad, ed., *New Challenges in the Modern Economic History of Indonesia,* Leiden: Programme of Indonesian Studies, 1993.
Proposals for expanding the activities of the ITRDC, New York, 1951.
Provencher, Ronald, *Mainland Southeast Asia: An Anthropological Approach*, Pacific Palisades, CA: Goodyear, 1975.
Puey Ungphakorn, 'The Economics of Tin Control', PhD dissertation, University of London, 1949.
Pulak, Cemal, 'The Copper and Tin Ingots from the Late Bronze Age Shipwreck at Uluburum', in Ünsal Yalçin, ed., *Der Anschnitt, Anatolian Metal I,* vol. 13, Bochum, 2000.
Quereajzu Calvo, Roberto, *Llallagua. Historia de una montaña*, La Paz: Los Amigos del Libro, 1984.
Ramin, Jacques, *Le problème des Cassitérides: et les sources de l'étain occidental depuis les temps protohistoriques jusqu'au début de notre ère*, Paris: Picard, 1965.
Randall-Maciver, David, 'The Rhodesia Ruins: Their Probable Origin and Significance', *Geographical Journal,* vol. 27, no. 4, April 1906.
Rawlins, J. S. D., 'French Enterprise in Malaya', *Journal of the Malaysian Branch of the Royal Asiatic Society,* vol. 39, December 1966.
Rawson, Jessica, 'The Ancestry of Chinese Bronze Vessels', in S. Lubar and W. D. Kingery, eds, *History from Things*, Washington, DC: Smithsonian Institution, 1993.

Reyer, E., *Zinn. Eine geologish-montanistisch-historische Monografie*, Berlin, 1881.

Richards, John F., *The Mughal Empire*, Cambridge: Cambridge University Press, 1995.

Rhohe, Randall, 'The Chinese and Hydraulic Mining in the Far West', *Mining History Journal,* 1994.

Robbins, Lionel, *The Great Depression*, London: Macmillan, 1934.

—— *Economic Planning and International Order*, London: Macmillan, 1937.

Robertson, W., *Tin. Its Production and Marketing*, Westport, CT: Greenwood, 1982.

Robinson, J.A., 'The International Tin Restriction Schemes with Special Emphasis on British Malaya', PhD dissertation, George Washington University, 1945.

Roddy, Peter, *The International Tin Trade*, London: Woodhead, 1995.

Rogers, Christopher D., 'Consumer Participation in the International Tin Agreements', *Malayan Economic Review*, vol. 14, 1969.

—— 'A Scheme for Enforcing the Tin Ceiling Price', *Intereconomics*, November 1970.

Rosewarne, Stuart, 'Capital Accumulation and the Export of Mining Capital before World War II', in E. L. Wheelwright and Ken Buckley, eds, *Essays in the Political Economy of Australian Capitalism*, Sydney: Australia and New Zealand Book Co., 1983.

Rowe, J. W. F., 'The Artificial Control of Raw Material Supplies',*Economic Journal*, vol. 40, no. 159, September 1930.

—— 'Keeping up the price of tin', *The Listener*, vol. 13, 1936.

—— *Markets and Men*, Cambridge: Cambridge University Press, 1936.

—— 'Some General Economic Aspects', in W. Y. Elliott, ed., *International Control in the Non-Ferrous Metals*, New York: Macmillan, 1937.

—— *Primary Commodities in International Trade*, Cambridge: Cambridge University Press, 1965.

Ryckmans, P., 'Le pays du Grands Lacs', *Le Miroir du Congo Belge*, Brussels: N. E. A, 1929.

Sabin, Andew, 'Tin availability – market economy countries', in Robin Amlot, ed., *The Market for Tin*, London: Tin Publications, 1986.

Şahoğlu, Vasif, 'The Anatolian Trade Network and the Izmir Region', *Oxford Journal of Archaeology*, vol. 24, no. 4, 2005.

Salmon, J. H. M., *A History of Gold Mining in New Zealand*, Wellington, 1963.

Sánchez Gomez, J., *De Minería, Metalúrgica y Comercio de Metales, 1450–1610*, Salamanca, 1989.

Scalisi, Philip, *Classic Mineral Localities of the World, Asia and Australia*, New York: Van Nostrand, 1983.

Schut, M., *Tinrestrictie en Tinprijs*, Haarlem: Nederlandsch Economisch Instituut, 1940.

Schmitz, Christopher, 'The Rise of Big Business in the World Copper Industry 1870–1930', *Economic History Review,* vol. 39, no. 3, August 1986.

Schonherr, Hansjoachim, *Anfange arbeitsrechtlicher Regelungen des Kapitalismus im erzgebirgischen Zinnbergbau des 15./16. Jahrhunderts*, Leipzig, 1983.

Shell, Colin, 'The Early Exploitation of Tin Deposits in South West England', in Michael Ryan, ed., *The Origins of Metallurgy in Atlantic Europe*, Dublin: Stationery Office, 1978.

Sinha, H., 'Jute futures in Calcutta', *Economica*, November 1929.

Sinclair, Keith, 'Hobson and Lenin in Johore: Colonial Office Policy towards British Investors and Concessionaires, 1878–1907', *Modern Asian Studies*, vol. 1, no. 4, 1967.

Skelton, Alex, '"Copper", "Nickel"', in W. Y. Elliott, ed., *International Control in the Non-Ferrous Metals*, New York: Macmillan, 1937.

Skilliter, S. A., *William Harborne and the Trade with Turkey*, Oxford: Oxford University Press, 1977.

Slessor, Robert, 'Chinese Non-Ferrous Metals', Australasian Institute of Mining and Metallurgy, *Proceedings*, New Series, no. 65, 31 March 1927.

Smith, Alan, 'Delagoa Bay and the Trade of South-Eastern Africa' in Richard Gray and David Birmingham, eds, *Pre-Colonial African Trade*: *Essays on Trade in Central and Eastern Africa before 1900*, London: Oxford University Press, 1970.

Smith, George Vinal, *The Dutch in Seventeenth-century Thailand*, DeKalb, IL: Northern Illinois University Press, 1977.

Smith, Nicol, *Burma Road*, Garden City, 1942.

Société générale métallurgique de Hoboken, 1908–1958, Brussels: Cuypers, 1957.

Sompop Manarunsgan, *Economic Development of Thailand, 1850–1950*, Bangkok: Institute of Asian Studies, 1989.

Spar, Debora, *The Cooperative Edge. The Internal Politics of International Cartels*, Ithaca: Cornell University Press, 1994.

Spraos, John, *Inequalising Trade?*, Oxford: Clarendon Press, 1983.
Staley, Eugene, *Raw Materials in Peace and War*, New York: Council on Foreign Relations, 1937.
Stokes, Ralph, *Malay tin-fields*, Singapore: Straits Times, 1906
Stuart-Fox, Martin, 'The French in Laos, 1887–1945', *Modern Asian Studies*, vol. 29, no. 1, February 1995.
Supaporn Ariyasajsiskul, 'The So-called Tin Monopoly in Ligor. The Limits of VOC Power vis-à-vis a Southern Thai Trading Polity', *Itinerario*, vol. 28, no. 3, 2003.
Suslow, Valerie, 'Cartel Contract Duration: Empirical Evidence from Interwar International Cartels', *Industrial and Corporate Change*, vol. 14, no. 5, 2005.
Sutherland-Harris, Nicola, 'Trade and the Rozwi Mambo' in Richard Gray and David Birmingham, eds, *Pre-Colonial African Trade*, New York: Oxford University Press, 1970.
Sutton, J. E. G., 'Igbo-Ukwu and the Nile', *African Archaeological Review*, vol. 18, no. 1, 2001.
Suwannathat, K., 'Thai Wartime Leadership Reconsidered: Phibun and Pridi', *Journal of Southeast Asian Studies,* vol. 27 no. 1, March 1996.
Swan, William, *Japanese Economic Activity in Siam. From the 1890s until the Outbreak of the Pacific War*, Gaya: Centre for Southeast Asian Studies, 1986.
Sweet, C. T., 'The Kyerwa Tin Deposits, Tanganyika', Institution of Mining and Metallurgy, *Transactions*, vol. 54, 1940.
Swinton, *I Remember*, London: Hutchinson, 1948.
—— *Sixty Years of Power*, London: Hutchinson, 1966.
Tarring, Trevor, *Trading in Metals*, London: Mining Books, 1983.
Taylor, J. W., 'A Nigerian Tin Trade in Antiquity?'*Oxford Journal of Archaeology,* vol. 1, no. 3, November 1982.
ter Braake, A. L., 'Rationalisatie der Banka Tinwinning', *Koloniale Studiën,* vol. 25, no. 2, 1941.
Thatcher, Dorothy, *Pai Naa*, London: White Lion, 1974.
Thoburn, John, *Multinationals, Mining and Development: A Study of the Tin Industry*, Westmead: Gower 1981.
—— *Tin in the World Economy*, Edinburgh: Edinburgh University Press, 1994.
Thompson, Carl, 'The heroic age of the tin can. Technology and ideology in British Arctic exploration, 1818–1835', in D. Killingray, *et al.*, eds, *Maritime Empires*, Woodbridge: Boydell and Brewer, 2004.
Thompson, Virginia, 'Siam: Manoeuvering towards Self-Sufficiency', *Far Eastern Survey,* vol. 7, no. 35, 21 December 1938.
Thorne, Stuart, *The History of Food Preservation*, Kirby Lonsdale: Parthenon, 1986.
Tin Buffer Stock Scheme, 1938/41, *Report and Accounts*, London: np, 1943.
Tin Producers' Association, *International Tin Control and Buffer Stocks*, London: Tin Producers' Association, 1944.
Tomlinson, Edward, *New Roads to Riches in the Other Americas*, New York: Scribners, 1939.
Torgasheff, Boris, 'Tin in the Far East', *Internationale Bergwirtschaft*, vol. 3, no. 12, December 1928.
Tregonning, K. C., *Straits Tin*, Singapore: Straits Times, n.d.
van de Wal, S. L., *Herinnergingen van Jhr. Mr. B. C. de Jonge*, Groningen: Walters-Noordhof, 1968.
van der Eng, Pierre, 'The Real Domestic Product of Indonesia, 1880–1989', *Explorations in Economic History*, vol. 29, no. 3, July 1992.
van Goor, Jurrien, 'A Hybrid State. The Dutch Economic and Political Network in Asia, ' in Claude Guillot, *et al.*, eds, *From the Mediterranean to the China Sea*, Wiesbaden: Harrasowitz, 1998.
van-Helten, Jean-Jacques, 'La France et L'Or des Boers: Some Aspects of French Investment in South Africa between 1890 and 1914', *African Affairs,* vol. 84, no. 335, April 1985.
—— and Geoffrey Jones, 'British Business in Malaya and Singapore since the 1870s', in R. P. T. Davenport-Hines and Geoffrey Jones, eds, *British Business in Asia since 1860*, Cambridge: Cambridge University Press, 1989.
Vichitvong Na Pombhejara, *Pridi Banomyong and the Making of Thailand's Modern History*, Bangkok, 2001.
Vogel, Hans, 'Bergbau auf Gold, Silver, Kupfer und Zinn in Yunnan bis zum Ende der Mongolenzeit', in Albert Lutz, ed., *Der Goldschatz der drei Pagoden*, Zurich: Museum Rietberg, 1991.
von Falkenhausen, Lothar, *Suspended Music. Chime Bells in the Culture of Bronze Age China*, Berkeley, CA: University of California Press, 1993.
Wagenführ, Horst, 'England in internationalen Kartellen', *Kartellrundschau*, 1940.

Wallace, D. R., *The Romance of Jute*, Calcutta, 1909.
Warburton, David A., *Macroeconomics from the Beginning, The* General Theory, *Ancient Markets, and the Rate of Interest*, Neuchâtel: Recherches et publications, 2003.
Waszkis, Helmut, *Dr. Moritz (Don Mauricio) Hochschild 1881–1965*, Frankfurt: Vervuert, 2001.
—— and Peter Waszkis, *The Story of Metal Trading*, London: Metal Bulletin, 2003.
Webster, Anthony, *Gentlemen Capitalists: British Imperialism in South East Asia, 1770–1890*, London: Tauris, 1998.
Weisgerber, Gerd and Jan Cierny, 'Tin for Ancient Anatolia?' *Der Anschnitt, Anatolian Metal* II.
Wheatley, Paul, *The Golden Khersonese: Studies in the Historical Geography of the Malay Peninsula before* A.D. *1500,* Westport, CT: Greenwood Press, 1973.
White, Nicholas, *Business, Government and the End of Empire: Malaya, 1942–1957*, Kuala Lumpur: Oxford University Press, 1996.
—— 'Gentlemanly capitalism and empire in the twentieth century: the forgotten case of Malaya, 1914–1965', in Raymond Dummett, ed., *Gentlemanly Capitalism and British Imperialism. The New Debate on Empire*, Harlow: Longman, 1999.
Wilkins, Mira and Harm Schroter, eds, *The Free Standing Company in the World Economy, 1830–1996*, Oxford: Oxford University Press, 1998.
Willbourn, E. S., 'The Geology and Mining Industries of Johore', *Journal of the Malaysian Branch of the Royal Asiatic Society,* vol. 6, pt. 4, November 1928.
Williams, Jack, *Cricket and England: A Cultural and Social History of the Inter-war Years*, London: Cass, 1999.
Williamson, David, 'The ITC collapse: background and consequences', in Robin Amlot, ed., *The Market for Tin*, London: Tin Publications, 1986.
Williamson, Philip, 'Financiers, the Gold Standard, and British Politics, 1925–1931', in John Turner, ed., *Businessmen and Politics. Studies of Business Activities of British Politicians, 1900–1945*, London: Heinemann, 1984.
Wollnik, Josef, *Zinn, Wandlungen in der Erzeugung und Verwendung des Zinns nach dem Weltkrieg*, Leipzig: Bibliographisches Institut, 1936.
Wong Lin Ken, *The Malayan Tin Industry to 1914*, Tucson: University of Arizona Press, 1965.
Wong, William E., *Mineral Wealth of China*, Shanghai, 1927.
Wurm, Clemens, *Business, Politics, and International Relations: Steel, Cotton, and International Cartels in British Politics, 1924–1939*, Cambridge: Cambridge University Press, 1993.
Yacob, Shakila, *The United States and the Malaysian Economy*, Abingdon: Routledge, 2007.
Yates, P. Lamartine, *Commodity Control*, London: Cape, 1943.
Yener, K. Aslihan and Pamela B. Vandiver, 'Tin Processing at Göltepe, an Early Bronze Age Site in Anatolia', *American Journal of Archaeology,* vol. 97, no. 2, April 1993.
—— *The Domestication of Metals: The Rise of Complex Metal Industries in Anatolia*, Leiden: Brill, 2000.
Yergin, Daniel, *The Prize. The Epic Quest for Oil, Money, and Power*, New York: Simon & Schuster, 1991.
Yip Yat Hoong, *The Development of the Tin Mining Industry in Malaya*, Kuala Lumpur: University of Malaya Press, 1966.
Yuen Choy Leng, 'Japanese Rubber and Iron Investments in Malaya, 1900–1940', *Journal of Southeast Asian Studies,* vol. 5, no. 1, March 1974.
Zaharah binti, Haji Mahmud, *Change in a Malay Sultanate: A Historical Geography of Kedah before 1939*, Kuala Lumpur: Oxford University Press, 1966.

III Newspapers, periodicals, annuals

American Metal Market
Bangkok Times
Bolivia
British Malaya
Bulletin d'études géologiques et minières coloniales
Chemical Engineering and Mining Review
Chronique des Mines Coloniales

Comercio y Minas
Economisch-Statistische Berichten
Economist
El Diario
Engineering and Mining Journal
Far Eastern Survey
Financial Times
Financial News
FMS Chamber of Mines Yearbook
Industrial Australian and Mining Standard
Internationale Bergwirtschaft
Investors' Chronicle
La Belgique Financière, Coloniale et Maritime
Malayan Tin and Rubber Journal
Metal Industry
Metal Bulletin
Metallwirtschaft
Mineral Industry
Minerals Yearbook
Mining Journal
Mining Magazine
Mining and Metallurgy
Mining World and Engineering Record
New York Times
Revista Financiera y Minera
Revista de Economía y Finanzas
Straits Times
South African Mining and Engineering Journal
Times
Tin
Transactions, Institution of Mining and Metallurgy
Wild Cat Monthly

IV Statistical sources

American Bureau of Metal Statistics, *Yearbook*, New York, annual.

Anglo-Oriental Mining Corporation, *Tin-World Statistics*, London, annual.

Department of Commerce and Statistics, *Statistical Yearbook of the Kingdom of Siam*, Bangkok, annual.

d'Avenel, Georges, *Histoire économique de la propriété, des salaires, des denrées et de tous les prix en général, depuis l'an 1200 jusqu'en l'an 1800*, New York: Franklin, 1968–1969.

Federated Malay States, Mines Department, *Monthly Bulletin of Statistics Relating to the Mining Industry,* Kuala Lumpur.

Garnsey,W.K., *Eastern tin dredging companies*, Sydney, 1937.

House of Commons, *Papers, Accounts relating to Tin*, 1815–1854.

—— *Papers, Annual Statement of the Trade of the United Kingdom*, 1855–1913.

Hsiao Liang-li, *China's Foreign Trade Statistics, 1884–1949*, Cambridge: Harvard University Press, 1974.

International Tin Council, *Statistical Yearbook, 1959*, London, 1959.

International Tin Research and Development Council, *Statistical Bulletin*, The Hague, monthly.

—— *Statistical Yearbook, 1938,* The Hague, 1938.

Kelly and Walsh, *Guide to Malayan Tin Companies*, Singapore, 1927.

London and Cambridge Economic Service in J. M. Keynes, *Collected Works*, London: Macmillan, 1983, vol. 12.

Metallgesellschaft, *Statistische Zusammenstellungen, 1928–1937*, Frankfurt, 1938.

Mining Yearbook, London: Skinner, annual.

Pring and Docker, *Guide to Eastern Tin Stocks for Australian Investors*, Sydney, 1927.
Quin's Metal Handbook and Statistics, London: Metal Information Bureau, annual.
Schmitz, Christopher J., *World Non-Ferrous Metal Production and Prices, 1700–1976*, London: Cass, 1979.
Swagler, R. M., 'An Analysis of Variations in Alternative Terms-of-Trade Measures. A Case Study of Bolivia, 1925–1965', PhD dissertation, Ohio State University, 1971.
Stock Exchange Official List of Making- up Prices, London.
The Record, Bangkok.
U.S. Bureau of the Census, *Historical Statistics of the United States*, Washington, DC: GPO, 1960.
Were, J. B., *Tin Shares as an Investment*, Melbourne, 1938.

V Internet sources

http://minerals.usgs.gov/minerals/pubs/of01-006/ (accessed 27 June 2005).
www.eh.net/hmit/ppowerbp/ (accessed 10 June 2005).
www.eh.net/hmit/ukgdp (accessed 20 June 2005).

Index